Roland Trimen, James Henry Bowker

South-African Butterflies:

A Monograph of the Extra-Tropical Species (Volume 3)

Roland Trimen, James Henry Bowker

South-African Butterflies:
A Monograph of the Extra-Tropical Species (Volume 3)

ISBN/EAN: 9783744756419

Printed in Europe, USA, Canada, Australia, Japan

Cover: Foto ©berggeist007 / pixelio.de

More available books at **www.hansebooks.com**

SOUTH-AFRICAN BUTTERFLIES:

A MONOGRAPH

OF THE

EXTRA-TROPICAL SPECIES.

BY

ROLAND TRIMEN, F.R.S., F.L.S., F.Z.S., F. Ent. S., &c.

CURATOR OF THE SOUTH-AFRICAN MUSEUM, CAPE TOWN;

ASSISTED BY

JAMES HENRY BOWKER, F.Z.S., F.R.G.S.

COLONEL (RETIRED) IN THE CAPE SERVICE,
LATE COMMANDANT OF FRONTIER ARMED AND MOUNTED POLICE,
GOVERNOR'S AGENT IN BASUTOLAND,
AND CHIEF COMMISSIONER AT THE DIAMOND FIELDS OF GRIQUALAND WEST.

VOL. III.

PAPILIONIDÆ AND HESPERIDÆ.

LONDON:
TRÜBNER & CO., LUDGATE HILL.
1889.

CONTENTS.

	PAGE
RHOPALOCERA .	1
FAMILY IV.—PAPILIONIDÆ	1
Sub-Family—PIERINÆ	3
,, PAPILIONINÆ	191
FAMILY V.—HESPERIDÆ	259
APPENDIX I.—ADDITIONAL SPECIES, RECENTLY ASCERTAINED TO BE SOUTH AFRICAN . .	377
APPENDIX II.—ADDITIONS AND CORRECTIONS	393
ERRATA	419
SYSTEMATIC INDEX TO VOL. III. . .	423
INDEX .	427
LIST OF SPECIES FIGURED IN THE PLATES	437

RHOPALOCERA.

Family IV.—PAPILIONIDÆ.

Papilionidæ, Leach, "Samouelle's Comp., p. 234 (1819)."
Papilionides and *Pierides*, Boisduval, Sp. Gen. Lep., i. pp. 171 and 402 (1836).
Papilionidæ, Swainson, "Phil. Mag., Ser. II. vol. i. p. 187 (1827);" and Hist. and Nat. Arr. Ins., p. 86 (1840).
Papilionidæ, Westwood, Intr. Mod. Class. Ins., ii. p. 347 (1840).
Papilionidæ and *Pieridæ*, Doubleday, Gen. Diurn. Lep., i. pp. 1 and 32 (1846–47).
Papilionidæ, Bates, Journ. Ent., 1861, p. 219.
Papilionidæ and *Pieridæ*, Trimen, Rhop. Afr. Aust., i. pp. 10 and 24 (1862).

IMAGO.—First pair of legs in both sexes as fully developed in all respects as the other pairs. Discoidal cell always closed in both fore and hind wings.

LARVA.—In *Pierinæ* cylindrical, tapering to each extremity, without processes or other appendages, usually with very short pubescence : in *Papilioninæ* stouter, often thickened towards anterior extremity, smooth, or with long tubercular prominences, possessing on second (prothoracic) segment an exsertible strongly-scented forked tentacle.

PUPA.—More or less angulated; head singly pointed in *Pierinæ*, bifid in *Papilioninæ*; attached erectly, obliquely, or horizontally, by the tail and by a girth round the middle.

The two Sub-Families, *Pierinæ* and *Papilioninæ*, which compose this Family, are readily distinguishable by the characters of the perfect insect. The *Pierinæ* have in the *fore-wings* the first disco-cellular nervule either obsolete or very short; the third disco-cellular and lower radial nervules disposed as in other butterflies; the branches of the sub-costal nervure varying in number from three to five; no internomedian nervule; and the internal nervure rarely present, and then very short and slender, and running into the submedian nervure : in the *hind-wings* the inner margins are convex, forming a groove or channel beneath the abdomen, and the internal nervure is long and strong; the precostal nervure is always simple. The palpi are of moderate size and porrected; the antennæ have a straight club. The legs are short, and there is no appendage to the tibia of the first pair; the tarsal claws are bifid, and usually provided with pulvilli and paronychia. The abdomen is slender, and nearly always of moderate length or rather short. The *Papilioninæ* have in the *fore-wings* the first disco-cellular nervule almost always well developed; the third disco-cellular nervule so disposed in relation to the median nervure that the lower radial appears to be a fourth median nervule; the sub-costal nervure always five-branched (except in *Parnassius* and *Hypermnestra*, where it is four-branched); the interno-median nervule present in four genera; and the internal nervure always present (except in *Doritis*), and running to a free termination on the inner margin; in the *hind-wings* the inner margins are concave (often folded back), leaving the abdomen perfectly free, and the internal nervure is wanting; the precostal nervure is branched, and forms with the costal nervure a small prediscoidal cell (in all but three genera). The palpi are usually very small and short, and closely appressed to the head; the antennæ have a curved club. The legs are long, and the fore-tibiæ have on the inner side a conspicuous projecting appendage, while the tarsal claws are simple and without appendages. The abdomen is thicker than usual, and enlarged posteriorly.

The full development of the first pair of legs in both sexes, and the median suspensory silken girth of the chrysalis, are the main characters of this Family, and together indicate a nearer relationship to the *Heterocera* than the three preceding Families exhibit. Of its two Sub-Families, the *Papilioninæ* better exhibit this relationship, and in the median appendage of their fore-tibiæ show a point of affinity to the *Hesperidæ*—the last Family of butterflies.

In numerical strength the *Papilionidæ* stand fourth of the five Families, about fourteen hundred species being recorded. Of these, the *Pierinæ* muster a large majority (about nine hundred species), and they are also much more prolific in recognised generic forms, having thirty-nine or forty, while the *Papilioninæ* have but thirteen. The latter contain, however, not only a far larger proportion of species remarkable for size, variety of form, and splendour of colouring, but also, in the genus *Ornithoptera*, the very princes of the butterfly world, gigantic in stature, and of the richest and most intense hues. The *Papilioninæ* have not the same world-wide range as the *Pierinæ*, being mostly natives of tropical and sub-tropical lands, and even *Parnassius* (alpine above all the genera) not extending within the Arctic circle. The *Pierinæ* are mostly of the middle size or rather smaller, and present a simple, but usually bright and conspicuous, colouring of white, yellow, or orange, tipped or bordered in many cases with black; some of the tropical and sub-tropical forms are, however, more variegated, and *Teracolus* and *Anthocharis* in most cases present a brilliant patch of orange, red, crimson, or violet at the tips of the fore-wings.

The larvæ of the two Sub-Families (as indicated above) differ considerably, the robuster caterpillar of the *Papilioninæ* alone possessing the strongly-scented extrusible forked tentacle on the back of the second segment.

In South Africa the *Papilionidæ* are not very extensively represented; the known *Pierinæ* numbering nine genera and sixty-one species, and the *Papilioninæ* one genus containing fifteen species. As compared with the Ethiopian region generally, this number falls short of only one genus (*Larinopoda* in *Pierinæ*); but it is very much scantier in species, the region yielding about a hundred and forty *Pierinæ* and sixty-two *Papilioninæ*.

Many of the most perfect cases of mimicry of butterflies of other tribes occur in this Family, as well as a certain number within the limits of the Family, Sub-Family, or even Genus itself. These will be found more particularly referred to under the several subdivisions and species concerned.

Sub-Family I.—PIERINÆ.

Pierinæ, Colianæ, and *Liciniianæ,* Swainson, Cab. Cyc., Hist. and Nat. Arr. Ins., pp. 88, 89 (1840).
Pierides, Boisd., Sp. Gen. Lep., i. p. 402 (1836).
Pieridæ, Dup., " Cat. Lep. Fr., p. 23 (1846)."
Pieridæ, Doubl., Gen. D. Lep., i. p. 32 (1847).
Pieridæ, Trim., Rhop. Afr. Aust., i. p. 24 (1862).
Pierinæ, Bates, Journ. Ent., 1864, p. 177.
Pierinæ, Butl. (Revision), Cist. Ent., i. p. 33 (1870).

IMAGO.—*Head* small or of moderate size, sometimes rather broad, always more or less hairy, and usually densely so clothed; *eyes* round, prominent, smooth; *haustellum* (*maxillæ*) long; *palpi* compressed, rather short, usually with rather bristly hair beneath, but sometimes with lengthened scales,—basal joint much developed, never shorter, and usually twice (sometimes thrice) as long as the second one,—terminal joint always small, seldom as long as the second, and frequently minute; *antennæ* usually of moderate length (very long in the genera *Leptalis* and *Euterpe*), with an elongate-ovate club, but in many cases shorter and thicker, with a very gradually-formed blunt or truncate club.

Thorax usually not large (in *Leptalis, Leucophasia, Terias,* and *Pontia* very short and slender), but moderately robust in some genera (*Callidryas, Gonepteryx, Hebomoia,* &c.), often with long dense silky hair above. *Wings* large (the hind-wings mostly larger in proportion to the fore-wings than in other groups,—in some species of *Leptalis* actually twice as large as the fore-wings), usually of moderate thickness, but in the slender-bodied genera very thin, and in the stout-bodied ones of considerable thickness; discoidal cell always closed (in *Leucophasia* extremely short). *Fore-wings* apically rather prominent, sometimes sub-acuminate, rarely (*Gonepteryx*) shortly and acutely falcate: subcostal nervure 3–5-branched,—the first nervule (occasionally uniting with costal nervure) always (except in *Leptalis* and *Leucophasia*) —and the second one very frequently—given off before extremity of discoidal cell; first disco-cellular nervule almost always wanting, the upper radial springing from subcostal nervure beyond (or sometimes just at) extremity of cell; internal nervure rarely developed (*Callidryas, Hebomoia, Gonepteryx*), and then very short and slender, running into submedian nervure. *Hind-wings* rounded, very rarely (*Gonepteryx*) angulated on first or second median nervule, or (a few species of *Callidryas*) produced into a long blunt projection at anal angle: costa usually much arched, and very prominent basally; hind-margin sometimes slightly or moderately dentate; inner margins convex, and forming a more or less complete groove under abdomen; radial nervule almost always originating nearer to subcostal than to median nervure; internal nervure always well-developed and often long. *Legs* rather short (except in *Leptalis*), usually slender, almost hairless except at base of femora beneath; tibiæ short, the terminal spurs of the middle and hind pairs very small; tarsi long, especially their first joint, and with bifid terminal claws.

Abdomen slender, arched, usually of moderate length or rather short (long in *Leucophasia*, and in *Leptalis* extending beyond hindwings).

LARVA.—Elongate, cylindrical, without segmental constriction, somewhat tapering anteriorly and posteriorly, without spines or other appendages, clothed with a more or less developed very short pubescence, rarely with longer hairs.

PUPA.—Rather elongate, usually with a more or less prominent median dorsal ridge (higher on thorax), and with the thorax and base of abdomen laterally angulated; often much curved backward, and with wing-covers forming a very convex ridge upon breast; head with a single acute projection in front directed forward.

The affinity of this Sub-Family to the *Papilioninæ* is clearly shown by the perfect condition of the first pair of legs, and the general form and silken girth of the chrysalis; but it differs widely in the neuration of the fore-wings, the form of antennæ and palpi, the convex (instead of concave) inner margin of the hind-wings, the absence of any middle spur or process on the tibiæ of the first pair of legs, the form of the caterpillar, and the single-pointed (instead of bifid) head of the chrysalis. Doubleday (*op. cit.*) pointed out that some *Pierinæ* approached very near to the *Lycænidæ*, instancing specially the genus *Terias;* and Mr. Butler has described and figured as one of the *Pierinæ*, *Larinopoda lycænoïdes*, a very curious West-African butterfly, which, with a general resemblance to *Pontia*, appears structurally to be in some respects intermediate between *Eronia* and the Lycænide genus *Deloneura*.[1] I have above (vol. ii. pp. 225. 227, and 234) called attention to the relation to the *Pierinæ* shown by the perfect fore-legs in both

---

[1] Mr. W. F. Kirby (*Rhop. Exot.*, Pt. ii., Oct. 1887) states that *Larinopoda* is now more correctly referred, both by Mr. Butler and others, to the *Lycænidæ*, being nearly related to *Liptena* and allies.

sexes of the Lycænidæ genera *Deloneura* (apparently), *Arrugia*, and *Lachnocnema*; and I understand that similar structure has been found in some American *Lycænidæ*.

Many of the most familiar forms among butterflies belong to the *Pierinæ*, such as the European "Garden Whites" (*Pieris*), "Orange-Tips" (*Anthocharis*), "Brimstones" (*Gonepteryx*), and "Clouded Yellows" (*Colias*), and throughout the globe members of this Sub-Family are conspicuous by their abundance, activity, and prevalent tints of white or yellow. For by far the greater part they are insects of medium size, many being rather under than over it—such as the genera *Terias*, *Pontia*, *Leucophasia*, *Anthocharis*, and *Elodina*. The American genus *Nathalis* contains only two or three very small species, and some kinds of *Terias* expand only 1¼ inches, but the dwarfs of the tribe are the South-American species of *Leucidia* (closely allied to *Terias*), which are less than an inch across the wings. The largest known *Pierinæ* (*Hebomoia* in the Oriental Region and *Amynthia* in the Neo-Tropical) do not expand above 4 inches.

As Mr. A. R. Wallace has shown in his excellent memoir "On the *Pieridæ* of the Indian and Australian Regions,"[1] this Sub-Family is specially noticeable for its general and uniform distribution over the earth, species of it being met with in all latitudes and at all elevations, and in every kind of station, whether forest-clad or open, sheltered or exposed. These butterflies are nevertheless far more numerous and varied in tropical countries, and most of the large and more richly-coloured forms are sylvan in their haunts. The extensive Indian and Australian genus *Thyca* affords a striking instance, its numerous members being conspicuous for brilliant red and yellow markings on a dark ground, the under side of their wings being usually more highly ornamented than the upper side.

As a rule, there is much disparity between the sexes, both as regards colouring and marking. Except in the few cases where the ♀ only directly mimics other butterflies, by far the greater divergence from the ordinary plain type of the Sub-Family is shown by the ♂ s; and the latter also sometimes possess special sexual badges, viz., a serrated costa in the fore-wings (*Prioneris* and *Callidryas*), and patches of elevated scales and brushes of silky hair on the wings (both in *Callidryas*, and the former to a slight extent in *Colias*).

The Ethiopian Region is not rich in *Pierinæ*; out of a roll of some forty genera and nine hundred species, it possesses but nine genera and about a hundred and forty species. Of these, only two genera, *Mylothris* and *Herpænia*, are peculiar to the region, but the species are nearly all confined to it. South Africa has yielded representatives of all the genera and sixty-one species, viz., *Pontia*, 1 sp., *Terias* 7, *Mylothris* 3, *Pieris* 14, *Herpænia* 1, *Teracolus* 29, *Colias* 1, *Eronia* 4, and *Callidryas* 1.

[1] *Trans. Ent. Soc. Lond.*, 3rd Series, vol. iv. p. 301 (1867).

The gregarious and onward-flying habits of the *Pierinæ* are well known, and reach their climax in the genus *Callidryas* (see *infra*, p. 185, and vol. i. p. 31), whose species have exceptional powers of flight, well indicated by their large, solid thorax and thick, strongly-veined wings.

Mimicry is strikingly exhibited in this Sub-Family, and it was mainly the study of the deceptively exact imitation of various South-American *Danainæ* by species of the Pierine genus *Leptalis* that enabled Mr. Bates to give to science the satisfactory explanation of the phenomenon to which I have adverted in vol. i. pp. 35–36. The ♀ s of *Perrhybris*, a genus very closely allied to *Pieris*, and a species of *Euterpe*, also mimic various *Danainæ*; and one of the latter genus (*E. Tereas*) closely imitates the ♀ *Papilio Zacynthus*. Mr. Wallace, in the paper above cited, commented on the imitation by the ♀ s of various Malayan species of *Eronia* of the common kinds of *Danais* in the same region, and further brought to notice that mimicry occurred within the limits of the *Pierinæ* themselves, specifying several cases in which the slow-flying and showy species of *Thyca* are simulated by species of the genera *Prioneris* and *Pieris*.

I am able to adduce two similar cases in South Africa, where the slow-flying *Mylothris Agathina* is nearly copied by both sexes of *Pieris Thysa* and by the female of *Eronia Argia*; while in Western Africa *Mylothris Poppea* (Cram.) is the model followed by *Pieris Rhodope* (Fab.).

The genus which is by far the best developed in South Africa is *Teracolus*, white or yellow butterflies with a patch of bright colour—orange, red, crimson, or violet—at the tip of the fore-wings. I have discussed at length (*infra*, p. 82) the various groups of this large and most difficult genus, which contains some of the most beautiful of all the *Pierinæ*.

Seasonal dimorphism—or palpable difference between the earlier and later broods of the same species—has been shown to occur in the well-known cases of *Pieris Napi* (Linn.), of which *P. Bryoniæ* (Ochs.) is the early or "winter" form, and *Anthocharis Ausonia* (Hübn.), of which *A. Belia* (Cram.) is the "winter" brood. The progress of observation of the life-history of butterflies has of late years gone to show that instances of this kind of dimorphism are not, as was supposed, limited to countries where a severe winter prevails, but occur (and are probably more numerous) in tropical and sub-tropical regions where the climate is divided into wet and dry seasons.[1] Three cases in South Africa certainly exist, viz., those of *Pieris Charina*, Boisd., *P. Thysa*, Hopff., and *P. Severina* (Cram.), in each of which the winter or dry-season brood on the South-East coast is considerably smaller, and in the two latter with enlarged black markings. Probable cases are that *Pieris alba* (Wallengr.) is the dry-season brood of *P. Pigea*,

---

[1] See Mr. L. de Nicéville's paper on Calcutta *Satyrinæ* in the *Journal of the Asiatic Society of Bengal*, vol. lv. p. 239 (1886).

Boisd., *P. Abyssinica*, Lucas, the same of *P. Gidica*, Godt., and *Teracolus Jobina*, Butl., the same of *T. speciosa* (Wallengr.), as, in each case, the smaller form appears only in the dry season and the larger one only in the wet season. These latter cases can only be proved by careful breeding from the eggs of one or both forms; at present, we have only the recorded dates of capture (and far too few of them) to go upon. When really systematic and extensive rearing of successive generations from the ova shall have been fully carried out and recorded by competent lepidopterists, it cannot be doubted that a great and most instructive rectification of our so-called "species" will ensue, not only in this Sub-Family, but throughout the Order.

Genus PONTIA.

Pontia (Fab., 1807,—part.), Boisd., Sp. Gen. Lep., i. p. 430 (1836).
Pontia, Doubl., Gen. Diurn. Lep., i. p. 40 (1847); Trim., Rhop. Afr. Aust., i. p. 25 (1862).
Nychitona, Butl., Cist. Ent., i. p. 41 (1870); Moore, Lep. Ceylon, i. p. 117 (1881).
Leptosia (Hübn., 1816,—part), Distant, Rhop. Malay., p. 287 (1882-86).

IMAGO.—*Head* wide, clothed above and in front with rather rough short hair; *eyes* large, smooth; *palpi* short, acute, the tips far apart, —only terminal joint porrected, the others appressed to the face and clothed inferiorly with long hairs; *antennæ* of moderate length, rather thick, with a gradually-formed but well-marked fusiform club.

Thorax very small and short, narrower than head, clothed with scales and short hairs; prothorax forming a marked neck. *Forewings* elongate, with costa much arched, apex very rounded, and hindmargin convex; costal nervure terminating at about two-thirds length of wing; subcostal nervure three-branched, the first nervule emitted far (about one-third), the second a little before extremity of discoidal cell; upper radial united to third subcostal nervule at some distance beyond, lower radial at extremity of cell; lower disco-cellular nervule long, strongly curved inwardly, joining third median nervule at a pronounced angle; discoidal cell rather broad, and three-fifths the length of wing; submedian nervure curved downward near base. *Hind-wings* rather bluntly obovate; costa very slightly curved beyond basal lobe; apex somewhat pronounced; hind-margin very convex; anal angle rounded; inner margin convex at base only, so that no channel or groove is formed beneath abdomen; costal nervure ending on costa not far beyond middle; subcostal nervules much curved, especially upper one, which is emitted a full third before extremity of discoidal cell; upper disco-cellular extremely short (so that radial nervule appears almost to spring from second subcostal), lower one long, angulated, oblique; discoidal cell about half length of wing; internal nervure strongly bent downward, terminating at a little beyond middle. *Legs* of moderate length, slender, clothed with scales; tibiæ slightly, tarsi very bristly; terminal spurs of middle and hind tibiæ very small.

Abdomen very slender, elongate, arched, laterally compressed.

This is a very well-marked genus, containing a few closely-allied species, and having a very wide range over the tropical and sub-tropical regions of the Old World, from Western Africa to North-Western Australia. Its extreme slenderness of structure, long abdomen, and contour of wings give it a strong resemblance to the Palæarctic genus *Leucophasia*—which it also resembles in its plain white livery, sometimes blackish-tipped in the fore-wings—but it is widely different in neuration, having only two instead of five subcostal nervules in the fore-wings, and very large instead of extremely small discoidal cells. On the whole, *Pontia* appears to be most nearly related to the small Australian genus *Elodina*, Felder, which presents very similar neuration as well as colouring, but has the apex of the fore-wings pointed, and the first subcostal nervule of the hind-wings originating near the extremity of the discoidal cell. Next to *Elodina*, the genus *Terias* comes nearest to *Pontia*.

In habits the *Pontiæ* greatly resemble the European *Leucophasiæ*. They frequent wooded places and flit feebly about bushes and herbage, preferring shade to sunshine.

Besides *P. Alcesta* (Cram.), which is the only species known to occur in South Africa, the Ethiopian Region yields *P. Medusa* (Cram.),[1] a larger form from the Gold Coast with greenish under-side markings, and two from Madagascar—one of which (*P. sylvicola*, Boisd.) is exceedingly close to *Alcesta*, while the other (of a singularly pure white) appears to be undescribed.

239. (1.) **Pontia Alcesta**, (Cramer).

Papilio Alcesta, Cram., Pap. Exot., iv. pl. ccclxxix. f. A (1782).
Papilio Narica, Fab., Ent. Syst., iii. 1, p. 187, n. 578 (1793).
Pontia Narica, Boisd., Sp. Gen. Lep., i. p. 432, n. 3 (1836); App. Voy.
  Deleg. Afr. Aust., p. 586 (1847).
Pontia Alcesta, Wallgrn., K. Sv. Vet.-Akad. Handl., 1857; Lep. Rhop.
  Caffr., p. 6.
Pontia Alcesta, Trim., Rhop. Afr. Aust., i. p. 26, n. 12 (1862).
„   „   Hopff., Peters' Reise Mossamb., Ins., p. 349.

PLATE X. fig. 1 (♀).

Exp. al., 1 in. 5–9 lin.

Pure-white; fore-wing with a blackish apical border and a blackish discal spot. Fore-wing: costal margin irregularly hatched and speckled with blackish to about middle, and sometimes very sparsely speckled beyond middle; apical border moderately wide at apex, but attenuated to a point at its commencement on costa at extremity of first subcostal nervure, and similarly at its hind-marginal termination (which

[1] *Pap. Exot.*, ii. pl. cl. fig. F (1779).

is usually a little above the extremity of third median nervule, but sometimes below it); inner edge of apical border irregularly excavate above and below upper radial nervule; rarely a minute blackish spot at extremity of second median nervule, and occasionally another on inter-nervular fold below it; discal spot rather large, sub-reniform, not far from hind-margin, its lower extremity on second median nervule and its upper one rather above third median nervule; circumference of this spot not sharply defined, but somewhat suffused. *Hind-wing:* without marking, except on hind-marginal edge, where there is a series of minute inter-nervular blackish linear marks (sometimes all but imper-ceptible), and occasionally traces of a series of similar minute marks at extremities of nervules. UNDER SIDE.—*Hind-wing and narrow costal, apical, and hind-marginal border of fore-wing more or less faintly tinged with straw-yellow, and sparsely hatched or freckled with short, thin, fuscous lineolæ. Fore-wing:* fuscous lineolæ extend below costal border, more especially near base, where they cover about a third of discoidal cell as far as median nervure; discal spot somewhat smaller and fainter than on upper side; along hind-marginal edge a series of minute but very distinct inter-nervular blackish spots, of which between first median nervule and submedian nervure there are two. *Hind-wing:* lineolæ pretty evenly distributed, but sparse generally beyond middle; about middle, and about midway between middle and hind-margin, the lineolæ tend more or less prominently to group themselves into two nearly straight and almost parallel obliquely transverse striæ, of which the outer is much more pronounced; a series of inter-nervular hind-marginal spots, quite as in fore-wing, but slightly larger.

The sexes do not differ except in size, and in the ♀ having a rather broader apical border on the upper side of the fore-wing, and the under side usually paler in tint and less freckled.

A dwarf ♂, taken near Pinetown, in Natal, by Colonel Bowker, expands only 1 in. 2½ lin. In another ♂, from the same locality, the discal blackish spot is on the upper side of the fore-wings much reduced, and on the under side obsolescent.

I do not think that the *Narica* of Fabricius, in which the apical blackish tip of the fore-wing is wanting, or almost wanting, is separable from *Alcesta*;[1] the four specimens from West Africa (Ashanti, Gold Coast) which I examined in the British Museum do not in other respects exhibit any special feature, except in the rather closer freckling of the under side of the hind-wing. I find that South-African specimens of *Alcesta* almost always exhibit a smaller but better-defined discal spot on the fore-wings, and a yellower tinge on the under side of the hind-wings and border of the fore-wings than are found in recognised West-African examples of the species. No South-African individual has come under my notice in which there was any failure of the apical fuscous border of the fore-wings.

Alcesta, as found in Natal, Zululand, Delagoa Bay, and Queriuba, and also in Western Africa, is very near *Xiphia*, Fab., inhabiting India, Ceylon, and part of the Indian Archipelago, but is distinguishable by having both the

---

[1] Cramer's very rough figure of the upper side apparently represents a ♂ with scarcely any apical border in the fore-wings. He notes the under side as wholly white, and gives the Guinea Coast as *habitat*.

blackish markings of the fore-wings smaller and not so dark, and the freckling of the under side less dense. The very closely-allied *Sylvicola*, Boisd. (of which *Nupta*, Butler, must, I think, be considered as a variation only), from Madagascar, may be separated from *Alcesta* by the comparative or almost complete failure of the discal spot, and by the duller, narrower, inwardly more suffused apical border.[1]

Although I was on the look-out for this butterfly when visiting Natal in the summer of 1867, I found it only on three occasions, frequenting the borders of woods on the coast, at the Umgeni and near Verulam. It has very much the flight and habits of the European *Leucophasia*, flitting slowly and feebly about the herbage in shady spots. Colonel Bowker has forwarded a good many examples, taken near D'Urban and Pinetown in March, April, and May; the few examples that I met with were on the wing at the end of February and the end of March.[2]

Mr. Druce notes (*Proc. Zool. Soc. Lond.*, 1875, p. 414) that the specimens of this butterfly brought from Angola by the late Mr. J. J. Monteiro were very small.

Localities of *Pontia Alcesta*.

I. South Africa.
   E. Natal.
      a. Coast Districts.—D'Urban. Verulam. Pinetown (*J. H. Bowker*).
   F. Zululand.—St. Lucia Bay (the late *Colonel H. Tower*).
   H. Delagoa Bay.—Lourenço Marques (*Mrs. Monteiro*).

II. Other African Regions.
   A. South Tropical.
      a. Western Coast.—"Angola (*J. J. Monteiro*)."—Druce. Congo and Loango (Chinchoxo).—Coll. Brit. Mus.
      b. Eastern Coast.—Querimba.—Coll. Brit. Mus. "Tchouaka (*Raffray*)."—Oberthür.
   B. North Tropical.
      a. Western Coast.—Gold Coast (Accra and Ashanti) and Sierra Leon.—Coll. Brit. Mus. "Lower Niger (W. A. Forbes)."—Godman and Salvin.

Genus TERIAS.

Terias, Swainson, Zool. Illustr., i. text to pl. 22 (1820-21).
Xanthidia, Boisduval, "Lep. Amer. Sept., p. 48 (1833)."
Terias, Boisd., Sp. Gen. Lep., i. p. 651 (1836).
  „   Doubl., Gen. D. Lep., i. p. 76 (1847).
  „   Trim., Rhop. Afr. Aust., i. p. 75 (1862).
  „   Butl., Cist. Ent., i. p. 44 (1870), and (Revision) Proc. Zool. Soc. Lond., 1871, p. 526.
  „   Moore, Lep. Ceylon, i. p. 118 (1881).
  „   Distant, Rhop. Malay., p. 302 (1882-86).

IMAGO.—*Head* small, clothed densely with short hair; *eyes* smooth, globose, large; *palpi* very short, slender, scarcely projecting beyond

---

[1] There is another form in Madagascar, of which two ♂s and a ♀ are in the South-African Museum, presented by Mr. E. L. Layard, who captured them, I believe, on the N.W. coast. It is of a singularly pure white, with the apical fuscous much reduced (especially in the two ♂s), but the discal spot almost as large and rounded as in *Xiphia*, and with the under side all but pure white, and its frecklings very faint and sparse.

[2] Two D'Urban examples, forwarded by Mr. A. D. Millar, are dated 17th September 1887.

forehead, compressed, clothed beneath with closely-appressed scaly hairs,—terminal joint very small, blunt, scaly; *antennæ* rather short and thick, with an elongate, very gradually formed, depressed club, blunt at the tip.

Thorax very short, slender, clothed with scales and short hair. *Wings* large and thin: *fore-wings* rather elongate, more or less truncate; costa strongly arched at base, but thence only slightly so; apex usually rounded or moderately pronounced, sometimes acuminate; hind-margin more or less convex; posterior angle usually well-marked, but sometimes rounded; inner margin slightly concave about middle; costal nervure much thicker than the others, ending at or a little beyond middle; four subcostal nervules,—the first originating about one-third, the second only a little, before extremity of discoidal cell,—the third and fourth forking at some distance beyond cell, not quite half-way to apex; upper radial nervule united to subcostal nervure about midway between end of cell and third subcostal nervule,—lower one originating at junction of disco-cellular nervules, of which the upper is shorter and less curved inwardly than the lower; discoidal cell rather broad, terminating somewhat truncately about middle. *Hind-wings* bluntly obovate, but sometimes with hind-margin sub-angulated at extremity of second median nervule; costa more or less strongly arched from base; anal angle mostly rather marked; inner margin moderately convex, so as to form a pretty complete but wide and shallow groove for abdomen; pre-costal nervure much atrophied, scarcely traceable; costal nervure arched, ending beyond middle (sometimes not very far before apex); first subcostal nervule arched, given off just before extremity of discoidal cell; upper disco-cellular nervule rather short, almost straight—lower one fully twice as long and obliquely angulated; discoidal cell rather short, but widening outwardly, terminating truncately before middle. *Legs* slender, scaly; fore and middle femora longer, hind-femora much shorter, than tibiæ; terminal tibial spurs minute; all tarsi longer than respective tibiæ, and thickly spinulose.

Abdomen slender, elongate (not quite as long as inner margin of hind-wings), much compressed laterally, slightly arched.

LARVA.—Of the ordinary Pierine form, towards extremities only slightly attenuated; clothed with a very short pubescence.

PUPA.—Slender, laterally compressed, much attenuated and sharply pointed at extremities; wing-covers very prominent ventrally, laterally flattened into a very convex keel.

These characters of larva and pupa are from figures of the early stages of *T. Hecabe*, (Linn.), in Horsefield and Moore's *Catalogue of Lepidopterous Insects in the Museum of the H. E. India Company* (1857), pl. i. ff. 11, 11*a*, and in Moore's *Lepidoptera of Ceylon*, 1881, pl. 45, f. 1*c*.; and (as regards the pupa only) from a figure of (presumably) a Japanese species given by Mr. A. G. Butler on pl. vi. of the *Trans. Entom. Soc. Lond.* for 1880, and the chrysalis skin of a Cape specimen of *T. Brigitta*, (Cram.). The Cingalese larva of *Hecabe* is depicted as pale yellowish-green with a dark-brown head, and the pupa as dark purplish-brown. The larvæ of this genus are stated by Doubleday (*Gen. D. Lep.*, i. p. 78) and Thwaites (*Lep. Ceylon*, i. p. 119) to feed on *Leguminosæ*.

Terias is a very extensive genus of closely-allied forms, occurring throughout the tropical and sub-tropical zones of both hemispheres, and in America extending as far northward as Virginia and Pennsylvania in the United States. It finds, however, by far its largest development in America (South and Central), whence about half the recorded species have been received; and in the second place comes the Indian region, yielding some thirty species. The Australian and African regions seem about equally poor in comparison, some twelve forms being known to inhabit each of them. I recognise seven forms as natives of South Africa, but am very doubtful as to the actual limits which can be defined as separating them in the "species" sense.

The difficulty of dealing satisfactorily with the numerous forms of *Terias* is admitted by all lepidopterists, and has been the subject of comment by E. Doubleday (*op. cit.*), Bates' (*Journ. of Ent.*, 1861, p. 245), Butler (*loc. cit.*, and *Ann. and Mag. Nat. Hist.*, 1886, p. 212), and Distant (*op. cit.*). Bates has pointed out the extreme similarity of certain forms inhabiting such widely separated countries as St. Domingo (West Indies) and the Malayan Archipelago; and as regards variation in one and the same locality, Butler (*Trans. Ent. Soc. Lond.*, 1880, p. 197, pl. vi.) has described and figured a series of seventeen (selected from no less than 154) examples taken at Nikko in Japan, exhibiting every gradation, "from the most heavily-bordered of the Japanese representative of *T. Hecabe* to the palest *T. Mandarina*, in which the border has practically disappeared." In connection with this remarkable case, it is most interesting to note that the late Mr. H. Pryer has recorded (*op. cit.*, 1882, p. 488) his having bred in Japan from eggs laid by *Mandarina* all the broader-bordered variations (11) figured by Butler, and also his having personally observed that *Mandarina* and the narrow-bordered variations that approach it are merely the autumnal brood of the same butterfly. Butler inclined to attribute this excessive variation to the crossing of various races of *Terias*,[1] but Mr. Pryer's experience certainly seems to show that it is a case of seasonal dimorphism in a transitional condition.

The butterflies of this genus are all below the medium size, and some of them very small; their normal colouring and pattern are very simple, consisting of some tint of yellow or orange (a few species are white), with a black border that is sometimes confined to the fore-wings. This black border altogether disappears in some forms, while in many it is very broad in the fore-wings. Several species, both in the Old

---

[1] A case of a ♂ *T. Zoë* having been captured *in copula* with a ♀ *T. Brigitta*, is noted below, p. 16.

and New Worlds, are further ornamented by a black stripe along the inner margin of the fore-wings (like the corresponding marking so prevalent in many females and some males of the genus *Teracolus*), and in the pretty American group represented by *T. Delia* and *T. Elathea*, Cram., this stripe is in the ♂ edged with orange on the inner margin itself.

These insects are for the most part of weak flight, and keep near the herbage, but those of them that are more partial to open ground are moderately active on the wing, and have the Pierine habit of flying onward instead of keeping about one spot. Doubleday (*op. cit.*, p. 78) observes that *T. Nicippe* (Cram.)—which in colouring and pattern strongly resembles one of the deeper-tinted species of *Colias*—in the United States flew abundantly in clover-fields in company with *Col. Philodice* and *Col. (Meganostoma) Caesonia*, and that in flight it was more like those swift butterflies than its own congeners.

The South-African forms are all yellow, with more or less developed black borders; the males having the ground-colour bright gamboge or chrome and the border deep black, while the females are of a paler, more sulphur tint, with the border more dusky. Three groups are typically represented by *Terias Brigitta* (Cram.), *T. floricola*, Boisd., and *T. Desjardinsii*, Boisd.: with the first is associated *T. Zoë*, Hopff.; with the second *T. Æthiopica*, mihi (= *Senegalensis*, Geyer, *nec* Boisd.), and *T. Butleri*, mihi; and with the third *T. regularis*, Butl.

The *Brigitta* group has the fore-wing border very broad, and the hind-wing one much narrower, or even (in the ♀) reduced to small nervular spots; the upper side of the ♀ is more or less irrorated basally with fuscous scales: the under side of both sexes is also faintly irrorated, the hind-wing being feebly marked with some transverse dusky streaks.

The *Floricola* group (not separable from the Indian *T. Hecabe* (Linn.), group) has the fore-wing border broad or very broad hind-marginally and more or less deeply bi-excavated or bi-sinuated on its inner edge, and the hind-wing border very narrow, or represented only by small spots or even dots; the upper side of the ♀ is not irrorated; the under side of both sexes is free from irroration, and its markings usually much more distinct, dentated irregularly, and ferruginous, there being commonly a conspicuous patch of the latter colour close to the apex of the fore-wings.

The *Desjardinsii* group has the fore-wing border moderately broad, narrow, or very narrow, with its inner edge more as in the *Brigitta* group, and the hind-wing border varying from narrow to purely macular; the under-side markings are intermediate in character between those of the two foregoing groups respectively; the hind-wings are bluntly angulated at extremity of second median nervule.[1]

[1] The allied forms from Central and South America separated from *Terias* by Mr. Butler in 1870, under the generic title of *Sphaenogona*, present a sharp angle or even prominent caudate projection at the same point of the hind-wings.

The species inhabiting South Africa are prevalent on the eastern side, and three of them, viz., *Brigitta, Zoë*, and *Regularis*, penetrate as far as the western central districts of the Cape Colony, but are scarce there.

240. (1.) Terias Brigitta, (Cramer).

♀ *Papilio Brigitta*, Cram., Pap. Exot., iv. pl. cccxxxi. ff. B, C (1782).
Pieris Brigitta, Godt., Ent. Meth., ix. p. 135, n. 53 (1819).
♂ ♀ *Terias Brigitta*, Boisd., Sp. Gen. Lep., i. p. 676, n. 38 (1836).
    "   "   Trim., Rhop. Afr. Aust., i. p. 80, n. 52 (1862).
Terias Rahel, Hopff. [part], Peters' Reise Mossamb., Ins., p. 368 (1862).
♂ *Terias Canduce*, Feld., Reise d. Novara, Lep., ii. p. 213, n. 228 (1865).
♀ (as ♂) *Terias Brigitta*, Staud., Exot. Schmett., i. pl. 16 (1884).

Exp. al., (♂) 1 in. 5½–7 lin.; (♀) 1 in. 5–9 lin.

♂ *Bright gamboge-yellow; fore-wing with a broad black border, hind-wing with a very narrow one, both dentate internally on nervules. Fore-wing:* base very narrowly speckled with blackish; on costa close to base some blackish speckling, soon giving place to a narrow costal black border, which extends to about middle and thence rapidly widens, to become very broad apically, whence it gradually narrows along hind-margin to posterior angle; hind-marginally this border is more or less completely interrupted, between first median nervule and sub-median nervure, by the yellow ground-colour; the yellow interrupting streak more or less obscured with blackish specks; on costal edge, near apex, three minute pale-yellowish inter-nervular marks indenting the black border. *Hind-wing:* hind-marginal narrow border commencing very faintly at extremity of costal nervure, widening about extremity of second subcostal nervule, and thence narrowing to a point at extremity of first median nervule (in one specimen this border is extremely narrow, and becomes merely macular on the nervules in its lower portion); inner margin as far as submedian nervure bordered with pale sulphur-yellowish; costa also narrowly edged with that tint; base very narrowly and (usually) faintly speckled with blackish. *Cilia* yellow, irregularly tinged with reddish (especially in fore-wing). UNDER SIDE.—*Rather dullish pale-yellow, with the hind-wing (chiefly marginally) and apical border of fore-wing more or less tinged with creamy-reddish*, for the most part very minutely speckled with fuscous atoms. *Fore-wing:* a very faint, slender, terminal disco-cellular striola, more or less interrupted in its middle portion; a faint (sometimes obsolete), transverse, short, sub-apical reddish-grey streak, from near costa to below second radial nervule. *Hind-wing:* a minute fuscous spot close to base below costal nervure; a transverse row of three rather larger reddish-grey spots near base, of which the middle one is in discoidal cell; a double very slender terminal disco-cellular striola, more distinct and reddish than that of fore-wing; beyond middle, three obliquely-transverse rather irregular reddish-grey streaks,—of which the first is short, just beyond middle, from costal nervure to first subcostal nervule,—the second, at some distance beyond middle, rather broad, from first subcostal nervule to submedian nervure, but very widely interrupted from third to first median nervules,—and the third, considerably beyond the second, short, dentate on median nervules. On hind-margin of both wings a series of minute black spots marks the extremities of the nervules.

♀ *Much paler, inclining to sulphur-yellow; black border not so dark and more restricted in fore-wing and wanting in hind-wing.* Forewing: costa rather faintly irrorated with dusky atoms from base to a point about one-third of its length; black border apically as broad as in ♂, its internal edge less regularly excavated, its costal commencement much more attenuated and farther from base, and its hind-marginal termination very abrupt on first median nervule; small yellow inter-nervular costal streaks better marked; rarely the hind-marginal border is interrupted by a yellow streak on second median nervule; a small (usually minute) black spot at extremity of submedian nervure. *Hind-wing:* a hind-marginal series of very small (sometimes minute, and more rarely towards apex of moderate size) black spots, at extremity of each nervule as far as first median. UNDER SIDE.—Hind-wing and costal-apical border of fore-wing of an uniform dull pale rufous-creamy, varying in depth of tint, its minute fuscous speckling more general than in ♂; markings usually duller and fainter than in ♂, sometimes very indistinct. The ♀ exhibits much variation in the tint of the under side, some examples being of quite a pallid cream-colour with only a slight reddish tinge, while in others the rufous colour is very pronounced.

Mr. Butler has recently (*Ann. and Mag. Nat. Hist.*, March 1886, p. 214) expressed the opinion that *Brigitta* is the Southern representative of *Candace*, Feld., from Abyssinia. An examination of the ♂ and ♀ of the latter in the British Museum has convinced me that the two are identical, *Candace* being simply rather more faintly marked and less reddish on the under side than usual.

Among its South-African congeners, *T. Brigitta* is distinguished in both sexes, but more especially in the female, by *the prevalent reddish tint of the under side, as well as of the cilia both above and below.*[1] In this respect it appears to approach *T. Herla*, Macl., from Northern Australia, but I have not seen the latter species, and can only judge by the published descriptions of it. Apart from the peculiarity mentioned, *Brigitta* differs in the following particulars from its close ally, *T. Zoë*, Hopff., viz. :—(♂) Ground-colour a deeper, warmer yellow ; black border of fore-wing narrower costally and at posterior angle—above which it is usually narrowly interrupted; border of hind-wing much narrower, and becoming more or less submacular and obsolete towards anal angle ; (♀) ground-colour a deeper, richer yellow, without fine dusky

---

[1] I think it not unlikely that *Terias Seruli*, Westw. (App. Oates's *Matabeleland*, &c., p. 342, 1881), is a variation of the ♀ *Brigitta*, in which the under-side markings are obsolete ; these markings being very faint in many examples of *Brigitta*, and in one specimen from Delagoa Bay (unusually red beneath) almost imperceptible.

irroration except on basal third of costa; black border of fore-wing darker, and of hind-wing altogether wanting or feebly represented by some hind-marginal spots. The close affinity of the two forms is nevertheless very apparent, and is illustrated by the circumstance of the capture *in copulâ* of a ♂ *Zoë* and a ♀ *Brigitta*, in April 1863, by Colonel Bowker, while collecting in Kaffraria Proper.

I was surprised not to meet with this *Terias* while collecting in Natal in 1867; it is evidently not uncommon there in most seasons. My correspondents in different parts of Eastern South Africa have noted nothing characteristic in the habits or haunts of *Brigitta*, but I have observed that the ♀ is much oftener forwarded than the ♂.

Mr. D'Urban found both sexes at King William's Town, and noted the abundance of the butterfly in February and April 1861. Colonel Bowker met with the ♀ in the same locality on May 1, 1874.

Localities of *Terias Brigitta*.

I. South Africa.
   B. Cape Colony.
      a. Western Districts.—Van Wyk's Vley, Carnarvon District (*E. G. Alston*).
      b. Eastern Districts.—King William's Town (*W. S. M. D'Urban* and *J.¹ H. Bowker*). Orange River, near Aliwal North (*J. H. Bowker*).
      c. Griqualand West.—Vaal River (*J. H. Bowker*).
   D. Kaffraria Proper.—Butterworth, Tsomo River, and Bashee River (*J. H. Bowker*).
   E. Natal.
      a. Coast Districts.—D'Urban (*J. H. Bowker* and *M. J. M‘Ken*).
      b. Upper Districts.—Estcourt (*J. M. Hutchinson*).
   F. Zululand.—St. Lucia Bay (the late *Colonel H. Tower*).
   G. Swaziland (the late *E. C. Buxton*; photogr.).
   H. Delagoa Bay.—Lourenço Marques (*Mrs. Monteiro*).
   K. Transvaal.—Potchefstroom District (*T. Ayres*). Marico River (*F. C. Selous*). Eureka, near Barberton (*C. F. Palmer*).

II. Other African Regions.
   A. South Tropical.
      a. Western Coast.—Damaraland (*C. J. Anderson*). "Angola (*J. J. Monteiro*)."—Druce. "Congo: Kinsembo (*H. Ansell*)."—Butler.
      b¹. Eastern Interior.—Mashunaland (*F. C. Selous*).
      bb. Madagascar.—Coll. Brit. Mus. (1886).
   B. North Tropical.
      a. Western Coast.—"Lower Niger (*W. A. Forbes*)."—Godman and Salvin. Gold Coast (*F. Bourke*). "Senegal."—Boisd.
      b. Eastern Coast.—Abyssinia: "Massaouah (*Raffray*)."—Oberthür. "South Abyssinia (*Zander*)."—Felder.

241. (2.) Terias Zoë, Hopffer.

♂ *Eurema pulchella*, Geyer, Forts. Hübn. Zütr. Exot. Schmett., p. 8, ff. 815–816 (1837).
♀ *Terias Zoë*, Hopff., "Monatsb. k. Akad. Wissensch. Berl., 1855, p. 640;" and Peters' Reise Mossamb., Ins., p. 369, pl. xxiii. ff. 10, 11 (1862).

♂ *Terias Drona*, Wallgrn., K. Sv. Vet.-Akad. Handl., 1857; Lep. Rhop. Caffr., p. 19.
♀ *Terias Zoë*, Wallgrn., *loc. cit.*
♂ *Terias Rahel*, Hopff. [part], Peters' Reise, &c., p. 368 (1862).
♂ ♀ „ „ and *Terias pulchella*, ♀, Trim., Rhop. Afr. Aust., i. p. 76, n. 49, and p. 78 (1862).

Exp. al., (♂) 1 in. 4–7 lin.; (♀) 1 in. 4–8 lin.[1]

♂ *Bright chrome-yellow*, of variable depth and brightness; *with strongly-marked black borders to both wings*. *Fore-wing:* base powdered with blackish; a black border from base along costa, gradually widening to apex, where it is very broad, and thence narrowing to anal angle, where it is about one line in breadth, the inner edge of its hind-marginal portion slightly scalloped, forming small acute projections on nervules. *Hind-wing:* blackish at base and along submedian nervure to a little before middle; a black band along hind-margin, widest in apical portion (but even there not or very slightly broader than the anal-angular portion of band in fore-wing), and gradually diminishing to anal angle, where it ends in a point. *Cilia* of fore-wing fuscous, slightly mixed with pale-yellow,—of hind-wing all pale-yellow. UNDER SIDE.— *Citron-yellow. Fore-wing:* at extremity of discoidal cell, a double narrow blackish line including the yellow disco-cellular nervule, occasionally almost obsolete; the broad border of upper side indicated by a greenish tint, sparsely irrorated with blackish atoms; in some specimens an indistinct, blackish, oblique stria near apex; small black dots at extremities of nervules; inner-marginal portion of wing paler than the rest. *Hind-wing:* more or less irrorated with blackish atoms; a transverse series of three blackish dots, sometimes yellow-centred, near base; an indistinct, blackish line, like that in fore-wing, at extremity of discoidal cell; beyond middle two or three faint-blackish, interrupted, obliquely transverse streaks (often scarcely perceptible),—the first on costa, only reaching subcostal nervure; the second, widely interrupted on median nervules, almost crossing the whole wing; the third, on median nervules; black dots on hind-margin as in fore-wing; occasionally a small rounded blackish spot close to base, between costal and subcostal nervules.

♀ *Very much paler, varying from pale lemon-yellow to primrose-yellow, more or less densely and generally irrorated with blackish atoms;* dark border in both wings of a duller black,—in hind-wing more or less suffused, and sometimes submacular, or actually broken into nervular spots contiguous only near apex. *Fore-wing:* blackish irroration densest in basal half, thinner (or sometimes wanting) in discal area; dark border commencing farther from base, and broadly interrupted with irrorated ground-colour between first median nervule and sub-median nervure, so that a well-marked terminal blackish spot is isolated

---

[1] A dwarf ♀, captured at Pinetown, Natal, by Colonel Bowker, expands only an inch across the wings.

at posterior angle. *Hind-wing:* blackish irroration usually denser than in fore-wing, except near costal margin; dark border very variable, sometimes broader or considerably broader than in ♂, rarely narrower, always ending at extremity of first median nervule. *Cilia* of fore-wing more mixed with pale-yellow generally, and almost wholly pale-yellow about posterior angle. UNDER SIDE.—*Much paler (in hind-wing commonly with a whitish cast), with irroration of blackish atoms usually more pronounced;* markings as in ♂.

I followed Doubleday in referring this butterfly to *Rahel*, Fab., in my work above quoted; but Fabricius' description (*Ent. Syst.*, iii. 1, p. 204, n. 367) is certainly not applicable to it, and Mr. Butler is apparently right in identifying the Bornean *T. Towlana*, Feld., with the Fabrician *Rahel*.
Goyer, who undoubtedly figures (*op. cit.*) a small ♂ from the Cape of Good Hope, refers it to the nearly-allied *T. pulchella*, Boisd., from Madagascar and Mauritius; but, as I hold these forms to be distinct,[1] I can only adopt the name of *Zoë*, given by Hopffer, who (*op. cit.*) describes and figures a ♀ from Querimba, East Africa, differing in no respect that I can detect from many South-African specimens before me.
M. Aurivillius kindly forwarded me a specimen of Wallengren's *Drona* (*op. cit.*), and I was thus enabled satisfactorily to determine it as the ordinary ♂ *Zoë*. From the well-known and widely-spread Indian *T. Drona* of Horsfield, the South-African species is readily distinguished, as far as the ♂ is concerned, by its very much deeper yellow ground-colour (the ♂ *Drona* being nearly as pale as the ♀ *Zoë*), narrower and much duller blackish border, and less basal blackish. The ♀s of the two forms are, however, exceedingly close to each other; but, as far as I have noticed, the Indian one has usually much fainter under-side markings.
The intimate alliance between *T. Zoë* and *T. Brigitta* has already been pointed out (*supra*, p. 15), as well as the fact of the capture of an ordinary ♂ of the former paired with a ♀ of the latter. I twice took the paired sexes of *Zoë*, on the same day (18th March 1867), in Natal, and have since received from that Colony two pairs captured *in coitu* by Colonel Bowker, and one by Mr. J. M. Hutchinson; in all these five cases, as well as in a sixth, where the paired sexes were captured in the Transvaal by Mr. W. Morant, the ♂s do not vary, but the ♀s exhibit the variations above described as regards both the development of the hind-wing border and the extent of fuscous irroration on the upper side.
This gaily-hued little butterfly was numerous about open ground throughout such parts of Natal as I visited in the summer of 1867. It is conspicuous on the wing, and, while active, is not swift, but keeps near the ground, and often settles on flowers. Colonel Bowker found it "very common all over the Trans-Kei territory in summer and autumn, frequenting open grassy country." I have taken it near Port Elizabeth, and also about Grahamstown and the neighbourhood, but it was rather scarce in both places. The season during which I observed the species was from the end of January to the beginning of April;

---

[1] The characters distinguishing *T. pulchella*, Boisd., though slight, appear to be remarkably constant—a condition of value in such a genus as *Terias*. They are (in the ♂) a much deeper yellow ground-colour on upper side, and a much broader border—near apex of hind-wing nearly twice as broad—of an intenser black, with the basal blackish of both wings very much darker; while (in the ♀) the border of the fore-wing, singularly enough, is conspicuously narrower than in the ♀ *Zoë*. On the under side both sexes of *Pulchella* are of a deeper yellow. In Mauritius, where I found this butterfly numerous, it appeared to be larger than the specimens usually received from Madagascar, and with the black border somewhat less broad.

but Mr. D'Urban noted its occurring in British Kaffraria from September to April, and Colonel Bowker took it at Delagoa Bay about the middle of the former month. *Zoë* has a more general range in South Africa than any of its congeners, but has not been recorded in the Cape Colony westward of Port Elizabeth on the south coast, or of the Carnarvon district inland.[1] Its distribution over the African continent seems to be very wide, but corresponding in the main to that of *Brigitta*.

Localities of *Terias Zoë*.

I. South Africa.
 B. Cape Colony.
   a. Western Districts.—Van Wyk's Vley, Carnarvon District (*G. E. Alston*).
   b. Eastern Districts.—Port Elizabeth. Uitenhage (*S. D. Bairstow*). Grahamstown. King William's Town (*W. D'Urban*). East London (*P. Borcherds*). Windvogelberg, Queenstown District (*Dr. Batho*). Burghersdorp, Albert District (*D. R. Kannemeyer*).
   c. Griqualand West.—" Kimberley."—H. L. Feltham. Vaal River (*J. H. Bowker* and *Mrs. Barber*).
 D. Kaffraria Proper.—Butterworth and Bashee River (*J. H. Bowker*).
 E. Natal.
   a. Coast Districts.—D'Urban. Verulam. Umvoti River. Mapumulo. Pinetown (*J. H. Bowker*).
   b. Upper Districts.—Hermansburg. Great Noodsberg. Greytown. Pietermaritzburg. Estcourt (*J. M. Hutchinson*). Rorke's Drift and Biggarsberg (*J. H. Bowker*).
 F. Zululand.—St. Lucia Bay (the late *Colonel H. Tower*). Napoleon Valley (*J. H. Bowker*).
 H. Delagoa Bay.—Lourenço Marques (*J. H. Bowker* and *Mrs. Monteiro*).
 K. Transvaal.—Potchefstroom (*W. Morant*). Potchefstroom and Lydenburg Districts (*T. Ayres*). Eureka, near Barberton (*C. F. Palmer*). Marico River (*F. C. Selous*).
 L. Bechuanaland.—Motito (the late *Rev. J. Frédoux*).

II. Other African Regions.
 A. South Tropical.
   a. Western Coast.—Damaraland (the late *C. J. Andersson*). "Angola (*Pogge*)."—Dewitz. "Chinchoxo (*Falkenstein*)."—Dewitz. Congo.—Coll. Brit. Mus.
   b. Eastern Coast.—"Quilimane."—Hopffer. "Zanzibar (*Raffray*)." —Oberthür.
 B. North Tropical.
   a. Western Coast.—"Lower Niger (*W. A. Forbes*)."—Godman and Salvin. Gold Coast (*F. Bourke*).
   b. Eastern Coast.—"Somaliland (*Thrupp*)."—Butler.

242. (3.) Terias Floricola, Boisduval.

Xanthidia Florirola, Boisd., Faune Ent. Madag., &c., p. 21, n. 2 (1883).
♂ ♀ *Terias Floricola*, Boisd., Sp. Gen. Lep., i. p. 671, n. 29 (1836).
♂ *Terias Ceres*, Butl., Ann. and Mag. Nat. Hist., March 1886, p. 218, pl. v. f. 3.

---

[1] Mr. G. Warden has since (June 1887) brought a specimen taken at Ookiep, in the Namaqualand District.

Exp. al., 1 in. 5–8 lin.

♂ *Bright chrome-yellow; fore-wing only with a narrow black border widened about the apex.* Fore-wing: border commencing at a point on costa just before extremity of second subcostal nervule, its apical portion moderately wide, but its hind-marginal portion rather abruptly becoming very narrow about end of third median nervule, and continuing so to posterior angle,—just above which it is sub-interrupted by some yellow scaling; the inner edge of this border emits more or less acute denticulations on nervules, and between these is more or less excavated, especially in its lower portion; costa with a black edging line of extreme tenuity. *Hind-wing:* a series of nine very small but (with the exception of the two last) very distinct black hind-marginal spots, all nervular except the eighth, which is between first median nervule and submedian nervure. UNDER SIDE.—*Very much paler yellow, with pale dull-ferruginous markings.* Fore-wing: two slender transverse linear marks in discoidal cell before middle, of which the first is near base and minute, and the second considerably larger and more or less angulated; at extremity of cell a small geminate marking of irregular form, with a more or less prominent superior projection outward, enclosing two whitish spots; a subapical, elongate, irregular marking, wider inferiorly, lying between costa and third median nervule; space between superior part of this marking and apex sparsely scaled with dull-ferruginous; a marginal series of nervular black spots throughout, minute on hind-margin, but larger (in fine specimens somewhat conspicuous) on costa; inner-marginal border whitish. *Hind-wing:* at extremity of discoidal cell a marking similar to that in fore-wing, but more irregular and more deeply insected; a sub-basal curved transverse row of three small whitish spots in ferruginous rings; close to base, between costal and subcostal nervures, a fourth spot (smaller and imperfect); a very indistinct, irregular, transverse, discal streak from costa to inner margin, broken completely on first subcostal, radial, and all three median nervules, and on submedian nervure; series of minute hind-marginal spots as on upper side.

♀ *Similar, rather paler.* Fore-wing: apical black border broader, prominently so at point (between radial and third median nervules) where hind-marginal narrowing begins. UNDER SIDE.—As in ♂, but the large subapical ferruginous marking of fore-wing rather broader and darker.

The foregoing description is made from specimens captured by myself in Mauritius. The very few South-African specimens (three ♂ s and five ♀ s) which I consider referable to *Floricola* are slightly larger, and of a paler yellow tint on the upper side; and in the ♀ the black apical border is a little broader and decidedly more prominent inwardly. On the under side the yellow is purer and brighter; the subapical marking of the fore-wing is broader and darker, and the other markings are in general better developed. In these respects they approach *Senegalensis*, Geyer (*nec* Boisd.), which is indeed a very close ally of

Floricola, distinguished by its broader, inwardly *angulated* apical border of fore-wing, and its very strongly marked under side.[1]

I have examined the ♂ Mauritian and South-African examples on which Mr. Butler founded his *T. Ceres*, as well as two ♀ s, ticketed "South Africa" and "Natal," which in the British Museum collection are associated with the two ♂ s. The interruption of the border of the fore-wing above the posterior angle, which Mr. Butler specially notes as a distinguishing character, is in neither ♂ complete.

The Madagascar specimens of *Floricola* in the South-African Museum are more like the South-African than the Mauritian examples, especially the ♀ s, one of which is of an unusual paleness, inclining to whitish.

This appears to be a rare form in South Africa. I took two ♀ s near D'Urban, Natal, in February and April 1867, and the late Colonel Tower brought one ♂ from St. Lucia Bay later in the same year. Colonel Bowker has sent only two examples (both from D'Urban), taken respectively in 1878 and 1879. My captures were made in the Botanic Garden, where several species of *Terias* frequented the flowers of *Vinca rosea*.

In Mauritius I found *Floricola* numerous, and generally distributed about grassy waste spaces, in company with its congener *Pulchella*.

Localities of *Terias floricola*.

I. South Africa.
  E. Natal.
    a. Coast Districts.—D'Urban.
  F. Zululand.—St. Lucia Bay (*Colonel H. Tower*).

II. Other African Regions.
  A. South Tropical.
    bb. Eastern Islands.—Madagascar: Murundava (*Grevé*). Mauritius. "Bourbon."—Boisduval.

243. (4.) Terias Æthiopica, *sp. nov.*

♂ *Eurema Senegalensis*, Geyer [*nec* Boisd.], Forts. Hübn., Zutr. Exot. Schmett., p. 41, ff. 969-970 (1837).

Exp. al., 1 in. 7-8½ lin.

Closely allied to *T. floricola*, Boisd. *Chrome-yellow ground-colour bright but rather paler; black border in fore-wing only, but broader superiorly; nervular hind-marginal spots in hind-wing larger.* Fore-wing: black border shaped much as in *Floricola*, but considerably broader apically, and presenting a marked angulated prominence inwardly on third median nervule (much as in *Hecabe*, Linn.), as well as a usually rather more pronounced small projection on first median nervule. *Hind-wing*: terminal nervular spots considerably enlarged and thinly elongated laterally. UNDER SIDE.—*Markings as in Floricola, but much more strongly marked, especially the subapical ferruginous patch of fore-wing, which is broad and dark (more particularly in ♀), and the irregular transverse discal streak of hind-wing.* Fore-wing: terminal disco-

[1] See *T. Æthiopica*, Trim., *infra*, No. 243.

cellular mark indistinctly geminate, larger; subapical patch more or less reproducing internal outline of upper side of black border as far as third median nervule, but in ♀ emitting downward from its outer side a narrow somewhat incurved projection as far as inter-nervular fold between third and second median nervules. *Hind-wing:* terminal disco-cellular mark much enlarged, not geminate, sub-reniform; discal transverse streak very distinct, and usually (more particularly in ♀) outwardly suffused in its middle portion between first subcostal and third median nervules; hind-marginal nervular spots not lengthened laterally as on upper side, and only a little larger than in *Floricola*.

I have reluctantly been compelled to re-name this *Terias*, because, though evidently the same as *Senegalensis*, Geyer, it is not the form to which the name of *Senegalensis* was given by Boisduval in 1836 (*Sp. Gen. Lep.*, i. p. 672). The latter author states expressly that the under side of his species was so faint in its markings as at first to appear to have none, whereas the *Terias* now under notice, of which Geyer figures the ♂, is remarkable for the great development of the under-side pattern, especially in the ♀. Boisduval's insect is evidently very close to *Hecabe*, L., and perhaps not separable from it.

In the above description *T. Æthiopica* is sufficiently distinguished from *T. floricola*, Boisd., but it comes even nearer to the Indian and Australian *T. Sari*, Horsfield (from which the Cingalese *T. simulata* and *T. uniformis* of Moore[1] cannot, I think, be separated). The only obvious distinctions of *Æthiopica* are on the upper side the very slight inward *inferior* prominence of the fore-wing border (which in *Sari* is very pronounced), and the hind-marginal marking of the hind-wing being restricted to small separate spots (instead of, as in *Sari*, a narrow continuous blackish border). This latter feature is, however, inclined to be unstable, for in one ♂ *Æthiopica* the thin, laterally elongated spots almost unite. A hind-wing border of variable width is a character of *T. Hecabe*, and thus *Sari* is nearer to the latter than *Æthiopica* is.

I have seen but few examples of this *Terias* from South Africa. A ♂ was sent by Colonel Bowker from Kaffraria Proper in 1866, and another labelled "Natal" was received at the South-African Museum in a collection acquired about the same date. In 1867 Colonel Tower gave me a ♀ captured at St. Lucia Bay, and the Museum received a series of the same sex taken in Natal by the late Mr. M. J. M'Ken. I did not meet with the insect during my collecting excursions in Natal.

Localities of *Terias Æthiopica*.

I. South Africa.
    D. Kaffraria Proper.—Tsomo River (*J. H. Bowker*)
    E. Natal.
        a. Coast Districts.—D'Urban (*M. J. M'Ken*).
    F. Zululand.—St. Lucia Bay (*Colonel H. Tower*).

II. Other African Regions.
    B. North Tropical.
        a. Western Coast.—" Senegal."—Geyer.

---

[1] *Lep. Ceylon*, i. p. 119, pl. 45, ff. 2, 2a, 2b; and p. 120, pl. 46, ff. 2, 2a, 2b (1881).

[2] Boisduval (*op. cit.* p. 670) gives *Sari* as a ♀ variety of *Hecabe*, but he has not been followed by Wallace (1867), Butler (1871), or Kirby (1871).

244. (5.) Terias Butleri, sp. nov.

Exp. al. (♂) 1 in. 7 lin.; (♀) 1 in. 9 lin.

♂ *Sulphur-yellow; fore-wing with rather broad, inwardly deeply excavated black border; hind-wing with a very narrow, inwardly slightly suffused border.* Fore-wing: costa irrorated finely with black for about one-third from base, thence rather strongly black-edged to first subcostal nervule, where the apical border begins to widen; upper part of border shaped as in *Æthiopica* as far as third median nervule, but thence much broader, the lower inward projection between first median nervule and posterior angle being at least twice as broad. Hind-wing: border extending from first subcostal to first submedian nervure, attenuated to each extremity, and crossed by yellow inter-nervular folds; at extremity of submedian nervure, and of fold between it and first median nervule, a minute black spot. UNDER SIDE.—*Almost the same tint of yellow; usual markings very indistinct; no trace in fore-wing of any subapical patch or streak.* Fore-wing: disco-cellular dot and streak faint and minute, and terminal markings very indistinct, small, and narrow. Hind-wing: small sub-basal annulets less indistinct; terminal disco-cellular mark rather narrow, interrupted on fold; discal irregular streak faint or nearly obsolete.

♀ *Considerably paler, sometimes inclining to whitish-yellow, with broader duller borders.* Fore-wing: a faint sparse minute fuscous irroration from base (in one example not so sparse, and extending over basal third of wing); border broader throughout, but more markedly so at inferior inward projection. Hind-wing: border variable in width, but always narrow, more suffused inwardly; in the example above mentioned a very sparse fuscous irroration for a little distance from base, chiefly in discoidal cell. UNDER SIDE.—As in ♂, but much paler, and markings even more indistinct, almost obliterated; in one example (not that above mentioned) a faint trace in fore-wing of a ferruginous subapical patch.

Unwilling as I am to add another to the numerous species-names of this bewildering genus, I see no help for it in the case of this South-African form, near as it is to *T. Hecabe* (Linn.). There is little else to depend on in this group of *Terias* except the form of the fore-wing border, and this in *T. Butleri* is intermediate between that exhibited by the West-African *T. Leonis*, Butl., and the pattern so strongly developed in *T. Hecabe*. *Leonis* itself is allied to *Æthiopica*, mihi, in its upper-side markings, but their rather greater development, combined with the feebly marked under side, link it to *Butleri*. The latter is readily distinguished from its South-African allies by its deeply excavated broad border in the fore-wing and distinct though narrow one in the hind-wing, as well as by its exceedingly indistinct marking on the under side.

It is with great pleasure that I name this butterfly after my friend Mr. Arthur G. Butler, of the British Museum, whose praiseworthy labours in this and other most difficult genera merit the thanks of all lepidopterists.

T. Butleri would appear to be a scarce species in South Africa, as the only examples that have come under my notice are a few captured by myself on the coast of Natal in February and March 1867.[1] There was, however, nothing in the habits of the butterfly to distinguish it from its congeners, and it is quite possible that it may be passed over among the rest in the localities they all frequent together.

Localities of *Terias Butleri*.

I. South Africa.
 E. Natal.
  a. Coast Districts.—D'Urban. Victoria County.

245. (6.) Terias Desjardinsii, Boisduval.

♂ *Xanthidia Desjardinsii*, Boisd., Faune Ent. Madag., &c., p. 22, pl. 2, f. 6 (1833).
♂ *Terias Desjardinsii*, Boisd., Sp. Gen. Lep., i. p. 671, n. 30 (1836).
♂ *Terias florirola*, Wallengr., K. Sv. Vet.-Akad. Handl., 1857; Lep. Rhop. Caffr., p. 19.
♂ *Terias Desjardinsii*, Trim., Rhop. Afr. Aust., i. p. 78, n. 51 (1862).
♀ ,, ,, Bates, Proc. Zool. Soc. Lond., 1863, p. 476, n. 2.
♀ *Terias aliena*, Butler (?), Ann. and Mag. Nat. Hist., 1880, p. 337, n. 20.
♂ ♀ *Eurema Desjardinsii*, Möschler, Verh. Zool.-Bot. Gesellsch. Wien, 1883, p. 281.

PLATE 10, fig. 2 (♂).

Exp. al., (♂) 1 in. 5½–8 lin.; (♀) 1 in. 6–8¼ lin.

♂ *Gamboge-yellow; fore-wing with a narrow or very narrow black border; hind-wing bluntly angulated at extremity of second median nervule, and with an extremely narrow black border, occasionally broken into nervular spots.* Fore-wing: on costa not far from base commences a thin black edging, gradually widening beyond middle to a border of variable width apically, which continues more narrowly along hind-margin to posterior angle, where it again slightly widens; inner edge of border irregularly and bluntly dentate superiorly, but emitting regular denticulations on median nervules and submedian nervure. *Hind-wing:* border little more than a slender edging, rather strongly dentated on nervules inwardly, diminishing to a point at extremity of first median nervule or of submedian nervure; rarely this border is reduced to nervular spots of moderate or even small size; inner-marginal border conspicuously paler, quite sulphur-yellow. *Cilia* of fore-wing greyish-brown mixed with rufous and at posterior angle with yellow,—of hind-wing, yellow mixed with rufous. UNDER SIDE.— *Sulphur-yellow, with pale-ferruginous markings. Fore-wing:* apical border rather widely tinged with pale-ferruginous, continued narrowly

[1] A ♀ specimen occurs in a collection made by Mr. A. W. Eriksson during 1884-85, in the country between the Transvaal and Gubulewayo, Matabeleland.

to a point at posterior angle; an oblique subapical streak of darker ferruginous, from fourth subcostal to third median nervule; in discoidal cell usually a small faintly-marked spot nearer base than extremity; terminal disco-cellular marking very narrow and indistinct except at summit. *Hind-wing*: ordinary basal spot and sub-basal spots usually distinct (the middle one of the three latter broken into three); terminal disco-cellular marking larger, wider, and better marked than that of fore-wing, partly interrupted on disco-cellular fold; discal ray indistinct in its upper and lower portions, but its central part (between first subcostal and third median nervules) suffused, angulated, and usually rather conspicuous. *Cilia* of both wings pale ferruginous-rufous, immediately preceded by nervular black dots.

♀ *Sulphur-yellow; fore-wing with border duller, much diminished hind-marginally; hind-wing with a series of black nervular dots only on hind-margin.* Fore-wing: a suffusion of gamboge-yellow over basi-inner-marginal area, rising over lower part of discoidal cell; blackish border broken into small separate spots on first median nervule and submedian nervure (the latter spot even wanting occasionally), and rarely so even as high as second and third median nervules; the apical part of border varying in width accordingly. *Hind-wing*: a slight suffusion of gamboge-yellow from base along median nervure. UNDER SIDE.—Like that of ♂, but paler, *except subapical oblique streak of fore-wing, and suffused central part of discal streak of hind-wing, which are much darker ferruginous, more sharply angulated, and very conspicuous.*

In both sexes the apex of fore-wing is more pronounced and acute than in the four species above described.

Among the drawings of some of Wallengren's type specimens kindly obtained for me by Mr. Chr. Aurivillius, there is a careful figure of that author's *T. floricola*, which proves to be, as I anticipated, the ♂ *Desjardinsii*.

As regards *T. aliena*, Butler, from Madagascar, I have seen the type in the British Museum; it is a very worn and faded specimen, and Mr. Butler agrees with me in thinking that it is in all probability only a small ♀ of *Desjardinsii*.

There can be no doubt that Möschler, who (*loc. cit.*) carefully describes both sexes, is in error in associating *Senegalensis*, Geyer, with *Desjardinsii*, the former being entirely different in such important characters as the shape of the wings and the form of the border of the fore-wings, and belonging in fact to the *Floricola* group.

The acuter apex of the fore-wings and the angulation of the hind-wings are alike in this species and in *T. regularis*, Butl., if, as I believe, I have correctly identified the latter, but *Desjardinsii* is readily distinguished by the very reduced border in both sexes. The specimen figured on Plate 10[1] exhibits the widest border yet observed in

---

[1] In this figure not nearly sufficient prominence is given to the principal markings of the under side, viz., the ferruginous subapical streak of fore-wing and upper discal streak of hind-wing, which are well marked in the specimen sent to the artist.

the ♂. The majority of the South-African ♂s have the border about as narrow as in the Malagasy type figured by Boisduval; but, on the other hand, there are some examples in which it is much more reduced, the extreme of narrowness being reached in a small ♂ from the Bashee River, and another from Lydenburg in the Transvaal. The ♀s vary similarly as far as the fore-wing border is concerned, but here the extreme of reduction breaks up the hind-marginal portion of the border into very nearly separated spots.

The ♀ of this *Terias* was sent to the South-African Museum by Colonel Bowker from Butterworth, Kaffraria Proper, as long ago as 1861, with the note that it frequented forest-clad country, while its allies (*T. Zoë* and *Brigitta*) were found quite in the open. It was not until some years later that ♂s and further ♀s were received from the same territory; and in 1865 I met with both sexes when landing for a day at D'Urban, Natal, towards the end of June. In 1867 I took the butterfly sparingly in the same locality during February and March, and also found it near Maritzburg in April. It was taken in Zululand during the same year by the late Colonel Tower. The best locality for it near D'Urban was the Botanic Garden, whence the late Mr. M'Ken also sent a good many specimens. On the wing it can be recognised at once by the narrowness of its black bordering.

Localities of *Terias Desjardinsii*.

I. South Africa.
   D. Kaffraria Proper.—Butterworth and Bashee River (*J. H. Bowker*).
   E. Natal.
      a. Coast Districts.—D'Urban. Avoca (*J. H. Bowker*).
      b. Upper Districts.—Maritzburg.
   F. Zululand.—St. Lucia Bay (the late *Colonel H. Tower*).
   K. Transvaal.—Lydenburg District (*T. Ayres*). Lydenburg (*A. F. Ortlepp*).

II. Other African Regions.
   A. South Tropical.
      bb. Eastern Islands.—Madagascar (*E. L. Layard*).

246. (7.) **Terias regularis**, Butler.

♀ *Terias Desjardinsii?* ♀, Trim., Rhop. Afr. Aust., i. p. 79 (1862).
♂ „ *regularis*, Butl., Ann. and Mag. Nat. Hist., 4th Ser., vol. xviii. p. 486 (1876).

Allied to *T. Desjardinsii*, Boisd.

Exp., (♂) 1 in. 6–8 lin.; ♀ 1 in. 6–8½ lin.

♂ *Gamboge-yellow; fore-wing rather broadly bordered with black, hind-wing about half as broadly bordered.* Fore-wing: base narrowly blackish; costal nervure as far as first subcostal nervule, and a space above and below it, irrorated with black atoms; black border almost as broad along costa as in *T. Zoë*, not so wide apically, but broader hind-marginally, and specially broader at posterior angle,—its inner edge rather more irregular superiorly and hollowed about its middle. *Hind-*

wing; black border well defined, about the same width as in *Zoë*, its inner edge forming sharper and more numerous dentations on nervules, its lower extremity becoming attenuated and suffused with yellowish scales, and terminating in a point on submedian nervure; inner margin rather widely bordered with sulphur-yellow. *Cilia* without any rufous tinge, but mixed with greyish-brown in fore-wing. UNDER SIDE.—*Sulphur-yellow; dark borders faintly indicated from upper side; markings generally arranged as in Desjardinsii, but dusky-grey instead of ferruginous, and very indistinct;* the subapical streak of fore-wing and upper discal streak of hind-wing usually altogether wanting, but sometimes faintly indicated. *Fore-wing:* no ferruginous tinge at apex or along hind-margin. *Hind-wing:* costal border finely and rather sparsely irrorated with blackish atoms.

♀ *Sulphur-yellow, more or less sparsely irrorated with blackish atoms; fore-wing border duller, but as broad, or nearly as broad, as in ♂, except just at posterior angle; hind-wing border reduced, varying from a narrow stripe, forming strong nervular projections inwardly, to a series of disconnected slender nervular marks. Fore-wing:* irroration most developed near base, along costa, and in upper part of discoidal cell; a slight suffusion of gamboge-yellow over lower basi-inner-marginal area, as in *Desjardinsii*. *Hind-wing:* irroration scantier than in fore-wing, usually confined a little space close to base, but sometimes extending over discoidal cell and below it. UNDER SIDE.—As in ♂, but paler, and the subapical streak of fore-wing and upper discal streak of hind-wing usually more or less distinct and tinged with ferruginous, but never strongly marked as in *Desjardinsii*.

I have not seen Mr. Butler's type of *Regularis*, which he described from an Abyssinian collection of insects, but from his description I feel almost certain that it is identical with the ♂ of the South-African butterfly here described. I also associate with *Regularis* two ♂ s from the coast of Tropical Western Africa, which are in the South-African Museum, and which only differ in the broader black border of the wings (in the hind-wing especially, where it ends truncately above submedian nervure), and their somewhat more rounded outline.

I wrote a description of this *Terias* in 1861, from a ♂ captured at King William's Town by Mr. W. S. M. D'Urban in January of that year, but did not then, or for some years afterwards, think it sufficiently distinct from *Desjardinsii* to be more than a variety of that species. In 1867 I had the pleasure of capturing a number of both sexes, both on the coast and in the inland districts of Natal, and was so fortunate as to take two pairs *in copulâ*,—one at Fort Buckingham, on the Tugela, on the 8th March, and the other near Greytown on the 11th. The species was on the wing in February, March, and April, and Colonel Bowker took one near Maritzburg early in May. When flying, it is not distinguishable from *T. Zoë*. Mr. R. Lightfoot, of Cape Town, has shown me a small ♂ received in a collection made near Plettenberg Bay, in the Knysna District, by Miss Newdigate in the year 1886. This is the only example I know of which has occurred so far to the south and west.

Localities of *Terias regularis*.

I. South Africa.
    B. Cape Colony.
        a. Western Districts.—Forest Hall, near Plettenberg Bay (*Miss Newdigate*).
        b. Eastern Districts.—King William's Town (*W. S. M. D'Urban*).
    E. Natal.
        a. Coast Districts.—D'Urban. Verulam. Mapumulo.
        b. Upper Districts.—Fort Buckingham, Greytown, and Umvoti District. Pietermaritzburg.
II. Other African Regions.
    B. North Tropical.
        a. Western Coast (*J. M. Pask*).
        b1. Eastern Interior.—" Atbara" River [Soudan].—A. G. Butler.

GENUS MYLOTHRIS.

Mylothris (Hübn., 1816, part.), Butler, Cist. Ent., i. p. 42 (1870).
Pieris, Godart [part.], Enc. Meth., ix. p. 10 (1819); *et Auct.*
Pieris, Trim. [part], Rhop. Afr. Aust., i. p. 26 (1862).

IMAGO.—*Head* small, clothed above and in front with rather long and sparse bristly hair; *palpi* slender, elongate, scaly above,—basal and middle joints with long and bristly hairs beneath,—terminal joint rather longer than middle one, porrect and ascendant, slightly thickened just before tip; *antennæ* of moderate length and thickness, with a short, abruptly-formed, flattened club, rounded at the tip.

Thorax short and rather slender, clothed densely with hair, which above is long and silky. *Wings* thin (in some species very delicate) and large. *Fore-wings* rather elongate, with the costa very slightly arched; apex not acute but pronounced; hind-margin almost straight or slightly convex; costal nervure ending at about three-fifths of wing from base; subcostal nervure three-branched,—first and second nervules given off not far apart, near extremity of discoidal cell—the first very short (sometimes in *M. Agathina* joining costal nervure, and in one ♂ specimen almost obsolete),—second long, ending not far before apex; upper radial united to subcostal nervure at origin of third nervule, lower radial from junction of two disco-cellular nervules; of the latter the lower is twice as long as the upper, but much less oblique and slightly angulated or merely curved; discoidal cell large, nearly three-fifths length of wing, broad and truncate at extremity. *Hind-wings* bluntly obovate; hind-margin convex, very slightly sinuate; inner margins convex, meeting below abdomen; costal nervure very short, terminating considerably before middle; first subcostal nervule emitted considerably before extremity of cell, much arched, ending at some distance before apex,—second nearly straight, ending at apex; disco-cellular nervules both oblique, the lower nearly or about twice as long as the upper, and sometimes slightly angulated; internal nervure ending about middle of inner margin. *Legs* moderately stout, scaly; femora with a little hair basally beneath, about equal in length to tibiae; tarsi rather longer than tibiae, both rather densely and strongly spinulose; terminal spurs of hind-tibiae very small, of middle ones obsolete.

Abdomen of moderate length, laterally compressed, much arched, larger posteriorly.

LARVA.—Of the ordinary Pierine form, rather attenuated posteriorly; clothed with longer hairs than usual in the Sub-Family. Foodplant *Loranthus* (in the case of *M. Agathina*).

PUPA.—Head with frontal process long and recurved upward; a dorsal series of prominent tubercles (larger on thorax) along middle line of back, and two laterally-projecting claw-shaped processes on each side of basal half of abdomen (*M. Agathina* and *Rüppellii*).

I follow Mr. Butler in thinking that this remarkable section of the old genus *Pieris* is generically separable, considering not only the three-branched subcostal neuration of the fore-wings, but the peculiar *facies* and pattern.[1] The curiously tuberculated chrysalis is also a point to be taken into consideration, as well as the slow flight and apparent absence of the onward-moving Pierine habit. The few known species (about seven or eight) are confined to the Ethiopian Region, one (*M. Phileris*, Boisd.) being peculiar to Madagascar; they are all closely allied. The three found in South Africa are *M. Agathina*, (Cram.), which apparently ranges over all Tropical Africa; *M. Trimenia*, Butl., and *M. Rüppellii*, Koch, which are found in Eastern Africa as far northward as Abyssinia. These butterflies are of very plain pattern, the ♂ s being white above with small hind-marginal black spots [2] (*Trimenia*, however, having lemon-yellow hind-wings, and *Rüppellii* a basal suffusion of orange-red in its fore-wings), while the ♀ s are more or less deeply and broadly tinged with ochreous-yellow (in *Trimenia* entirely confined to the hind-wings), or with fuscous-grey, and usually have the hind-marginal spots rather enlarged. On the under side, *Rüppellii* is almost the same as on the upper side, but *Agathina* and *Trimenia* have the hind-wings and the apex of fore-wings ochre-yellow, the former possessing a conspicuous orange-red basal suffusion in the fore-wings.

It is of much interest to note that, like the allied slow-flying but more richly coloured species of the allied Oriental and Australian group, *Thyca*, Wallengr.,[3] some kinds of *Mylothris* are the subjects of mimicry

---

[1] In *Proc. Zool. Soc. Lond.*, 1872, pp. 37, 38, Mr. Butler extended *Mylothris* to include the South-American *P. Pyrrha*, Fab., and allies; but they do not seem at all closely related to the African species with which they are associated, and are usually separated by lepidopterists under the genus *Perrhybris*. The ♀ s of this group are noted for mimicry of various *Heliconinae*.

[2] In the West-African *M. Chloris*, (Fab.), the outer three-fifths of the whole area of the hind-wings is uniformly blackish on both surfaces. This species was included in my former work on the strength of Boisduval's giving "Natal" as one of its localities; but I now omit it, seeing that no South-African example appears to be known.

[3] See Wallace, *Trans. Ent. Soc. Lond.*, 3d Ser., iv. pp. 309, 344, and 383.

by Pierinæ of related genera. The imitation of *M. Agathina* by *Pieris Thysa*, Hopff., is deceptively close in both sexes; and *M. Poppea*, (Cram.), is similarly copied by *P. Rhodope*, (Fab.), on the West Coast. *M. Agathina* is also mimicked by the ♀ *Eronia Argia*, (Fab.).

247. (1.) Mylothris Agathina, (Cramer).

♂ *Papilio Agathina*, Cram., Pap. Exot., iii. pl. ccxxxvii. ff. D, E (1782).
♂ *Pieris Agathina*, Boisd., Sp. Gen. Lep., i. p. 511, n. 106 (1836).
♂ ♀ *Pieris Agathina*, Hopff., Peters' Reise Mossamb., Ins., p. 351, t. xxi. ff. 11, 12 [♀] (1862).
♂ ♀ " " Trim. [part], Rhop. Afr. Aust., i. p. 28, n. 15 (1862).
♂ ♀ " " Staud., Exot. Schmett., i. pl. 19 (1884).

Exp. al., (♂) 2 in. 6-8 lin.; (♀) 2 in. 6-9 lin.

♂ *White, with black marginal spots. Fore-wing*: costa narrowly, apex more widely, black; four small black rounded spots on hind-margin at ends of nervules. *Hind-wing*: six black spots on hind-margin at extremities of nervules. UNDER SIDE.—*Markings similar to those of upper side. Fore-wing: apex yellow-ochreous;* on its hind-marginal edge are three small spots continuous of the row along hind-margin; *base broadly suffused with deep orange-red. Hind-wing: entirely pale yellow-ochreous;* hind-marginal spots rounder than on upper side; portion of costa above costal nervure orange-red.

♀ *Varies from dull whitish-ochreous to dull orange-ochreous; spots on hind-margins larger than in ♂*. UNDER SIDE.—As in ♂, but ground-colour of *fore-wing* usually dull yellowish instead of white, and its apex and whole of hind-wing of a deeper yellow-ochreous.

In the ♂ the lowest hind-marginal spot in the fore-wing is at the extremity of the first median nervule, and is much smaller than the rest, being occasionally a mere dot; while in the hind-wing the two uppermost, respectively on second subcostal and radial nervules, are considerably less than the other four.

In the ♀ the spot on first median nervule of fore-wing is proportionately larger than in the ♂, there is an additional spot (very small) at extremity of submedian nervure, and there is more or less tendency in all the spots to extend acutely in the direction of base. The spots of the hind-wing, although sometimes much larger than in ♂, do not exhibit this tendency.

In both sexes the neuration of the under side is more or less clouded with whitish where traversing the ochre-yellow ground-colour, and the costal edging of the fore-wing is much more densely white-scaled than on the upper side.[1]

The ♀ s exhibit on the upper side gradations, from being only

---

[1] Oberthür records that specimens taken in Abyssinia by M. Raffray were smaller, and with less developed black markings than those found in Natal. A ♂ example captured by Colonel Bowker at Quilimane presents the same differences.

slightly tinged with yellowish (and having the fore-wings almost as white as in the ♂) to having the whole surface of a dull orange-ochreous (deepening into reddish at bases). The most richly and deeply tinted specimens that I have seen are from Delagoa Bay.

LARVA.—Transversely barred with alternate dull red and blackish bands speckled with yellow; and clothed generally with fine grey hair of some length; a yellowish-white lower lateral stripe on each side, from second to last segment immediately above the legs. Head black, varied with yellow down the middle. Length, $\frac{3}{4}$ inch. (From notes and drawings by Mr. J. P. Mansel Weale of specimens from near King William's Town.)—See PLATE 2, ff. 3.

PUPA.—White, more or less tinged with cream-colour in parts, and curiously marked with black. Head with a long frontal horn, curved upward, cream-coloured. *Thorax* cream-coloured dorsally, but with a broad black marking along the middle; a small anterior acute black tubercle on each side, and on median ridge a series of three white, black-edged, broad, blunt, tubercular processes, slanting forward. Wing-covers black with a greenish tinge. Abdomen dorsally white and black, the latter forming a large lozenge-shaped marking (widest on seventh segment) acuminate anteriorly on fifth and posteriorly on ninth segment; on each side a row of small black spots; below these a broad black stripe; along median ridge a series of seven small white black-edged tubercles, of which the second, third, and fourth are blunter and larger than the rest; both the sixth and seventh segments bearing on each side a large, broad, acute, slightly forward-curved, tooth-like white projection; anal extremity very pointed. Length, $\frac{2}{3}$ inch. Attached by anal extremity and thoracic silken girth to web of silk spread on a leaf. (From notes and drawings by Mr. J. P. Mansel Weale, and drawings by Mrs. Barber, of specimens from near King William's Town; and drawings by Captain H. C. Harford of a specimen found at Pinetown, Natal.)—See PLATE 2, ff. 3*a*.

The singular pupa was sent to Mrs. Barber in December 1868 by Miss Fanny Bowker, who discovered it near King William's Town, and the drawings reproduced in Plate 2 were received by me from Mrs. Barber during the same month. Captain Harford's drawings reached me the following year, and Mr. Mansel Weale's in 1873. Mr. Mansel Weale discovered the larva, and wrote on 20th March 1873: "I have found *Loranthus olecefolius* swarming with the larvæ of *Agathina*; they follow each other like a regiment in line, or like the Processionary Moth."

The pupa, from its black-and-white colouring, and particularly from its attachment to a leaf covered just about it with white silk, very probably presents, at a little distance, the appearance of a bird's dropping;[1] but on a closer inspection the dorsal aspect is by no means

---

[1] Mr. Weale wrote in February 1877: "The chrysalides both of *Agathina* and *Poppea* (=*Rüppellii*) very much resemble bird-droppings with mistletoe seeds intermixed."

unlike that of a spider of the *Gasteracantha* group. From every point of view, it looks altogether unlike what it really is—the chrysalis of a butterfly—and no doubt derives protection from its disguise.

Few African butterflies are better known or more widely spread over the continent than *Agathina*. It is rather a heavy flyer, haunting wooded spots, and is very conspicuous on the wing. Like many others of the Sub-Family, it is of somewhat gregarious habits, but I never observed in it any tendency to the direct onward flight in one direction so characteristic of many *Pierinæ*. It keeps much about low trees, and often settles on the leaves.

In South Africa Proper it is confined to the Eastern Territories, not being recorded in the Cape Colony from farther westward than the Albany District. It becomes numerous about King William's Town and in the Trans-Kei country, and is abundant on the coast of Natal. The perfect insect appears to be "out" all the year round. I have personally taken it in the months of January, March, April, June, and August, and have ample records of its capture in May, October, November, and December. Colonel Bowker sent me the paired sexes taken on the Bashee River in May 1863, at King William's Town in 1872, and also at D'Urban, Natal, on the 11th October 1881.

Localities of *Mylothris Agathina*.

I. South Africa.
  B. Cape Colony.
    b. Eastern Districts.—Grahamstown (*M. E. Barber* and *H. J. Atherstone*). Kowie River, Bathurst District (*J. L. Fry*). Windvogelberg, Queenstown District (*Dr. Batho*). King William's Town. "East London and Keiskama River mouth."—W. S. M. D'Urban.
  D. Kaffraria Proper.—Butterworth and Bashee River (*J. H. Bowker*).
  E. Natal.
    a. Coast Districts.—D'Urban. Avoca (*J. H. Bowker*). Itongati River. "Lower Umkomazi."—J. H. Bowker.
    b. Upper Districts.—Maritzburg (*S. Windham*). Estcourt (*J. M. Hutchinson*).
  F. Zululand.—St. Lucia Bay (the late *Colonel H. Tower*).
  G. "Swaziland."—The late E. C. Buxton.
  H. Delagoa Bay.—Lourenço Marques (*Mrs. Monteiro*).
  K. Transvaal.—Potchefstroom (*V. E. Noren*).

II. Other African Regions.
  A. South Tropical.
    a. Western Coast.—Damaraland (*J. A. Bell*). "Angola (*Pogge*)."—Dewitz. "Congo: Kinsembo (*H. Ansell*)."—A. G. Butler. "Chinchoxo (Falkenstein)."—Dewitz.
    b. Eastern Coast.—Quilimane (*J. H. Bowker*). Zambesi (*Rev. H. Rowley*). "Querimba."—Hopffer.
    *b*1. Eastern Interior.—Shashani River, near Bulawayo; and Inyambare River (*F. C. Selous*).
  B. North Tropical.
    a. Western Coast.—Sierra Leone.—Coll. Brit. Mus.
    b. Eastern Coast.—"Abyssinia (*Raffray* and *Antinori*)."—Oberthür.

248. (2.) Mylothris Trimenia, Butler.

♂ ♀ *Pieris Agathina,* Vars. B. and C., Trim., Rhop. Afr. Aust., i. p. 29 (1862); and pl. 2, f. 2 [♂] (1866).
Pieris Trimenia, Butl., Cist. Ent., p. 13 (1869).
Mylothris Trimenia, Butl., Proc. Zool. Soc. Lond., 1872, p. 36.
♂ *Tachyris Trimenia,* Staud., Exot. Schmett., i. pl. 17 (1884).

Exp. al., (♂) 2 in. 3–4 lin.; (♀) 2 in. 1–3 lin.

♂ *Fore-wing white, hind-wing lemon-yellow; a common hind-marginal row of nervular black spots.* Fore-wing: base very narrowly clouded with black; costa with narrow but well-defined black margin from base to a little before apex, where it widens considerably, forming an apical blackish tip as in *Agathina*; hind-marginal nervular spots variable in size, more or less narrowed and pointed baseward,—the spot at end of lower radial nervule joined to apical blackish, and the lowest spot (at end of submedian nervure) very small. *Hind-wing:* hind-marginal spots varying in size, more rounded than those in fore-wing, one more in number than the corresponding series in *Agathina*,—the additional (seventh) spot being at the end of first subcostal nervule; the middle (third and fourth) spots of this row smaller than the rest; base narrowly blackish about origins of nervures. UNDER SIDE.— *Hind-wing and apex of fore-wing citron-yellow tinged with ochreous.* Fore-wing: costal black almost obliterated by white scaling; apical black only represented by three small spots at extremities of nervules; other hind-marginal spots much smaller than on upper side,—that on submedian nervure minute or wanting; no black at base, but only a slight tinge of ochrey-yellow. *Hind-wing:* costa at base and as far as middle edged with orange-yellow; hind-margin also narrowly tinged with the same colour; hind-marginal spots rather larger and better defined than on upper side.

♀ *Similar to ♂, but hind-wing paler or deeper creamy yellow-ochreous.* Fore-wing: no black at base, but a very faint tinge of yellowish; hind-marginal spots considerably larger (especially that on submedian nervure), sometimes almost contiguous as far as first median nervule. *Hind-wing:* spots rounder, and usually slightly larger; no black at base. UNDER SIDE.—As in ♂, but all the yellow colouring deeper and more ochreous, being of about the same tint as in ♂ *Agathina*.

Thorax clothed on the breast with *orange-yellow* pubescence.

This near ally of *Agathina* is easily known by its constantly yellow hind-wings (lemon-yellow in the ♂, ochre-yellow in the ♀) and pure white fore-wings in both sexes. The conspicuous orange-red flush at the base of the fore-wings on the under side which characterises *Agathina* is wanting in *Trimenia*; and the latter constantly possesses an additional spot (at the extremity of the first subcostal nervule) in the hind-wing, and has the under side of the thorax clothed with orange-yellow (instead of white) hair.

Mr. W. S. M. D'Urban first brought this beautiful species to my notice, forwarding specimens taken near King William's Town in 1860. Colonel Bowker subsequently met with it in the Trans-Kei; and I had the pleasure of observing it in Natal (Fort Buckingham or Tunjumbili) early in March 1867. It is quite a woodland butterfly, but seems to prefer forests at a considerable elevation.[1] I was struck with the resemblance which living specimens, when on the wing, bore to faded yellow leaves drifting before the breeze, and Mr. J. P. Mansel Weale, writing to me in the same year, independently recorded the same impression, conveyed to him by the notice of examples occurring in the woods near Bedford in the Cape Colony. In 1877, Mr. Weale, writing from Breidbach, near King William's Town, expressed his belief that the larva of *Trimenia* would be found to feed on *Loranthus prunifolius*, parasitic on *Schotia latifolia*.

Localities of *Mylothris Trimenia*.

I. South Africa.
   B. Cape Colony.
      b. Eastern Districts.—Bedford (*J. P. Mansel Weale*). King William's Town (*W. S. M. D'Urban, M. E. Barber*, and *J. H. Bowker*).
   D. Kaffraria Proper.—Tsomo and Bashee Rivers (*J. H. Bowker*).
   E. Natal.
      a. Coast Districts.—D'Urban (*A. D. Millar*). Pinetown (*J. H. Bowker*).
      b. Upper Districts.—Tunjumbili.

II. Other African Regions.
   B. North Tropical.
      *b*1. Eastern Interior.—Abyssinia: "Shoa (Antinori)."—Oberthür.

249. (3.) Mylothris Rüppellii, Koch.

Pieris Rüppellii, Koch, "Indo-Austr. Lep. Fauna, p. 88 (1865)."
♂ „ „ Feld., Reise d. Novara, Lep., ii. p. 167, n. 146 (1865).
Var. ♂ ♀, *Pieris Poppea*, Trim., Rhop. Afr. Aust., ii. p. 321, n. 215 (1866).
Var. ♂ ♀, *Pieris Haemus*, Trim., Trans. Ent. Soc. Lond., 1879, p. 342.
Pieris Rueppelii, Oberth., Etudes d'Ent., iii. p. 16, pl. i. f. 2 (1878); and Ann. Mus. Civ. Genova, xv. p. 149 (1880).

PLATE X. fig. 3 (♂), 3*a* (♀).

Exp. al., (♂) 2 in. 1–7 lin.; (♀) 2 in. $2\frac{1}{2}$–$6\frac{1}{2}$ lin.

♂ *White, with black hind-marginal spots on nervures. Fore-wing*: base slightly irrorated with fuscous; *a broad basal suffusion of orange-red* spreading over rather more than half of discoidal cell; costa edged with fuscous, which is widest (and white-scaled) near base, but very narrow about middle of wing; apex rather widely bordered with fuscous; four hind-marginal spots, of which the first sometimes joins apical black, and the fourth (at end of first median nervule) is always minute. *Hind-wing*: base slightly fuscous; a faint orange suffusion,

---

[1] Mr. Alfred D. Millar has sent me a ♂ taken near D'Urban on 11th September 1887, and informs me that he met with five others during that month, but that the species is rare both in that vicinity and about Pinetown.

fading outwardly into yellow, in basal region; six hind-marginal spots of small size, of which the first (at end of second subcostal nervule) is minute, and the second and sixth smaller than the three others. UNDER SIDE.—*Fore-wing:* no basal fuscous, and only some faint fuscous scales along edge of costa; orange-red suffusion brighter and spreading nearer to extremity of discoidal cell; in place of apical fuscous three small marginal nervular spots, of which the first is a little *before* apex at end of second subcostal nervule. *Hind-wing:* white, very faintly tinged with yellowish near base and inner margin; costa at and for a little distance from base bordered with orange-red; hind-marginal spots (especially the first) larger and rounder than on upper side.

♀ *More or less universally suffused with pale creamy-ochreous*, into which the basal orange-red (which is much duller than in ♂) gradually fades; on the discs the nervures are all more or less clouded with whitish; hind-marginal spots much larger than in ♂. *Fore-wing:* apical fuscous crossed by two or three ochreous inter-nervular rays; a minute spot at extremity of submedian nervure. UNDER SIDE.—In the paler specimens almost as white as in ♂, but in the darker ones more or less tinted with creamy-yellow generally. *Fore-wing:* orange-red suffusion fills discoidal cell, and faintly extends inner-marginally almost to posterior angle; hind-marginal spots *not* enlarged as on upper side, but blacker and rounded.

This beautiful *Pieris* is a very near relation of the West African *Poppea*, Cram.,[1] and of the Malagasy *Phileris*, Boisd.; it is, perhaps, not quite so closely allied to *Agathina*, Cram. As far as the ♂ s are concerned, the *very conspicuous orange-red basal suffusion* of the forewings *on the upper side* readily distinguishes *Rüppellii* from all the three species named; *Rhodope* and *Phileris* having only a very faint and much smaller orange-yellow tinge, while in *Agathina* there is none at all. The fainter basal suffusion in the hind-wings is also wanting in all the three allied forms. On the *under side*, *Rüppellii* has none of the rich ochre-yellow of *Agathina*, but its white is less pure than that of *Poppea* and *Phileris;* the hind-marginal spots, however, are smaller than in the two latter, and agree very closely with those of *Agathina;* and the basi-costal orange-red of the hind-wings is bounded (as in *Agathina*) by the costal nervure, instead of extending suffusedly beyond it, as in *Poppea;* while the wide suffusion from the base of the fore-

---

[1] Dewitz (*Nov. Act. K. Leop.-Carol.-Deutsch. Akad. Naturf.*, xli. p. 185, 1879) and Möschler (*Verh. K.K. Zool.-Bot. Gesellsch. Wien*, 1883, pp. 270-77) have independently been at pains to point out the characters which in reality widely separate such to all appearance closely allied species as *Rhodope*, Fab., and *Poppea*, Cram., which authors generally have treated as identical. Not only has *Rhodope* constantly eight hind-marginal spots (instead of six) in the hind-wing, as well as other minor differences, but *the subcostal nervure of the fore-wing has an additional (fourth) branch*. This latter distinction is actually more of *generic* than specific value. Möschler (*loc. cit.*, p. 270) is, however, quite in error in placing the ♀ of my *Poppea* (= ♀ *Hanus*, described in 1879) as that of *Rhodope*, Fab., its neuration agreeing with that of the ♂ with which I associated it, as well as with the neuration of the true *Poppea*, Cram., *Agathina*, Cram., and *Phileris*, Boisd.

wings is but little wider, and if anything less bright, than in *Agathina*, though very much brighter and broader than in the other two species.

With respect to the ♀ s, that sex of *P. Phileris* is not known to me;[1] but as regards the remaining species, *Rüppellii* is more like *Agathina* than *Poppea* in either of its forms. It is best distinguished from *Agathina* by (1) the broader red suffusion from base; (2) the white clouding of the nervures on disc; (3) the broader apical fuscous and larger hind-marginal spots in the fore-wings; and on the under side by (5) the much paler and less yellow hind-wings and apex of fore-wings, especially white on the hind-marginal edge. Both forms of *Poppea*, but particularly that in which the fore-wings are white, are specially characterised by the enlarged hind-marginal spots of the *upper side*, which in the fore-wings form broad, elongated nervular rays; but on the under side the corresponding spots are quite as in *Rüppellii*, excepting that they are slightly larger.

In both sexes of *Rüppellii* the wings are thinner and weaker in structure than in *Agathina*, but not so delicate and semi-transparent as in *Poppea* and *Phileris*.

It was not until I saw M. Oberthür's figure above quoted that I knew how very close my species *Hæmus*, described in 1879, was to the *Rüppellii* of Koch, which, from Felder's remarks (*op. cit.*), I had supposed to be most probably an Abyssinian variety of *Agathina*. Judging from that figure (of the ♂) and from Felder's description of another ♂, the South-African form, though it can scarcely be separated as a species from *Rüppellii*, constitutes a variety which is of larger size, better developed hind-marginal spots and fore-wing apical blackish, rather more restricted basal red in fore-wing on upper side,[2] and whiter under side. Neither Oberthür nor Felder describe the ♀ *Rüppellii*. Two ♂ s sent by Mr. Selous from Mashunaland in 1883 are considerably nearer the Abyssinian type than specimens from South Africa proper, being of smaller size, and with the black marginal markings much reduced,—most of the spots (especially the upper three in the hind-wing) being minute or obsolete.

PUPA.—" Satiny-white, with a pale-green tinge dorsally; projections on wing-covers tipped with ochreous-yellow and black; faintly mottled with grey ventrally. In shape closely resembles pupa of *Agathina*."—J. P. Mansel Weale, *in epist.*, February 1877.

Mr. Weale wrote that he suspected the larva to feed on *Loranthus Dregei*, parasitic on two kinds of *Acacia*, *Combretum*, and *Schotia*, but had not succeeded in identifying it, although he had found the pupa.

[1] The butterfly figured as the ♀ *Phileris* by Boisduval (*Faune Ent. de Madag.*, &c., pl. 2, f. 5) is an entirely different species, and has been named *Pieris Grandidieri* by Mabille.

[2] In plate 10, fig. 3, the basal red of the fore-wing on the upper side is not quite wide enough on the median nervure, while the basal yellow of the hind-wing is a little too much extended over the discoidal cell.

Mr. W. S. M. D'Urban was the first to discover this very beautiful insect in South Africa, having captured two specimens at King William's Town in May 1861. In January 1878, I saw an example in the same locality, flitting about flowers in Colonel Bowker's garden. On the Bashee, in Kaffraria Proper, Colonel Bowker met with many specimens, including some females. At Breidbach, near King William's Town, Mr. Weale observed that this species was more prevalent in wet weather, while *Ayathina* appeared more when it was dry. I am not aware of the occurrence of the species in Natal. A specimen from the Zambesi was in the collection of the late Mr. Hewitson in 1867, and, as above mentioned, Mr. Selous sent two examples from Mashunaland in 1883. Oberthür notes it as very rare in Abyssinia, the Marquis Antinori having brought from thence only a single specimen.

Localities of *Mylothris Rüppellii*.

I. South Africa.
  B. Cape Colony.
    b. Eastern Districts.—King William's Town (*W. S. M. D'Urban* and *J. P. Mansel Weale*). Kei River Mouth (*J. H. Bowker*).
  D. Kaffraria Proper.—Bashee River (*J. H. Bowker*).
II. Other African Regions.
  A. South Tropical.
    b1. Eastern Interior.—Zambesi.—Hewitson Coll. Mashunaland (*F. C. Selous*).
  B. North Tropical.
    b1. Eastern Interior.—Abyssinia: "Southern (*Rüppell*)."—Felder; "Shoa (*Antinori*)."—Oberthür.

Genus PIERIS.

Pieris, Schrank [part], "Fauna Boica, ii. 1, pp. 152, 164 (1801)."
  „   Latreille [part], "Hist. Nat. Crust. et Ins., xiv. p. 111 (1805)."
Pontia, Fab., "Illiger's Mag., vi. p. 283 (1807)."
Pieris, Boisduval [part]. Sp. Gen. Lep., i. p. 434 (1836).
  „   Doubleday [part], Gen. Diurn. Lep., i. p. 42 (1847).
Pinacopteryx, Wallengren, K. Sv. Akad.-Handl., 1857; Lep. Rhop. Caffr., p. 7.
Belenois and *Synchloë* (Hübn. 1816), Butler, Cist. Ent., i. pp. 50, 51 (1870).

IMAGO.—*Head* of moderate size, more or less densely clothed with bristly hairs; *palpi* slender, compressed, rather long,—basal joint as long as (or sometimes rather longer than) middle one; both scaly above, and densely clothed with long bristly hair beneath,—terminal joint scaly, slender, usually about as long as, or a little shorter than, middle joint, moderately acuminate or slightly blunted, porrect (sometimes half hidden by long hair of middle joint); *antennæ* of moderate length and thickness, with the club short, rather abruptly formed, more or less flattened, and rounded at tip.

Thorax of moderate length and robustness, densely clothed with hair, which is longer and more silky above. *Fore-wings* of moderate length; costa scarcely arched beyond basal curve; apex always marked, and sometimes prominent and acute; hind-margin straight, or very slightly concave about middle; inner margin usually slightly concave about middle; subcostal nervure four-branched,—first nervule given off some distance before extremity of discoidal cell, short, often joining costal nervure near its extremity (*Calypso* group),—second long, ending not far before apex, given off rather nearer to first than to extremity of cell,—third subapical, short or very short, ending at apex or just before it (in *Brassicæ* group exceedingly short or even [*P. Daplidice* and *P. Hellica*] wanting),—fourth ending at or a little below apex; upper radial nervule united with subcostal nervure at some distance beyond end of cell; upper disco-cellular nervule very oblique, usually considerably more than half as long as lower one, which is transverse and slightly curved; discoidal cell rather broad, truncate at extremity, more than half length of wing. *Hind-wings* broad, rounded, sometimes rather elongate anal-angularly; costa moderately or slightly arched; hind-margin very convex, entire or slightly sinuated; anal angle rounded; inner margins meeting, and forming a tolerably deep groove almost to end of abdomen; first subcostal nervule much arched, given off a long way before end of cell; disco-cellular nervules both oblique, the upper one much shorter (sometimes less than half as long), the lower one slightly curved. *Legs* moderately long and thick; femora with fine hair beneath; fore-tibiæ much shorter than femora; middle and hind tibiæ about as long as femora, their terminal spurs moderately developed; tibiæ and tarsi finely spinulose.

Abdomen curved, compressed, moderately long and slender, rather larger terminally, tufted with silky hair at base on back, and more thinly so along middle line beneath; anal valves in ♂ sometimes with a short terminal curved spine (*Calypso* group); in other species, a tuft of long bristly hairs beneath, just at base of each anal valve.

LARVA.—Rather elongate, shortly pubescent, slightly attenuated at extremities; head small.

PUPA.—Rather slender; much attenuated posteriorly; a more or less elevated ridge along middle line of back, and another on each side of first three segments of abdomen; an acute projection in front of head; and a prominent tubercle at middle of dorso-thoracic ridge, and on each lateral abdominal ridge on second segment.

Boisduval's list of this genus numbered (1836) no fewer than 166 species; Doubleday's (1847), 175. Wallace, after separating from it (1867) the extensive sections *Thysa*, *Tachyris*, and *Prioneris*, gave the number of described species as 171; and if (as I consider preferable) *Tachyris* be not separated generically, the total will be raised to 237. Mr. Butler's revision of the genus was first given (1870) in *Cistula Entomologica*, vol. i. pp. 39–52, and, as amended (1872) in *Proc. Zool. Soc. Lond.*, p. 27, he still further divided the old genus *Pieris* into fifteen genera, containing altogether 341 species, but restricted the name of *Pieris* to a very limited group of Tropical American species (fifteen), forming Section I. of Boisduval's subdivision of the old genus, and for which Hübner's name *Perrhybris* has been generally adopted. I cannot find that his genera *Appias* (= *Tachyris*, Wallace), *Belenois*, *Pontia*, and *Synchloë* are satisfactorily separable; these I regard as constituting the bulk of the genus *Pieris*, and they contain, according to Mr. Butler, 179 described species. Of those genera of Mr. Butler's which include African species, I think that *Mylothris* and *Herpænia* are deserving of adoption, considering their peculiarities of structure and pattern.

Pieris, as here regarded, is of almost universal distribution, New Zealand being the only extensive land area at any considerable distance from the poles which has no known representative of the genus. Taking the number of recorded species at about 230, it is noticeable that the Oriental and Australian regions greatly preponderate in richness and variety of forms, upwards of a hundred, or not far from half the genus, being native there, and a considerable number of these apparently inhabiting both regions. The Neo-Tropical Region yields between fifty and sixty species, and the Ethiopian about thirty-seven. The great Palæarctic Region and the Nearctic are both exceedingly poor, possessing respectively fourteen and eleven species only.

In South Africa fourteen species are known to occur; they belong to four groups, of which *P. Saba*, (Fab.), *P. Pigea*, Boisd., *P. Calypso*, Drury, and *P. Daplidice*, (Linn.), may be regarded as the respective representatives, viz.:—

GROUP 1.—*Saba*, Fab., representative.

Sexes extraordinarily different. ♂ white, with an incomplete narrow apical hind-marginal border in fore-wings; under side with hind-wings and apex of fore-wings creamy. ♀ with very broad black hind-marginal borders to both fore and hind wings, and with basi-disco-cellular area of fore-wings usually also black; under side white with similar but much fainter blackish markings. The ♂ has an inferior abdominal tuft of bristly hairs,—the characteristic of Wallace's genus *Tachyris*.

(1 species.)

GROUP 2.—*Pigea*, Boisd., representative.

Sexes moderately dissimilar. ♂ greenish-white (in one species, *P. Spilleri*, Staud., sulphur-yellow), with very small and inconspicuous hind-marginal blackish nervular spots (scarcely apparent except at and near apex of fore-wings, where they are enlarged); under side with hind-wings and apex of fore-wings faintly tinged with yellowish or greenish, sometimes speckled with fuscous and with traces of discal fuscous spots in hind-wings. ♀ usually more or less tinged with ochre-yellow, especially in hind-wings, and with one or more discal fuscous spots in fore-wings; hind-marginal nervular spots larger, especially in fore-wings; under side more deeply coloured with ochre-yellow.

(5 species: *Pigea*, Boisd., *Alba*, Wallengr., *Simana*, Hopff., *Charina*, Boisd., *Spilleri*, Staud.)

GROUP 3.—*Calypso*, Drury, representative.

Sexes very dissimilar. ♂ white, with moderately broad white-spotted black borders, better developed in fore-wings, which have also a conspicuous terminal disco-cellular black spot or oblique bar; under side with hind-wings and apex of fore-wings of some tint of yellow, with similar blackish markings, but usually with black neuration and discal spots. ♀ more or less tinged with yellow, especially in hind-wings (which are sometimes deep ochre-yellow), and with all black markings much enlarged.

In one species (*Mesentina*, Cram.) the apex of fore-wings is rather produced and pointed, and in two others (*Gidica*, Boisd., and *Abyssinica*, Lucas) it is strongly so characterised.

(7 species: *Thysa*, Hopff., *Ogygia*, Trim., *Zochalia*, Boisd., *Mesentina*, Cram., *Gidica*, Godt., *Abyssinica*, Lucas, *Severina*, Cram.)

GROUP 4.—*Daplidice*, Linn., representative.

Sexes much alike. White, with black markings much as in last section, but in ♀ stronger, and including an additional inferior discal spot in fore-wings. Under side with hind-wings and apex of fore-wings varied with white and dull-greenish and yellow-tinged markings.

In *Daplidice* and *Hellica*, Linn., the third subcostal nervule of fore-wings is wanting.

(1 species: *Hellica*, Linn.)

The caterpillars of this genus appear for the most part to be attached to *Cruciferæ* and *Capparideæ*; those of the European *P. Brassicæ*, *Rapæ*, &c., and of the allied North-American *P. oleracea*, are notorious for their injuries to cultivated cabbages, turnips, and related plants.

250. (1.) Pieris Saba, (Fabricius).

♀ *Papilio Saba*, Fab., "Sp. Ins., p. 46, n. 199 (1781);" and Ent. Syst., iii. 1, p. 201, n. 627 (1793).
♀ *Papilio Epaphia*, Cram., Pap. Exot., iii. pl. ccvii. ff. D, E (1782).
♀ *Papilio Hypatia*, Dru., Ill. Nat. Hist., iii. pl. xxxii. ff. 5, 6 (1782).
♀ *Pieris Higinia*, Godt., Enc. Meth., ix. p. 133, n. 45 (1819).
♂ *Pieris Orbona*, Boisd., Faune Ent. Madag., &c., p. 18, pl. 1, f. 3 (1833).
♀ and (as ♂) Var. ♀, *Pieris Malatha*, Boisd., *loc. cit.*, ff. 4, 5.
♂ ♀ *Pieris Orbona*, Boisd., Sp. Gen. Lep., i. p. 497, n. 89 (1836).
♂ ? *Pieris Saba*, Doubl., Gen. D. Lep., i. p. 46 (1846).
♂ ♀ „ „ Hopff., Peters' Reise Mossamb., Ins., p. 353 (1862).
♂ ♀ „ „ Trim., Proc. Ent. Soc. Lond., 1881, p. vii. pl. ix. ff. 3, 4.
♂ ♀, ♀ Var. *Pieris Saba*, Grandid., Hist. Phys., &c., Madag., xix., ii. Atlas 1, pl. 36, ff. 1-8 (1885).

Exp. al., (♂) 2 in. 1–5 lin.; (♀) 2 in. 1–5½ lin.

♂ *White; fore-wing with a rather narrow blackish apical and hind-marginal border. Fore-wing:* space between costa and costal nervure rather closely irrorated with blackish; costa with a linear black edging from before middle to end of second subcostal nervule; at the latter point commences the apical hind-marginal blackish border, which is narrow, and rather suffused and ill-defined on its inner edge (where it forms more or less marked nervular dentations), extending to between second and first median nervules; usually a small blackish mark at extremity of first median nervule. *Hind-wing:* sometimes without marking, but usually with a series of very small, faint, nervular blackish marks along hind-margin, of which the first (at extremity of first subcostal nervule) is much longer and more pronounced than the rest. UNDER SIDE.—*Hind-wing and apical border of fore-wing faintly tinged with rather dull cream-colour. Fore-wing:* no blackish border; in discoidal cell a narrow basal suffusion of chrome-yellow; some scarcely perceptible indications of very small hind-marginal nervular spots. *Hind-wing:* costa from base to a point considerably before middle narrowly edged with chrome-yellow; in some specimens, traces of some very small hind-marginal nervular spots, those at extremities of first subcostal and second median nervules more apparent than the rest.

♀ *Blackish border of fore-wing very broad, and a very broad blackish hind-marginal border in hind-wing; basal area of fore-wing very broadly blackish; white ground thus reduced to a discal band in fore-wing, and a broad basi-central space in hind-wing. Fore-wing:* basal blackish extending from costa to inner margin, completely and exactly filling discoidal cell to its extremity, narrower and somewhat suffused below cell; very broad apical hind-marginal border widest on costa (where it extends sometimes to a little before middle), somewhat abruptly narrowed near and at posterior angle, its inner edge regularly dentated on median nervules,—the dentation on third median nervule often more or less completely united to outermost extremity of black filling discoidal cell; in blackish border, and towards its outer part, a subapical oblique row of three inter-nervular, elongate, rather ill-defined white spots (of which the lowest is rarely obsolete), between fourth subcostal and third median nervules. *Hind-wing:* base narrowly and usually faintly irrorated with blackish; hind-marginal broad border somewhat variable in width, its inner edge irregular and more or less suffused, but with a more or less marked dentation or prominence on second subcostal nervule. UNDER SIDE.—*Hind-wing and apex of fore-wing greyish-white with a very slight opalescent tinge and a faint marginal edging of yellowish; hind-marginal dark borders much duller than on upper side, especially in hind-wing. Fore-wing:* basal blackish much fainter, obsolete below discoidal cell; yellow basal suffusion much more extensive than in ♂, covering about two-thirds of cell, and reaching also to costal edge; greyish-white apical patch representing on its inner side the white spots of the upper side. *Hind-wing:* yellow costal edging extending to middle or a little beyond; hind-marginal border diffusedly narrowed in its lower half, and outwardly marked vaguely with a greyish-white shade along its upper half; sometimes a ray of chrome-yellow along submedian nervure from base.

VARIETY of ♀.—(Var. *Flavida*, Grandidier.)—More or less tinged with lemon-yellow, especially near margins: hind-marginal blackish border much narrower (especially in hind-wing) and more sharply dentated inwardly; basal blackish of fore-wing all wanting except a narrow faint basi-costal border, which scarcely enters discoidal cell (except very slightly and diffusedly at base and along its upper edge), and terminates somewhat truncately a little before extremity of cell. *Fore-wing*: subapical spots in border yellowish. *Hind-wing*: basal irroration wanting or very slight and restricted; hind-marginal border less than half as wide as in typical ♀, its inner edge much better defined, and prominently (in one example acutely) dentating the ground-colour on nervules. UNDER SIDE.—Hind-wing and apex of fore-wing more creamy in tint; blackish border considerably narrower and duller in fore-wing, and scarcely perceptible in hind-wing. *Fore-wing*: no cellular blackish (except, in one specimen, the very faintest trace along upper part of cell).

In this butterfly the dissimilarity of the sexes is so extreme that one cannot wonder at Boisduval's treating them originally (1833) as distinct species. The variety of the ♀ just described was, however, at first regarded by that author as the ♂ of the typical ♀, which he named *Malatha*; but this was rectified in the *Species Général* (1836), which recognised that *Orbona* and *Malatha* were ♂ and ♀ of one species, and noted the so-called ♂ of *Malatha* as a form of the ♀.

I have examined the type of *Saba* in the Banksian Collection at the British Museum. It is a small but broadly black-marked ♀ from "Sierra Leone." The ♀ figured by Cramer as *Epaphia* (stated to be from the same locality) is very broadly black-marked, and is represented as possessing an orange basal suffusion in discoidal cell of the fore-wing, and also a broad orange suffusion (outwardly fading into yellow) between the nervures in the basal half of the hind-wing. Drury's *Hypatia* is also a "Sierra Leone" ♀; it is larger than Cramer's specimen, and is figured as having a tinge of yellow over the basi-inner-marginal part of both wings on the upper side, and the border of the hind-wing narrower than in Cramer's figure; while, on the under side, the suffusion of the cell in the fore-wing is pale ochreous-yellow, and there is scarcely a trace of the orange and yellow rays in the hind-wing. Godart's description of a ♀, likewise from Sierra Leone, accords better with Drury's than with Cramer's figures. Neither of the figures just mentioned exhibit the junction in the fore-wing of the disco-cellular with the hind-marginal blackish (on third median nervule), which usually occurs in the Natalian and Delagoa Bay specimens of the typical ♀.

I do not remember to have seen, nor have I found any record of, any ♀ examples linking the variety *Flavida* with the typical ♀; but the form and position of the small dusky costal stripe in the former reminds one of the similar marking in the ♀ *Papilio Meriones*, Feld., of Madagascar, and indicates (with the well-developed but not very broad hind-marginal borders) the initial stage of the extraordinarily developed dark markings which characterise the typical form. I have suggested (*Proc. Ent. Soc. Lond.*, 1881, p. viii.) that the ♀ *Saba* may be modified in mimicry of a common and evidently protected slow-flying diurnal moth, *Nyctemera apicalis*, Walker, which has a very wide African range, and possesses a near ally in Madagascar.[1]

As I have noted (*loc. cit.*, p. vii.), Colonel Bowker captured the paired sexes near the Umgeni, Natal, in January 1881, and I believe that this is the only instance of such capture recorded in the case of this species. The two specimens in question are figured in *Trans. Ent. Soc. Lond.*, pl. ix. (1881), the ♂ being an individual with the marginal markings rather more pronounced than usual, and the ♀ one of strongly developed black markings throughout.

The late Colonel Tower first made this species known to me as a South-African native, having taken the ♂ at St. Lucia Bay. In 1872 the late Mr. E. C. Buxton met with two of each sex on the Natal Coast (D'Urban and Verulam) during October.

The butterfly appears to be far from common in Natal. I did not meet with it during my visit in 1867, and Colonel Bowker has sent only five examples. From Delagoa Bay, Mrs. Monteiro has sent several of both sexes, including two of the ♀ Var. *Flavida*. A single ♀ of this variety occurred in a collection made in Mashunaland, near the Zambesi, in 1882, by Mr. F. C. Selous.

The distribution of *Saba* in the Ethiopian Region is very wide, embracing apparently a very large portion of the South-Tropical tracts and much of the North-Tropical coast on the Western side. In Madagascar it would appear to be rather numerous, being of frequent occurrence in collections sent from thence. I have not found any notice of the particular haunts or habits of the species. Colonel Bowker's Natalian specimens were taken in the month of January.

Localities of *Pieris Saba*.

I. South-Africa.
E. Natal.
a. Coast Districts.—D'Urban (the late *E. C. Buxton* and *J. H. Bowker*). Umgeni (*J. H. Bowker*). Verulam (the late *E. C. Buxton*).
b. Upper Districts.—Maritzburg (*A. S. Windham*).
F. Zululand.—St. Lucia Bay (the late *Colonel H. Tower*).
H. Delagoa Bay.—Lorenço Marques (*Mrs. Monteiro*).

II. Other African Regions.
A. South Tropical.
a. Western Coast.—" Angola (*Pogge*)."—Dewitz.
b. Eastern Coast.—" Querimba."—Hopffer. " Zanzibar."—Kirby, Cat. Hewits. Coll.

---

[1] In marking, the nearest ally of the ♀ *Saba* is the ♀ of the Malayan and Sumatran *Cynis*, Hewits., figured by Mr. Distant (*Rhop. Malay.*, pl. xxvi. f. 6) as *Udaina Cynis*. In this ♀ the white is even more reduced in the fore wings, the subapical spots being wanting, and the lower part of the discal band much narrower; the black is, however, represented by dull fuscous-brown, which is ill-defined in the hind-wings. The ♂ *Cynis* has a broad costal, apical, and hind-marginal black border in the fore-wings.

b1. Eastern Interior.—Mashunaland (*F. C. Selous*). "Nyassa."—Kirby, Cat. Hewits. Coll.

bb. Eastern Islands.—Madagascar (*De Robillard*). "Sainte-Marie et Madagascar."—Boisduval.

B. North Tropical.

a. Western Coast.—"Fernando Po; Mungo and Victoria, Cameroons (*Buchholz*)."—Plötz. Old Calabar.—Coll. Brit. Mus. "Lower Niger (*W. A. Forbes*)."—Godman and Salvin. Abomey.—Coll. Brit. Mus. Cape Coast Castle and Elmina (*Bourke*). Sierra Leone.—Coll. Brit. Mus. "Senegal."—Boisduval.

251. (2.) Pieris Thysa, Hopffer.

♂ ♀ *Pieris Thysa*, Hopff., "Monatsb. K. Akad. Wissensch. Berl., 1855, p. 639, n. 1;" and Peters' Reise Mossamb., Ins., p. 349, t. xxi. ff. 7, 8 [♂], 9, 10 [♀] (1862).

♂ *Pieris Agathina*, Var. A., Trim., Rhop. Afr. Aust., i. p. 29 (1862).

Var. ♂, *Belenois Sabrata*, Butl., Trans. Ent. Soc. Lond., 1870, p. 526.

Exp. al., (♂) 2 in. 5½–7 lin.; 2 in. 3–11 lin.

♂ *White, with large hind-marginal nervular black spots, united in fore-wing to form an inwardly strongly dentated border, but all separate in hind-wing; a submarginal row of small inter-nervular black spots, more developed in hind-wing.* Fore-wing: costal margin with a narrow black edging, scaled with white at and near base, and rather abruptly widened a little before apex; hind-marginal macular border continuous as far as second median nervule, its component spots more or less trigonate and inwardly acute; a separate spot at extremity of first median nervule, and a very small narrow one at extremity of submedian nervure; spots of submarginal row four, rounded,—three between costa and lower radial nervule, very close to (or even touching) apical black,—the fourth (often indistinct) apart from the rest, between third and second median nervules; base with a sub-metallic gloss, and with the under side red faintly indicated. *Hind-wing:* six hind-marginal spots of moderate size, rather widely separated, rounded, from second subcostal nervule to submedian nervure,—the third, fourth, and fifth larger than the first and second, and the sixth much the smallest of all and sublinear; very rarely the trace of a seventh spot at extremity of first subcostal nervule; seven small spots in submarginal row, between costal and submedian nervures,—of which the seventh is always, and the fourth and sixth usually, very indistinct; a faint ochrey-yellow edging on inner margin for a little distance from anal angle. UNDER SIDE.—*Hind-wing and small apical patch of fore-wing creamy ochreous-yellow; hind-marginal spots much smaller and narrower, but submarginal spots larger, darker, and better defined than on upper side.* Fore-wing: a narrow basal suffusion of orange-red, outwardly radiating slightly on the white ground, but not reaching quite to middle of discoidal cell; of the seven hind-marginal spots, the upper four (bordering apical yellow) are very small, or even obsolete, but the remaining three moderately developed, the lowest being larger than on upper side. *Hind-wing:* on costa a narrow scarlet-red edging for a little distance from base; submarginal spots inclining to be pointed inwardly, especially the fourth, which is smaller and thinner than all the rest; hind-marginal spots greatly reduced, sublinear, crossed by yellow nervules, the upper three obsolescent or sometimes obsolete. *Cilia* white, widely interrupted with black at ends of nervules.

♀ *Hind-wing and (usually) greater part of fore-wing paler or deeper (rather dull) yellow-ochreous; black markings, especially hind-marginal ones of hind-wing, in some examples much larger than in* ♂. *Fore-wing:* basal reddish suffusion very much broader and sprinkled with grey scales,—filling about three-fourths of discoidal cell; disc sometimes white, only submarginally tinged with yellow-ochreous,—sometimes all faintly tinged with that colour, but more deeply so submarginally; lowest spot in submarginal row usually wanting; hind-marginal series of spots variable in development, but lowest spot always larger than in ♂. *Hind-wing:* a wide basal reddish-grey speckled suffusion, reaching from costa to submedian nervure and almost filling discoidal cell; submarginal spots (except the last) all larger and darker than in ♂; hind-marginal spots in one specimen of the same form as, and no larger than, in ♂, but in two others so enlarged as to be all but contiguous, and sub-rhomboidal in form. UNDER SIDE.—*As in* ♂, *but scarlet suffusion at base of fore-wing very much deeper in tint, and so much enlarged as to occupy three-fourths of discoidal cell and extend considerably below it;* neuration whitish where on ochreous-yellow ground.

This species, in both sexes, is subject to variation in size, and in the development of the hind-marginal spots; and the ♀ also varies much in the proportion and extent of the general suffusion of dull yellow-ochreous in the fore-wing. Dewitz records that in examples from Chinchoxo, on the Loango Coast, the ♂ has no basal yellow on the under side of the fore-wings, and the ♀ has the marginal markings much more extended than in the Querimba examples described by Hopffer. The variety named *Sabrata* by Butler is distinguished by its small size, and by the brighter, richer colouring of the under side of the ♂: it is represented in the British Museum collection (1886) by two ♂ s and a ♀ from Zanzibar.

Thysa is so remarkably like *Agathina,* Cram., in colouring and marking, that its real alliance with the group represented by *Calypso,* Drury, is apt to be overlooked; and up to 1866, when I had seen but one specimen (in the collection of the British Museum), I was misled into regarding it as a variety of *Agathina.* There can be little or no doubt that it directly mimics the common and widely prevalent butterfly just named.[1] Though recorded over rather a wide range in Africa south of the Equator, it appears to be everywhere scarce. Hopffer (*op. cit.*) states that the Peters Expedition brought a few examples from Querimba. I met with a solitary ♀ in Natal on 22nd

---

[1] It is very noticeable how closely the ♀ follows all the variations of tint exhibited by the ♀ *M. Agathina.*

February 1867; it was flying pretty actively in a wood at the Umhlanga, and was netted by the late Mr. M. J. M'Ken, with whom I was on a collecting excursion. Mr. W. Morant lent me in 1869, for description, a ♂ captured near Pinetown, and Colonel Bowker sent a perfect specimen of the same sex taken by him in that locality in June 1884, as well as a pair which he took nearer D'Urban in 1881. Single examples have also reached me from Maritzburg and St. Lucia Bay, as well as three forwarded from Delagoa Bay by Mrs. Monteiro. Möschler (*Verh. K.K. Zool.-Bot. Gesellsch. Wien*, 1883, p. 10) notes the receipt of a ♀ from Baziya, east of the Bashee River in Kaffraria Proper; and this is the southern limit of the range of the insect, as far as I can discover. It is not improbable that *Thysa* is less rare than it appears to be, as it might very often be passed over by collectors as the common *M. Agathina*.[1]

Localities of *Pieris Thysa*.

I. South Africa.
    D. Kaffraria Proper.—"Baziya (*Baur* and *Hartmann*)."—Möschler.
    E. Natal.
        a. Coast Districts.—D'Urban (*J. H. Bowker*). Umhlanga. Pinetown (*W. Morant* and *J. H. Bowker*).
        b. Upper Districts.—Maritzburg (*S. Windham*).
    F. Zululand.—St. Lucia Bay (the late *Colonel H. Tower*).
    H. Delagoa Bay.—Lourenço Marques (*Mrs. Monteiro*).

II. Other African Regions.
    A. South Tropical.
        a. Western Coast.—"Angola (*J. J. Monteiro*)."—H. Druce. Congo.—Coll. Brit. Mus. "Chinchoxo (*Falkenstein*)."—Dewitz.
        b. Eastern Coast.—"Querimba."—Hopffer. Zanzibar [Var. *Sabrata*, Butl.].—Coll. Brit. Mus.

252. (3.) Pieris Pigea, Boisduval.

    ♀ *Pieris Pigea*, Boisd., Sp. Gen. Lep., i. p. 523, n. 124 (1836).
    ♀ „ „ Wallgrn., K. Sv. Vet.-Akad. Handl., 1857; Lep. Rhop. Caffr., p. 7 [*nec* ♂].
    ♂ *Pinacopteryx Simana*, Wallgrn., *op. cit.*, p. 10.
    ♀ *Pieris Pigea*, Trim., Rhop. Afr. Aust., i. p. 29, n. 16 (1862) [*nec* ♂].
    ♂ *Belenois inana*, Butl., Trans. Ent. Soc. Lond., 1870, p. 526.

PLATE X. fig. 5 (♂), fig. 5*a* (♀).

Exp. al., (♂) 1 in. 10 lin.—2 in. 2½ lin.; (♀) 1 in. 11 lin.—2 in. 2 lin.

♂ *Glossy greenish-white.* Fore-wing: a very narrow blackish mark at base; costa edged throughout by a black line; a hind-marginal

[1] No fewer than eleven specimens have been sent to me by Mr. A. D. Millar, who took them in the vicinity of D'Urban, Natal. From the dates of capture kindly furnished by Mr. Millar, I find that nine of these (six ♂s, three ♀s) were taken in August and September 1887; they are all smaller than usual, and have the hind-marginal black markings decidedly reduced. The remaining two are a pair taken *in copulâ* on the 29th January 1888; they are of large size, and both (but more especially the ♀) have the hind-marginal markings strongly developed. The ♀, indeed, has these markings (more particularly in the hind-wings) larger and more extended inwardly than in any other example that I have seen; her fore-wings are almost pure white on the disc, but the reddish basal suffusion is densely clouded with grey, and there is a separate patch of grey scales near posterior angle.

I suspect that the larger and more strongly marked specimens represent what is the summer or wet-season brood. The ♀ that I took in February was also very large, and had the dark markings much developed.

series of seven very small but distinct black nervular spots from third
subcostal nervule (at apex) to first median nervule; about apex some
black scaling, forming a very narrow border, uniting more or less
incompletely the hind-marginal spots one to three or (rarely) four; an
exceedingly slender black hind-marginal edging line from apex to third
median nervule. *Hind-wing:* a hind-marginal series of seven minute
black spots at extremities of nervules, of which those on median nervules are best developed. UNDER SIDE.—*Hind-wing and apex of fore-
wing (rather widely) very pale dull-yellowish. Fore-wing:* costa narrowly
tinged with dull-yellowish; rarely a faint or very faint small discal
fuscous spot between third and second median nervules; apical pale-
yellowish very narrowly continued along hind-margin to about first
median nervule; hind-marginal spots either wanting entirely or extremely minute. *Hind-wing:* costa narrowly edged with ochre-yellow
from close to base almost to extremity of costal nervure; a discal row
of five more or less indistinct small fuscous inter-nervular spots (of
which the third, between third and second median nervules, is best
expressed), between second subcostal nervule and submedian nervure;
hind-marginal spots as on upper side.

♀ *Fore-wing tinged with pale ochreous-yellow on margins; hind-wing
wholly pale ochreous-yellow; dark markings larger generally than in ♂.
Fore-wing:* ochreous-yellow tinge sometimes more or less faintly per-
vading the whole area, but always more developed marginally, especially
about apex and narrowly along hind-margin; costal margin from base
usually more or less thinly sprinkled with fuscous; *a conspicuous fuscous
spot on disc between third and second median nervules,* and usually an
ill-defined diffused fuscous mark between first median nervule and submedian nervure (sometimes vaguely extended to posterior angle by
fuscous scaling below submedian nervure); rarely the faint trace of a
third, much smaller spot, between second and first median nervules;
nervular hind-marginal spots larger (sometimes very much larger) and
longer, and including an additional one, smaller than the rest, at
extremity of submedian nervure. *Hind-wing:* neuration marked with
whitish; hind-marginal spots always larger, and usually conspicuous;
usually the more or less indistinct traces of a curved discal row of
small fuscous spots between second subcostal nervule and submedian
nervure. UNDER SIDE.—*Hind-wing and apical and marginal yellow of
fore-wing rather deeper in tint than on upper side, and altogether different
from the colouring of the ♂;*[1] hind-marginal spots rarely much larger
than in ♂, and in fore-wing sometimes wanting. *Fore-wing:* upper
discal spot darker than above; lower one better defined, very much
smaller, not diffused; trace of intermediate spot seldom present. *Hind-
wing:* six inter-nervular spots in curved discal row constant in number,
small, somewhat diffused, and ill-defined; costal edging from base
orange-red, wider than in ♂, and faintly continued beyond middle.

[1] In the figure of the ♂, pl. x. fig. 5, the yellow of the under side is too deep and
decided in tint.

A specimen of the ♀ sent from Pinetown, Natal, in 1869 by Mr. W. Morant is remarkable for having the hind-wings white except for a faint ochreous-yellow suffusion in basal third, and a very narrow ochreous-yellow hind-marginal border like that of the fore-wings; while the yellowish of the under side is but little darker than in the ♂. A similar ♀, but much worn, was taken (paired with the ♂) near D'Urban by Colonel Bowker in 1879.

I had not seen the ♂ *Pigea* in 1862, but mistakenly took for it the ♀ of the closely-allied but smaller *P. alba*, Wallengr. Colonel Bowker sent only ♀s from Kaffraria Proper, but I met with both sexes on the Natal coast in 1867.

Wallengren (*loc. cit.*) describes what purports to be "*P. Simana* (Hpfr.) ♀;" but his description is undoubtedly that of a ♂, and agrees thoroughly with the characters of the ♂ *Pigea*; moreover, in some rough sketches of types sent by this author to Mr. W. F. Kirby, and kindly lent by the latter for my inspection, "*P. Simana*, Hpfr.," is undoubtedly the representation of a ♂, and to all appearance that of *Pigea*.

Colonel Bowker has sent me four pairs of the sexes captured *in copulâ*; three of these were received in 1879, and one in 1881. The ♂s of these pairs differ in size only, but one of the ♀s (as noted above) is abnormally pale.

I found this species rather numerous near D'Urban and Verulam, on the coast of Natal, in February, March, and April 1867. It was active on the wing, keeping to open places about the borders of woods, and reminding one (except for the deeper yellow of the females) of the abundant *Pieris Rapæ* of Europe. It ranges to Kinsembo in Congoland, and is reported by M. Oberthür to have been received from widely separated localities in Tropical Eastern Africa.

Localities of *Pieris Pigea*.

I. South Africa.
D. Kaffraria Proper.—Bashee River (*J. H. Bowker*).
E. Natal.
  a. Coast Districts.—D'Urban. Verulam. Pinetown (*W. Morant*).
K. Transvaal.—Potchefstroom (*T. Ayres*).

II. Other African Regions.
A. South Tropical.
  a. Western Coast.—Congoland: "Kinsembo (*H. Ansell*)."—Butler.
B. North Tropical.
  b. Eastern Coast.—Red Sea: "Massowah (*Raffray*)."—Oberthür.
  *b*1. Eastern Interior.—"Lake Tsana (*Raffray*)."—Oberthür. "Shoa (*Antinori*)."—Oberthür.

253. (4.) Pieris alba, (Wallengren).

♂ *Pinacopteryx alba*, Wallgrn., K. Sv. Vet.-Akad. Handl., 1857; Lep. Rhop. Caffr., p. 10, n. 7.
♀ *Pieris Pigea*, ♂, Trim., Rhop. Afr. Aust., i. p. 29, n. 16 (1862).

Exp. al., (♂) 1 in. 10½ lin.—2 in. 1 lin.; (♀) 1 in. 10 lin.—2 in. 0½ lin.

Closely allied to *Pigea*. Boisd., but smaller.

♂ Glossy greenish-white. *Fore-wing:* a very narrow basal black mark; costa with a very fine linear black edging throughout; this edging faintly continued from apex along hind-margin to about extremity of second median nervule; about apex some very limited blackish scaling immediately precedes black edging; spots at extremities of nervules either wanting altogether or exceedingly minute. *Hind-wing:* usually an exceedingly minute spot at extremity of each nervule and of submedian nervure. UNDER SIDE.—*Hind-wing and apex of fore-wing glistening greyish-creamy with a slight pinkish tinge. Fore-wing:* tint of apical area extending and gradually narrowing along costa and along hind-margin to between second and first median nervules; no hind-marginal nervular spots. *Hind-wing:* a narrow orange edging on costa as in *Pigea;* sometimes the traces of an indistinct discal series of grey spots; no hind-marginal nervular spots.

♀ Dull-white, tinged with pale-yellowish on margins (more prominently in hind-wing, which is rarely all tinged with yellow). *Fore-wing:* costa with a pale-brownish border above costal nervure; apical blackish usually wider than in ♂ and more continuous than in *Pigea;* hind-marginal blackish also better developed; nervular spots very small (that at extremity of submedian nervure minute), but almost always distinct; a fuscous spot (sometimes ill-defined or occasionally almost obsolete) between third and second median nervules. *Hind-wing:* hind-marginal spots very small or minute, but almost always present. UNDER SIDE.—*Hind-wing and apex of fore-wing varying from almost the same tint as in ♂ to pale, deep, or very deep ochre-yellow;* hind-marginal spots wanting altogether in fore-wing, very minute (rarely wanting) in hind-wing. *Fore-wing:* discal fuscous spot always present, rarely indistinct. *Hind-wing:* discal row of spots situated as in *Pigea*, but seldom well-defined, sometimes partly (and in one specimen almost wholly) obsolete.

In addition to its smaller size and (in the ♂ s and paler ♀ s) remarkably glistening, satin-like under side, *Alba* presents in both sexes a more acute apical outline in the fore-wings, and in the ♀ a much less development of yellow on the upper side of the hind-wings. I think it clear, from the description given by Wallengren making no mention of the discal spot on the fore-wings on either surface, that only ♂ examples reached that author. The ♀ s of the ochre-yellow coloration beneath are more frequently met with than those of the paler tint resembling the ♂ s.

I did not find *P. Alba* during my summer sojourn in Natal; but the late Mr. M'Ken sent down from D'Urban many specimens of both sexes in the years 1869 and 1870, and Colonel Bowker has forwarded several from the same locality, including the paired sexes captured on 22d August 1878. The only month of capture noted on Colonel Bowker's examples is August, and I am disposed to think that *Alba* makes its appearance on the wing only in the winter or dry season.[1]

Some dated specimens since received from Mr. Alfred D. Millar tend to confirm this view, a ♂ having been captured on the 15th August and three ♂ s and a ♀ on the 17th September 1887. It is not impossible that *Alba* may be the winter brood of *Pigea*.

Localities of *Pieris Alba*.

I. South Africa.
   E. Natal.
      a. Coast Districts.—D'Urban (*M. J. M'Ken, J. H. Bowker*). D'Urban District (*T. Ayres*). Malvern (*J. H. Bowker*). Pinetown (*W. Morant*). Mouth of Tugela River (*J. H. Bowker*).

254. (5.) Pieris Simana, Hopffer.

♂ ♀ *Pieris Simana*, Hopff., "Monatsb. K. Akad. Wissensch. Berl., 1855, p. 640, n. 13;" and Peters' Reise Mossamb., Ins., p. 354, t. xxiii. ff. 3, 4 [♂], 5, 6 [♀] (1862).

♂ *Pieris Pigea*, Wallengr., K. Sv. Vet.-Akad. Handl.; Lep. Rhop. Caffr., p. 7 [*nec* ♀].

Exp. al., (♂) 1 in. 9½–11 lin.; (♀) 1 in. 11 lin.

Very closely allied to *Charina*, Boisd.

♂ *Greenish-white with a bright pearly lustre; basal and marginal blackish markings quite as in Charina.* UNDER SIDE.—*Almost without marking, as white as upper side; the hind-wing and apex of fore-wing very rarely with the faintest tinge of yellowish. Fore-wing*: terminal nervular black dots exceedingly minute; terminal disco-cellular fuscous dot very minute; in Querimba specimen (*apud* Hopffer), a good-sized rounded, discal, fuscous spot, like that in ♀, between third and second median nervules. *Hind-wing*: irroration either wholly wanting or extremely faint and limited; spots of basal area and discal row very vaguely and only partly represented, the more constant and less indistinct being the spots below median nervure and the costal spot of discal row; terminal disco-cellular spot distinct; costa at base with more distinct orange-yellow edging line.

♀ *More yellowish than ♂, and rather more so than ♀ Charina; blackish markings generally more developed than in Charina. Fore-wing*: basal dusky suffusion wider; spots of costal streak beyond middle confluent, not separable; hind-marginal spots enlarged and confluent into a quite continuous, rather wide border, with acute inward nervular projections. *Hind-wing*: hind-marginal nervular spots larger than in *Charina*, sub-sagittiform, acutely-pointed outwardly. UNDER SIDE.—

---

[1] Colonel Bowker has since forwarded three pairs taken *in copulâ* at Malvern, near D'Urban, in May 1888. Two of the ♀ s are of the pale, more glossy, under-side colouring, while the third presents a deep ochre-yellow hue, and has the upper side of the hind-wings generally but unequally tinted with pale ochre-yellow.

Hind-wing and apex of fore-wing more or less yellowish, but without, or almost without, any irroration. *Fore-wing:* costal bar beyond middle obsolete; a faint yellowish suffusion in basal area; discal blackish spot well-marked. *Hind-wing:* markings as indistinct as in ♂; orange edging on costa from base more developed, extending almost to end of costal nervure.

Wallengren's note (*loc. cit.*) of "*posticis maris utrinque albis*" renders it clear that he had not before him in association with the ♀ *Pigea*, Boisd., the actual ♂ of that species (in which the under side of the hind-wings is pale dull-yellowish), but the ♂ *Simana*, Hopff.,[1] to which alone among the males of the South-African species of this group the character in question belongs.

The larger size of both sexes, the white under-side of the ♂, and almost spotless and unirrorated nature of that surface in both ♂ and ♀, and the greater development of the blackish markings on the upper side of the ♀, are the principal features distinguishing *Simana* from *Charina*. The singular character which Hopffer gives of the presence in the ♂ of the discal blackish spot between third and second median nervules on the under side of the fore-wing is entirely wanting in the seven Natalian ♂ s before me. I feel doubtful whether this form should be considered as more than a variety of *Charina*, Boisd.; hitherto I have seen no examples linking it to the specimens of *Charina* which are little irrorated on the under side.

D'Urban, in Natal, is the only locality known to me for this butterfly within South-African limits. The first ♀ example I saw was one taken by Mr. W. Morant in November 1869; before then I had referred two ♂ s which I captured in February 1867 to *Charina*. Several ♂ s and a ♀ were received at the South-African Museum in 1870 from the late Mr. M. J. M'Ken; and in November 1881 Colonel Bowker sent me a pair taken *in copulâ*. The ♂ of this pair has only the very faintest traces of a few spots on the under side of the hind-wing, and the ♀ is slightly yellowish-tinged on the upper side, but distinctly pale-yellowish on the under side.[2]

Localities of *Picris Simana*.

I. South Africa.
  E. Natal.
    a. Coast Districts.—D'Urban.

II. Other African Regions.
  A. South Tropical.
    b. Eastern Coast.—Querimba.

---

[1] Wallengren has, in fact (*vide supra*, p. 46), referred the true ♂ *Pigea* to *Simana*, Hopff., ♀.

[2] Altogether similar paired sexes, captured on 10th December 1887 at D'Urban, have been kindly forwarded to me by Mr. A. D. Millar. Colonel Bowker has also sent a further pair, with the ♀ rather whiter beneath, taken together at Malvern, near D'Urban, in May 1888.

255. (6.) Pieris Charina, Boisduval.

♂ ♀ *Pieris Charina*, Boisd., Sp. Gen. Lep., i. p. 525, n. 128 (1836).
♂ ♀ „ „ Trim., Rhop. Afr. Aust., i. p. 30, n. 17 (1862).

PLATE X. fig. 4 (♂).

Exp. al., (♂) 1 in. $8\frac{1}{2}$–$10\frac{1}{2}$ lin.; (♀) 1 in. $7\frac{1}{2}$–9 lin.

♂ *Greenish-white, with a bright pearly lustre, most brilliant near bases of wing.* Fore-wing: base blackish; costa narrowly edged with blackish, wider so close to apex; on hind-margin a row of black dots, lessening towards anal angle, one at extremity of each nervule, but none at extremity of submedian nervure. *Hind-wing:* a series of seven minute black dots (rarely absent) along hind-margin, at extremities of nervules; base more or less blackish.

UNDER SIDE.—*Variable.* Fore-wing: a very small black dot at extremity of discoidal cell; *costa narrowly, apex broadly, tinged with very pale dull-yellowish, sprinkled more or less thickly with dark-grey dots.* Hind-wing: entirely of the same pale-yellowish, thickly sprinkled *with grey atoms;* costa at base tinged with *pale bright green*, and sometimes faintly edged with yellow; a distinct black dot at extremity of discoidal cell; some dark-grey ill-defined spots near base, arranged transversely; beyond middle, parallel to hind-margin, a band of about seven darker spots, commencing on costa with a rather conspicuous larger mark; hind-marginal dots larger than those in *fore-wing*. (The grey irroration of the under side is often very thinly sprinkled, and is frequently altogether wanting in the central portion of the hind-wing.)

♀ *Rather more yellowish than ♂, but with the same pearly lustre, which is indeed considerably more extended in basal area of fore-wing.* Fore-wing: a more or less distinct disco-cellular dot; beyond middle a row of brownish spots parallel to hind-margin, consisting of three small spots forming a streak from costa almost to lower radial nervule, and a rounded larger spot between third and second median nervules; occasionally a fifth indistinct diffused spot between first median nervule and inner margin; *spots on hind-margin very much larger than in ♂, contiguous, somewhat rhomboidal in form (in some instances almost forming a marginal stripe)*, the last spot on submedian nervure. Hind-wing: dots on hind-margin larger than in ♂, but not one-fourth the size of those in fore-wing; occasionally a row of spots beyond middle, similar to that on under side of ♂, but not so distinct. UNDER SIDE.— As in ♂, but often with hind-wing and apex of fore-wing of a more decided yellow tint. Fore-wing: the spots beyond middle faintly marked; hind-marginal dots barely larger than in ♂; base faintly tinged with very pale greenish-yellow.

In Boisduval's description of this species, the examples in which the under-side irroration is wanting or slight are typical, while those densely irrorated constitute his *Variété* B.

There are many intermediate grades in both sexes, but I noticed at Knysna, in the Cape Colony, that while in the late spring and early summer the densely-irrorated form prevailed, in the autumn it was almost always the rather larger sparsely-irrorated specimens that were met with. A densely-irrorated ♂, received from Colonel Bowker, was taken at King William's Town, however, as late as 1st May; but in January and February 1870 all the examples I captured near Grahamstown were sparsely irrorated. In August 1865 I took a ♀ intermediate in irroration near D'Urban, Natal.[1]

Charina is well distinguished from *Pigea*, Boisd., and *Alba*, Wallengr., by its smaller size and much more lustrous surface gloss in both sexes, and by the costal macular streak beyond the middle of the fore-wings in the ♀, while its under side in both sexes, apart from the variable dark irroration, presents a dark terminal disco-cellular spot and several other dark marks in the basal area. There is no tendency to yellow on the upper side of the hind-wing of the ♀ *Charina*, and on the under side, when a yellow tint prevails, it is exceedingly pale and dull.

The figure (PLATE X. fig. 4) represents a ♂ of the densely-irrorated form, but does not give the under side with sufficient accuracy, the irrorations being much too rufous as well as the spots, and the latter on the disc being run together into a streak instead of being kept separate.

Wherever woods extend on the eastern side of South Africa this butterfly seems to be found in some abundance. It flies actively, and at Knysna I often observed a good many specimens congregated about flowering bushes. On the 6th February 1870 I saw a large number, in company with even greater flights of *P. Gidica* and *Severina*, flying for some hours over the long hill at Highlands, near Grahamstown; though all going onward in the same direction, many of them occasionally stopped to visit flowers, and then resumed their journey. Mrs. Barber informed me that at the beginning of November 1867 immense flights of these three species, with the addition of many *P. Zochalia*, passed over Highlands for three consecutive days in the same direction, notwithstanding several changes of wind during that time.

I have not found any record of *Charina's* occurrence beyond South Africa proper, but the very closely-related *Simana*, Hopffer, which is perhaps only a variety of *Charina*, inhabits Tropical East Africa.

Localities of *Pieris Charina*.

I. South Africa.
  B. Cape Colony.
    a. Western Districts.—Knysna. Plettenberg Bay.
    b. Eastern Districts. Port Elizabeth (*W. D'Urban*). Uitenhage (*S. D. Bairstow*). Grahamstown. Kowie River (*J. L. Fry*), and Kleinemond River (*H. J. Atherston*). Bathurst District. King William's Town (*W. D'Urban* and *J. H. Bowker*).
  D. Kaffraria Proper.—Butterworth and Bashee River (*J. H. Bowker*).

---

[1] Two ♂s from this locality, captured by Mr. A. D. Millar on 17th September 1887, are also of this intermediate degree of under-side irroration.

E. Natal.
  a. Coast Districts.—D'Urban. "Lower Umkomazi."—J. H. Bowker.
  b. Upper Districts. — Greytown. Estcourt (*J. M. Hutchinson*).
    Colenso (*W. Morant*).
F. Zululand.—Coll. Brit. Mus.

256. (7.) Pieris Spilleri, Staudinger.

♂ ♀ *Pieris Spilleri*, Staud., Entom. Nachrichten, 1884, iv. p. 52; and
          Exot. Schmett. i. pl. 18 (1884) [♂].
♂ ♀   „   „   Spiller, Entomologist, 1884, p. 62.
? VAR. ♂, *Pieris Gallenga*, H. G. Smith, Ann. and Mag. Nat. Hist., 5th
    Ser., xix. p. 62 (1887).[1]

Exp. al., (♂) 1 in. 7–8 lin.; (♀) 1 in. 6–8 lin.

♂ *Bright sulphur-yellow; fore-wing with a narrow fuscous apical edging. Fore-wing:* base very narrowly but strongly marked with fuscous; costa very thinly edged with a black line; apical edging beginning a little before apex, and continued along part of hind-margin by three nervular spots, of which the last, smallest, and least distinct is on second median nervule. *Hind-wing:* without marking of any kind. UNDER SIDE.—*Almost uniform sulphur-yellow, rather paler than on upper side* (especially towards inner margin of fore-wing); *or with the hind-wing and apical area of fore-wing of a slightly deeper yellow. Hind-wing:* on disc some indistinct or all but obsolete traces of small faint, fuscous spots, most apparent between median nervules.

♀ *Yellow less bright than in* ♂. *Fore-wing:* costa not with a bounding black line, but with a pale-reddish tinge, especially near base; apical hind-marginal border duller and less distinct. UNDER SIDE.—*Hind-wing and costal and apical hind-marginal area of fore-wing pale ochreous-yellow with a slight reddish tinge. Fore-wing:* base very slightly tinged with orange-yellow, extending along costa. *Hind-wing:* costa at base thinly edged with orange-yellow; four discal spots better marked than in ♂, but very indistinct.

Second form of ♀.—*Fore-wing dull-whitish; hind-wing dull-yellowish. Fore-wing:* base with a very faint yellowish basal suffusion; apical hind-marginal border much fainter and duller than in ♂, but much broader (especially at apex), and prolonged as far as posterior angle. *Hind-wing:* a pale orange-yellow suffusion from base, filling discoidal cell and extending for some distance below and beyond it; hind-margin sulphur-yellow; a row of 5–6 indistinct small fuscous spots at extremities of nervules. UNDER SIDE.—*Paler and clearer than upper side; both wings white, edged almost throughout with sulphur-yellow, which is brightest on costa near base and in fore-wing about apex. Fore-*

---

[1] Mr. H. G. Smith kindly showed me two Delagoa Bay specimens presenting the peculiarities on which he afterwards founded the new species *Gallenga*. They were smaller than usual (*exp. al.* 1 in. 4½ lin.), with the apical edging of fore-wing reddish-brown and very narrow, and the under side of the hind-wing and of the apex of the fore-wing tinged with reddish.

wing: basal suffusion distinct, pale orange-yellow. *Hind-wing:* discal spots larger, but indistinct.

Dr. Staudinger, who received six specimens from Mr. Spiller, notes that, of the two females among them, one was as yellow as the males, while the other was of the form just described. Mr. Spiller (*loc. cit.*) does not mention this difference in the ♀s; but he writes: "Described from twelve specimens taken in Natal, six of which are in my own collection, and the remainder in the possession of Dr. Staudinger."

I agree with Dr. Staudinger in placing this very distinct species in the *Charina* and *Pigea* group of *Pieris*, and think it on the whole nearer to *Pigea*; especially in respect of the second form of ♀, which, except for the want of discal spots in the fore-wing, is not unlike a miniature ♀ of the latter butterfly. The bright yellow of the ♂ (exactly that of the ♂ "Brimstone" butterfly (*Gonepteryx Rhamni*) of Europe), which also characterises one form of the ♀, at once distinguishes *P. Spilleri* from its near congeners.

Through the kindness of Mr. H. Grose Smith I had, in August 1884, the opportunity of examining a ♂ and yellow ♀ received by him from Delagoa Bay, and almost simultaneously the South-African Museum received from the same locality a ♂ and dull-tinted ♀ taken by Mrs. Monteiro.

Mr. Spiller (*loc. cit.*) writes: "This species is evidently very rare in Natal; its flight is rapid, and cannot be confounded with the similarly-coloured species of the genus *Terias*, these latter insects being feeble flyers." He does not state in the place quoted the locality of the species; but, in a previous communication to the same journal (*Entomologist*, 1882, p. 6), he mentions "a lovely canary-coloured *Pieris*, which I met with frequently in the woods on the Zululand border,"—which I presume was the insect under notice.

Colonel Bowker, on 27th June 1888, met with this species on the coast of Natal, "about half a mile from the sea, between the Tongaati and Umhloti rivers." He sent me six ♂s and a yellow ♀, with the note that the ♂s were numerous and active, and flying in company with *Terias Brigitta*, from which they could be distinguished by their canary-yellow tint. In the net this species was fragile in a very marked degree. It frequented the bush only, coming out of the thickets, flying down the edge for some distance, and then re-entering the cover; the flight was low, about two or three feet from the ground.

Colonel Bowker subsequently found a ♂ in a collection made near D'Urban in 1877, and notes that two were noticed on the wing at the Umkomazi by Mr. F. Barber on 7th July 1888. He is of opinion that it is a common winter butterfly along the Natal coast, but has hitherto escaped notice owing to its resemblance to the species of *Terias*.

Localities of *Pieris Spilleri*.

I. South Africa.
    E. Natal.—? Tugela River.
        a. Coast Districts.—Between Tongaati and Umhloti (*J. H. Bowker*). "D'Urban and Umkomazi."—J. H. Bowker.
    H. Delagoa Bay.—Lourenço Marques (*Mrs. Monteiro*).

257. (8.) Pieris Ogygia, Trimen.

♂ ♀ *Pieris Ogygia*, Trim., Trans. Ent. Soc. Lond., 1883, p. 356.

PLATE XI. fig. 2 (♀).

Exp. al., (♂) 2 in. 1 lin.; (♀) 2 in. 6½ lin.

♂ *White, with black markings*. *Fore-wing*: costa edged very narrowly with black; base with a slight pearly gloss; at extremity of cell an oblique angulated transverse streak, commencing rather widely just below costa, narrowing much to angle (where it is narrowest), and thence abruptly broad to its termination; a rather narrow apical and hind-marginal border (widest at apex), rather sharply indenting the white on nervules, ending abruptly on first median nervule; at extremity of submedian nervure a small spot, quite separate from hind-marginal border; immediately before border, four rather small spots, of which three form an oblique row from costa (the first touching apical border), and the fourth is between second and third median nervules and rather indistinct. *Hind-wing*: on hind-margin six nervular spots, all (except that at end of submedian nervure, which is small and almost linear) large, subovate, well separated. UNDER SIDE.—*Hind-wing and apex of fore-wing pale chrome-yellow; nervures of hind-wing universally black*. *Fore-wing*: in upper part of cell, for a little distance from base, a flush of orange; disco-cellular terminal stripe fainter than on upper side, commencing a little farther from costa; four submarginal spots distinct; apical yellow extending narrowly to about middle of hind-margin; a hind-marginal row of seven small inwardly-acuminate black nervular spots. *Hind-wing*: costa, from base to before middle edged with orange-red; a submarginal row of seven inter-nervular, small, subsagittate black spots, of which the middle or fourth one is very small; hind-marginal spots all smaller than on upper side (except that at the end of submedian nervure, which is larger), narrow, flattened, sub-rhomboidal, well separated; a faint blackish dot on each side of first median nervule near its origin; on fold between median and submedian nervures, for a little distance from base, a very faint orange streak.

♀ *Similar to ♂, but considerably larger*. *Hind-wing*: a submarginal row of five small blackish spots, corresponding to the sub-sagittate ones of the under side, but wanting the first and last spots. UNDER SIDE.—*Fore-wing*: basal flush of orange wider. *Hind-wing*: a black line marks middle part of fold between median and submedian nervures.

This species partakes of the characters of both *P. Calypso*, Drury, and *Zochalia*, Boisd.; it is, perhaps, on the whole, more nearly allied to the latter, especially as regards the ♀. *Ogygia* is distinguished from *Zochalia* by having in the *fore-wing* (1) a narrower, more angulated, disco-cellular streak, and (2) a much narrower apical and hind-marginal border, enclosing no white spots; and in the *hind-wing* (3) much rounder hind-marginal spots, not acuminate inwardly (or united by the festooned line often found in *Zochalia*). On the under side (4) the yellow colouring is much brighter; (5) the *fore-wing* has a basal flush of orange, and (6) only four separated submarginal spots instead of a continuous band; while in the *hind-wing* (7) the nervures are much more narrowly black, and (8) the cellular striæ and festooned submarginal line are wanting.

From *Calypso* the butterfly is readily separated by its smaller size and much narrower black border of the fore-wings, as well as, on the under side, by its black nervures and very much smaller submarginal black spots of the hind-wing, and the want in the same wing of the conspicuous terminal disco-cellular black spot. In the ♀ there is no resemblance to that of *Calypso*, which has a broad dusky border, and ground suffused with yellow and grey,—much like those presented by the darker ♀ s of *Gidica*, Boisd.

The male above described was sent to me by Mr. Walter Morant in 1869, and I then made a description and drawing of the specimen, which the captor believed was taken near Pinetown, in Natal, where he was resident at the time. The ♀ here described and figured was taken by the late Mr. M. J. M'Ken at D'Urban in 1866; but being unset, and, I believe, on the same pin with several other butterflies, had escaped notice in a collection received by the South-African Museum.

It occurs to me as not impossible that this may be the butterfly referred to *Calypso* by Boisduval in the Appendix to Delegorgue's Travels (p. 586), with the *habitat* of Port Natal,—the ♀ *Ogygia* having been taken for the ♂ *Calypso*. No South-African specimens of the latter species have ever come under my notice.

Localities of *Pieris Ogygia*.

I. South Africa.
  E. Natal.
    a. Coast Districts.—D'Urban (*M. J. M'Ken*). Pinetown (*W. Morant*).

258. (9.) **Pieris Zochalia**, Boisduval.

Pieris Zochalia, Boisd., Sp. Gen. Lep. i. p. 506, n. 100 (1836).
  ♂ ♀   ,,   Trim., Rhop. Afr. Aust., i. p. 37, n. 22 (1862).

PLATE X. fig. 6 (♀).

Exp. al., (♂) 2 in.; (♀) 1 in. 11 lin.–2 in. 3 lin.

♂ *White, with black markings. Fore-wing*: costa from base narrowly black-edged; basal half of wing with a *brilliant pearly gloss*; a short black terminal disco-cellular bar from costa, contracted in its middle portion; a deep black band on hind-margin, narrowing to a point at anal angle, irregularly excavate on its inner edge, and generally containing five sub-triangular white spots, of which the fourth is occasionally almost obsolete. *Hind-wing*: generally a very short, thin black line or streak at extremity of discoidal cell; on hind-margin a row of six sub-rhomboidal black spots, generally united by a very faint regularly festooned blackish streak, which touches their inner extremity; base slightly glossed with a pearly lustre. UNDER SIDE. —*Fore-wing*: similar in marking; marginal band much larger than on upper side, of a *pale greyish-yellow*, only separated by the black-clouded nervules. *Hind-wing*: *entirely pale greyish-yellow; all the nervules broadly defined with black*; festooned submarginal streak much darker and well developed, and presenting an additional separate portion (beyond middle, but before the main portion) between costa and second subcostal nervule; spots on hind-margin much thinner than on upper side, sublinear; costa at base edged with bright-yellow; a bifid black longitudinal streak in discoidal cell, and a simple similar streak between median and submedian nervures, becoming bright-yellow near base; a short transverse diffused black mark from extremity of discoidal cell to near middle of submedian nervure, and a similar shorter mark between median and submedian nervures before middle.

♀ *All black markings broader than in ♂. Hind-wing: usually creamy ochre-yellow. Fore-wing*: white spots in hind-marginal border always smaller and diffused, sometimes tinged with yellowish, occasionally almost obsolete; costal transverse bar very variable in breadth; when broadest, not, or very slightly, constricted in the midde. *Hind-wing*: submarginal festooned streak usually complete and well-marked, as on under side, but variable in development; hind-marginal spots sometimes so large as almost to touch each other. UNDER SIDE.— Hind-wing and apical hind-marginal border of fore-wing usually of a clearer yellow than in ♂, inclining to lemon-yellow. *Fore-wing*: a rather narrow basal suffusion of orange-yellow. *Hind-wing*: a short faint edging of orange-yellow on costa a little before apex.

I have captured three unusually small individuals of this species, viz., a ♂ (exp. 1 in. 9 lin.) and a ♀ (exp. only 1 in. 5½ lin.) at Knysna, and another ♀ (exp. 1 in. 8 lin.) at Port Elizabeth, Cape Colony.

The ♀, as a rule, has the upper side of hind-wings ochre-yellow, and that of the fore-wings sometimes has a very faint tinge of lemon-yellow. The two ♀ s in my collection, which resemble the ♂ in having the upper side of the hind-wings white, I took at Knysna.

Specimens of both sexes from the Eastern districts of the Cape Colony, Kaffraria, and Natal are usually a little larger than the typical form from Knysna district, and they almost always have the yellow of the under side clearer and more decided, with an inclination to the tint of sulphur. A ♂ taken by Colonel Bowker at Malvern, near D'Urban, Natal, in August 1885, has the under-side yellowish dull and pale, as in the type form, but on the upper side the white hind-wings have a moderately wide border of pale sulphur-yellow all along the hind-margin. Another ♂ (without label of locality, but taken, I believe, in Kaffraria by Colonel Bowker) is singular in having the hind-marginal spots of the hind-wings so greatly reduced on the *upper side* that they are smaller and more linear in form than the corresponding spots are on the under side.

In the pattern of the black borders, and the size and shape of the white spots which they enclose, the ♂ *Zochalia* most resembles on the upper side the ♂ *Severina*, Cram., but is at once distinguished by the terminal disco-cellular bar from costa, which is very black and sharply defined, and nearly as broad as in the ♀ *Mesentina*, Cram. In both sexes the width of the basal pearly gloss in the fore-wings is a distinctive feature, as well as the thin not diffused blackish neuration of the under side of the hind-wings. The ♀ differs greatly from the same sex of both *Severina* and *Mesentina* in the hind-marginal border of the hind-wings, which, instead of being very broad, dusky, and almost spotless, is almost as narrow and lightly festoon-edged inwardly as in the ♂. The tendency to have the upper side of the hind-wings only yellow—and that of a deep clear tint—is also a special character of the ♀ *Zochalia* among its immediate congeners.

This species appears to be peculiar to South Africa, and to be more prevalent in the Cape Colony than elsewhere. I found it numerous in wooded places at Knysna and Plettenberg Bay, and not uncommon near Grahamstown. In Natal I did not meet with it, and only a few examples have reached me from that Colony and Kaffraria Proper. A single specimen was sent from the Transvaal by Mr. T. Ayres. It is quite sylvan in its haunts, never being found far from woods, but chiefly delighting in flying pretty briskly along their edges, or in open places on their outskirts, settling often on flowers. I have captured it from the end of September to the middle of February, and once took a specimen (at Knysna) as late as the 7th May.

Localities of *Pieris Zochalia*.

I. South Africa.
   B. Cape Colony.
      a. Western Districts.—Knysna and Plettenberg Bay.
      b. Eastern Districts.—Port Elizabeth. Grahamstown. King William's Town (*W. S. M. D'Urban*).
   D. Kaffraria Proper.—Mouth of Kei River and Bashee River (*J. H. Bowker*).
   E. Natal.
      a. Coast Districts.—D'Urban (*J. H. Bowker*).
   F. "Zululand."—Boisduval.
   K. Transvaal.—Potchefstroom District (*T. Ayres*). Eureka, near Barberton (*C. F. Palmer*).

259. (10.) Pieris Mesentina, (Cramer).

♂ *Papilio Mesentina*, Cram., Pap. Exot., iii. pl. cclxx. ff. A, B (1782).
♂ *Papilio Aurota*, Fab., Ent. Syst., iii. 1, p. 197, n. 614 (1793).
♂ ♀ *Pieris Mesentina*, Godt., Enc. Meth., ix. p. 130, n. 34 (1819).
♂ ♀ „ „ Boisd., Sp. Gen. Lep., i. p. 501, n. 95 (1836).

♀ *Pinacopteryx Syrinx*, Wallengr., Wien. Ent. Monatschr., iv. p. 34, n. 4 (1860).
♂ *Pieris Mesentina*, Hopff., Peters' Reise Mossamb., Ins., p. 352 (1862).
♂ ♀ ,, ,, Trim., Rhop. Afr. Aust., i. p. 35, n. 21 (1862).
♀ *Pieris Lordaca*, Walk., Entomologist, 1870, p. 48.
Larva and Pupa (*Indian*), Horsf. and Moore, Cat. Lep. E. I. C. Mus., i. pl. xii. ff. 9, 9a (1857).

Exp. al., (♂) 2 in. 0–3 lin.; (♀) 1 in. 10 lin.—2 in. $3\frac{1}{2}$ lin.

♂ *White, with blackish borders.* *Fore-wing*: costa narrowly edged with black, and bordered with greyish above discoidal cell; a curved black transverse streak, thickest at its lower extremity, at end of cell; a black hind-marginal border, with a very irregular inner edge, narrowing to a point at posterior angle, and containing in its upper half six elongate somewhat triangular white marks; base with a slight pearly gloss. *Hind-wing*: a hind-marginal black border, containing incompletely (owing to more or less obsolescent inner edge) four large subovate white spots; base with a pearly gloss. UNDER SIDE.—*Similar in marking; hind-wing faintly tinged with yellowish*. *Fore-wing*: spots in hind-marginal border larger, better defined; a black streak between costal and subcostal nervures from base to upper end of disco-cellular terminal transverse streak. *Hind-wing*: all the neuration clouded with blackish; costa and a longitudinal streak between median and submedian nervures tinged from base with chrome-yellow; a yellow edging before middle of inner margin; a blackish transverse stripe from extremity of discoidal cell to longitudinal streak just mentioned, and another from end of costal nervure to second subcostal nervule; hind-marginal border completely enclosing five spots of the ground-colour.

♀ *Varying from yellowish-white to dull-yellowish; markings much as in ♂, but all broader.* *Fore-wing*: curved bar at extremity of cell connected with base by a black costal stripe; spots in hind-marginal border small and ill-defined, or sometimes wanting. *Hind-wing*: a linear transverse mark at extremity of cell; beyond middle a blackish subcostal marking, sometimes united to hind-marginal border; four somewhat rounded spots in border, sometimes obliterated. UNDER SIDE.—*Hind-wing and apical spots of fore-wing whitish, more or less tinged and clouded with chrome-yellow*. *Fore-wing*: a basal suffusion of rather dull chrome-yellow, very variable in extent, sometimes almost filling discoidal cell and tinging costa beyond middle. *Hind-wing*: neuration thickly black-clouded; spots in hind-marginal border varied with yellow.

The typical *Mesentina*, figured by Cramer, is the ordinary form of continental India, in which the ♂ has the black markings generally more strongly represented, and both sexes (but especially the ♂) exhibit a deeper yellow on the under side of the hind-wings and apex of fore-wings. The latter is, however, a decidedly variable character in Indian specimens (though I have not seen any quite so pale as the African ♂ s, in which the colouring is almost white), and appears, from the observations of Colonel Yerbury and others, to be largely a seasonal modification, the spring brood being the more deeply coloured.[1] Mr. Butler (*P. Z. S. Lond.*, 1886, p. 374) has separated the darker brood as *Belenois auriginea*.

As regards the name of *Lordaca* applied by the late Mr. F. Walker to specimens from the African side of the Red Sea, no lepidopterist can doubt that his description is strictly applicable to the ordinary ♀ of the African *Mesentina*. The same must be remarked concerning *Pinacopteryx Syrinx* of Wallengren, from Damaraland, which was referred by its author in 1872 (*K. Sv. Vetensk.-Akad. Förhandl.*, p. 44) to *P. Gidica*, Godt., as a probable variety; but in 1875 (*op. cit.*, p. 90) to *P. Severina*, (Cram.),—the previous reference to *Gidica* being ascribed "*lapso typographico*."

I had lately (in October 1886) the opportunity of examining the fine series, Asiatic and African, of this butterfly in the British Museum collection, and noticed that the smallest specimens were from Damascus, Huswah (Aden), and Somaliland; a ♀ from Madagascar was also smaller than usual.[2] The forty examples were separated into six sets, of which the second only was named *Mesentina* (three Indian specimens), the third (ten Indian specimens) " *? Lordaca*, Walk.," and the fourth (eight Asiatic and five African specimens) *Lordaca*, Walk. The sixth set (eight African) bore the name of *Agrippina*, Feld., but the latter, as I have pointed out (p. 70 *infra*), is really a slight variation of the ♂ *Severina*, Cram. I could not discover any satisfactory characters by which this instructive series could be regarded as forming more than one species.

The ♂, as pointed out by Oberthür (*Ann. Mus. Civ. Genov.*, xv. 1880, p. 150), exhibits much variation in the development of the curved black streak at the extremity of the discoidal cell of the forewings, especially on the upper side, where the thin superior part of the streak is often more or less evanescent, or even wanting altogether in some examples.

The ♀, besides the variation in ground-colour above described, varies considerably in the development of the black markings generally; as regards the clouded neuration of the under side of the hind-wings, the most strongly marked specimen I have seen is one taken at Delagoa Bay by Mrs. Monteiro.

LARVA.—Pale-yellow, greenish on the back; a broad, brownish, lateral stripe from head to tail; head pinkish; a few short hairs near head and along the sides.

---

[1] The Ceylon representative, *Taprobana*, Moore, seems to be constantly darker than the Indian *Mesentina*. Even in the ♂ I have found no case in which the white spots of the dark borders are not greatly reduced or partly obsolete on the upper side, and a similar deficiency is observable in both sexes as regards the under side. The yellow of the under side is also remarkably deep and rich, often inclining to orange, in both sexes.

[2] A dwarf ♀, taken near Grahamstown, Cape Colony, by Mrs. Barber, is only an inch and a half across the expanded wings.

PUPA.—Sharply angulated, slender, head beaked. Light-brown, varied with darker-brown; a white, pink-spotted, longitudinal stripe on abdominal segments. Represented as suspended to the stalk of some plant.

The above descriptions of larva and pupa are made from figures in Plate xii. (f. 9, 9a.) of Horsfield and Moore's Catalogue of Lepidoptera in the East India Company's Museum, vol. i. The food-plant of the larva is not stated, nor is its locality given.

Colonel Yerbury, quoted by Mr. Butler in *Proc. Zool. Soc. Lond.*, 1884, p. 492, and 1886, p. 376, notes that near Aden he had reared caterpillars of this butterfly ("*Lordaca*, Walk.") on *Capparis galeata*, and that in Western India they feed on a *Capparis* with dark-red blossom (*C. horrida*).

Mr. W. D. Gooch's notes and outline sketches of the earlier stages of *Mesentina* near D'Urban, in Natal, agree very fairly with the figures above described, and may be thus summarised, viz.:—

LARVA.—Yellowish olive-green on back, marked down the middle with a double dark-brownish line; on each side a deep citrine-green stripe, bearing on each segment a minute yellow spot,—these lateral stripes inflect a little on eleventh segment, and join dorsally at their extremities on twelfth segment; below lateral stripes yellowish-green; just above legs with whitish-grey pubescence, inclining to form a tuft on each segment; on second segment two longer subdorsal tufts of similar hair projecting above the head. Head bright reddish-brown.

PUPA.—Very light-brownish, dorsally flecked with dark-brown; edges of wing-covers and part of neuration dark; angular projections on each side of dorsal base of abdomen black; lateral streaks of abdomen, and line along median dorsal carina of thorax white. Form quite like that of the pupa of *Severina*, Cram.

This well-known species has a wide range over all the Ethiopian region (except, apparently, the tropical north-west forest sub-region), and over South-West Asia, from Syria to Calcutta. In South Africa it seems to be far more numerous in the uplands of the interior than on the coast. In Natal during the summer of 1867 I met with only four specimens; and not many examples have reached me in collections made in that Colony. Colonel Bowker described it as very numerous all over Basutoland, and Mr. H. L. Feltham informs me that in Griqualand West it is by far the most abundant species of the genus. Two stragglers of this butterfly have been recorded by me as visiting Cape Town,—the first taken in the Museum enclosure in April 1873, and a second, closely observed by myself for some time, on 14th April 1878, about flowers in the Botanic Gardens. Several times, however, in the later summer I have seen on Table Mountain a "White" hurrying past, which, although I could not identify it, was clearly not the only resident species, *P. Hellica*, and very probably was *Mesentina*. The species has occurred in all the collections I have seen from Damaraland; in a small one formed by Mr. John A. Bell there were as many as thirty-eight specimens of it. Boisduval (*op. cit.*, p. 502) notes that in some parts of Africa this butterfly at certain seasons migrates in innumerable hosts, but he gives no authority for the statement. Colonel Bowker noticed in Basutoland that numbers of *Mesentina* flew in an eastward direction.

Localities of *Pieris Mesentina*.

I. South Africa.
  B. Cape Colony.
    a. Western Districts.—Cape Town [occasional visitor]. Ookiep, Namaqualand District (*G. Warden*).
    b. Eastern Districts.—Grahamstown (*Mrs. Barber*). King William's Town (*Mrs. Tyrwhitt Drake*). Murraysburg (*J. J. Muskett*). Colesberg (*A. F. Ortlepp*).
    c. Griqualand West.—Vaal River (*J. H. Bowker* and *H. L. Feltham*). Kimberley (*H. L. Feltham*).
    d. Basutoland.—Koro Koro and Maseru (*J. H. Bowker*).
  D. Kaffraria Proper.—Bashee River (*J. H. Bowker*).
  E. Natal.
    a. Coast Districts.—D'Urban. Verulam. Mapumulo.
    b. Upper Districts.—Estcourt (*J. M. Hutchinson*). Rorke's Drift (*J. H. Bowker*).
  F. Zululand.—Isandlhwana and Napoleon Valley (*J. H. Bowker*).
  H. Delagoa Bay.—Lourenço Marques (*Mrs. Monteiro*).
  K. Transvaal.—Potchefstroom (*W. Morant*). Potchefstroom District (*T. Ayres*). Limpopo and Marico Rivers (*F. C. Selous* and *A. W. Eriksson*).
  L. Bechuanaland.—Motito (the late *Rev. J. Frédoux*).

II. Other African Regions.
  A. South Tropical.
    a. Western Coast.—Damaraland (the late *C. J. Andersson, J. A. Bell*, and *W. C. Palgrave*). "Angola (*Pogge*)."—Dewitz. Congo.—Coll. Brit. Mus.
    b. Eastern Coast.—"Querimba."—Hopffer.
    b1. Eastern Interior.—Lotsani and Makloutse Rivers (*F. C. Selous*). "Bamangwato, Tati, Gubulewayo, Inyati, and Gwailo River (*Oates*)."—Westwood. Zambesi.—"Victoria Falls (*Oates*)."—Westwood; "Tette."—Hopffer. Lake Nyassa.—Coll. Brit. Mus.
    bb. Eastern Islands.—Madagascar (*J. Caldwell*).
  B. North Tropical.
    b. Eastern Coast.—"Somaliland (*Thrupp*)."—Butler. Tajora[1] (*J. K. Lord*). Red Sea: "Harkeko, Raffa (*J. K. Lord*)."—Walker.
    b1. Eastern Interior.—Abyssinia: "Shoa (*Antinori*)."—Oberthür.

IV. Asia.—Syria:—Damascus.—Coll. Brit. Mus. Bagdad.—Coll. Brit. Mus. Arabia:—Aden (Huswah).—Coll. Brit. Mus. Western India.—"Kurrachee."—Colonel C. Swinhoe; Campbellpore and Chitta Pahar (*Yerbury*).—Coll. Brit. Mus. Afghanistan: Kandahar, Bolan Pass, Quetta.—Coll. Brit. Mus. Punjaub.—Coll. Brit. Mus. Bengal: "Calcutta."—De Nicéville. Barrackpore.—Coll. Brit. Mus. "Madras."—Moore.

---

[1] Through the kindness of Professor R. Meldola I have received a ♂ from Mr. Lord's collection taken in this locality. It is rather smaller than usual, and has the dusky neuration of the under side very faintly marked.

260. (11.) **Pieris Gidica,** Godart.

♂ *Pieris Gidica,* Godt., Enc. Meth., ix. p. 131, n. 37 (1819).
♂ ♀ *Pieris Gidica,* Boisd., Sp. Gen. Lep., i. p. 503, n. 97 (1836).
♀ *Pinacopteryx Westwoodi,* Wallgrn., K. Sv. Vet.-Akad. Handl., 1857; Lep., Rhop. Caffr., p. 9, n. 4.
♀ *Pinacopteryx Doubledayi,* Wallgrn., *loc. cit.*, p. 8, n. 2.
♂ ♀ *Pieris Gidica,* Trim., Rhop. Afr. Aust., i. p. 34, n. 20 (1866).

PLATE XI. fig. 1 (♂).

Exp. al., (♂) 2 in. 0–5 lin.; (♀) 2 in. 0–5½ lin.

♂ *White, with blackish markings. Fore-wing:* apex produced and pointed; costa edged with blackish, close to base greyish; base slightly diffused with greyish, and with a pearly gloss; at extremity of discoidal cell, a black, angulated streak, united to blackish costal edging, and thickest at its lower extremity; on hind-margin a blackish band, rather wider at apex, and ending on first median nervule, enclosing six very small white spots, and united by blackish nervules to an inner blacker, more irregular transverse band ending on second median nervure. *Hind-wing:* on hind-margin a linear black edging, and from four to six moderately-sized, inwardly-acute black spots at extremities of nervules; base greyish, with a slight pearly gloss; a few indistinct traces of the dark wavy streaks on under side.

UNDER SIDE.—*Fore-wing:* similar in marking; in discoidal cell a short longitudinal dusky ray from base; the spots in hind-marginal band enlarged, confluent with white between hind-marginal and sub-marginal bands, and slightly tinged with yellowish. *Hind-wing:* pale-yellowish, the nervures marked blackly upon it; costal, median, and submedian nervures clouded with blackish; two transverse rows of irregular blackish angulated streaks, one before, the other beyond, middle,—the inner one much interrupted, and joining the dark clouding on median nervure,—the outer one composed of thin lunular markings; neither row extending beyond submedian nervure; hind-marginal spots larger and somewhat squarer than on upper side, some of them united by fainter, curved streaks from their inner extremities, which touch the points of the lunular marks of the *outer transverse row;* costa edged at base with bright-yellow.

♀ *Varies from whitish to dull yellowish, blackish markings broader than in* ♂. *Fore-wing:* apex not so produced as in ♂; base broadly suffused with greyish nearly to extremity of discoidal cell, and glossed with a *violaceous lustre;* a short broad stripe of black from costa, at extremity of discoidal cell; submarginal band joined to hind-marginal one, the narrow space enclosed between the two composing three or four yellowish spots, often indistinct; an ill-defined blackish spot beyond middle, between first median nervule and inner margin. *Hind-wing:* from base a dusky-greyish cloud along both sides of median nervure, and slightly along costa; a black elongate mark at extremity of discoidal cell; hind-marginal spots united by broad arched streaks, so as to form a series of *festoon-like* markings; before them a more or less connected row of arched streaks (as in under side of ♂) touching them. UNDER SIDE.—*Very similar to ♂, but the markings broader. Fore-wing:* paler than on upper side; base only light-greyish, but in the whiter specimens with a pale-yellow tinge as far as extremity of cell. *Hind-wing:* deeper in tint than in ♂; blackish markings often more or less diffused.

The ♂ varies but little, except in size; but the black markings generally are more pronounced in some specimens, and in a few of the smaller examples the upper part of the terminal disco-cellular streak on the upper side of the fore-wings is obsolescent or wanting. The ♀, on the other hand, is highly variable, not only in size, but in ground-colour and development of markings. Examples of the medium pale-yellowish tint, with all the marginal markings defined with more or less clearness, are most prevalent; but specimens not rarely occur in which the yellow is much deeper on both upper and under side, and the marginal markings are on the upper side in both wings confluent into a broad dark border without (or with only the traces of) the usual spots, while the basal clouding is broader and darker. Females in which the ground-colour is whitish or nearly white are the scarcest; in one of three, which I took near Grahamstown, the borders and basal suffusion are almost as strongly marked as in the yellower examples just mentioned.[1]

I captured the paired sexes near Grahamstown on the 6th, and at Uitenhage on the 23d February 1870, and Colonel Bowker took them near D'Urban, Natal, in November 1881. The ♂ of Colonel Bowker's pair is remarkable for presenting on the upper side of the fore-wings the coalescence of the outer and inner series of white spots in the dark border, usually found only on the under side,—the confluent spots are, however, much irrorated with fuscous scales.

The longer and much more pointed fore-wings and inferiorly elongated hind-wings well distinguish the ♂s of *Gidica* and *Abyssinica* from their allies in South Africa, and, in a less degree, the ♀s also. The ♂ *Gidica* is a very rapid flyer, and its swift irregular course over and among the trees and underwood of its sylvan haunts makes it by no means an easy capture on the wing. It constantly visits flowers, however, especially those of *Calodendron capense* (the so-called "Wild Chestnut") and of *Plumbago capensis*, and is then taken without much difficulty. The species is very numerous in the wooded parts of South Africa, but does not make its appearance until the warm weather is well advanced. At Plettenberg Bay, near Grahamstown, and on the Natal coast, I found it abundant at the end of January and through February in different years, and in the last-named district up to the beginning of April. Mr. W. S.

---

[1] My determination of Wallengren's *Pinacopteryx Westwoodi* and *P. Doubledayi* as respectively ♂ and ♀ of *Gidica* was confirmed by some rough drawings of his type specimens shown to me by W. F. Kirby. "*P. Westwoodi*, Wlgrn.," in the drawings is certainly the ♂ *Gidica*; "*P. Doubledayi*, Wlgrn., ♂," is a ♀ *Gidica* in which the fore-wings are whitish; and "*P. Doubledayi*, Wlgrn., ♀," is a yellow ♀ *Gidica*.

D'Urban met with it about King William's Town from November to May, and I took a specimen at Grahamstown as late as the 25th May.

Though known to occur in various very distant points in Tropical Africa, *Gidica* does not appear to flourish anywhere out of South Africa proper. Three examples have reached me from Damaraland, and a few have been taken in Abyssinia; but its place over the greater part of the tropical interior seems to be taken by the abundant *Mesentina*.

Localities of *Pieris Gidica*.

I. South Africa.
 B. Cape Colony.
  a. Western Districts.—Knysna. Plettenberg Bay.
  b. Eastern Districts.—Port Elizabeth (*W. S. M. D'Urban*). Uitenhage. Grahamstown. Kowie River (*J. L. Fry*), and Tharfield (*Miss M. Bowker*). Bathurst District. Keiskamma Mouth and King William's Town (*W. S. M. D'Urban*). Queenstown (*W. S. M. D'Urban*).
 D. Kaffraria Proper.—Bashee River (*J. H. Bowker*).
 E. Natal.
  a. Coast Districts.—D'Urban. Verulam. Umvoti.
  b. Upper Districts.—Estcourt (*J. M. Hutchinson*). Rorke's Drift (*J. H. Bowker*).

II. Other African Regions.
 A. South Tropical.
  a. Western Coast.—Damaraland (*J. A. Bell*).
  b. Eastern Coast.—Zambesi River (*Rev. H. Rowley*).
 B. North Tropical.
  b1. Eastern Interior.—Abyssinia: "Shoa" (*Antinori*).—Oberthür. "Atbara River" [Soudan].—Butler.

261. (12.) Pieris abyssinica, Lucas.

♂ ♀ *Pieris abyssinica*, Luc., "Rev. et Mag. Zool., 2nd Ser., iv. p. 328 (1852)."
♂ ♀ *Pieris Gidica*, Var. (Allica, Boisd. MS.), Oberth., Études d'Ent., iii. p. 16 (1878).

Exp. al., (♂) 2 in. 0–4 lin.; (♀) 1 in. 10 lin.—2 in. 2½ lin.

Very closely allied to *Gidica*, Godt.

♂ *White, with blackish markings; pattern of markings quite as in Gidica, but blackish, duller, inclining to brown. Fore-wing:* apical hind-marginal border not so clearly defined, rather diffused, its outer row of whitish spots more or less obsolescent. UNDER SIDE.—*Hindwing and apical area of fore-wing more or less tinged with pale dull-reddish, on which the darker markings are less distinct and much browner than in Gidica, and sometimes obsolescent. Hind-wing: a longitudinal whitish ray* (conspicuous in the more reddish-tinged examples) from base almost to hind-margin, traversing discoidal cell, bounded inferiorly by a dark-brown ray along the line of disco-cellular fold and the extension of that fold beyond extremity of cell; basal edging of costa bright-orange; neuration clouded much as in *Gidica*, but itself with a pale-reddish tinge.

♀ *Slightly tinged with pale-yellow, or sometimes nearly white; basal clouding much more restricted than in Gidica* ♀. *Fore-wing:* apical hind-marginal border as in the more lightly marked specimens of *Gidica*; no blackish spot on lower discal area beyond middle. *Hind-wing:* hind-marginal nervular spots separate from each other, or only very imperfectly united by indistinct arched streaks; inner row of arched streaks represented by separated cuneiform fragments only. UNDER SIDE.—As in ♂, but darker in ground-colour, and with the whitish markings of hind-wing more prominent.

As a rule, the fore-wings (especially in the ♂) are more produced, sometimes even inclining to be subfalcate; the size of this butterfly is also on the whole smaller than that of *Gidica*. A dwarf ♂ from Estcourt, Natal, expands only 1 in. 10 lin. A ♂ from Zumbo, on the Zambesi River, is remarkable for having only very faint traces of the usually well-marked hind-marginal spots on the hind-wings.

Though widely distributed in South Africa, *Abyssinica* is very scarce; and I have seen only ten authenticated specimens from the sub-region. If it were restricted to a particular territory, one would be disposed to regard it as a local variety of *Gidica*; and if it were not so rare, it might be taken for possibly a seasonal form of that species.[1] I captured a single example at Knysna early in May 1859; two have been sent by Mrs. Barber from near Grahamstown, and two were taken at D'Urban, Natal, by the late Mr. M'Ken. In Swaziland the late Mr. E. C. Buxton (who sent me photographs of this and many other butterflies of his capturing) seems to have met with the insect pretty frequently.

Localities of *Picris Abyssinica*.

I. South Africa.
  B. Cape Colony.
    a. Western Districts.—Knysna.
    b. Eastern Districts.—Grahamstown (*Mrs. Barber*).
  E. Natal.
    a. Coast Districts.—D'Urban (the late *M. J. M'Ken*).
    b. Upper Districts.—Estcourt (*J. M. Hutchinson*).
  G. "Swaziland."—The late *E. C. Buxton*.

II. Other African Regions.
  A. South Tropical.
    b. Eastern Coast.—Mombas: "Endara (Kersten)."—Gerstäcker.
    *b*1. Eastern Interior.—Zambesi River: Zumbo (*F. C. Selous*).
  B. North Tropical.
    *b*1. Eastern Interior.—Abyssinia: "Lake Tsana (*Raffray*)."—Oberthür. Soudan: "Atbara River."—Butler; "Khartoum."—Gerstäcker.

---

[1] Six specimens (five ♂s and a ♀), with dates of capture—from 3d to 17th September 1887—have been kindly forwarded to me from D'Urban, Natal, by Mr. A. D. Millar, who inclines to think this form the dry-season brood of *Gidica*. The latter certainly seems restricted in appearance to the summer and autumn months.

262. (13.) Pieris Severina, (Cramer).

♀ *Papilio Severina*, Cram., Pap. Exot., iv. pl. 338, ff. G, H (1782).
Pieris Severina, Godt., Enc. Meth., ix. p. 131, n. 36 (1819).
♂ ♀ ,, Boisd., Sp. Gen. Lep., i. p. 507, n. 101 (1836); App. Voy. Deleg. Afr. Aust., p. 586 (1847).
♂ *Pinacopteryx Mesentina*, Wallengr., K. Sv. Vet.-Akad. Handl., 1857; Lep. Rhop. Caffr., p. 9, n. 3.[1]
♀ *Pinacopteryx Severina*, Wallengr., *loc. cit.*, p. 8, n. 2.
♂ ♀ *Pieris Severina*, Trim., Rhop. Afr. Aust., i. p. 32, n. 19 (1862).
♂ ♀ ,, ,, Hopff., Peters' Reise Mossamb., Ins., p. 352 (1862).
♂ *Pieris Agrippina*, Feld., Reise Novara, Lep., ii. p. 173 n. 159 (1865).
♂ ♀, *Pieris Severina*, Staud., Exot. Schmett., i. pl. 18 (1884).
VAR. ♂ ♀, *Pieris Boguensis*, Feld., *loc. cit.*, n. 160.

Exp. al., (♂) 1 in. 10½ lin.—2 in.; (♀) 1 in. 11 lin.—2 in. 1 lin.

♂ *White, with white-spotted black hind-margins.* Fore-wing: costa with a linear black edging; a black border on hind-margin, broad at apex, and narrowing to a point at anal angle,—its inner edge irregularly dentate and excavate; a row of elongate white spots, from four to seven, in this border, lessening in size towards posterior angle, but not extending below first median nervule; a small sublinear black spot at extremity of discoidal cell. *Hind-wing:* a black border of moderate width on hind-margin, enclosing four rather large white spots,—of which rarely all are indistinct except the first; in some specimens an irregular black transverse mark on costa before apex; very rarely a thin blackish line at extremity of discoidal cell. UNDER SIDE.—*Main markings like those of upper side, but hind-marginal borders not so black, tinged with brown; hind-wing and marginal spots in fore-wing lemon-yellow or greyish-yellow,—the former with neuration almost always more or less clouded with black.* Fore-wing: base with a faint yellow tinge; rarely a projection from costal black edging towards terminal disco-cellular spot; hind-marginal spots larger and more sharply defined. *Hind-wing:* costa near base edged with chrome-yellow; pre-apical costal black mark conspicuous; a similar but diffused transverse black mark from about origin of second median nervule to about middle of submedian nervure, crossed by a longitudinal black streak from near base of median nervure to near anal angle; hind-marginal border containing five spots of the ground-colour, the additional spot being close to anal angle; near base, a short longitudinal chrome-yellow streak between median and submedian nervures, and another close to inner margin. *Cilia* black, interrupted with white between nervules.

♀ *Varying from very pale whitish-yellow to pale ochreous-yellow; hind-marginal black borders very broad.* Fore-wing: only two or three

---

[1] This reference was confirmed by a figure in the rough drawings of some of Wallengren's type specimens shown to me by Mr. W. F. Kirby, which, under the name of "*Pinacopteryx Mesentina*, Cram., *var.*," unquestionably represented a ♂ *Severina*.

small yellow spots in hind-marginal border, close to apex; terminal disco-cellular black spot rounder and usually larger than in ♂; a black projection from costal edging (as on under side of ♂),—very rarely extending almost to disco-cellular spot. *Hind-wing*: costal spot near apex united to hind-marginal border, which inwardly radiates more or less on nervules, and contains a single yellowish spot close to apex; disco-cellular terminal line rarely distinct. UNDER SIDE.—*Varying from lemon-yellow to deep chrome-yellow; hind-wing often inclining to whitish. Fore-wing*: some tint of yellow between the two just mentioned, suffused from the base with orange or orange-yellow; spots in hind-marginal border yellow, usually seven. *Hind-wing*: neuration more strongly and generally black-clouded; six or seven spots in hind-marginal border.

VARIETY A. (*P. Boguensis*, Felder), ♂ and ♀.—*Fore-wing*: in both sexes, instead of merely a small terminal disco-cellular spot, a well-marked (sometimes rather broad), short, oblique, blackish bar from costa, where it is wider and somewhat diffused.

LARVA.—Dull reddish-sandy on back, with a median longitudinal streak of violaceous-grey. On each side succeeds a wide ferruginous-brown stripe, followed by a narrow pale-yellow one, fringed inferiorly by thinly-set white hairs of moderate length. All the dorsal surface except the median streak shagreened with minute elevated whitish or yellowish dots arranged in transverse lines; also across the back of each segment (rather before its middle) a series of much larger widely-separated elevated and acuminate yellow dots,—six on the second, third, and fourth segments, and four on each of the others,—followed (a little beyond middle of each segment) by two similar dots. Ventral surface, including legs, pro-legs, and under part of head, pale greyish-green. Head above ferruginous, in front inferiorly yellowish; shagreened like dorsal surface, and with two or three larger acuminate spots on each side; inferiorly and laterally with some thinly-set white hairs like those on body. Length, 9 lin.

In its earlier stages, down to a length of only $4\frac{1}{2}$ lin., the larva is dorsally much tinged with greenish.

PUPA.—Pale-brownish with a reddish tinge, or pale-creamy with a greenish tinge, superiorly more or less speckled with blackish. Cephalic process, dorsal ridge of thorax, and acute projections on each side of back of second abdominal segment, outlined with black. Blackish irroration in brownish specimens extending quite across back of each segment, only leaving clear a narrow incision-bar; but in creamy examples restricted so as to form four longitudinal series of small spots. Mixed with this irroration on each segment are three or four transverse series of minute whitish spots, and also towards the front of each segment a transverse row of larger elevated yellow spots (four on each abdominal segment, six or more on each thoracic one). Beneath creamy-white or greenish-white; wing-covers streaked with blackish along and near inner margin and also near hind-margin.

These descriptions of *larva* and *pupa* are made from numerous living specimens received from Colonel Bowker in August 1887, having been collected by him near D'Urban, Natal. The larvæ had almost finished their supply of food by the time that they arrived, and I liberated them all except one which was suspended for pupation, and from which I obtained a ♀ *Severina* on the 9th September. Seven of the pupæ received had the date of pupation attached, and I thus ascertained that the duration of the chrysalis state was from fourteen to seventeen days. The eight examples (three ♀, five ♂) that I reared from these Natalian pupæ, as well as several others reared at the same time by Colonel Bowker and afterwards forwarded to me for comparison, were all of the rather smaller form, with duller-tinted under side marked by heavy blackish neuration, proper to the winter or dry season ; but one of the ♂ s that I reared was of the *Boguensis* variety, with a very completely developed oblique costal bar marking the extremity of the discoidal cell.

The typical ♂ varies to some extent in the width of the black borders on the upper side, as well as in the size and distinctness of the white spots which they contain; in one example (from Delagoa Bay) the inner part of the border of the hind-wings is so feebly developed that these spots are scarcely separated from the white ground.[1] On the under side the tint of the hind-wing is sometimes of a duller, greyer tinge, and in these examples the neuration is strongly and generally fuscous-clouded ; while in the specimens which have this surface pale and bright the nervures are often almost free of clouding, more especially on the disc. Two rather small ♂ s from the Limpopo River exhibit the latter character in a very marked degree, and in that respect resemble the very closely allied *P. Creona*, (Cram.), of West Africa.

In a ♂ that I captured near Grahamstown, the basal pale-yellow suffusion of the fore-wing on the under side is abnormally developed, filling the discoidal cell and spreading beyond it along the costa. The same peculiarity exists to a much less extent in two other ♂ s,—one from Kaffraria Proper and the other from Natal.

The typical ♀ on the upper side presents a variable width in the borders, and the ground of the fore-wings sometimes (and of the hind-wings very rarely) is nearly white. On the under side of the hind-wings the clouding of the neuration is less variable than in the ♂, being commonly well developed.

An example from Barberton, Transvaal, received from Mr. J. P. Cloete in March 1888, is of a remarkably deep rich yellow above, with the dark border abnormally wide,—in the fore-wings almost touching

---

[1] It is evidently on a similar ♂ from Port Natal that Felder (*op. cit.*) has founded his *P. Agrippina*. The feature in question is a character of the African ♂ s of *Mesentina*, Cr.

the disco-cellular spot, and in the hind-wings without the usual enclosed spot. This ♀ was accompanied by a ♂ of large size, exhibiting the *Agrippina* character above described (see note, p. 70).

Dr. Felder, from whom I received a Bogos ♂ of *Boguensis*, expressed the opinion (*op. cit.*, p. 174) that this form was very probably but a local variety of *Severina*. I have since, however, taken a ♀ near Grahamstown and received a ♂ and two ♀ s from D'Urban, Natal. The marked feature of the costal bar of the fore-wings imparts to it something of the aspect of *Mesentina*; and it is noteworthy that in the ordinary ♀ s of *Severina* (especially on the under side), a fragmentary, or rarely complete, narrow costal bar is present.

It is very doubtful whether *Severina* is entitled to be held a distinct species from *Creona*, Cram. (*op. cit.*, i. t. xcv. ff. C, D [♂], E, F [♀]). This Tropical-African form is, however, constantly smaller, and presents in both sexes proportionately wider borders (that of the hind-wings in the ♂ s having all the spots it contains on the upper side obsolete except the apical one), and on the under side unclouded neuration in the hind-wings, as well as broader and brighter longitudinal orange-yellow streaks. The ♂ has the disco-cellular spot of the fore-wing better developed, but the white spots in the border much reduced; the ♀ has a rather wide basal fuscous suffusion in both wings (much as in the ♀ *Gidica*, Godt.).

The figures and description of *Pieris Elisa*, Vollenhoven (Pollen and Van Dam's *Recherches sur la Faune de Madagascar*, &c., Part v. p. 12, pl. 2, ff. 3, 1877), are respectively so defective and insufficient, that I am unable with certainty to make out whether the Mayotte specimens represented and described are actually referable to *Severina* or not; but I think that they most probably are so referable, the figure marked "3 ♂" being apparently a ♂ of the slight variation above referred to as *Agrippina*, Felder, and that marked "3 ♀" to all appearance a ♂ of the variety *Boguensis*, Feld.[1]

Colonel Bowker has sent me four pairs of this butterfly captured by him *in copula*, viz., in 1873 a pair from Fort Warden, on the Kei River; in 1879 a pair, and in November 1881 two pairs from D'Urban, Natal. In the ♂ s of the Kei River pair the clouded neuration of the under side of the hind-wings is marked only near base and hind-marginal border; in that of the D'Urban pair, 1879, it is almost obsolete; and in those of the D'Urban pairs of 1881 it is well developed except on the middle of the disc. The ♀ of the Kei River pair is of average size, very yellow on the upper side, and has the central part of the under side of the hind-wings whitish streaked with yellow,—the dark neuration being very strongly marked; that of the D'Urban pair of 1879 is small, whitish on the upper side, and with the neuration

---

[1] It is remarkable that, next to *Creona*, Cram., the nearest known ally of *Severina* is the Australian *Teutonia*, Fab., which is considerably larger, and with much broader upper-side borders in the ♂, and a costal bar (like that of Var. *Boguensis*) in the fore-wings of the ♀; while the neuration of the hind-wings on the under side is very strongly black-clouded.

of the under side of the hind-wings dark clouded throughout; and those of the 1881 pairs are both yellowish on the upper side (with the costal streak of the fore-wing in one thinly prolonged almost to the disco-cellular spot), while on the under side one of them has the dark neuration of the hind-wings only marked near the hind-marginal border, but the other has it strongly developed throughout (in this latter specimen the narrow costal bar of the fore-wing is complete).

A ♂ and ♀ which I captured at D'Urban, Natal, in June 1865, and also a ♂ taken at the beginning of the following August, are smaller and with duller-tinted, more dusky-veined under sides than those which I afterwards took in that locality and near Grahamstown in the summer months of January, February, and March.[1] The ♀ referred to is almost white on the upper side; in this respect resembling a considerably larger ♀ from South-West Madagascar, which is in the South-African Museum.

This is a common and widely-spread butterfly over all the wooded parts of South Africa; its most westerly known locality is Knysna, on the coast of the Cape Colony. Although prevalent always about the edges of woods, it by no means confines itself to them, but flies actively over open ground at hand, often stopping to visit flowers. The ♂ s are swifter than the ♀ s, but both sexes were in the flights of *Pierinæ* which I witnessed on 6th February 1870 at Highlands, near Grahamstown, and I captured several females on that occasion.

Localities of *Pieris Severina*.

I. South Africa.
  B. Cape Colony.
    a. Western Districts.—Knysna (*Miss Wentworth*).
    b. Eastern Districts.—Port Elizabeth. Uitenhage (*S. D. Bairstow*). Grahamstown. Mouth of Kowie River (*J. L. Fry*). King William's Town (*W. S. M. D'Urban*). Windvogelberg, Queenstown District (the late *Major G. E. Bulger*). Kei River (*J. H. Bowker*).
  D. Kaffraria Proper.—Butterworth and Bashee River (*J. H. Bowker*).
  E. Natal.
    a. Coast Districts.—D'Urban. Pinetown (*J. H. Bowker*). Verulam. Umvoti. Mouth of Tugela (*J. H. Bowker*). Mapumulo.
    b. Upper Districts.—Greytown. Maritzburg (*S. Windham*). Estcourt (*J. M. Hutchinson*).
  F. Zululand.—Special locality not noted (*G. F. Angas*).—Coll. Brit. Mus.
  H. Delagoa Bay.—Lourenço Marques (*Mrs. Monteiro*).
  K. Transvaal.—Potchefstroom District (*T. Ayres*). Limpopo River (*F. C. Selous*). Barberton (*J. P. Cloete* and *C. F. Palmer*).

---

[1] Mr. Alfred D. Millar, of D'Urban, at my request, was good enough to capture and date numerous examples of this (as well as of other *Pierinæ*) during 1887-88, and to send them to me. Five ♂ s and four ♀ s taken on 15th August, and a ♂ taken on the 22d, are all of the duller more dusky-veined form; a ♂ taken on 22d September is larger and brighter beneath, showing some approach to the summer or wet season form; and three ♂ s taken on 3d December and a ♂ and ♀ taken on 29th January are all of the bright pale-yellow under side, with the neuration lightly and unequally blackish.

II. Other African Regions.
  A. South Tropical.
    a. Western Coast.—Angola: "Loanda (*R. Meldola*)."—A. G. Butler. Congo: "Kinsembo (*H. Ansell*)."—A. G. Butler.
    b. Eastern Coast.—Zambesi (*Rev. H. Rowley*). "Querimba."—Hopffer. Zanzibar.—Coll. Brit. Mus. "Endara, Mombas (Kersten)."—Gerstäcker. Kilima Njaro.—Coll. Brit. Mus.
    b1. Eastern Interior.—Makloutse and Tati Rivers (*F. C. Selous*). Mashunaland (*F. C. Selous*).
    bb. Eastern Islands.—Madagascar: Murundava River (*Grevé*). "Mayotte (*Pollen* and *Van Dam*)."—Vollenhoven.
  B. North Tropical.
    a. Western Coast.—"Accra and Aburi (*Weigle*)."—Möschler.[1]
    b. Eastern Coast.—Red Sea: Bogos (*Hansal*). [Var. *Boguensis*, Feld.]
    b1. Eastern Interior.—White Nile.—Coll. Brit. Mus. [Var. *Boguensis*, Feld.]

263. (14.) Pieris Hellica, (Linnæus).

♂ *Papilio Helice*, Linn., Mus. Lud. Ulr. Reg., p. 243, n. 62 (1764); and *P. Hellica*, Syst. Nat., i. 2, p. 760, n. 78 (1767).
♀ *Papilio Daplidice*, Cram., Pap. Exot., ii. pl. clxxi. ff. c, D. (1779).
♂ ♀ *Mancipium vorax Hellica*, Hübn., Samml. Exot. Schmett., i. t. 141, ff. 1, 2 [♂], 3, 4 [♀] (? 1806).
♂ ♀ *Papilio Raphani*, Esp.,[2] (Europ.) Schmett., Suppl., Bd. i. t. cxxiii. Cont. 78, ff. 3, 4 (1805).
Pieris Hellica, Boisd., Sp. Gen. Lep., i. p. 546, n. 156 (1836).
♂ ♀ *Pieris Hellica*, Trim., Rhop. Afr. Aust., i. p. 39, n. 24 (1862).
♂ *Synchloë Hellica*, Crowley, Trans. Ent. Soc. Lond., 1887, pl. iii. ff. 4, 5.

Exp. al., (♂) 1 in. 10 lin.—2 in.; (♀) 1 in. 9½ lin.—2 in.

♂ *White, with blackish markings. Fore-wing:* base narrowly greyish; costa narrowly edged with blackish; at extremity of discoidal cell a rather broad, short, blackish stripe, extending above subcostal nervure, but not touching costal edge; a broad, blackish border on hind-margin, abruptly terminating on second median nervule, and containing four rounded white spots, the third of which, from costa, is almost always connected with the white ground-colour of wing, dividing the band into two portions. *Hind-wing:* on hind-margin a row of small black spots, one at extremity of, and inwardly projecting along, each nervule.
UNDER SIDE.—*Fore-wing:* as on upper side, but disco-cellular stripe extended to costa by some fainter blackish colouring; *apical portion of hind-marginal band tinged and dusted with chrome-yellow. Hind-wing: nervures much clouded with yellow-dusted grey,* leaving a central, transverse, curved, macular band of white, and almost enclosing a row of five or six white spots on hind-margin; *costa at base, and most of the spaces between clouded nervures, tinged with bright chrome-yellow.*

---

[1] From Möschler's note on several ♂s and a ♀ (Abhandl. Senckenberg. Naturf. Gesellsch., 1887, p. 53), these Gold Coast specimens appear to belong to the form *Creona*, Cram.

[2] Esper's earlier figure of *P. Raphani* (op. cit., i. pt. 2, t. lxxxiv.) appears not to represent *Hellica*, Linn., but some allied species. Mr. W. F. Kirby (*Cat. D. Lep.*, 1871, p. 451) regards it as intended to represent a variety of *Daplidice*, Linn.

♀ *Similar to* ♂, *but with additional black markings*. *Fore-wing*: costa with a greyish border as far as disco-cellular band, which is much broader than in ♂; hind-marginal band slightly broader and blacker than in ♂,—its third white spot *always quite enclosed* in the black; two additional blackish spots near posterior angle, between first median nervule and inner margin, the lower one much smaller and fainter, and forming a continuation of the upper one. *Hind-wing*: a blackish band along hind-margin, commencing with a dark mark on costa (adjoining spot on inner margin of fore-wing), becoming obsolete towards anal angle, and containing four or five rather large ovate white spots. UNDER SIDE.—*Fore-wing*: only the *upper* of the two additional blackish spots present. *Hind-wing*: the clouding of the nervules wider than in ♂, in some strongly marked specimens so much so as to leave only a row of white spots in place of the central white band.

A dwarfed ♂, taken at Burghersdorp by Dr. Kannemeyer, measures only 1 in. $3\frac{1}{2}$ lin. across the expanded fore-wings.

LARVA.—Light-green, darker on inferior surface. A median dorsal violaceous stripe; and on each side a broader, less defined, deeper-greenish stripe mixed with violaceous, succeeded by a conspicuous pale-yellowish spiracular band. On each segment numerous black dots arranged in four transverse lines on back and sides (other scattered black dots on lower part of sides), and also four orange spots, situated anteriorly, two of which immediately precede the spiracles. Head black-dotted; spiracles conspicuously black. A few short hairs about body generally, numerous short hairs on head.

" Food-plants near Grahamstown, *Sisymbrium Capense* (and probably *S. lyratum*) and *Lepidium sativum*."—M. E. Barber.

PUPA.—Above yellow, sprinkled with black dots, beneath pale-green. A median dorsal pale-violet stripe; narrow thoracic ridge marked with a red line.

The first pupa I observed (found on 10th December) changed in colouring, four days afterwards, to light violet-grey, with a pale-yellow stripe along each side of the abdomen; the imago did not appear before the 20th December. The second pupa I reared retained its yellow and green colours throughout from the 24th April to the 8th May, when the imago emerged. One pupa was attached to the wall of a house, the other to a grass stem.

A very near ally of *Hellica* was brought from Kilima-Njaro by Mr. H. H. Johnston, and has lately been described and figured as *Synchloë Johnstonii* by Mr. P. Crowley (*Trans. Ent. Soc. Lond.*, 1887, p. 35, pl. iii. ♂, ff. 1, 3; ♀, f. 2). I made notes on two ♂s and a ♀ of this form in the British Museum collection (October 1886). It is at once distinguishable by its longer wings, the fore-wings being also acuter apically; and the ♂ has the black border of the fore-wings on the upper side internally unbroken, with the enclosed white spots smaller than in *Hellica*. On the under side, the neuration is more heavily clouded—the nervures themselves being widely greyish, bounded on both sides with fuscous;—but in the hind-wings the yellow and white discal inter-nervular markings are brighter and more completely unite the hind-marginal spots (which are elongated) to the median series.

Next to *Johnstonii*, the nearest known form to *Hellica* is the Arabian *Glauconome*, Klug,[1] which differs, however, on the under side in the greener clouding that occupies much more of the field,—especially in the basal area of the hind-wings, where the white is reduced to two good-sized oval spots (one in discoidal cell, the other just above it),—and in having the neuration itself yellow, while the inter-nervular yellow rays so conspicuous in *Hellica* (especially those uniting the median and marginal white spots of the hind-wings) are wanting. In these respects, and in the hind-marginal markings on the upper side of the hind-wings, *Glauconome* is nearer to the well-known *Daplidice*, Linn, than to *Hellica*; and the same may be said of *Glauconome's* close ally, *Iranica*, Bienert, of which I have examined specimens from Afghanistan (Candahar) in the British Museum.[2]

Hellica is one of the commonest and most abundant butterflies in South Africa; over nearly all districts it seems to be found throughout the year, but is most prevalent in the spring and summer months. It frequents open ground generally, and is the only "White" resident in the Cape Peninsula. On the wing it is by no means swift, keeping near the ground, and often settling on flowers. On the coast of Natal I did not meet with it, but it was not uncommon on the higher levels inland, particularly near Maritzburg and Greytown. I have not met with any example in collections from Tropical localities, nor have I found any record of the species' occurrence out of South Africa proper.[3]

Localities of *Pieris Hellica*.

1. South Africa.
    B. Cape Colony.
        a. Western Districts.—Cape Town. Hout Bay. Stellenbosch. Paarl. Wellington. Malmesbury. Ceres and Vogel Vley, Tulbagh District. Worcester. Robertson. Genadendal, Caledon District (*G. Hettarsch*). Mossel Bay. Knysna. Plettenberg Bay. Oudtshoorn (*Adams*). Van Wyk's Vley, Carnarvon District (*E. G. Alston*). Ookiep and Klipfontein (*L. Péringuey*), and Amenous, Namaqualand District.
        b. Eastern Districts.—Port Elizabeth. Uitenhage. Grahamstown. Tharfield, Bathurst District (*Miss M. L. Bowker*). East London (*P. Borcherds*). King William's Town (*W. S. M. D'Urban*). Windvogelberg, Queenstown District (*Dr. Batho*). Burghersdorp, Albert District (*D. R. Kannemeyer*). Colesberg (*A. F. Ortlepp*). Murraysburg (*J. J. Muskett*).

---

[1] Symb. Phys., dec. I. pl. vii. ff. 18, 19 (1829).

[2] For a note on the differences between *Iranica* and *Glauconome*, see Butler, Proc. Zool. Soc. Lond., 1884, p. 492.

[3] The late Mr. F. Walker gave (*Entomologist*, 1870, p. 49) localities both east and west of the Red Sea for *Hellica*; but there can be no doubt that the species he received from that region must have been the closely-allied *Glauconome*, Klug.

  c. Griqualand West.—Kimberley.
  d. Basutoland.—Maseru (*J. H. Bowker*).
 C. Orange Free State.—Locality not noted (*C. Hart*).
 D. Kaffraria Proper.—Butterworth and Bashee River (*J. H. Bowker*).
 E. Natal.
  b. Upper Districts.—Karkloof (*W. Morant*). Greytown. Maritzburg. Estcourt (*J. M. Hutchinson*).
 K. Transvaal.—Potchefstroom District (*T. Ayres*).

GENUS HERPÆNIA.

Herpœnia, Butler, Cist. Ent., i. p. 52 (1870); Proc. Zool. Soc. Lond., 1872, p. 67.
Pieris (Auct.—part.), Trimen, Rhop. Afr. Aust., i. pp. 26 and 40 (1862).

 IMAGO.—Nearly allied to *Pieris*. *Palpi* much shorter, especially terminal joint; *antennœ* shorter, more slender. *Fore-wings* apically rather prominent (especially in ♂); posterior angle well marked; first and second subcostal nervules given off not far apart, at some distance before extremity of discoidal cell; upper radial nervule longer, united to subcostal nervure at a point considerably nearer to extremity of cell; upper disco-cellular nervule exceedingly short, slanting inward,—lower one very long, strongly curved inward in middle; discoidal cell thus terminating very truncately; second and third median nervules nearer to each other than second and first. *Hind-wings* more elongated superiorly, shorter and sub-truncate inferiorly; anal angle rather pronounced; disco-cellular nervules almost as in fore-wings. *Abdomen* more elongate, particularly in ♂, where it is very slender and is only a little shorter than inner margins of hind-wings.
 The characters here given (of which the more important were pointed out *loc. cit.* by Mr. Butler) serve well to separate this genus; and its very remarkable and peculiar pattern is a character which emphasises the structural ones. The creamy-white or yellowish ground-colour is boldly and conspicuously broken up by not only very broad hind-marginal black borders enclosing spots of the ground-colour, but also by a very broad longitudinal black bar in the fore-wings (occupying base and discoidal cell, and joining the hind-marginal border about its middle), and a narrower obliquely transverse sub-basal black bar in the hind-wings (the superior extremity of which is linked to apical commencement of hind-marginal border by a costal and a subcostal short black ray). Both structure and pattern indicate approach to the genus *Teracolus*; in the fore-wings the disposition of the subcostal nervules being the same, and that of the upper radial and disco-cellular nervules almost the same; and the singular arrangement of wing-markings is not far different from that found in the ♀ *T. Evenina*, Wallengr.
 Mr. Butler has given names to several slight modifications of what,

after examination of a large number of examples from various quarters, I can but consider the only known species of this genus, viz., *H. Eriphia*, Godt. This butterfly has a very wide range in Tropical Africa, and in the south is of pretty general distribution, except in the south-west of the Cape Colony.

264. (1.) Herpænia Eriphia, (Godart).

Pieris Eriphia, Godt., Enc. Meth., ix. p. 157, n. 134 (1819).
♂ *Pontia Tritogenia*, Klug, Symb. Phys., t. viii. ff. 17, 18 (1829).
Pieris Tritogenia, Boisd., Sp. Gen. Lep., i. p. 513, n. 110 (1836).
♂ *Dismorphia Tritogenia*, Geyer, Forts. Hübn. Zutr. Exot. Schmett., p. 11, ff. 829-830 (1837).
♀ *Pinacopteryx Eriphia*, Wallgrn., K. Sv. Vet.-Akad. Handl., 1857; Lep. Rhop. Caffr., p. 10, n. 5.
Pieris Eriphia, Trim., Rhop. Afr. Aust., i. p. 40, n. 25 (1862).
Anthocharis Tritogenia, Hopff., Peters' Reise Mossamb., Ins., p. 356 (1862).
Herpænia Tritogenia, Butl., Cist. Ent., i. p. 52 (1870).
Var. *Herpænia Melanarge*, Butl., Proc. Zool. Soc. Lond., 1885, p. 774.

Exp. al., (♂) 1 in. 9 lin.—2 in. 2½ lin.; (♀) 1 in. 10 lin.—2 in. 4 lin.

♂ *White or creamy-white, with broad black stripes and borders; about equal areas of black and white.* Fore-wing: a broad black band from base occupies all discoidal cell, extending considerably below and beyond it, and united by a narrower extension to a broad border on hind-margin; the latter containing five variously-shaped white spots (of which that next costa is very small, the second, third, and fifth of moderate size, the fourth large and ovate), and so *widened on inner margin* as to occupy the outer half of it; sometimes in black discoidal cell a paler or whitish longitudinal streak. Hind-wing: a transverse black stripe at base; before middle a black stripe, with two or three irregular indentations on its edges, straight across wing from costa near apex to about middle of inner margin; from commencement of this band all along hind-margin to anal angle a black band—abruptly widened, and with a straight inner edge parallel to the band before middle, from discoidal nervule to inner margin—containing six good-sized white spots. Cilia yellowish-white. UNDER SIDE.—*Fore-wing*: black markings clouded with white near inner margin, and along upper part of discoidal cell, dull and brownish tinted in central portion, and *replaced by grey-dusted ochreous at apex*; in discoidal cell, a broad longitudinal yellowish-white stripe. Hind-wing: costa, especially near base, edged with orange-yellow; *basal, median, and hind-marginal bands dark-grey-dusted ochreous*; median band more irregular and macular than on upper side; spots on hind-margin larger, but not so clearly defined; neuration yellowish-white.

♀ *Similar to ♂, but commonly with the ground-colour tinged with sulphur-yellow; the black markings duller, and sometimes tinged with brown.* UNDER SIDE.—*Fore-wing*: blackish markings not clouded with white; white markings generally more or less tinged with sulphur-yellow. *Hind-wing*: all the white markings usually more or less tinged with sulphur-yellow along margins of wing.

As above noted, the size of this species is very variable. I have observed that, although large examples are found throughout the South-African range, all the smallest specimens come from the drier (usually upland and interior) tracts.

VARIETY A.—♂ and ♀ (*Melanarge*, Butl.).

♂ *Ground-colour decidedly creamy;* along hind-margins a dull ochrey-reddish tinge tinging both black border (especially about apex of fore-wing) and outer part of its creamy spots. UNDER SIDE.—*Hind-wing and apical hind-marginal area of fore-wing suffused with dull ochrey-reddish (in the paler portions with a tinge of pink) dusted with dark-grey;* the dark bands and border very ill-defined reddish-brown, in some specimens scarcely distinguishable.

♀ (Two examples). Ground-colour pale sulphur-yellow. UNDER SIDE.—As in ♂, but the ochrey tint paler, not so red, and with the position of the typical white markings vaguely indicated by some whitish clouding. *Fore-wing*: pale markings sulphur-yellow.

(*Hab.*—Natal, Delagoa Bay, Transvaal, Griqualand West, Damaraland, and Somaliland.)

From Mr. Butler's description (*loc. cit.*) I think that there can be no doubt that his *Melanarge*, of which he describes three male examples from Somaliland, is identical with the Variety A. just described.[1] I should also refer to the same variety his *Lacteipennis* (*Ann. and Mag. Nat. Hist.*, 4th Ser., xviii. p. 489, 1876), from Abyssinia, notwithstanding its unusually small size (*exp. al.* 1 in. 7 lin.), if it were not for his description of the hind-wings as having " several submarginal black spots, sagittate (with the points upwards), towards apex," which looks as if the ordinary white-spotted black border of the upper side were wanting on those wings.

I have not been able to discover any character by which *Tritogenia*, Klug, can be separated from *Eriphia*, Godt. The description of the latter only differs in giving four instead of six white spots in the black border of the hind-wing; but these spots vary a good deal in size and distinctness, and Godart probably did not include the first large costal spot (which, indeed, is but narrowly separated from the central white field); and I have seen several examples in which the sixth (last) spot is almost obsolete. Mr. Butler, in the paper last cited, observes: "*Eriphia*, which I have examined from Angola, is a larger and more creamy-coloured insect than *H. Tritogenia* (with which it has been united), . . the markings are not quite the same on

---

[1] Under the name of *Herpænia iterata*, Mr. Butler has recorded some examples from Kilima-Njaro (F. J. Jackson), stated to differ from *H. melanarge* only in its considerably larger size (55 mm. = about 2 in. 2 lin.), and in the broader subbasal black belt and larger white marginal spots in the hind-wings (*Proc. Zool. Soc. Lond.*, 1888, p. 96).

the hind-wings." He does not, however, specify what differences exist.

Geyer (*op. cit.*) figures a ♂ from " South Africa," which agrees very well with specimens taken near Grahamstown and in Kaffraria Proper, but in one point differs from every other example that I have seen, viz., in the narrow but complete separation of the disco-cellular longitudinal black band from the hind-marginal black border on the upper side of both fore-wings.

Boisduval, describing northern specimens under the name *Tritogenia*, mentions that the ♀ has a whiter ground-colour than the ♂. This is quite the exception in southern examples. I took one such ♀ near Grahamstown in 1870, but all the others I have seen exhibit a more or less decided inclination to pale-yellow.

Hopffer (*op. cit.*) notes that the ground-colour varies as much in the Mozambique specimens as in those from Nubia and Senegal, and mentions one example of the variety with the brown under-side colouring.

The general resemblance of this pied butterfly to the "Marbled White" *Satyrinæ* of Europe (*Melanargia Galathea*, Linn., and allies) has been noticed by many lepidopterists. The form and position of the longitudinal band of the fore-wings and the straight transverse ante-median band of the hind-wings are very singular, and the only other species in which similar (though smaller and duller) markings occur is *Teracolus Evenina*, Wallengr., ♀.

I had the pleasure of observing this beautiful insect in life for the first time at the end of January 1870, near Grahamstown, and during the following month met with it not uncommonly in that neighbourhood. It frequented steep hillsides on the borders of woods, but never entered the shade of the woods themselves; delighting in the *Scabiosa* flowers, which were abundant in such stations. It is very conspicuous on the wing, and is easily captured, being rather slow of flight and settling frequently. I afterwards saw the species near Uitenhage. There is no part of South Africa proper where this butterfly appears to be abundant, but it is evidently numerous in the South-Tropical belt between 23° and 20°, twenty-five examples occurring in Mr. J. A. Bell's small collection made in Damaraland, and nineteen in a series collected by Mr. A. W. Eriksson between the northern limit of the Transvaal country and Gubulewayo, in Matabeleland. Mr. H. Barber informed me that two specimens he sent from Matabeleland were captured while drinking at a hole dug to collect water.

Localities of *Herpænia Eriphia*.

I. South Africa.
   B. Cape Colony.
      a. Western Districts.—Swellendam (*W. Cairncross*). Knysna (*Miss Wentworth*). Spectakel (*L. Péringuey*) and Ookiep (*G. Warden*), Namaqualand District. Prieska, Orange River, Victoria West District (*F. Purcell*).
      b. Eastern Districts.—Uitenhage. Grahamstown. Kleinemond River, Bathurst District (*H. J. Atherstone*). Murraysburg (*J. J. Muskett*).

c. Griqualand West.—Vaal River (*J. H. Bowker*). "Kimberley."—
H. L. Feltham.
d. Basutoland.—Koro Koro and Headwaters of Orange River (*J. H. Bowker*).
C. Orange Free State.—Hebron (*W. Morant*).
D. Kaffraria Proper.—Bashee River (*J. H. Bowker*).
E. Natal.
  a. Coast Districts.—D'Urban (*M. J. M'Ken*). Pinetown (*H. J. Harford*.—Var. A.). "Tugela River."—E. C. Buxton.
  b. Upper Districts.—Estcourt (*J. M. Hutchinson*).
G. "Swaziland."—E. C. Buxton.
H. Delagoa Bay.—Lourenço Marques (Mrs. Monteiro.—Type and Var. A.).
K. Transvaal.—Potchefstroom District and Origstadt, Lydenburg District (*T. Ayres*.—Type and Var. A). Marico River (*F. C. Selous*.—Type and Var. A.).
L. Bechuanaland. Kamhanni Pass and Klibbolikhonni (the late *J. W. Burchell*).

II. Other African Regions.
A. South Tropical.
  a. Western Coast.—Damaraland (the late *C. J. Andersson, J. A. Bell*, and *J. J. Christie*.—Type and Var. A.). "Angola (*Monteiro*)."—Druce. "Angola: Loanda (*R. Meldola*)."—Butler. "Congo: Kinsembo (H. Ansell)."—Butler.
  b. Eastern Coast.—Zambesi (*Rev. H. Rowley*).—Coll. Hope Oxon. "Tette (Zambesi) and Querimba."—Hopffer.—Type and Var. A. Mombasa : "Lake Jipé (O. *Kersten*)."—Gerstäcker.
  b1. Eastern Interior.—Matabeleland : Sinquasi (*H. Barber*). Between North Transvaal and Gubulewayo (*A. W. Eriksson*).
B. North Tropical.
  a. Western Coast.—"Senegal."—Boisduval.
  b. Eastern Coast.—"Somaliland (Thrupp)."—Butler.—Var. A. Red Sea : "Massowah (*Raffray*)."—Oberthür.
  b1. Eastern Interior.—Abyssinia : "Shoa (*Antinori*)."—Oberthür. "Kordofan (*Vienna Museum*)."—Möschler. Soudan : "Atbara River."—Butler. White Nile.—Coll. Hewitson. Dongola : "Ambukohl."—Klug.

GENUS TERACOLUS.

Teracolus, Swains., Zool. Illustr., 2nd Ser., vol. iii. text to pl. 115 (1833).
Anthocharis, Boisd. [part], Sp. Gen. Lep., i. p. 556, and *Idmais*, Boisd., p. 584 (1836).
Anthocharis (Sect. iii., *Callosune*) and *Idmais*, E. Doubl., Gen. D. Lep., i. pp. 57 and 59 (1847).
Anthopsyche and *Ptychopteryx*, Wallengr., K. Svensk. Vetensk.-Akad. Handl., 1857 ; Lep. Rhop. Caffr., pp. 10 and 17.
Thespia, Wallengr., K. Sv. Vet.-Akad. Förhandl., 1858, p. 77.
Anthocharis and *Idmais*, Trim., Rhop. Afr. Aust., i. pp. 42 and 60 (1862).
Teracolus, Butl., Cist. Ent., i. p. 47 (1869) ; and [Revision] Proc. Zool. Soc. Lond., 1876, p. 126.

IMAGO.—*Head* rather small, densely hairy ; *eyes* large, globose, smooth ; *palpi* rather short, slender, laterally compressed, ascendant,

—basal and middle joints densely clothed beneath with projecting hairs of unequal length,—terminal joint small, thin, rather acute, directed forward, clothed with short appressed hairs; *antennæ* short and rather thick, with a somewhat abruptly-formed, broad, flattened club, blunt at the tip.

Thorax rather robust or moderately slender, clothed superiorly with close silky hair (longer laterally and posteriorly), and inferiorly with a shorter, denser down. *Fore-wings* with costa but slightly arched; apex varying from rather acute to rounded; hind-margin entire, rarely prominent in apical portion, sometimes convex in middle (especially in ♀); inner margin almost always nearly straight, or but slightly hollowed about middle (but in the *Fausta* group prominently convex in ♂); costal nervure extending to about middle; first and second subcostal nervules originating not far apart at some distance before extremity of discoidal cell, and usually closely approximate throughout,—the first often almost touching costal nervure,—third nervule short, originating nearer to apex than to extremity of cell, and usually terminating at apex, but sometimes very short and ending on costa; upper radial nervule united to subcostal just at extremity of discoidal cell; upper disco-cellular nervule short (in *Daira* group extremely short) curved inwardly,—lower one three times as long, sub-angulated inwardly, moderately oblique inferiorly. *Hind-wing*: costa very prominently convex at base, but not strongly arched generally (except in *Subfasciatus* and the *Fausta* group); hind-margin entire or but very slightly sinuated; costal nervure rather short, terminating at a little beyond middle; second subcostal nervule usually originating before, or considerably before, extremity of discoidal cell (but in several groups *at* that extremity, and in the *Daira* group slightly *beyond* it); lower disco-cellular nervule long, sub-angulated, oblique, joining third median nervule at some distance from latter's origin; internal nervure extending to beyond middle; inner margins very prominent basally, and meeting so as to hide basal half or more of abdomen; anal angle rounded. *Legs* of moderate length and thickness, scaly; femora with short hair beneath; tibiæ of fore and middle legs shorter, of hind-legs considerably longer, than femur,—their terminal spurs short; tarsi long, spinulose beneath.

Abdomen slender, rather short.

LARVA.—Resembling that of *Pieris*, but stouter anteriorly, and segmental incisions deeper; some short scattered hairs, sparsely distributed, but more numerous laterally; dorsal surface with marked transverse rugæ, and two transverse rows of tubercular spots on each segment.

PUPA.—Head acuminate somewhat abruptly, but rather wide and with eye-covers prominent. Thorax dorsally moderately convex, laterally very slightly sub-angulated at bases of wing-covers. Wing-covers exceedingly prominent ventrally, highly convex in outline, but laterally compressed keel-fashion. Abdomen slender, straight, dorsally flattened, its anal extremity bluntly bifid. Silken girth in depression about base of abdomen, only free from a point near middle of each wing-cover.

The above characters of larva and pupa are taken from specimens of those stages of *T. Anterippe* (Boisd.) received alive from Colonel Bowker, and also from spirit specimens of another *Teracolus* (*T. Achine*, Cram., or *T. Garisa*, Wallengr.) obtained from Mr. W. D. Gooch. A pupa of *T. Pleione*, (Klug), in the British Museum, is of a somewhat stouter build, but does not materially differ.

The pupa of *Teracolus* is at once distinguishable from that of *Anthocharis* by its much shorter and more abruptly (instead of very gradually) acuminate head, and much more bulging wing-covers. Its outline and shape combine most of the characters of *Anthocharis* and *Colias*, while the prominent keeled wing-covers resemble those of *Gonepteryx*.

The alliance of this genus with *Pieris* is apparent, the chief distinguishing characters being the shorter palpi and antennæ and the differing neuration of the fore-wings, in which the first and second subcostal nervules originate close together at some distance before the extremity of the discoidal cell, and the upper radial nervule is united to the subcostal nervure just at the extremity of that cell. Its affinity to the genus *Colias* is more remote (although in pattern and colouring of the under side the *Amata* group has a strong resemblance to it, and the robust type species, *T. subfasciatus*, is not unlike it in appearance), the antennæ and palpi being totally different, as well as the subcostal neuration of the fore-wings.

It is not without reluctance that I adopt Swainson's name of *Teracolus*[1] for this genus; but, as I agree with Mr. Butler that Swainson's type (*Subfasciatus*), though presenting several special minor characters, cannot be generically separated either from *Idmais*, Boisd., or *Callosune*, Doubl., the law of priority demands this course.

Neither structure nor pattern is by any means uniform in this large and difficult group. Besides the variations pointed out in my diagnosis, numerous divergences occur as regards the thickness of both antennæ and wings and the system of coloration in the latter. After careful investigation of a large number of species, I consider that the genus may with advantage be arranged in nineteen sections,[2] as follows, viz.:—

SECTION I.—Representative: *Subfasciatus*, Sws.

General structure robust; wings thick. Antennæ rather short and thick, with broad blunt club. Fore-wings acute in both sexes; hind-wings

---

[1] The derivation of this name given in Agassiz's *Nomenclator Zoologicus*, viz., "τέρας, miraculum; κόλος, mutilus," seems altogether fanciful; and it is almost certain, from Swainson's text, that the founder coined the term as a combination of *Terias* and *Colias*. In Swainson's figure of the neuration the first subcostal nervule is omitted.

[2] The sections which appear to have no representative in South Africa proper are enclosed within square brackets.

with a *fringe of hairs on costa near base*.[1] First and second subcostal nervules of fore-wings closely approximate; hind-wings with discoidal cell more than half their length, *costa and costal nervure strongly arched*, second subcostal nervule originating some distance before extremity of discoidal cell.

♂ *Sulphur-yellow, with apical patch of fore-wing almost concolorous, but with an incomplete internal broad black border.* ♀ *paler with orange-red apical patch.* Under side dull greenish-white striolated with dusky-grey.

1 Species.

Range.—South Tropical Africa and Extra-Tropical South Africa as far as 30° lat.

SECTION II.—Representative: *Agoye*, Wallengr.

Wings thin; fore-wings apically prominent, with apex subacute in both sexes; hind-wings not much arched costally. *Junction (in fore-wings) of upper radial nervule with subcostal nervure farther from base, so that upper disco-cellular nervule is strongly curved and extended outwardly, forming a very acute upper angle to discoidal cell*; hind-wings with discoidal cell half their length, second subcostal nervule originating much nearer extremity of discoidal cell than in Section 1.

White, with small ochrey orange-yellow or yellow-ochreous apical patch in fore-wings, in ♂ *black-bordered inwardly, is* ♀ *much duller and broadly blackish or brownish-bordered on both sides.* Under side whitish or faint-yellowish, almost devoid of marking.

Nervures in ♂ *more or less black on upper side.*

2 Species.—*Agoye*, Wallengr.; *Bowkeri*, Trim.

Range.—South Tropical Africa and Extra-Tropical South Africa as far as 30° lat.

SECTION III.—Representative: *Eris*, Klug.

Structure of Section I., but less robust, and shape of fore-wings as in Section II.

♂ *White, with moderate-sized dull yellowish-ochreous apical patch, slightly glossed with violet, bordered outwardly with brown and inwardly with black; an intensely black inner-marginal band in fore-wings and narrower costal band in hind-wings:* under side white or yellowish-white. ♀ *White, yellowish-white, or sulphur-yellow; pale part of apical patch reduced to some separate elongate whitish or yellowish spots;* inner-marginal band of fore-wings much narrower and duller; costal band of hind-wings obsolete; under side yellow-ochreous.

3 Species.—*Eris*, Klug; *Fatma*, Feld.; *Coliagenes*, Butl.

Range.—All Africa (except Extra-Tropical North Africa west of Egypt) and Arabia.

---

[1] This fringe of hairs (which occurs in both sexes) is quite peculiar to *T. subfasciatus*, no other species in the entire genus possessing it. This character recurs, however, in *Eronia Leda*.

SECTION IV.—Representative: *Ione*, Godt.

Structure of Sections I. and III., but *fore-wings apically not so prominent as in Sections II. and III., nor so acute at apex itself as in Section I.;* and hind-margin in ♀ inclining to convexity; *club of antennæ less abruptly formed and not so broad at tip;* costa of hind-wings not much arched, discoidal cell less than half their length.[1]

♂ *White, with glittering-violet apical patch, blackish-bordered externally* (except in *Eunoma*, Hopff.) *and internally* (except in *Eunoma* and *Hetæra*, Gerst.); *neuration generally black in parts;* under side white or pale-yellowish, with faint traces of a dark discal ray in hind-wings. ♀ *White, yellowish-white, or sulphur-yellow; apical patch violet or orange-red, broadly black-bordered and traversed by black spots, or black with a median row of white spots;* bases more or less clouded with blackish; terminal parts of longitudinal blackish bands usually well-marked in both fore and hind wings; under side yellower or redder than in ♂, with a more or less complete elbowed discal dark ray in hind-wings.

This Section contains (with Section V.) the largest species of the Genus.

7 Species.—*Ione*, Godt.; *Speciosa*, Wallengr.; *Jobina*, Butl.; *Phlegyas*, Butl.; *Regina*, Trim.; *Hetæra*, Gerst.; *Eunoma*, Hopff.

Range.—Tropical Africa and Extra-Tropical South Africa as far as 30° lat.

[SECTION V.—Representative: *Mananhari*, Ward.

Structure of Section IV., but fore-wings with apex more rounded, and discoidal cell of hind-wings about half their length.

♂ *Sulphur-yellow, with rather narrow plain black apical patch.* ♀ *Yellowish-white, with rather wide black hind-marginal border to both fore and hind wings.* Under side yellow, marked much as in *Speciosa* (Sect. IV.).

1 Species.

Range.—Madagascar.]

[SECTION VI.—Representative: *Halimede*, Klug.

Structure as in Section IV., but hind-wings with discoidal cell half their length.

White, with an apical blackish border, but no patch; *an ochreous-yellow or orange-yellow suffusion over inner-marginal (and sometimes also disco-cellular) area of fore-wings and costal (and sometimes disco-cellular and lower discal area of hind-wings;* usually a basal grey patch in fore-wings. ♀ *with a common discal series of black spots, larger in fore-wings.* Under side white or whitish, with discal series of spots very small in ♂ and small in ♀.

---

[1] In *T. Regina*, Trim., the lower radial nervule of fore-wings originates much nearer to the upper one, so that the upper disco-cellular nervule is almost as short as in Section xiv.

4 Species.—*Halimede*, Klug; *Cœlestis*, Swinhoe; *Heliocaustus*, Butl.; *Pleione*, Klug.

Range.—Eastern North-Tropical Africa and Southern Arabia.]

[SECTION VII.—Representative: *Fausta*, Oliv.

Neuration of wings generally and strong costal arch of hind-wings as in Section I., but discoidal cell of hind-wings less than half their length. *Fore-wings apically as in Section IV.*, but (in ♂) *inner margin prominently convex from base to a little beyond middle, and immediately above the sinuated submedian nervure a narrow elongate sac before middle. Reddish-isabelline or creamy salmon-red, with incomplete concolorous apical patch*, bordered on both sides with black, *the outer (hind-marginal) border with spots of the ground-colour.* Under side very pale reddish-creamy or yellowish-creamy, deeper towards bases; ♂ almost without markings, ♀ with ill-defined macular discal dark-grey ray.

4 Species.—*Fausta*, Oliv.; *Faustina*, Feld.; *Fulvia*, Wall.; *Vi*, Swinhoe.

Range.—Syria, Arabia, N.W., W., and S. India.

(Part of Boisduval's Genus *Idmais*—" Groupe I.")]

[SECTION VIII.—Representative: *Amata*, Fab.

Fore-wings elongate, slightly prominent apically, apex itself not acute. Neuration ordinary, but *hind-wings with second subcostal nervule originating only just before extremity of discoidal cell*, which is rather less than half the length of the wing. *Creamy salmon-red or salmon-red and white (♂), yellowish or white*—sometimes basally flushed with salmon-red—(♀), *with broad black hind-marginal border enclosing one row or two rows of spots of the ground-colour;* disco-cellular terminal black spot in fore-wings much enlarged, usually united to base by a costal or subcostal blackish stripe. Under side resembling that of genus *Colias, greenish-yellow or greenish*, the borders scarcely indicated by pale-greyish; a discal row of fuscous spots, mostly indistinct in ♂, but better defined and tinged with red in ♀.

9 Species.—*Amata*, Fab.; *Calais*, Cram.; *Phisadia*, Godt.; *Dynamene*, Klug; *Vestalis*, Butl.; *Rorus*, Swinhoe; *Pidus*, Swinh.; *Puellaris*, Butl.; *Castalis*, Staud.

Range.—Tropical Africa, Southern Arabia, Madagascar, N.W., W., and S. India and Ceylon.

(Part of Boisduval's Genus *Idmais*—" Groupe II.")]

SECTION IX.—Representative: *Danaë*, Fab.

Structure and neuration of Section IV.; but *hind-wings with discoidal cell in ♂ half, in ♀ less than half, their length.* ♂ *White, with very large vivid crimson-red (sometimes violet-shot) apical patch, inwardly with a broad black border; bases often suffused with grey;* under side white, yellowish, or reddish-creamy, with *a common discal series of dark (often sub-ocellate) spots.* ♀ *White or some shade of yellow with all the blackish markings (especially the basal suffusion) enlarged; coloured portion of apical patch usually paler and duller, intersected by a macular blackish ray, and sometimes reduced to a mere row of dull-reddish or even ochrey-whitish rays in the middle of the black;* under side deeper in tint than in ♂,—the discal spots larger.

11 Species.—*Danaë,* Fab.; *Eupompe,* Klug; *Antenpompe,* Feld.; *Dedecora,* Feld.; *Annæ,* Wallengr.; *Wallengrenii,* Butl.; *Dulcis,* Butl.; *Subroseus,* Swinh.; *Taplini,* Swinh.; *Miles,* Butl.; *Walkeri,* Butl.

Range.—Tropical Africa and Extra-Tropical South Africa, Arabia, India, and Ceylon.

SECTION X.—Representative: *Eucharis,* Fab.

Fore-wings shaped as in Section IV., &c., but *first and second subcostal nervules not so close together, and third much shorter,—ending on costa at some little distance before apex.* Hind-wings (in both sexes) with discoidal cell less than half their length; second subcostal nervule originating not far before extremity of discoidal cell.

White or yellowish-white or sulphur-yellow, with large broad orange apical patch, outwardly (and sometimes very faintly inwardly) black-bordered; patch paler in ♀ and traversed by the upper part of a common discal series of blackish spots. UNDER SIDE.—Yellow or reddish, with discal markings indistinctly reproduced in ♀.

6 Species.—*Eucharis,* Fab.; *Evarne,* Klug; *Auxo,* Lucas; *Topha,* Wallengr.; *Citreus,* Butl.; *Liagore,* Klug (excl. " ♀ ").

Range.—Tropical Africa and Extra-Tropical South Africa. India, and Ceylon.

[SECTION XI.—Representative: *Etrida,* Boisd.

Structure not examined; apparently as in Section X.

White, with small orange apical patch, widely black-bordered both externally and internally, especially in ♀, which has bases rather widely suffused with blackish. UNDER SIDE.—White with yellow borders, or yellowish generally, with a discal series of blackish or brownish spots, not well-marked in ♂, but better developed in ♀.

4 Species.—*Etrida,* Boisd.; *Pernotatus,* Butl.; *Casimirus,* Butl.; *Bimbura,* Moore.

Range.—India.]

[SECTION XII.—Representative: *Evanthe,* Boisd.

Structure slight, wings very thin. Fore-wings with rounded apex and rather convex hind-margin. Hind-wings with second subcostal nervule originating at extremity of upper part of discoidal cell, which is less than half the length of the wing.

White, with a moderate-sized apical patch,—in the ♂ orange with a narrow outer black border and a broad diffused inner one (interrupted between subcostal nervure and lower radial nervule),—in the ♀ blackish with a dull ochre-orange central bar or wholly blackish. UNDER SIDE.—White or yellowish, markedly striolated with gray.

2 Species.—*Evanthe*, Boisd. ; *Pseudevanthe*, Butl.

Range.—Madagascar and Western India.]

SECTION XIII.—Representative : *Evenina*, Wallengr.

Fore-wings shaped as in Section IV., but neuration of hind-wings as in Section XII.

♂ *White, with a rather large orange apical patch, very narrowly black-bordered externally, but diffusedly or not at all internally; an ill-defined greyish and black inner-marginal stripe in fore-wings, and a blackish costal stripe in hind-wings.* ♀ *with orange patch paler, traversed by a submacular black streak; irregular basal blackish clouding, in fore-wings extending from base to extremity of discoidal cell and beyond and beneath it, in hind-wings not reaching to middle.* Under side varying from white to creamy or pinkish-creamy; in hind-wings a central dusky longitudinal ray (sometimes obsolete).

1 Species.—*Evenina*, Wallengr.

Range.—South-Tropical Africa and Extra-Tropical South Africa as far as 30° lat.

SECTION XIV.—Representative : *Daira*, Klug.

Fore-wings elongate, in ♂ apically prominent, in ♀ with convex hind-margin ; first and second subcostal nervules more widely separated ; upper radial nervule originating at upper acute angle of extremity of discoidal cell, almost as in Section II.; lower radial nervule originating very near subcostal nervure, so that upper disco-cellular nervule is very short. Hind-wings with *second subcostal nervule originating slightly beyond extremity of discoidal cell, which is very short,—much less than half length of wing.*

♂ *White, with good-sized orange apical patch, outwardly black-bordered, inwardly without any black or with a black mark or spot inferiorly: longitudinal black stripes in both wings very unstable,* wholly wanting in some species, but in others very broadly developed,—in hind-wings lower portion of black discal ray often more or less united with hind-marginal black marks. ♀ *with apical patch usually traversed or inwardly bordered by a black macular ray,*—in other respects very like ♂. Under side white or whitish, more or less varied with yellow or reddish-creamy,—usually with a very faint reproduction of the upper-side dark markings in the hind-wings.

11 Species.—*Daira*, Klug ; *Thruppi*, Butl. ; *Verburii*, Swinh. ; *Xanthus*, Swinh. ; *Antigone*, Boisd. ; *Phlegetonia*, Boisd. ; *Odysseus*, Swinh. ; *Sarcus*, Swinh. ; *Nouna*, Lucas ; *Eragore*, Klug ; *Demagore*, Feld.

Range.—Whole of Africa, Southern Arabia.

SECTION XV.—Representative : *Achine*, Cram.

Fore-wings shaped as in Section IV., but first and second subcostal nervules farther apart ; *hind-wings with second subcostal nervule origi-*

nating at upper part of extremity of discoidal cell, which is less than half length of wing.

♂ White, with bright-red, rosy-glossed, moderate-sized apical patch, bordered outwardly by black, which radiates broadly on nervules, and inwardly by narrower black (sometimes evanescent or obsolete); ordinary longitudinal blackish stripes on inner margin of fore-wings and costa of hind-wings usually present, moderately developed. ♀ with red of apical patch duller, much more widely black-bordered outwardly, and traversed or inwardly bordered by a broad black ray; longitudinal stripes (especially that on inner margin of fore-wings) very much broader, and hind-marginal border and discal ray often widened so as to meet. Under side white, whitish, or tinged with creamy-yellow or creamy-reddish, traversed in ♀ by a common blackish or brownish discal ray; neuration sometimes black.

6 Species.—*Achine*, Cram.; *Garisa*, Wallengr.; *Anterippe*, Boisd.; *Simplex*, Butl.; *Lais*, Butl.; *Halyattes*, Butl.

Range.—Tropical Africa and Extra-Tropical South Africa.

SECTION XVI.—Representative: *Evippe*, Linn.

Like Group XV., but fore-wings with hind-margin in ♂ convex, in ♀ very convex or almost bluntly elbowed about third and second median nervules.

♂ White, with bright-red (not rosy-glossed) curved apical patch, outwardly bordered broadly by black, which does not deeply or broadly radiate on nervules, and inwardly by black varying from very broad to very narrow; black longitudinal stripes sometimes wanting, sometimes moderately developed, or very heavily marked; discal ray (lower part) of hind-wings equally variable. ♀ White, yellowish, or yellow, with apical red much duller, narrower, more curved,—sometimes wanting, being replaced by black; longitudinal black stripes (as well as discal ray in hind-wings) always present, sometimes exceedingly broad. Under side white, yellowish, or faint pinkish-creamy; discal ray usually well-marked, especially in ♀.

5 Species.—*Evippe*, Linn.; *Omphale*, Godt.; *Theogone*, Boisd.; *Loandicus*, Butl.; *Microcale*, Butl.

Range.—Tropical Africa and Extra-Tropical South Africa.

SECTION XVII.—Representative: *Celimene*, Lucas.

Fore-wings in ♂ apically considerably produced; apex itself moderately acute; hind-margin superiorly convex and inferiorly concave; hind-wings with anal angular portion slightly produced. ♀ with these characters much less pronounced. Neuration as in Section XIV., except that in hind-wings second subcostal nervule originates a little before extremity of discoidal cell.

♂ White, with very large purplish-lake, violet-glossed apical patch, bordered outwardly and inwardly with black, traversed by a black ray, and crossed by black nervules; hind-margin of hind-wings spotted or bordered with black. ♀ *with apical patch black (sometimes partly marked very faintly with purplish), with a central row of luke-red or white spots.* Under side creamy or pale-yellow, with neuration beyond basal area more or less ferruginous-red ; apical patch inwardly rose-red, outwardly like hind-wings ; base of fore-wings with a red or orange suffusion.

4 Species.—*Celimene*, Lucas ; *Pholoë*, Wallengr. ; *Zoë*, Grandidier ; *Praeclarus*, Butl.

Range.—Tropical Africa ; Extra-Tropical South Africa as far as 27° lat. ; Madagascar.

[SECTION XVIII.—Representative : *Eulimene*, Klug.

Fore-wings moderately produced apically ; apex itself rather rounded.

White, with rather small ochreous-yellow (in ♂ shot with pink) apical patch, outwardly very narrowly, inwardly broadly, black bordered ; terminal disco-cellular black spot very large. *Under side white, with neuration mostly orange-red ;* hind-margins tinged with sulphur-yellow ; *an orange-red longitudinal stripe along costa of fore-wings ;* a discal series of black spots, larger in ♀. Sexes very similar ; all black markings stronger in ♀.

[Characters taken from Klug's figures.]

1 Species.—*Eulimene*, Klug.

Range.—Eastern North-Tropical Africa.]

SECTION XIX.—Representative : *Chrysonome*, Klug.

Fore-wings as in Section IV., but hind-wings with second subcostal nervule originating at, or only a little distance before, extremity of discoidal cell, which is considerably less than half length of wing.

Creamy ochreous-yellow with white bases, black hind-marginal borders spotted with ground-colour, and a sinuated black discal ray in fore-wings (sometimes in hind-wings also) ; or sulphur-yellow with black apical patch and hind-marginal borders similarly spotted. Under side whitish *or sulphur-yellow (field of fore-wings sometimes deep ochreous-yellow), with more or less broad ochreous-ferruginous neuration and transverse (sometimes submacular bars of the same colour.* Sexes very similar.

8 Species.—*Chrysonome*, Klug ; *Vesta*, Reiche ; *Velleda*, Lucas ; *Amelia*, Lucas ; *Gaudens*, Butl. ; *Aurigineus*, Butl. ; *Hanningtoni*, Butl. ; *Protomedia*, Klug.

Range.—All Africa except Extra-Tropical Northern to west of Egypt, and Extra-Tropical Southern to south of 30° lat.

It will be seen from the above analysis that twelve of the nineteen sections are represented in South Africa, and the same number in North-Tropical Africa, while South-Tropical Africa is richer than both by one section. Twelve sections are peculiar to Africa, including

Arabia, to which four of them extend; while three others are common to Africa, Arabia, and India. One section (*Etrida* group) is peculiar to India, and one (*Mananhari* group) to Madagascar; while another (*Eranthe* group) is common and peculiar to those two countries. India possesses representatives of six sections and Madagascar of four. Of the ninety species here enumerated, fifty-five appear to be peculiar to Continental Africa, or sixty-nine to the entire Ethiopian Region, including Arabia and Madagascar, which latter has yielded four endemic forms, while nine appear to be limited to Arabia. In South-Africa I recognise twenty-nine species, of which five only are not recorded as occurring beyond the tropical limit.

In these statements it must be noted that the numbers can only be regarded as generally indicative of the actual distribution. There is, perhaps, no genus of butterflies more puzzling to deal with than *Teracolus*, owing mainly to the multitude of closely-allied forms, the disparity in pattern and coloration exhibited by the sexes, and the instability of colouring and markings in the females. It is probable that no two lepidopterists would even approach agreement in discriminating the known species, and the mass of the genus must remain in a very unsatisfactory state until careful breeding of successive generations from the ova can be systematically applied to its elucidation. Even in South-Africa, as far as I can learn, only two or three species have been reared from the larvæ; but some little aid has been afforded by the record of the capture of the sexes *in copulâ*.

In Mr. Butler's "Revision" in 1876, forty-one South-African species were enumerated, and of these nineteen were for the first time described. I have had the advantage of examining the types of these species in the very large series of *Teracolus* contained in the collection of the British Museum, and comparing them with my own series from South Africa, and arrived at the conclusion that only five of them presented characters warranting specific separation from previously described forms.

The beauty of most of these butterflies is very remarkable, especially in the ♂s, where the brilliantly-tinted tips of the fore-wings, usually relieved by black edging, contrasts with a pure-white or pale-yellow field. The *Ione* group perhaps carries off the palm of loveliness, the lustre of the glittering violet tips (in *T. Regina* quite metallic) being unequalled in any other group, though the intense rich crimson of the tips of *T. Danaë* and its near allies is almost equally splendid. Bright-red, orange-red, orange, and yellow of various shades are the colours that ornament the wing-tips of most of the groups, while the *Célimène* section has an immense purple patch of a lustre inferior only to that in the *Ione* section.

As will be seen, however, in the account of the characters of the several sections given above, there are five groups, viz., those of *Mananhari*, *Halimede*, *Fausta*, *Amata*, and *Chrysonome*, in which the apex of the fore-wing is not rendered conspicuous by any bright contrasting colour. All these groups except the first are specially prevalent in the barren or even desert tracts of Eastern North-Tropical Africa, Southern Arabia, and Western India; but it must at the same time be noted that a good many of the brightly-tipped species of other groups inhabit the same countries.

The under side, where not white, is singularly colourless, apart from various tints more or less approaching the isabelline; and it is only in the *Celimene*, *Eulimene*, and *Chrysonome* sections that a more lively coloration, characterised by red neuration, is found.

In South Africa, undoubtedly the tract most productive of this genus is that lying between the tropical boundary and 30° lat.; indeed, the *Subfasciatus*, *Agoye*, *Ione*, *Evenina*, and *Vesta* groups do not appear to range any farther to the southward, and the *Celimene* group stops short about lat. 28°. As far as my records go, all the twelve South-African groups are represented in Transvaal (eighteen species known); nine in Cape Colony, including the territories of Griqualand West and Basutoland (seventeen species); eight in Natal (eighteen species); eight in Swaziland (thirteen species); eight in Delagoa Bay (twelve species); and six in Kaffraria Proper (eleven species). The comparative richness of the Cape Colony is due to its receiving several interior forms along its north-western and northern border, viz., *Subfasciatus*, *Evenina*, *Microcale*, and *Lais*, and possessing one (*Bowkeri*) peculiar to that tract.

Besides the last-named butterfly, only four others seem to be peculiar to South Africa, viz., *Speciosa*, Wallengr.; *Jobina*, Butl.; *Halyattes*, Butl.; and *Topha*, Wallengr.; and it is not improbable that further exploration will prove that even these few also inhabit the tropical region.

In the Cape Colony *Teracolus* thins out westward, only two species —small varieties of *Evenina*, Wallengr., and *Omphale*, Godt.—having been taken in Little Namaqualand, and the latter form in the south-west extending as far as Robertson. The Knysna district, besides the variety of *Omphale*, has hitherto yielded only *Achine*, Cram., and *Antevippe*, Boisd.; but the eastern districts are far more productive, possessing, in addition to the three last-named species, nine others, viz., *Eris*, Klug; *Annæ*, Wallengr.; *Wallengrenii*, Butl.; *Auxo*, Lucas; *Topha*, Wallengr.; *Gavisa*, Wallengr.; *Theogone*, Boisd.; *Phlegetonia*, Boisd.; and *Antigone*, Boisd. All the species that I have seen in life flit actively along the edges or about the outskirts of woods, with the exception of *Eris*, which flies with great rapidity over more open country. Even swifter than this last is (I learn from Colonel Bowker) the beautiful *Subfasciatus*—the type of the genus—which occurs in Griqualand West.

265. (1.) **Teracolus subfasciatus**, Swainson.

♂ *Teracolus subfasciatus*, Swains., Zool. Illustr., 2nd Ser., iii. pl. 115 (1833).
♂ *Anthocharis subfasciata*, Boisd., Sp. Gen. Lep., i. p. 567, n. 12 (1836).
♂ ♀ *Ptychopteryx Bohemani*, Wallengr., K. Sv. Vet.-Akad. Handl., 1857; Lep. Rhop. Caffr., p. 18.
Thespia Bohemanni, Wallengr., K. Vet.-Akad. Förh., 1858, p. 77.
♂ ♀ *Anthocharis subfasciata*, Trim., Rhop. Afr. Aust., i. p. 58, n. 38 (1862); and ii. p. 331 (1866).
Ptychopteryx Bohemani, Butl., Lep. Exot., p. 45 (1870).
♂ *Teracolus subfasciatus*, Staud., Exot. Schmett., i. pl. 23 (1884).

Exp al., (♂) 2 in. 1–3½ lin.; (♀) 2 in. 1–4½ lin.

♂ *Sulphur-yellow with deeper, somewhat ochre-tinged yellow apical patch.* Fore-wing: a small, elongate, black disco-cellular spot; costa narrowly black-edged from near base, paling and narrowing into a dull-fuscous apical border, which, outwardly bordering ochre-tinged patch, extends along hind-margin nearly to end of first median nervule; apical patch small, of the ground-colour, but suffused with greyish-ochreous, crossed by four fuscous nervules,—*inwardly bordered by a broad black bar from costa*, ending abruptly on or a little above third median nervule, where a few dusky scales (in some specimens) form an indistinct line between it and fuscous border. *Hind-wing: paler;* a fuscous indistinct spot on costa beyond middle. UNDER SIDE.— *Greenish-white; hind-wing and apex of fore-wing hatched closely with lines of ochrey-grey.* Fore-wing: cellular spot wanting; bar from costa faintly showing. *Hind-wing:* sometimes a thin, reddish line on edge of costa near base; hatching denser along costa, forming a dusky border; *cellular fold, from base to hind-margin, inferiorly marked by an ochrey-grey stripe*.

♀ *Much paler, or sometimes nearly white; apical patch brighter or duller orange.* Fore-wing: cellular spot sometimes ill-defined; fuscous border broader, tinged with ferruginous; costal edging reddish-grey; bar from costa reduced to a narrower fuscous submacular stripe, within which the orange sometimes more or less extends; along inner margin some faint, dusky irroration, very rarely developed into a rather wide longitudinal border. *Hind-wing as in ♂.* UNDER SIDE.—Similar; hatchings darker. *Fore-wing:* apical orange faintly indicated. *Hind-wing:* costal hatching and cellular fold-streak more strongly marked.

The ♂ is remarkably constant in colouring and marking, but the ♀, in addition to the variable features just mentioned, occasionally (two examples from Lydenburg District in Transvaal, and one from Makloutse River, north of Bamangwato) has the under side of the hind-wings and of the apical area of the fore-wings tinged with delicate creamy-pinkish. In these three ♀s the apical orange of the upper side of the fore-wings is brighter than usual, and extends before the reduced bar on its inner side.

On the convex part of the costa of the hind-wings there is (as

Wallengren has pointed out) a fringe of moderately long, delicate, silky-white hairs.

This very distinct and handsome species was discovered by Burchell in 1812, towards the northern limit of his South-African journeyings. Through Professor Westwood's kindness I was able to examine the original specimens, and to refer to Burchell's manuscript list of localities of the species in his collection. It appeared from the latter that three of the four examples were captured at "Chue Spring," and the fourth at "Little Klip." Burchell gives the latitude and longitude of the first of these localities, and I am thus able to determine its position on recent maps as about Honing Vley, in British Bechuanaland. Specimens from Motito, in the same tract of country, were sent to me in 1866.

The range of *Subfasciatus* lies between 20° and 30° S. lat., and it is known to occur from Damaraland on the Western Coast as far east as the Tati River in Matabeleland. Colonel Bowker took a good many specimens on the Vaal River, Griqualand West, and also sent a single example captured at Hope Town, on the left bank of the Orange River. The latter is its most southern locality known to me; and it certainly appears to be most numerous in the tropical portion of its range, Mr. J. A. Bell having brought no fewer than thirty-four specimens in his small collection formed in Damaraland. Colonel Bowker describes the butterfly as a swift flyer; he found it on the wing at Klipdrift (Barkly) in March and April, and took the Hope Town individual on 1st May 1871. Mr. H. L. L. Feltham informs me that he occasionally sees specimens in Kimberley, and took some in the month of December of the years 1884 and 1885. Wallengren records [1] Mr. Person's note that in Southern Transvaal the butterfly occurs in March and April.

Localities of *Teracolus subfasciatus*.

I. South Africa.
   B. Cape Colony.
      b. Eastern Districts.—Hope Town (*J. H. Bowker*).
      c. Griqualand West.—Klipdrift (Barkly) (*J. H. Bowker*). "Kimberley."—H. L. Feltham.
   K. Transvaal.—Potchefstroom (*W. Morant* and *T. Ayres*). Pretoria (*T. Ayres*). Limpopo River (*F. C. Selous*). Lydenburg District (*T. Ayres*).
   L. Bechuanaland.—Chue Spring (the late *W. J. Burchell*). Motito (the late *Rev. J. Frédoux*).

II. Other African Regions.
   A. South Tropical.
      a. Western Coast.—Damaraland (*J. A. Bell*).
      b1. Eastern Interior.—Makloutse and Tati Rivers (*F. C. Selous*). Matabeleland (*H. Barber*).

266. (2.) Teracolus Eris, Klug.

   ♂ *Pontia Eris*, Klug, Symb. Phys., t. vi. ff. 15, 16 (1829).
   ♂ ♀ *Pieris Eris*, Boisd., Sp. Gen. Lep., i. p. 514. n. 111 (1836).
   ♂ ♀ *Anthocharis Eris*, Reiche, Ferr. et Gal. Voy. Abyss., iii. pl. 460, p. 31, ff. 1-3 (1849).

---

[1] K. Vet.-Akad. Förh., 1875, p. 91.

♂ ♀ *Anthocharis Eris*, Trimen, Rhop. Afr. Aust., i. p. 59, n. 39 (1862).
♂ ♀ „ „ Hopff., Peters' Reise Mossamb., Ins., p. 356 (1862).
♀ *Teracolus Abyssinicus*, Butl., Ann. and Mag. Nat. Hist., 4th Ser., xviii. p. 486 (1876).
♂ ♀ *Idmais Eris* and *I. Maimuna*, Kirby, Proc. R. Dubl. Soc., 1880, pp. 46, 47.
♂ *Idmais Eris*, Staud., Exot. Schmett., i. pl. 23 (1884).

Exp. al., (♂) 1 in. 11 lin.—2 in. 1 lin.; ♀ 1 in. 11 lin.—2 in. 1 lin.

♂ *White, with dull yellowish-ochreous more or less violet-glossed apical patch. Fore-wing:* costa dusted with blackish; in some specimens a small black spot at extremity of discoidal cell; ochreous at apex divided into five elongate marks by the brown-clouded nervules, externally bordered by a brown violet-glossed edging, internally by a black stripe; the latter stripe becomes merged on third median nervule with a *glossy, deep black, inner-marginal band*, widening from base, and becoming very broad and *tending upward* beyond first median nervule, but extending very close or *quite to anal angle* and lower portion of hind-margin. *Hind-wing:* a duller black band, of considerable width, covering basal fourth or third of discoidal cell, and bounded inferiorly by second subcostal nervule, extends along near costa, ending suddenly, *with a concave excavation*, just before apex; in some specimens small black spots along hind-margin at ends of nervules, as far as second or first median nervule; blackish clouding from base extending a little between median and submedian nervures. UNDER SIDE.—*Whitish, or yellowish-white. Fore-wing:* apex faintly tinted with *yellow*,—both it and inner-marginal band being indicated by a faint greyish tinge; between first and third median nervules, not far from hind-margin, are two rounded blackish spots (in some specimens a third smaller spot below them, immediately beneath first median nervule). *Hind-wing:* slightly more yellowish than fore-wing; *spots* on costa edged with chrome-yellow near base.

♀ *Not as white, often more or less tinged with yellow, rarely pale lemon-yellow. Fore-wing: apical patch less distinct, narrower, macular, usually whitish, without violaceous gloss, sometimes tinged with ochre-yellow*, its outer margin rusty-brownish, its inner border blackish; the latter stripe continued by two blackish spots to first median nervule, where the lowermost of the two spots almost touches extremity of inner-marginal band, which is much narrower, duller, and shorter than in ♂, leaving anal angular area wholly white; disco-cellular spot larger, rounder. *Hind-wing:* beyond middle, traces of a *blackish, macular, angulated stripe*, only distinct at commencement on costa beyond middle; base rather widely dusted with fuscous. UNDER SIDE.—*Fore-wing:* tinged with pale lemon-yellow; apex pale yellow-ochreous, bordered inwardly as on *upper side*, but very faintly; below last blackish spot and first median nervule, a faint blackish mark defines extremity of inner-marginal band. *Hind-wing:* pale yellow-ochreous, a small brownish disco-cellular spot; a dark mark on costa, beyond middle, commences an indistinct, transverse, angulated row of brownish spots.

VARIETY A. (♂ and ♀).—♂ not differing on upper side; ♀ with the black markings usually more or less reduced, and with the apical patch of the fore-wing outwardly much suffused with dull-rusty brownish. UNDER SIDE.—*Hind-wing and apical area of fore-wing, in both sexes, pale dull creamy-reddish.*

(*Hab.*—Eastern Districts of Cape Colony, Natal, Transvaal, and Eastern South-Tropical Interior.)[1]

Examples intermediate between the variety and typical *Eris* occur, more particularly in the ♀, in which the under side presents some paler or more decided tinge of reddish. The most pronounced of these are three unusually small specimens, two ♂s and a ♀, brought from Damaraland by the late Mr. C. J. Andersson (*exp. al.* ♂ 1 in. 7½ lin. and 8½ lin. respectively; ♀ 1 in. 9½ lin.), but rather larger examples taken by Mr. A. W. Eriksson in the North-West Transvaal are almost as decidedly tinged, and so are two ♂s from the Albany district of the Cape Colony, and one from the Trans-Kei territory. Three ♂s captured by Mr. John L. Fry on the Maklontse River, North Bamangwato country, on the 20th May 1887, are all differently tinted beneath, one being slightly yellowish, another slightly tinged with reddish-brown, and the third dull brownish-creamy. The last mentioned has much the darkest under-side that I have seen in this species.

As will be seen from the above description, *Eris* is a decidedly variable species in both sexes, but especially in the ♀. The Dongolan type, as figured by Klug,[2] is a ♂ having the black longitudinal bands broader than in any South-African examples that I have seen; the band of the fore-wings leaving no trace of the white spot close to hind-margin between second and third median nervules, and that of the hind-wings at its extremity projecting downward very considerably beyond second subcostal nervule. The fore-wing band in South-African specimens usually leaves, besides the white spot just mentioned[3] (which is, however, very small in some Transvaal examples and a Delagoa Bay specimen, and is only just perceptible in one from Natal), two more or less apparent white marks on the hind-margin between the spot in question and the posterior angle; but these vary to a mere

---

[1] A ♀ of the variety from Grahamstown is labelled in the British Museum collection (September 1886) " *T. Johnstoni*, Butler;" but I am not aware that any description of it has been published.

[2] Reiche (*op. cit.*) figures an Abyssinian ♂ very like Klug's, but with the apical patch of the fore-wings darker and of a redder tinge, and a ♀ rather yellowish on the upper side and inclining to argillaceous on the under side.

[3] After examining the Angolan specimens of Mr. Kirby's *Maimuna* in the British Museum, and carefully considering his description (*loc. cit.*), I am unable to regard his new species as a recognisable one. The ♂ in the British Museum has the white spot in the hind-marginal border of the fore-wings well marked, and, like the ♀, agrees with the majority of South-African examples. The same remark applies to a ♂ and ♀ in the same collection from Victoria Nyanza, referred to *Maimuna* by Mr. A. G. Butler (*Ann. and Mag. Nat. Hist.*, xii. p. 101, 1883).

trace, and in one Delagoan and two Natalian specimens are totally wanting. The hind-wing band is almost always bounded inferiorly by the second subcostal nervule to its termination, but the terminal portion in one Transvaal specimen projects very slightly, and in a Natalian individual considerably below the nervule. The under side of Klug's figure is white (as in a good many South-African examples), but the dark bands of the upper side are depicted as showing more plainly through the wings than in any specimens which have come under my notice.

As regards the nervular hind-marginal spots on the upper side of the hind-wings, which are distinct but linear in Klug's figure, I find that, while it is unusual for them to occur in South-African specimens, they are found in those examples which more nearly approach Klug's type (in one Delagoa Bay specimen they are rather large and conspicuous). On the other hand, the terminal disco-cellular spot of the fore-wings (which is wanting in Klug's type and in the more heavily-banded South-African examples) is most developed in specimens which bear the largest hind-marginal white markings on the black band.

The ♀, besides varying in ground-colour,[1] exhibits great diversity in the development of the black band and of the apical hind-marginal border of the fore-wings, and in the colouring and pattern of the apical patch. In one from Natal, the inner-marginal band is scarcely narrower (except for a very large white spot at the posterior angle) than in the more lightly marked ♂ s; while the opposite extreme is met with in an example I captured near Grahamstown, where the band is exceedingly narrow, and the usual black connection with the apical patch is only represented by two quite separate very small blackish spots.[2] In this example the apical patch is externally broadly tinted with pale dull ferruginous, and the pale enclosed spots are almost obsolete, and it thus on the upper side approaches the ♀ of Variety A., above described. Reiche's figure of an Abyssinian ♀ depicts the under side as inclining to argillaceous.

In the British Museum collection I noted (October 1886) a remarkably large ♀ labelled *T. opalescens*, Butler, and ticketed "Delagoa Bay." This example is white, with the inner-marginal black of the fore-wings well developed, and the hind-marginal border rather broader than usual,—the white spots in the latter (with the exception of the two largest) being almost obsolete; in the hind-wings both the

---

[1] The pale lemon-yellow coloured ♀ seems to be rare (its occurrence was noted by Boisduval in 1836). I have seen only three specimens, respectively from Kaffraria Proper, Natal, and Angola, and a coloured photograph of a fourth example which was taken by the late Mr. E. C. Buxton either in Natal or Swaziland. The Natal specimen, taken by Mr. J. M. Hutchinson in Weenen County, belongs to Var. A., having the under-side reddish very pronounced.

[2] Felder's *Idmais Fatma*, from Kordofan (*Reise du Novara*, Lep., ii. p. 189, t. xxv. f. 3, 1865), which I have not seen, may perhaps be an unusually small and faintly marked ♀ of this kind; the description and figure show that the inner-marginal band is wholly wanting

discal and hind-marginal fuscous spots are well developed. The under side of this ♀ is pale-yellowish, with the usual spots pronounced. Nothing but its large size seemed to warrant its separation from the variable series of *Eris* ♀.

The ♂ of this species presents a very striking and peculiar appearance, owing to the intense highly-glossy black (and great width exteriorly) of the inner-marginal band of the fore-wings, and also to the unusual colouring of the apical patch, which combines to a considerable extent the characters of the same marking in *T. Subfasciatus* and *T. Ione*, Godt. The ♀ also has an aspect unlike that of the same sex of any of its congeners, but shows some approach both to *Ione* and to the desert group of which *T. Fausta*, Oliv., may be regarded as the type.

I know of no part of South Africa where this widely-ranging African species is at all numerous. I saw a few examples in the "Thorn" country near Greytown, in Natal; they were ♂ s, and flew with great rapidity. This was in March 1867, and the only other opportunity I had of seeing the butterfly in life was in January and February 1870, in the Albany district of the Cape Colony, where I captured one of each sex. Mr. W. Morant sent me some specimens from near Potchefstroom, with the note that they were taken on 25th February about a stony kopje, and flew very strongly. Mr. D'Urban found the species near King William's Town in March and April.

Like *T. Subfasciatus*, this butterfly seems to be more numerous in the Tropical parts of Africa, but it has also an immensely wider distribution northwards, having been recorded even from Cairo. There were fifteen specimens in the small collection made in Damaraland by Mr. J. A. Bell. As Gerstäcker remarks, the range of this species and of *H. Eriphia* appears to be about the same, but I do not know of the occurrence of the latter to the north of Dongola.

Localities of *Teracolus Eris*.

I. South Africa.
  B. Cape Colony.
    b. Eastern Districts.—Grahamstown and Zwaartwater Kloof. Albany District (Sub-typ. and Var. A.). Tharfield, Bathurst District (*Miss M. L. Bowker*—Sub-typ.). Uitenhage (*S. D. Bairstow*—Var. A.). Bedford (*J. P. Mansel Weale*—Var. A.). Seymour, Stockenstrom District (*W. C. Scully*—Sub-typ.). King William's Town (*W. S. M. D'Urban*—Sub-typ. and Var. A.). Windvogelberg, Queenstown District (*Mrs. Barber*).
    c. Griqualand West.—Vaal River (*J. H. Bowker*—Sub-typ.).
  D. Kaffraria Proper.—Bashee River (*J. H. Bowker*—Typ.).
  E. Natal.
    b. Upper Districts.—Greytown. Weenen County (*J. M. Hutchinson*—Typ. and Var. A.). Between Tugela and Mooi Rivers (*J. H. Bowker*).
  G. "Swaziland."—E. C. Buxton.
  H. Delagoa Bay.—Lourenço Marques (*Mrs. Monteiro*).
  K. Transvaal.—Potchefstroom (*W. Morant* and *T. Ayres*). Marico and Limpopo Rivers (*F. C. Selous* and *A. W. Eriksson*—Typ. and Var. A.). Origstadt Valley, Lydenburg District (*T. Ayres*—Var. A.).

II. Other African Regions.
  A. South Tropical.
    a. Western Coast.—Damaraland (*J. A. Bell* and the late *C. J. Andersson*—Typ. and Sub-Typ.). Angola: Ambriz (*J. J. Monteiro*).—Coll. Brit. Mus.
    b. Eastern Coast.—Zambesi River (*Rev. H. Rowley*). "Querimba."—Hopffer. Mombasa: "Lake Jipè (*Kersten*)."—Gerstäcker.
    b1. Eastern Interior.—"Tati and Ramaqueban Rivers (*Oates*)."—Westwood. Bamangwato and Matabeleland (*H. Barber*). Makloutse River (*J. L. Fry*—Sub-typ. and Var. A.). "Lake Nyassa."—Kirby, Cat. Hewits. Coll. Victoria Nyanza (*Rev. J. Hannington*).—Coll. Brit. Mus.
  B. North Tropical.
    a. Western Coast.—"Senegal."—Boisduval.
    b1. Eastern Interior.—"Abyssinia."—Reiche and Hopffer. Soudan: "Atbara River."—Butler. Dongola: "Ambukohl."—Klug. "Nubia."—Boisduval.
  C. Extra-Tropical North Africa.—"Cairo—(*J. K. Lord*)."—Walker.
IV. Asia.—"Arabia."—Boisduval.

267. (3.) Teracolus Agoye, Wallengren.

♂ *Anthopsyche Agoye*, Wallengr., K. Sv. Vet.-Akad. Handl., 1857; Lep. Rhop. Caffr., p. 15, n. 11.
♂ *Anthocharis Eosphorus*, Trim., Trans. Ent. Soc. Lond., 3rd Ser., i. p. 523, n. 3 (1863).
♂ ♀ *Anthocharis Agoye*, Trim., Rhop. Afr. Aust., ii. p. 325, n. 219 (1866).

Exp. al., (♂) 1 in. 7–8 lin.; (♀) 1 in. 9 lin.

♂ *White, with neuration (except near bases) finely marked with black; apical patch of fore-wing small, pale-ochrey orange-yellow, edged inwardly with rather diffused black.* Fore-wing: irregularly sprinkled with black scales, most thickly over basal half and along costal border; no terminal disco-cellular spot; apical patch extending to edges of wing, exteriorly very faintly tinged with brownish, interiorly bordered by a diffused black stripe, which is more or less obsolete a little below its middle part, but extends to hind-marginal termination of second median nervule. Hind-wing: basal half and costal border to apex rather thinly sprinkled with black scales. UNDER SIDE.—*White; hind-wing and apex of fore-wing more glossy, very faintly tinged with yellowish; no spots or markings.* Fore-wing: apical patch very faintly showing through from upper side. Hind-wing: costa narrowly edged with orange-yellow from base to beyond middle.

♀ *Less purely white; neuration not black, except rather faintly near extremity of some of fore-wing nervules.* Fore-wing: dull orange-yellow of apical patch much more restricted than in ♂, being outwardly bounded very broadly with dull-brownish (which diminishes to a point on hind-margin about or rather below first median nervule), and inwardly by a dull-brownish bar, broader and shorter than the black one in the ♂, and ending on lower radial nervule; a small fuscous terminal disco-cellular spot; blackish irroration limited to costa and discoidal cell. UNDER SIDE.—*Hind-wing, and basi-costal and apical hind-marginal areas of fore-wing, very pale straw-yellow;* in both wings a fuscous terminal disco-cellular spot, larger in hind-wing.

In a ♂ from the Lydenburg District of the Transvaal the under side of the hind-wings and apex of fore-wings is tinted with pale creamy-pinkish. A ♀ taken on the Limpopo by Mr. Selous exhibits on the upper side of the hind-wings a very faint small fuscous mark on the costa beyond the middle.

Among the drawings of some of Wallengren's types, kindly procured for me by Mr. Chr. Aurivillius, was a careful figure of the ♂ *Agoye*, thoroughly agreeing with the former's description (*loc. cit.*). The species is a very distinct one, easily recognised by the small size and peculiar ochreous-orange tint of the apical patch, to which in the ♂ are added the fine black neuration and irroration in both fore and hind wings.

This is a rare butterfly in collections, and was for many years known to me only by a series of seventeen specimens (only one ♀) brought from Damaraland by Mr. J. A. Bell, and by two or three other examples from the same country in the collection of the late Mr. C. J. Andersson. These examples had been pressed flat in a book, and did not properly exhibit all the characters of the species; but I have recently (in 1883 and 1885) received in good order a few individuals of both sexes from the North-West Transvaal and farther northward, from which the above description has been made.

I have no particulars as to the habits of *Agoye*, or as to its special stations. In geographical range, though known to occur from Damaraland on the west to the Lydenburg District on the east, it seems to be very restricted in latitude,—its place south of the 26th parallel being apparently taken by the nearly allied *Bowkeri*, mihi.

Localities of *Teracolus Agoye*.

I. South Africa.
    K. Transvaal.—Marico and Limpopo Rivers (*F. C. Selous*). Lydenburg District (*T. Ayres*).

II. Other African Regions.
    A. South Tropical.
        a. Western Coast.—Damaraland (*J. A. Bell* and the late *J. C. Andersson*).
        b1. Eastern Interior.—Bamangwato: Tauwani River (*F. C. Selous*). Between Transvaal Border and Gubulewayo (*A. W. Eriksson*).

268. (4.) Teracolus Bowkeri, Trimen.

♂ *Callosune Agoye*, Trim., Trans. Ent. Soc. Lond., 1870, p. 381.
♂ ♀ *Teracolus Bowkeri*, Trim., *op. cit.*, 1883, p. 358.

PLATE XI. fig. 4 (♂).[1]

Exp. al., (♂) 1 in. $6-7\frac{1}{2}$ lin.; (♀) 1 in. $7-7\frac{1}{2}$ lin.

♂ *White, with yellow-ochreous apical patch in fore-wing.* Fore-wing: base and costa sparsely irrorated with black; apical patch internally irregularly bordered with black, which is broad in its middle part (and sometimes also in its upper part), but attenuated and usually ill-defined towards its extremities; a narrow ray of clear yellow-ochreous immediately beyond the black, but the rest of apical patch tinged with greyish. *Hind-wing:* base irrorated with black, more widely and rather more closely than in fore-wing; a longitudinal ray of black irroration on costa beyond middle, its outer extremity sometimes strongly marked; subcostal nervules very rarely thinly defined with black; in some specimens a few black atoms scattered about disc. UNDER SIDE.—*Hind-wing and apical area of fore-wing very faintly tinged with yellowish.* Hind-wing: costa very narrowly edged with chrome-yellow from base to a little beyond middle; a general very fine and very sparse irroration of dusky atoms.

♀ *Fore-wing: apical patch fuscous-brownish, darker inwardly, traversed mesially by a dull yellow-ochreous ray,* which is sometimes suffused and ill-defined; base more widely irrorated than in male. *Hind-wing:* beyond middle, from costa, a transverse row of three ill-defined dull-fuscous spots, the first of which represents the termination of the black costal irroration in the male. UNDER SIDE.—*Yellowish colouring much more decided than in male.* Hind-wing: fine dusky irroration closer than in male.

VARIETY A.—(♂ and ♀). Considerably smaller: *exp. al.* (3 ♂ s) 1 in. 3-5 lin.; (2 ♀ s) 1 in. $2\frac{1}{2}-4\frac{1}{2}$ lin.

♂ Costal and basal irroration wanting in fore-wing and all but absent in hind-wing; inner black edging of apical patch of fore-wing narrower. UNDER SIDE.—Tinge of yellow more decided.

♀ Basal irroration wanting, as well as faint fuscous spots of hind-wing; apical patch of fore-wing redder, its dusky-brown borders much reduced; the external one almost obsolete. UNDER SIDE.—Yellowish deeper, in one specimen tinged with creamy-pinkish.

(*Hab.*—Namaqualand District, Cape Colony.)

This near ally of *T. Agoye*, Wallengr., is, in the ♂, distinguishable

---

[1] This figure is no credit to chromo-lithography, and gives a very incorrect idea of the species. The apical patch is much too dull and dusky in tint, and the ground-colour generally, instead of being pure white, as in nature, is represented with various yellow and greenish shades. On the under side, also, the apical shade is much too dusky, and the hind-wing too much stained with yellowish. I write while comparing with the figure the actual specimen from which it was drawn.

by its larger, paler, less warmly-tinted apical patch; total, or almost total, absence of black nervules in both wings; want of copious black irroration in fore-wing, and presence of strongly-marked irroration stripe on costa of hind-wing: the under side is somewhat more yellowish.[1] The ♀ has a much larger apical patch, owing to the breadth and extension inferiorly of the fuscous-brown on its inner border; but wants the disco-cellular terminal spot presented by *Agoye* ♀ in both wings: the under side is duller in tint, and wants both the disco-cellular spots and the tinge of pale yellow at the base of the fore-wing.

As above noted, I referred the first example seen of this form to *Agoye*, Wallengr., but the subsequent receipt of individuals of both sexes has shown me that it is quite distinct. Colonel Bowker, after whom I have named the butterfly, captured the first specimen in Basutoland. In 1871 he sent a second ♂ from Hope Town, on the Orange River, and, later in the same year, a ♂ and four ♀s from the Vaal River. On the 6th September 1872 I took a single ♂ at Kolberg, in Griqualand West, on the flowers of a fine species of *Cineraria* growing on a rocky hill. This example flew rather slowly.[2] Mrs. Barber also forwarded, in 1879, a ♂ from the Vaal River. Mr. L. Péringuey, who discovered the Variety A. above noted, informs me that he met with it only at Spectakel; the specimens taken were flying not rapidly about a rocky "kopje" of some elevation, in company with some other congeners, on the 11th November 1855.

Localities of *Teracolus Bowkeri*.

I. South Africa.
   B. Cape Colony.
      a. Western Districts.—Spectakel, Namaqualand District (*L. Péringuey*).
      b. Eastern Districts.—Hope Town (*J. H. Bowker*).
      c. Griqualand West.—Kolberg. Kimberley (*H. L. L. Feltham*). Klipdrift, Vaal River (*J. H. Bowker* and *M. E. Barber*).
      d. Basutoland.—Koro-Koro (*J. H. Bowker*).

269. (5.) Teracolus Ione, (Godart).

♂ *Pieris Ione*, Godt., Enc. Meth., ix. p. 140, n. 74 (1819).
♂ " " Boisd., Sp. Gen. Lep., i. p. 515, n. 112 (1836).
♂ ♀ *Anthocharis Ione*, Reiche, Ferr. and Gal., Voy. Abyss., iii. pl. 30, ff. 1, 2 [♂], 5, 6 [♀], 7 [♀ var.] (1849).
♂ ♀ *Anthocharis Ione*, Hopff., Peters' Reise Mossamb., Ins., pl. xxi. ff. 1, 2 [♂], 3, 4 [♀], 5, 6 [♀ var.] (1862).

---

[1] The example taken by me in Griqualand West has a slight tinge of creamy-pinkish on the under side.

[2] Mr. H. L. Langley Feltham, from whom I have lately received a ♀ with unusually dark and well-developed apical marking (taken at Kimberley on 20th November 1887), informs me that this butterfly has in his experience occurred but very sparingly in and about Kimberley. He gives the following dates, viz.: 1884—May 4th, ♀; December 14th, ♂; 1885—February 7th, ♂; April 15th, ♂; 1886—July 10th, ♂; 1887—April 10th and 22d, ♂; 11th, ♀; November, 4th, ♂; 20th, ♀.

♂ *Euchloë Jalone*, Butl., Cist. Ent., i. p. 14, n. 1 (1869).
♂ *Teracolus imperator*, Butl., Proc. Zool. Soc. Lond., 1876, p. 132, n. 20.
♂ *Callosune Ione*, Westw., App. Oates' Matabeleland, &c., pl. 338 (1881).
♂ ♀ *Callosune Jalone*, Staud., Exot. Schmett., i. pl. 23 (1884).

Exp. al., (♂) 2 in. $2\frac{1}{2}$–$4\frac{1}{2}$ lin.; (♀) 2 in. $2\frac{1}{2}$–5 lin.

♂ *White, with glittering-violet apical patch edged on both sides with blackish.* Fore-wing: base rather widely and closely sprinkled with blackish scales, especially along and near inner margin; costa also similarly sprinkled but becoming black-edged before middle; apical violet forming a moderately wide band, lessening in width towards its extremities (especially inferiorly), from second subcostal to between third and second median nervules, and divided by slender black crossing nervules into five unequal parts (besides a minute sixth part between third and fourth subcostal nervules); inner black edge of violet moderately broad, complete throughout, but rather diffused inwardly, extending from costa a little beyond middle to a point at extremity of first median nervule on hind-margin; outer edge broader, especially at apex, and (except on nervules) strongly clouded with pale grey, the dark indentations of the violet on nervules very slight; a small, thin, elongate, slightly curved terminal disco-cellular black spot; neuration only black beyond middle, except submedian nervure, which is heavily marked in black from considerably before middle. *Hind-wing*: neuration conspicuously black-marked throughout; in some specimens the black is barely or slightly thickened at extremity of nervules, but in most it is developed into a conspicuous series of hind-marginal spots with diffused edges; fold between median nervure and its first nervule and submedian nervure marked with a distinct black line, and some blackish irroration over the same inter-nervular space and lower basal part of discoidal cell. UNDER SIDE.— *White throughout, or with a very slight tinge of pale yellowish or pale pinkish-creamy over hind-wing and apical area of fore-wing.* Fore-wing: disco-cellular spot distinct; a very faint indication of the position of the violet patch of upper side. Hind-wing: costa edged narrowly with orange-yellow for a little distance from base (rarely as far as middle or a little beyond it); a terminal disco-cellular black spot, smaller than that in fore-wing; beyond middle a blackish mark on costa is the first and usually the most distinct trace remaining of a discal macular ray, normally angulated on third median nervule,—but its course seldom continuously marked even so far, and beyond that point never indicated except by two widely separated, small, often indistinct spots.

♀ *White (rarely more or less yellow); apical patch of fore-wing much broader than in ♂, orange-red, spotted and broadly bordered with black.* Fore-wing: basal clouding much denser, darker, and superiorly much broader than in ♂; disco-cellular spot much larger and rounder; between first median nervule and submedian nervure a large quadrate discal black spot, sometimes enlarged and confluent inwardly with blackish clouding of basal area ; orange-red of apical patch considerably wider than violet in ♂, sometimes more or less stained with crimson in parts, divided into five less distinctly, and traversed mesially by a series of five black spots,—the lowest of which is united with inner black border of patch between second and third median nervules ; inner black border of patch broad, rather diffused, irregularly excavated between nervules ; outer border broad, even, only slightly greyish-tinged outwardly, continued to posterior angle. *Hind-wing* : basal clouding variable in density and extent, but usually darker and more developed than in ♂ ; rather beyond middle on costa a more or less pronounced black mark (sometimes broad and enlarged inwardly) commences a very imperfect and short narrow discal ray ; nervular black terminal spots enlarged and confluent into a broad continuous hind-marginal border, emitting acute nervular dentations inwardly. UNDER SIDE.— *Hind-wing and apical area of fore-wing very pale greyish-yellow, faintly stippled with grey scales and short lineolæ. Fore-wing* : basal clouding much fainter, mixed with white ; apical orange-red very faintly represented, but the spots traversing it distinct, more or less sagittiform, the series having an additional (sixth) spot between second and first median nervules, and its two uppermost spots ferruginous ; large lower discal spot considerably smaller than on upper side, and divided longitudinally by white fold ; two or three small inter-nervular hind-marginal fuscous spots at and near posterior angle. *Hind-wing* : disco-cellular spot indistinct ; discal macular ray ferruginous, well-marked as far as third median nervule, and sometimes (with three widely separated spots) beyond it ; hind-marginal border faintly indicated, by a greyer shade.

Second (*Dimorphic*) *Form of* ♀.—Like ♀ just described, *except in apical patch of fore-wing, which is considerably narrower, and black, traversed mesially by a series of five rounded white spots*, of which the first is much the smallest, and the fifth much smaller than the intermediate three.

VARIETY A.—(♂) *Exp. al.*, 2 in. 3–4 lin.

Basal irroration much more restricted in both wings. *Fore-wing* : *apical violet considerably narrower*, its uppermost division much reduced, and its lowest (between third and second median nervules) either a mere trace or obsolete. UNDER SIDE.—*Black neuration almost as well developed as on upper side, but finer*,—in hind-wing (and to a much less extent in fore-wing) thickened to form small spots along hind-margin. *Hind-wing* : transverse discal ray well marked, blackish.

(*Hab.*—D'Urban County, Natal.)

It is not practicable to determine with certainty the exact form of ♂ upon which Godart (*loc. cit.*) founded his *Pieris Ione*, his description being too brief, and no locality being given ; but as he describes the under side as white, and as it is improbable that he should have had before him in the year 1818 any of the more locally restricted southern forms, I consider it judicious to regard as the typical *Ione* the form I have above described, which has a very wide Tropical-African range, extending northward to the White Nile on the east and to Senegal on the west.

I have examined the types of *Jalone* and *Imperator*, Butl., in the British Museum, with the result that I do not consider them separable as species, or even as marked varieties from *Ione*. The former is a ♂ from the White Nile, with pinkish-tinged under side, and with the neuration and other black markings of the upper side (especially the macular terminations of the hind-wing nervules) less strongly marked than in Reiche's figures (*op. cit.*) of the Abyssinian ♂. *Imperator*, on the other hand, is a ♂ from Senegal, rather more strongly marked than Reiche's ♂, particularly as regards the dusky discal ray of the under side of the hind-wings; and with it are associated (September 1886)—I think rightly—two ♂s and two deeply yellow-tinged ♀s from Mamboio in Eastern Africa, and a very slightly yellow-tinged ♀ from Madagascar. These two continental ♀s have a very brilliant appearance from their sulphur-yellow ground-colour and crimson-glossed red apical patch, and have the black markings strongly developed, while the Malagasy ♀ more resembles the ordinary pattern of *Ione* with orange-red apex.

Hopffer's ♂ *Ione* (*fig. op. cit.*) agrees more closely with Butler's *Imperator* than with *Jalone*, but has the dusky discal ray of the hind-wings only feebly represented by four faint brownish marks. Reiche's Abyssinian ♂ and Hopffer's Mozambique ♂ agree very nearly, the main difference being that the discal ray just mentioned is better marked in the former. In four Transvaal ♂s the ray in question is even more obsolescent than in Hopffer's figure. As regards the ♀, Reiche's and Hopffer's figures agree in the feeble development (in both forms of the sex) of the dusky clouding over the basal areas; but on the under side Reiche's figures exhibit a deeper and warmer colouring, the ♀ with red apex being pale creamy-ochreous, and that with black apex yellow, and both having the macular discal rays broad and diffused. In these features of the under side Reiche's figures approach the Natalian *Speciosus*, Wallengr.

The ♀s (red-tipped) that I have seen from the Transvaal and Delagoa Bay, while heavily clouded on the upper side, have the under side even paler and with fainter markings than Hopffer's figures. The ♂ accompanying the Delagoan ♀ is, like Butler's *Jalone*, pinkish-tinted on the under side.

The *Variety* A. of *Ione* above described is only known to me by a few ♂ examples from Natal, viz., one in the Hewitson collection (in 1867), another taken by the late Mr. M·Ken in 1869, and three captured by Colonel Bowker in December 1884. Their slightly smaller size and more restricted apical violet approximate them to the ♂ *Speciosus* inhabiting the same district; but the black neuration of the almost pure-white under side shows some resemblance to *Phlegyas*,

Butl. I have not seen any ♀ that appears certainly referable to this variety, but one or two of the ♀s referred to *Speciosus* exhibit some traces of similar black neuration towards the hind-margin of both fore and hind wings on the under side.

Localities of *Teracolus Ione*.

I. South Africa.
  E. Natal.
    a. Coast Districts.—D'Urban (the late *M. J. M'Ken*—Var. A.). Pinetown (*Colonel J. H. Bowker*—Var. A.).
    H. Delagoa Bay.—Lourenço Marques (*Mrs. Monteiro*).
    K. Transvaal.—Limpopo and Marico Rivers (*F. C. Selous*). Lydenburg (*A. F. Ortlepp*). Lydenburg District (*T. Ayres*).

II. Other African Regions.
  A. South Tropical.
    b. Eastern Coast.—Zambesi River (*Rev. H. Rowley*). "Tette."—Hopffer. "Querimba."—Hopffer. "Bagamoyo and Zanzibar (*Raffray*)."—Oberthür. Zanzibar.—Coll. Hewitson. Mamboio.—Coll. Brit. Mus.
    b1. Eastern Interior.—"Tati (*Oates*)."—Westwood.
    bb. Eastern Islands.—Madagascar.—Coll. Brit. Mus.
  B. North Tropical.
    a. Western Coast.—Senegal (the late *E. C. Buxton*).—Coll. Brit. Mus.
    b1. Eastern Interior.—Abyssinia: "Shoa (*Antinori*)."—Oberthür. White Nile (*Druce*).—Coll. Brit. Mus.

270. (6.) **Teracolus Speciosus**, (Wallengren).

♂ ♀ *Anthocharis Ione*, Boisd., App. Voy. Deleg. Afr. Aust., p. 587 (1847).
♂ *Anthocharis Erone*, Angas, Kafirs. Illustr., pl. xxx. f. 3 (1849).
♂ *Anthopsyche Speciosa*, Wallengr., K. Sv. Vet.-Akad. Handl., 1857; Lep. Rhop. Caffr., p. 16, n. 14.
♂ ♀ *Anthocharis Ione*, Trim., Rhop. Afr. Aust., i. p. 43, n. 26 (1862).
Teracolus Ione, Butl. [part.], Proc. Zool. Soc. Lond., 1876, p. 132.

Exp. al., (♂) 2 in. 2–4 lin.; (♀) 2 in. 2–4 lin.

♂ *White, with lustrous-violet apical marking broadly bordered with black.* Fore-wing: the nervures defined with black to same extent as in *Ione*, Godt.; base and costa greyish; a black terminal disco-cellular dot; violet band at apex narrow, divided into four by the nervules (the uppermost division much smaller than the rest, or sometimes obsolete), margined both outwardly and inwardly with a broad black band, the outer band narrowing to a thin streak at posterior angle. *Hind-wing*: base greyish; nervures all black, with moderately large diffused black spots on their hind-marginal extremities. UNDER SIDE.—*Fore-wing*: costa narrowly yellowish; *apical patch varying from pale straw-yellow to pale chrome-yellow*; disco-cellular spot larger than on upper side. *Hind-wing: same tint as apical patch of fore-wing*; nervures not black; a distinct spot at extremity of discoidal cell;

costa from base to a little before middle edged with orange-yellow; an elongate blackish spot on costa a little beyond middle; spots at extremities of nervules indistinct or almost obsolete, but nervules themselves black or blackish close to hind-margin.

♀ *White of a more yellowish tint than in ♂; the apical patch bright-orange. Fore-wing:* disco-cellular spot larger and rounder than in ♂; orange apical band divided lengthwise by a row of four rather large wedge-shaped black spots, the black bands bordering it broader than in ♂, and extending to posterior angle; a large subquadrate blackish spot on disc, immediately above submedian nervure; basal clouding much darker and more extended than in ♂, usually filling basal half of cell, and extending rather widely along inner margin as far or nearly as far as black quadrate spot. *Hind-wing:* on costa beyond middle a blackish streak commences, narrowing and gradually disappearing towards centre of wing; hind-margin broadly black, emitting deep acute dentations on nervules. UNDER SIDE.—*More deeply tinted than in ♂, hind-wing and apical patch of fore-wing externally dull lemon-yellow. Fore-wing:* inner portion of apical patch tinged with ochrey-orange and traversed (as on upper side) by a series of black spots; disco-cellular and inner-marginal spots as on upper side. *Hind-wing:* a narrow black spot, immediately surmounted by a white one at extremity of cell; transverse blackish stripe from costa beyond middle strongly marked as far as third median nervule, and beyond its extremity a blackish spot below that nervule; a second blackish spot below first median nervule.

Second (Dimorphic) Form of ♀.—Orange at apex of fore-wing wholly wanting, but to some extent replaced by three or four rather small, rounded, separate whitish spots; black markings generally broader and somewhat diffused on their edges.

This form is in the male sex readily separable from its near allies by the great development and intensity of the apical black of the fore-wings, which forms a broad inner margin to the violet band, much reducing the width and more or less the length of the latter. The outer black margin of the violet also wholly wants (or has only very faint traces of) the pale-grey clouding conspicuous in *Ione*. The decidedly yellow colouring of the under side is further a very characteristic feature, and is pronounced in both sexes. The two forms of ♀ present on the upper side no salient points of distinction from those of *Ione* (varying in the dusky clouding of basal areas, and to a less extent in yellowish tinting, much in the same way), but in the red-tipped ♀ the black spots traversing the red seem to be invariably much larger and nearer to its inner edge, while the red itself occupies a smaller space, scarcely extending below third median nervule. In several ♀s (of both forms) the angulated blackish discal ray of the under side of the hind-wings is completed by a variably-developed spot between second and first median nervules.

Speciosus is a remarkably local form, being apparently restricted to the coast belt of Natal, and, as far as I am aware, not having hitherto been recorded out of D'Urban County. Within these narrow limits it is very numerous, and I met with it in abundance about D'Urban at the end of January, all through February, and again at the end of March and beginning of April 1867. Colonel Bowker has taken it freely in December also. The red-tipped form of ♀ is much less frequently met with than the other. I fell in with three specimens only during my visit, and Colonel Bowker has also noted its scarcity as compared with the black-and-white-tipped ♀.

The lovely ♂ is a very active and even rapid flyer, but the ♀ is much slower in her movements. Both sexes are fond of flowers, and I captured the finest specimens I obtained on those of *Vinca rosea* and of *Lantana* in the Botanic Gardens on the Berea Hill. On the 1st February I observed and netted a ♂ and a red-tipped ♀ playing together close to the ground. I did not meet with the species anywhere away from the neighbourhood of D'Urban.

Localities of *Teracolus speciosus*.

1. South Africa.
   E. Natal.
      a. Coast Districts.—D'Urban. Pinetown (*J. H. Bowker*). "Lower Umkomazi."—J. H. Bowker.

271. (7.) Teracolus Jobina, Butler.

♂ ♀ *Euchloë Jobina*, Butl., Cist. Ent., i. p. 14, n. 2 (1869).
♂ ♀ *Teracolus Jobina*, Butl., Lep. Exot., p. 116, pl. xliii. f. 3 [♂] (1872).
♂ ♀ Var. *Callosune Jobina*, Staud., Exot. Schmett., i. pl. 23 (1884).

Exp. al., (♂) 1 in. 10½ lin.—2 in. 1 lin.; (♀) 1 in. 11 lin.—2 in. 1½ lin.

Closely allied to *Speciosus*, Wallengr.

♂ *White, with shining-violet apical patch of fore-wing comparatively larger than in Speciosus, and less widely black-edged on both sides.* Fore-wing: apical violet shaped much as in *Speciosus*, and divided similarly, but wider, and with a minute fifth portion between third and fourth subcostal nervules almost always present;[1] basal grey irroration very limited; on costa sometimes a creamy-reddish tinge from base for a little distance; terminal disco-cellular spot very faint or obsolete altogether. *Hind-wing*: usually subcostal nervules only black-marked (and those often only towards their termination); rarely the other nervules are black near hind-margin, and with an inclination to a spot at the extremity. *Cilia* in fore-wing creamy-reddish, except a white portion near posterior angle; in hind-wing more or less white in apical half, but the rest creamy-reddish. UNDER SIDE.—*Hind-wing and apical patch of fore-wing pale reddish-creamy with a tinge of pink,—the former generally but sparsely marked with short transverse grey striolæ.* Fore-wing: costa narrowly edged with creamy-reddish; terminal disco-cellular spot small but dis-

---

[1] This minute almost linear bit of violet is sometimes found in *Speciosus*.

distinct. *Hind-wing:* disco-cellular spot dusky, diffused; discal ray only represented by a short brownish streak from costa to second subcostal nervule; orange costal edging almost obsolete.

♀ *White, somewhat yellow-tinged generally or from bases only; apical patch of fore-wing bright orange-red,* wider than in *Speciosus,* and emitting rays along subcostal nervure and two radial nervules as far as extremity of discoidal cell, the inner black border being very much narrowed and obsolescent in its middle part. *Fore-wing :* terminal disco-cellular spot very small; basal clouding narrow and faint; discal spot very much reduced, diffused, or obsolescent; cuneiform black spots transversing apical red not so near its inner edge as in *Speciosus.* *Hind-wing:* basal clouding very faint and restricted; discal ray obsolescent except on costa; hind-marginal blackish border very much narrower and diffused, partly or entirely broken into separate spots. UNDER SIDE.—*Hind-wing and apical patch of fore-wing not pink-tinged, but the former rendered darker than in ♂ by closer and stronger striolation,* and the latter flushed with reddish and traversed by the usual series of blackish spots. *Hind-wing:* disco-cellular terminal spot small but distinct, immediately surmounted by a rather conspicuous whitish spot; discal ray dark-brownish rather strongly marked as far as third median nervule.

Dimorphic Form of ♀.—*Apical patch of fore-wing black, as in corresponding form of ♀ Speciosus, but white spots of series traversing it comparatively larger and not so widely separated.* UNDER SIDE.—Duller in tint; discal ray of hind-wing with macular traces of its inferior continuation.

In one ♀ of this form the discal spot of the fore-wings beyond middle and the hind-marginal spots of the hind-wings are exceedingly reduced, and in another those markings are so faint and minute as to be scarcely perceptible.

The characters given serve very well to distinguish this small violet-tipped *Teracolus* from *Speciosus,* but it must be noted that examples occur (I have before me four ♂s and a ♀ of each form) which show certain features intermediate between those of *Jobina* and *Speciosus.* These ♂s are a little larger than *Jobina* proper, and have the black borders of the apical violet broader, and the cilia of the hind-wings white throughout; and both sexes have the under-side colouring much yellower, with scantier striolation of the hind-wings. Two of these ♂s and the two ♀s were taken by the late Mr. M. J. M'Ken at D'Urban, Natal, late in April 1867.

Jobina seems to occur solely as a winter (or dry-season) butterfly. Apart from the non-typical individuals just mentioned as captured in April (which are, however, much nearer to true *Jobina* than to *Speciosus*), all the specimens whose dates of capture are known to me were taken in May, June, July, and August. I never saw this small form during my summer visit, which ended on April 9th; nor, on the other hand, am I aware of *Speciosus'* appearing on the wing except in the summer or wet season. It seems not impossible that the two butterflies may turn out to be summer and winter broods of the same species,[1] but this could only be proved by careful breeding from the egg.

As far as my records go, *Jobina* has a wider range than *Speciosus*, Colonel Bowker having met with it as far northward as the mouth of the Tugela River, and (judging from photographs and MS. sent to me), the late Mr. E. C. Buxton having taken it in Swaziland.

Localities of *Teracolus Jobina*.

I. South Africa.
  E. Natal.
    a. Coast Districts.—D'Urban (*J. H. Bowker* and the late *M. J. M'Ken*). Pinetown and Northdene (*J. H. Bowker*). Verulam and Mouth of Tugela River (*J. H. Bowker*).
  G. "Swaziland."—The late E. C. Buxton.

272. (8.) Teracolus Phlegyas, Butler.

♂ *Anthocharis Ione*, Var., Reiche, Ferr. et Gal., Voy. Abyss., pl. xxx. ff. 3, 4 (1849).[2]
♂ ♀ *Anthocharis Phlegyas*, Butl., Proc. Zool. Soc. Lond., 1865, p. 431, pl. xxv. ff. 3, 3a.
VAR. ♂ ♀, *Teracolus Buxtoni*, Butl., op. cit., 1876, p. 130, n. 17.
♀ *Callosune Buxtoni*, Westw., App. Oates' Matabeleland, p. 340, pl. E, ff. 7, 8 (1881).

Exp. al., (♂) 2 in. 0–2 lin.; (♀) 2 in. 3 lin.

♂ *White, with rather small lustrous-violet apical patch in fore-wing, not widely bordered with fuscous externally, and very narrowly so bordered internally. Fore-wing:* no disco-cellular terminal spot; apical violet five-partite by dark nervules, its outer dusky border not extending below second median nervule. *Hind-wing:* neuration conspicuously black, but no spots at ends of nervules. UNDER SIDE.—*White; hind-wing and apical area of fore-wing with black neuration. Hind-wing:* costa edged with orange-yellow from base to before middle; discal ray from costa very faint and ill-defined.

♀ *Apical patch orange-red extending to below second median nervule, externally with a rather wide brownish border. Fore-wing:* a small terminal disco-cellular spot; base and costa tinged with yellowish; inner edge of apical red without blackish except close to costa, the series of blackish spots traversing it rather small and ill-defined; discal spot near posterior angle indistinct. *Hind-wing:* hind-marginal nervular spots of moderate size, diffused. UNDER SIDE.—*Hind-wing and apical area of fore-wing very pale creamy-reddish, the latter flushed with pale orange-red inwardly. Fore-wing:* terminal disco-cellular spot distinct; traversing spots rather faint, ill-defined. *Hind-wing:* discal ray

---

[1] This is the opinion of Mr. A. D. Millar, an observer of long residence at D'Urban. The dated specimens of *Jobina* which he has kindly sent to me are two ♂s captured on 22d August, one ♂ on 22d September, and two ♂s on 24th.

[2] In fig. 4 the under side of the fore-wing is represented as closely dusted with fuscous, but this is probably from some defect in the engraving of the plate.

brown, macular, but complete, the spots between third median nervule and submedian nervure being represented.

VARIETY A., ♂ and ♀ (*Buxtoni*, Butl.).

♂ Apical patch of fore-wing rather larger, its outer border being broader, and its inner one (though narrow and diffused) more developed and complete throughout. *Fore-wing*: terminal disco-cellular spot sometimes present though minute. UNDER SIDE.—*Without black neuration;* hind-wing and apex of fore-wing with an extremely slight tinge of yellowish or reddish,—the former rarely with some scattered brownish-grey striolation, chiefly on margins; terminal disco-cellular spot small but distinct in both wings.

♀ Apical patch broader, brighter in tint, the orange radiating inwardly on nervules as far as extremity of discoidal cell, and outwardly strongly suffusing its brownish border. *Fore-wing*: discal spot near posterior angle sometimes wanting. *Hind-wing*: hind-marginal spots very much reduced or obsolete. UNDER SIDE.—*Hind-wing and apical area of fore-wing finely striolated with reddish brown. Hind-wing*: area generally from base as far as discal ray (which is much diffused and not macular), tinged with pale-brownish.

Three ♂s of this variety, respectively from Damaraland, the Transvaal, and the Zambesi have the under side and the cilia of fore-wing tinted with creamy-pinkish as in *Jobina*, Butler.

(*Hab.*—Damaraland, Tropical Interior South Africa, Transvaal, Swaziland.)

After examining the types in the British Museum, I could not find sufficient grounds for separating as species *Buxtoni* and *Phlegyas*. The variety (in the ♂) is nearer than *Phlegyas* proper to typical *Ione;* but, on the other hand, the black under-side neuration of *Phlegyas* approximates it to the Natalian Variety A. of *Ione* above described. The ♀ of the variety nearly resembles the ♀ *Jobina*, Butler, especially on the under side, but presents a much more extended field of apical red on the upper side. In size *Phlegyas* (including *Buxtoni*) is intermediate between *Ione* and *Jobina*.

The typical *Phlegyas* inhabits the Soudan (White Nile) and Abyssinia, and the variety *Buxtoni* is known to inhabit the tropical belt from the Zambesi southward, and on the eastern side to penetrate as far as the Transvaal and Swaziland. I have not seen any but red-tipped ♀s referable to this species. Westwood's figures (*op. cit.*) are somewhat larger and more strongly marked than any specimen I have seen, and indeed than, from his own description, would appear to have been among Mr. Oates's examples. The late Mr. E. C. Buxton sent me coloured photographs of Swaziland specimens of the butterfly.

Localities of *Teracolus Phlegyas*.

1. South Africa.
    H. Swaziland (*E. C. Buxton.*—Var.).
    K. Transvaal.—Marico and Limpopo Rivers (*F. C. Selous.*—Var.).
       Lydenburg District (*T. Ayres.*—Var.).

II. Other African Regions.
  A. South Tropical.
    a. Western Coast.—Damaraland (*H. Hutchinson* and *W. C. Palgrave*—Var.).
    b1. Eastern Interior. — Makloutse River (*F. C. Selous* — Var.). Zambesi River: Zumbo (*F. C. Selous*—Var.).
  B. North Tropical.
    b1. Eastern Interior.—Soudan: "White Nile (*Petherick*)."—Butler. "Abyssinia (*Ferret* and *Galinier*)."—Reiche.

273. (9.) Teracolus Regina, (Trimen).

♂ ♀ *Anthocharis Regina*, Trim., Trans. Ent. Soc. Lond., 3rd Ser., i. p. 520, n. 1 (1863).
♀ *Callosune Regina*, Westw., App. Oates' Matabeleland, p. 339, pl. E, ff. 9, 10 (1881).
Var. A. (♂), *Anthopsyche Ione*, Wallengr., K. Sv. Vet.-Akad. Handl.; Lep. Rhop. Caffr., p. 15.

Plate XI. fig. 3 (♀).

Exp. al., (♂) 2 in. 4½–8 lin.; (♀) 2 in. 4½–8½ lin.

♂ *Pure white, with very large and brilliant blue-shot violet-lake apical patch. Fore-wing:* a linear black, terminal disco-cellular dot; costa usually very faintly grey-dusted nearly to middle, where blackish edging of apical patch commences; apical patch composed of six elongate portions (defined by more or less strongly-marked black nervules), and lying between second subcostal and second median nervules, very narrowly and faintly edged with blackish internally, narrowly along costa to apex, and rather more widely externally from apex to first median nervule (slightly indenting the violet on each nervule); neuration beyond and below discoidal cell (and submedian nervure from a little before middle) black; base very slightly dusted with faint greyish. *Hind-wing:* usually spotless, occasionally with a hind-marginal series of small blackish nervular spots; first and second subcostal and discoidal nervules and third median nervule black for some distance from near their origins; an obsolescent greyish basal irroration. Under side.—*White, with a faint tinge of yellowish or pinkish in hind-wing; apical patch of fore-wing pinkish-grey, with a faint flush of pale orange-yellow. Fore-wing:* terminal disco-cellular spot distinct; occasionally a series of four or five very faint fuscous spots traversing apical patch. *Hind-wing:* costa edged from base to about middle with more or less bright yellow; a minute terminal disco-cellular black dot, usually edged with yellow superiorly; discal ray rarely represented by a series of small widely-separated fuscous spots; extremities of nervules rarely marked with blackish.

♀ *White, without black neuration; apical patch of the same colour as in ♂, but with a less brilliant blue gloss, and traversed mesially by a macular black stripe (sometimes so diffused as almost to obliterate inner* portion of violet),—its blackish edgings broader and darker. *Fore-wing*: disco-cellular spot larger, ovate or round; near posterior angle two small blackish diffused spots, continuous of the series traversing violet patch. *Hind-wing*: blackish hind-marginal spots not so rare as in ♂. Base of both wings (but especially that of fore-wing) with broader and darker greyish irroration. UNDER SIDE.—*Hind wing and apex of fore-wing* dull creamy-ochreous (sometimes tinged with reddish), hatched faintly and minutely with grey; disco-cellular spots well-developed, that in hind-wing yellow with an inferior black dot; transverse discal series of fuscous spots usually well marked, that on hind-wing elbowed at third median nervule, and continued thence to inner margin; no spots at extremities of nervules.

VARIETY A. (♂ and ♀).

♂ Blackish nervules and edgings of apical patch more strongly marked, the latter broader; bases more widely and closely irrorated with greyish; hind-marginal spots in hind-wing well marked, and five nervules (first subcostal to second median) finely black throughout their length. UNDER SIDE.—In both wings, nervules black close to hind-margin, and spots of discal series blackish, but usually very small.

♀ *Apical patch* obscure-white tinged with violet, the intersecting black macular ray very broad, almost obliterating inner portion of violet. *Hind-wing*: hind-marginal spots large, black, inwardly prolonged and acuminate on nervules. UNDER SIDE.—Hind-wing and apex of forewing strongly tinged with chrome-yellow; spots of discal series very dark and distinct, separate, but larger than in ♂; extremities of nervules more widely blackish than in ♂.

Dimorphic Form of ♀.—Apical patch black, broader throughout (and especially towards posterior angle) traversed mesially by a row of six rather small, outwardly narrowed white spots; upper of two discal spots near posterior angle merged in hind-marginal black. *Hind-wing*: hind-marginal spots greatly widened, forming a border narrowly interrupted with white on inter-nervular folds. UNDER SIDE.—As in ♀ just described, but the spots of discal series in both wings much enlarged.

(*Hab.*—Damaraland, North-West Transvaal, and Eastern Tropical South Africa.)

Some variation exists in both sexes of the typical form of *Regina*, individuals occurring of smaller size than usual, with the under side more decidedly reddish-tinged, and with the cilia also creamy-reddish. The most strongly marked ♂ s of this description that I have seen were from the Eastern Transvaal; two examples had the hind-wings quite pinkish-creamy, with faint dusky-grey striolation. I have not seen any ♀ in which the black macular ray traversing the violet apical patch is so nearly obsolete as in Westwood's figure (*op. cit.*) of a Tati specimen, but several ♀ s from Damaraland and the Transvaal exhibit considerable reduction of the ray. It is a ♀ of this kind that I have selected for my illustration. (Plate 11, fig. 3.)

The Variety A. above described is linked to the type by an intermediate ♂ taken by Mr. A. W. Eriksson in the region between the north-west limits of the Transvaal and South Matabeleland.

By means of an exquisite water-colour drawing of Wallengren's type, obtained for me by the kindness of Mr. Aurivillius, I have identified that author's *Ione* with my Variety A. of *Regina*; the markings being only a little less developed, and approaching those of Mr. Eriksson's example just mentioned. A second example was sent for my inspection by Mr. Aurivillius with the following note, viz.: "Referred by Wallengren to his *Ione*, but distinct from his type and description." This was a small worn example of my typical *Regina* ♂.

The very closely allied *Hetæra*, Gerst. (*Gliederth.-Fauna d. Sansibar-Gebietes*, 1873, p. 365, t. xv. f. 2), founded on a single ♂ from Endara, near Mombas, seems to be distinguished from *Regina* by its rather larger size and redder apical patch,—the latter being also entirely devoid of any trace of black on its inner edge.

This magnificent species excels in beauty all its near congeners, and is, moreover, the largest of the genus, with the exception of the doubtfully distinct *Hetæra*. The apical patch of the ♂ is nearly twice as broad as that of *Ione*, Godt., and of different colouring, the most brilliant metallic pale (almost glaucous) blue shifting over a ground of purplish-lake. The ♀, moreover, stands alone in presenting an apical violet space partaking to a large extent of the splendour proper to that of the ♂. The Variety A. is undoubtedly the finest form of the species, but one of the two forms of ♀ that are referable to it exhibits only traces of the purple of the typical ♀, and the other none whatever.

Like so many of its congeners, *Regina* appears to find its "metropolis" between the Zambesi and the limit of the Southern Tropic, and to penetrate but a little distance into the extra-tropical tracts. Mr. T. Ayres, however, informs me that he met with the species "in numbers for a very short time in December 1875," among the mountains in the Lydenburg District of the Transvaal, and several examples (including one ♀ of Var. A.) taken near the junction of the Marico and Limpopo Rivers have reached me from Mr. Selous and Mr. Eriksson. Mr. John A. Bell, who in 1862 made me first acquainted with the butterfly, brought down from Damaraland no fewer than sixty-seven specimens, and informed me that it was most abundant on the Botletle, one of the chief streams connected with Lake Ngami. On the eastern side of the interior, the Makloutze River and Tati seem to be favoured stations of *Regina*, Mr. Oates having noted it from the latter, and Mr. Selous and Mr. John L. Fry having each sent me ticketed specimens from both localities. Mr. Fry's examples from Makloutze River were taken on the 20th May 1887, and a ♂ of Variety A. from Tati on the 23d January. He informs me that at the former place the butterfly was numerous on the purple flowers of a species of *Cineraria*.

Localities of *Teracolus Regina*.

I. South Africa.

    K. Transvaal.—Marico and Limpopo Rivers (*F. C. Selous*—Typ. and Var. A.). Junction of Marico and Limpopo (*A. W. Eriksson*). Lydenburg District (*T. Ayres*).

II. Other African Regions.
  A. South Tropical.
    a. Western Coast.—Damaraland (*J. A. Bell* [Typ. and Var. A.], *H. Hutchinson, W. C. Palgrave,* and *J. J. Christie*).
    b1. Eastern Interior.—Makloutze River (*F. C. Selous* and *J. L. Fry*). Tati (*F. C. Selous* and *J. L. Fry*—Typ. and Var. A.).

274. (10.) Teracolus Eunoma, Hopffer.

♂ *Pieris Eunoma,* Hopff., "Monatsber. K. Akad. d. Wissensch. Berlin, 1855, p. 640," and Peters' Reise nach Mossamb., Zool., v., p. 353, t. xxiii. ff. 1, 2 (1862).

Exp. al., 2 in. $3\frac{1}{2}$ lin.

♂ *White; apical patch of fore-wing reduced to two small separated elongate purplish-lake markings between subcostal nervure and lower radial nervule; bases very finely and narrowly greyish-sprinkled.* Fore-wing: a small, ill-defined terminal disco-cellular blackish spot; apical margin narrowly bordered with brownish, internally dentate on nervules, on hind-margin not extending below third median nervule; close to costa, below second subcostal nervule, a small elongate faint blackish mark. *Hind-wing:* without marking of any kind. UNDER SIDE.—*Hind-wing and narrow costal and apical border of fore-wing pale ochrey-yellow. Fore-wing:* terminal disco-cellular spot better defined than on upper side. *Hind-wing:* costal margin edged with orange from base to about middle; a minute terminal disco-cellular blackish spot.

This curious species, in which the purple apical marking is so extraordinarily limited, and almost all the other ordinary markings of the group are wanting, was founded by Hopffer on a single ♂ example from Inhambane. It seems to be a butterfly of great rarity, the only other example recorded being a ♂, ticketed "Zanzibar," in the Hewitson Collection, where I noted it in 1867. I did not make a description of the latter, not at the time bearing in mind that the locality of the type lay just outside the Southern Tropic. The diagnosis given above is therefore taken from Hopffer's figures in the work quoted.

Localities of *Teracolus Eunoma*.

I. South Africa.
  1. "Inhambane."—Hopffer.
II. Other African Regions.
  A. South Tropical.
    b. Eastern Coast.—Zanzibar.—Coll. Hewitson.

275. (11.) Teracolus Annæ, (Wallengren).

♂ *Anthocharis Danaë,* Doubl., Gen. D. Lep., pl. vii. f. 2 (1847).
  „   „   Boisd., App. Voy. Deleg. Afr. Aust., p. 587 (1847).
♂ *Thestias Annæ,* Wallengr., K. Sv. Vet.-Akad. Handl., 1857; Lep. Rhop. Caffr., p. 16, n. 1.

♂ ♀ *Anthocharis Danaë*, Trim., Rhop. Afr. Aust., i. p. 44, n. 27 (1862).
Teracolus cinerescens, Butl., Cist. Ent., i. p. 172, n. 53 (1873), and Proc. Zool. Soc. Lond., 1876, p. 155. n. 99.
♂ ♀ *Callosune cinerescens*, Staud., Exot. Schmett., i. pl. 23 (1884).

Exp. al., (♂) 2 in. 0–3 lin.; (♀) 1 in. 11 lin.—2 in. 1 lin.

♂ *White, with very large broad crimson-red apical patch, divided into eight unequal parts by fine black neuration. Fore-wing:* costa black-edged; *basal area to beyond middle of discoidal cell and of inner margin suffused with dark-grey;* apical patch usually glossed with violet in some lights, externally bordered narrowly with black, which radiates for a little distance on nervules, and internally with a broad blackish border, diffused on its inner edge, touching extremity of discoidal cell, and extending to posterior angle; a terminal lunular disco-cellular black spot; between first median nervule and submedian nervure, almost touching internal border of red patch, a very faint indistinct blackish mark. *Hind-wing:* broadly suffused with dark-grey as far as extremity of discoidal cell (where there is a faint orange spot), and also as far as anal angle, where it is paler; on costa, beyond middle, a transverse short blackish streak, from which a broad blackish band, radiating for some distance on nervules, and usually crossed by whitish inter-nervular streaks, extends along hind-margin, narrowing to anal angle. UNDER SIDE.—*Fore-wing:* white; basal grey much paler than on upper side, its outer edge defined by a blackish streak; terminal disco-cellular spot larger and rounder; apical patch outwardly creamy-white, inwardly pale orange-red, traversed on its inner side by a sinuate row of six blackish spots; spot below first median nervule blacker and more distinct, and continuous of the sinuate row of spots; nervules near hind-margin clouded with blackish, which forms terminal spots on the three next anal angle. *Hind-wing:* very slightly tinged with yellowish; costa edged with orange-yellow from base to a little before middle; grey suffusion as extensive as on upper side, but very much paler; a rather large, orange-red, incompletely black-ringed spot at extremity of cell; beyond middle, from costa to submedian nervure, a sub-angulated series of seven red-centred blackish spots; neuration clouded with blackish between this series of spots and hind-margin. *In both wings* a very fine black line edging hind-margin.

♀ *Fore-wing:* suffusion from base much darker than in ♂, extending farther towards middle; disco-cellular spot larger, ovate; *apical marking smaller, dull-red, inclining to orange, and without violet lustre,* broadly black-bordered externally, and divided by a transverse, sinuate band of contiguous black spots, similar to those on under side of ♂; spot below third median nervule large, black, quadrate, often joined to inner border of apical marking. *Hind-wing:* dark suffusion not extending farther than in ♂, but of the same brownish-black as in fore-wing; disco-cellular orange spot hardly perceptible; costal mark beyond middle continued across wing by a row of contiguous blackish spots, which become merged in the blackish bands on inner margin and hind-margin (the latter of which is broader than in ♂) near anal angle. UNDER SIDE.—Like that of ♂. *Fore-wing:* apical marking yellower, and more broadly red inwardly, than in ♂; the black spots dividing it larger, sometimes contiguous; basal grey mixed with lemon-yellow. *Hind-wing:* yellowish, darker than in ♂; row of spots beyond middle larger. *Cilia* of fore-wing reddish-grey, but white or whitish near posterior angle; of hind-wing white, mixed with reddish in ♀.

VARIETY A. ♂ and ♀.—Usually smaller; *exp. al.*, (♂) 1 in. $9\frac{1}{2}$-$10\frac{1}{2}$ lin.; (♀) 1 in. $8\frac{1}{2}$ lin.—2 in.

♂ Basal grey much fainter and less developed, especially in fore-wing, where it is sometimes almost obsolete; hind-marginal border of hind-wing broken up into completely separate spots, except near apex.

♀ More or less tinged with lemon or ochreous-yellow; basal suffusion not nearly so dark, mixed with ochreous-yellow scales; apical patch of fore-wing usually larger and of brighter red, its inner border being narrower and in the upper part more or less diffused; hind-marginal border of hind-wing narrower, radiating little or not at all on nervules.

Under side in both sexes of a deeper creamy-tint in hind-wing and at apex of fore-wing: spots of discal series (especially in hind-wing and in ♀) more rounded and sub-ocellate, with centres more or less glistening.

(*Hab.*—Kaffraria Proper, Natal, North-West Transvaal, and Eastern Tropical South Africa.)

Of this variety, which links the typical *Annæ* to *Wallengrenii*, Butl., I possess a dwarf ♀, taken in some part of Kaffraria by Mr. H. J. Atherstone, which expands only 1 in. 6 lin.

Wallengren's reference of his *Annæ* to the Oriental genus *Thestias*, and his description of its apical patch as "rufo-fulva," for long puzzled me; and it was not until 1881, when Mr. Aurivillius kindly sent me a typical specimen from the Stockholm Museum, that I was enabled to identify it as the large African form of "*Danaë*," figured by Doubleday and mentioned by Boisduval in 1847, and separated by Mr. Butler in 1873 as *Cinerescens*.

Annæ is well distinguished from the Indian *Danaë*, Fab., by its larger size, much greater development of the basal dusky clouding in both wings, and hind-marginal blackish border in the hind-wing. In the ♂, too, the apical patch is redder (not so thoroughly crimson), with a broader blackish border on its inner edge, while in the ♀ the same marking is altogether different alike in its duller tint and the far greater development of its dark borders and intersecting macular ray. On the under side there is a closer resemblance, but *Annæ* has all the markings stronger, a wider red flush on inner side of the apical patch, and in the fore-wing a distinct blackish streak bounding the basal grey. The true *Danaë* is intermediate in characters between

Annæ and *Eupompe*, Klug, recorded from North-East Africa and Arabia.

The typical ♂ presents some variation in the dusky-grey basal clouding, but in only one specimen (from Kaffraria Proper) is there a marked diminution of it, and that in the fore-wings only. A ♂ from Damaraland approaches the individual just mentioned to some extent, and also presents a less developed and submacular border in the hind-wings. The latter peculiarity (approaching the Variety A.) occurs in an otherwise strongly-marked and very typical ♂ from the North-West Transvaal. The ♀ varies more in the development of the marginal blackish (which in the hind-wings of one Kaffrarian example is so great as to obliterate entirely the usual white band or spots beyond the discal blackish series), and in the tint of the apical patch. In one example from the North-West Transvaal this last feature is much suffused with fuscous, and the colour of the very diminished pale portion is very dull pale ochrey-yellowish.

In Natal I only once fell in with this strikingly handsome butterfly, viz., on 23d February, about a hill-top near Verulam, where two or three examples were flying rapidly about flowers. Mr. Gooch (*Entomologist*, 1880, p. 274) notes the capture of a fine series five miles inland from D'Urban.[1] Colonel Bowker has sent but a few examples from the same neighbourhood; but in former years he forwarded many fine specimens from the Bashee River in Kaffraria,[2] and in 1873 dispatched to me the paired sexes captured at Fort Warden, on the west bank of the Kei. To the northward of the Transvaal, and in the Tropical belt beyond, as well as in Damaraland, the type-form and the variety appear to be both numerous.

Localities of *Teracolus Annæ*.

I. South Africa.
  B. Cape Colony.
    b. Eastern Districts.—Fort Warden, Kei River (*J. H. Bowker*). St. John's River Mouth (Sir H. Barkly—Var. A.).
  D. Kaffraria Proper.—Bashee and Tsomo Rivers (*J. H. Bowker*).
  E. Natal.
    a. Coast Districts.—D'Urban (*J. H. Bowker* and the late *M. J. M'Ken*—Typ. and Var. A.).
    b. Upper Districts.—Rorke's Drift (*J. H. Bowker*).
  K. Transvaal.—Limpopo and Marico Rivers (*F. C. Selous*). Junction of those rivers (*A. W. Eriksson*—Typ. and Var. A.).

---

[1] Mr. A. D. Millar, in referring to some fine specimens taken in December 1887, informs me that during that month *Annæ* is plentiful in certain localities about D'Urban, but, as a rule, is restricted to those localities. He noted a ♀ laying eggs on a climbing plant.

[2] One of these specimens is very remarkable, as exhibiting to some extent the characters of both sexes. The ground-colour is as white as usual in the ♂; the basal clouding is irregularly but for its larger part made grey, as in the ♂, by close white scaling; the border of the hind-wings is like that of the ♂, but the discal row of black spots as in the ♀; and, finally, while the left apical patch is coloured and marked as in the ♀, the right one is (except for the narrow strip above subcostal nervule) entirely like that in the ♂.

II. Other African Regions.
  A. South Tropical.
    a. Western Coast.—Damaraland (*J. A. Bell* and the late *C. J. Andersson*).
    b. Eastern Coast.—Zanzibar (Coll. Hewitson—Var. A.).
    *b*1. Eastern Interior.—Bamangwato: Kama's Country (*H. Barber*). Tchakani and Makloutze Rivers (*F. C. Selous* and *John L. Fry* —Typ. and Var. A.).

276. (12.) Teracolus Wallengrenii, Butler.

♂ *Anthopsyche Danaë*, Wallengr.,[1] K. Sv. Vet.-Akad. Handl., 1857; Lep. Rhop. Caffr., p. 14.
♀ *Anthopsyche Eupompe*, Wallengr.,[1] *loc. cit.*
♂ *Anthocharis Danaï*, Trim. [part],?Rhop. Afr. Aust., i. p. 45 (1862), and [*Anth. Eupompe*] ii. p. 330 (1866).
♂ ♀ *Teracolus Wallengrenii*, Butl., Proc. Zool. Soc. Lond., 1876, p. 157, n. 105.
♀ *Callosune Wallengrenii*, Westw., App. Oates' Matabeleland, p. 341, pl. E, ff. 3, 4 (1881).

Exp. al., (♂) 1 in. 6½–11 lin.; (♀) 1 in. 5–10 lin.
Nearly allied to *Annæ*, Wallengr., Var. A.

♂ *White, with large crimson-red, slightly violaceous-glossed apical patch, shaped and bordered as in Annæ; basal grey very narrow and faint in fore-wing, and equally faint (though extending near inner margin to beyond middle) in hind-wing, and without any defined edge externally.* Hind-wing: hind-marginal blackish border represented only by a series of small or very small spots on nervules, often more or less acuminate inwardly. *Cilia* of fore-wing reddish, except for a short white portion about posterior angle; of hind-wing white, except for a reddish portion about and near anal angle. UNDER SIDE.—*Hind-wing and apical area of fore-wing pale reddish-creamy, very finely and closely dusted with grey.* Fore-wing: apical patch decidedly yellow-tinged over its inner half,— the intersecting blackish spots small, ill-defined, but with more or less indistinct submetallic brassy centres, and partly surrounded by some diffused dull-red elongate rays; rarely some greyish clouding in basal area. Hind-wing: terminal disco-cellular spot reniform, submetallic brassy (or rarely silvery) ringed with blackish; seven spots of discal row similar, but with darker rings mixed with reddish or ferruginous scales (the two lowest spots small and dull); basi-costal orange edging not well-defined; a minute black spot at extremity of each nervule.

♀ *Apical red patch brighter, paler, and larger than in Annæ; the spots intersecting it smaller and more separate, and the dark borders narrower,—the inner border being also obsolete as regards its upper part;* basal grey often not more developed than in ♂, but variable,—in some examples almost as strong as in *Annæ*, Var. A. Hind-wing: spots of

---

[1] Wallengren's descriptions are sufficient to warrant the references here given; but I have been able to confirm them by the loan from Mr. Aurivillius of typical specimens belonging to the State Museum in Stockholm.

discal series usually smaller and narrower; spots of hind-marginal series usually small, diffused and separate, but occasionally so much enlarged as almost to touch each other. UNDER SIDE.—As in ♂, but spots of discal series in both wings, and terminal disco-cellular spot on hind-wing, larger, and more distinctly silvery-centred. *Fore-wing*: reddish streaks on inner side of apical patch enlarged and confluent; a hind-marginal series of minute nervular black spots.

Wallengrenii appears to represent in the south the species named *Eupompe* by Klug (*Symb. Phys.*, pl. vi. ff. 11-14), which that author records from Abyssinia, Dongola, and Arabia, and which Geyer (*Zutr. Ex. Schmett. Hübn.*, ff. 991-992) also figures from Senegal. In the typical *Eupompe*, the carmine red of the apical patch is in both sexes more vivid and pure than in *Wallengrenii*, and the under side of the hind-wings is white, and either without or with quite minute discal spots. The ♀ is represented by Klug as having the basal grey broadly and strongly developed, as well as the large nervular spots of the hind-wings; but the discal series of the latter wings are faint ochreous-yellow, while on the under side the wide subapical red of the fore-wing and the discal spots of the hind-wing (orange-yellow with imperfect thin blackish rings) are marked features. I possess, however (through the kindness of Professor Meldola), specimens from Harkeko, on the Red Sea, collected by Mr. J. K. Lord, which, in the decidedly yellowish under side of the hind-wing in both sexes, exhibit a tendency in the direction of *Wallengrenii*; the two ♀s of this variation are much smaller and with much less developed basal clouding than Klug's figure, and one ♀ entirely wants the apical red in the fore-wings.

As shown by the measurements given above, *Wallengrenii* is extremely variable in size. Judging from coloured photographs sent to me by the late Mr. E. C. Buxton, some of the specimens taken in Swaziland are even smaller than the minimum I have given. On the other hand, a ♂ from Delagoa Bay surpasses the maximum I have given, expanding as widely as the largest ♂ *Annae*. Nevertheless, the characters emphasised in my description of *Wallengrenii* suffice to distinguish the butterfly even from the Variety A. of *Annae*, which most nearly approaches it.

Localities of *Teracolus Wallengrenii*.

1. South Africa.
    B. Cape Colony.
        b. Eastern Districts.—Kingscote, Keiskamma River (*W. S. M. D'Urban*).
    D. Kaffraria Proper.—Special locality not noted (*J. H. Bowker*).
    E. Natal.
        a. Coast Districts.—D'Urban (the late *M. J. M'Ken*). Verulam (*J. H. Bowker*). Victoria County (*W. Morant*).
        b. Upper Districts.—Weenen County (*J. M. Hutchinson*).
    G. "Swaziland."—The late *E. C. Buxton*.
    H. Delagoa Bay.—Lourenço Marques (*Mrs. Monteiro*).
    K. Transvaal.—Marico River (*F. C. Selous*). Lydenburg District (*T. Ayres*).

II. Other African Regions.
  A. South Tropical.
    *b*1. Eastern Interior.—" Tauwani River (*Oates*)."—Westwood.

277. (13.) **Teracolus Auxo,** (Lucas).

♂ ♀ *Anthocharis Evarne*, Boisd., App. Voy. Deleg. Afr. Aust., p. 587 (1847).
♂ *Anthocharis Auxo*, Lucas, " Rev. et Mag. Zool., 1852, p. 422."
♂ ♀ *Anthopsyche Evarne*, Wallengr., K. Sv. Vet.-Akad. Handl., 1857; Lep. Rhop. Caffr., p. 14, n. 9.
♂ *Anthocharis Evarne*, Trim., Rhop. Afr. Aust., i. p. 55, n. 36 (1862).
♂ ♀ *Callosune Auxo*, Staud., Exot. Schmett., i. pl. 23 (1884).
? ♂ *Teracolus Syrtinus*, Butl., Proc. Zool. Soc. Lond., 1876, p. 163, n. 124.

Exp. al., (♂) 1 in. 7–10 lin.; (♀) 1 in. 7–10 lin.

♂ *Bright pale sulphur-yellow; apical patch large, orange, very slightly glossed with pink, occupying nearly half area of fore-wing.* Fore-wing: base very narrowly blackish; costa edged with black and narrowly speckled with blackish; apical patch internally sometimes with an ill-defined faint blackish edging, externally with a well-marked black edging, widest at apex, on nervules acutely indenting orange, and narrowing to lower extremity of patch, beneath which (immediately above posterior angle) it suddenly widens, forming a blackish spot. *Hind-wing:* a hind-marginal series of small indistinct blackish nervular spots, in some specimens very minute. UNDER SIDE.—*Hind-wing and apical area of fore-wing chrome-yellow.* Fore-wing: apical patch indicated by a flush of deeper yellow and very faint greyish edging; along hind-margin an inter-nervular series of thin black dots. *Hind-wing:* a terminal disco-cellular blackish dot, usually rather indistinct; inter-nervular series of dots as in fore-wing.

♀ *Sometimes paler than ♂, occasionally nearly white; apical patch sometimes rather paler than in ♂, traversed by part of a discal series of large well-marked blackish spots.* Fore-wing: base rather widely but sparsely dusted with blackish; a conspicuous well-defined terminal disco-cellular spot, very variable in size, reniform or subovate in two examples, acuminate superiorly; outer border of orange patch browner than in ♂, emitting longer nervular rays; discal ray composed of three more or less united smaller spots, obliquely placed between dark costal border and lower radial nervule,—two united spots (of which the upper is much the larger) between third and first median nervules, and two (of which the upper is subquadrate and the largest in the series) a little before the two last-named spots, between first median nervule and inner margin; between these three groups of spots in the discal series there is usually a thin indistinct connecting line of blackish scales. *Hind-wing:* a broken discal macular ray of blackish spots, widely and completely interrupted between second subcostal and third median nervules, angulated on the latter nervule, and thence consisting of two or three separate spots, and reaching either to first median nervule or to submedian nervure; basal blackish irroration very sparse, but inferiorly extending widely; blackish spots of hind-marginal series always greatly larger than in ♂, and acuminate inwardly on nervules,— sometimes so much developed as to form a broad border only interrupted by crossing lines of the ground-colour on inter-nervular folds. UNDER SIDE.—*Hind-wing and apical area of fore-wing usually rather duller and more ochreous than in ♂. Fore-wing:* inner-marginal area whitish; terminal disco-cellular spot not so black or so sharply defined as on upper side,—the nervule traversing it often yellowish-white; discal macular series indistinctly marked except the two largest spots. *Hind-wing:* discal macular ray only indicated by some very faint dull reddish-grey marks; a good-sized terminal disco-cellular rounded whitish spot in a thin reddish-grey ring.

Dimorphic Form of ♀.—*Orange of apical patch obsolete*, the only traces of it being some scales on nervules and inter-nervular folds; ground-colour yellowish-white; discal blackish markings smaller and more separate than usual.

(*Hab.*—Kaffraria Proper and Natal.)

VARIETY A. ♂ and ♀.—♂ Yellowish-white except for a tinge of sulphur-yellow along inner edge of apical patch. *Fore-wing:* costal edge very finely black, its border scarcely irrorated with grey; apical hind-marginal border more bluntly indenting the orange on nervules, its lower extremity forming a smaller blackish mark below the orange. *Hind-wing:* spotless; in one example only with the faintest possible indication of minute hind-marginal nervular spots. UNDER SIDE.— Much paler than in type; inner-marginal area of fore-wing white. *Fore-wing:* base tinged with sulphur-yellow. *Hind-wing:* in two specimens, some small faint reddish-brown spots, far apart, indicate position of a discal ray.

♀ Ground-colour as in ♂; markings as in type, except that hind-marginal spots of hind-wing appear to be always rather small and widely separate. UNDER SIDE.—Paler than in type.

One ♂ and two ♀s of this variety to some extent approximate to *Topha*, Wallengr., having the under side of the hind-wing and apical area of fore-wing tinted with pale reddish-creamy, which in the ♀s is speckled with dark-grey.

(*Hab.*—North-West Transvaal and Eastern Tropical South Africa.)

The type-specimens of *Syrtinus*, Butl., in the collection of the British Museum appeared to me to be pale, rather worn ♂s of *Auxo*, slightly approaching the variety just described. They were ticketed "Senegal" and "Lake Nyassa."

Auxo differs from *Evarne*, Klug,[1] which is itself a very near ally of the Indian *Eucharis*, Fab., in its general yellow ground-colour, which in both sexes of *Evarne* is confined to the fore-wings, and is there very much paler, especially in the ♂. The ♂ *Auxo* has the apical

[1] *Symb. Phys.*, pl. vi. ff. 1-4. *Hab.* "Ambukohl" (Dongola).

orange better defined inwardly (and sometimes faintly blackish edged), while its outer blackish border is broader, and forms a broad mark immediately above posterior angle, which is quite wanting in *Evarne;* and, on the under side, sulphur-yellow and deep-yellow take the place of the white area of fore-wing and the yellowish-white of the hind-wing. I have not seen a ♀ *Evarne*, but, judging from Klug's figures, it has much smaller black markings than the ♀ *Auxo*, and wants the whole of the discal spots of the hind-wings except the costal one (which is exceedingly small), while the pale-yellowish under side has the markings very faint and reduced throughout.[1]

Colonel Bowker forwarded in 1873 the paired sexes, taken at Fort Warden on the Kei River: the ♀ of this pair was almost white, with the apical colouring pale salmon-reddish, and the black markings moderately developed. From Kaffraria Proper he had previously sent a good series of both sexes, quite agreeing with the Natalian typical form; and among the few more recently sent by him from Natal is a fine example of the dimorphic ♀ wanting the orange at the apex of the fore-wings. During my stay in Natal I noticed but few of this gaily tinted species; those that I captured flew with moderate activity, and settled often on flowers. These individuals were taken in February and March.[2] Westward of the Kei River, *Auxo* appears to be scarce, being mostly replaced by *Topha*, Wallengr.; but Mr. J. P. Mansel Weale has recorded (*Trans. Ent. Soc. Lond.*, 1877, p. 274) the occurrence of a number of specimens in a spot near King William's Town which had, during the preceding summer, yielded only *Topha*. Mr. Weale notes the appearance of the species from December to April.

The Variety A. above described seems to be numerous in the interior on both sides of the tropical boundary, specimens having been sent from those tracts by Mr. Selous, Mr. Eriksson, and Mr. Fry.

Localities of *Teracolus Auxo*.

I. South Africa.
  B. Cape Colony.
    b. Eastern Districts.—Keiskamma Hoek (*H. J. Atherstone*). "King William's Town."—J. P. Mansel Weale. Fort Warden, Kei River (*J. H. Bowker*).

---

[1] Another even nearer ally of *Auxo* occurs on the White Nile and in Abyssinia, viz., *Xantherarne*, Butl. (*loc. cit.*, n. 123). The ♂ has a larger orange patch, without any trace of inner dark edging, and the ♀ has very strong black markings; while the apex of the fore-wings is in both sexes much rounder. *T. citreus*, Butl. (*loc. cit.*, n. 120), from the White Nile and Hor Tamanib—I have received from Professor Meldola a pair from the latter locality—is a smaller form, with the under side in both sexes (but more strongly in the ♀) pale reddish-creamy reticulated with grey hatchings. The *Evarne* of Geyer (*Forts. Hübn. Zutr. Exot. Schmett.*, ff. 993-994), from Senegal, is evidently a variety of the ♀ of this form, in which the apical orange is wanting. In the ♀ I have from Hor Tamanib there is scarcely the faintest tinge of that colour.

[2] Mr. A. D. Millar notes the abundance of the species in certain localities near D'Urban during December 1887. It frequents the same species of creeper that is affected by *T. Annae*. Three ♂s and two ♀s were kindly forwarded by Mr. Millar.

D. Kaffraria Proper.—Tsomo and Bashee Rivers (*J. H. Bowker*). St. John's River Mouth (*Sir H. Barkly*).
E. Natal.
  a. Coast Districts.—D'Urban. Verulam. Umvoti.
H. Delagoa Bay.-- Lourenço Marques (*Mrs. Monteiro*).
K. Transvaal.—Limpopo and Marico Rivers (*F. C. Selous*). Lydenburg District (*T. Ayres*).

II. Other African Regions.
A. South Tropical.
  b1. Eastern Interior.—Tauwani River and Tchakani Vley (*F. C. Selous*—Var. A.). Between North-West Transvaal and South Matabeleland (*A. W. Eriksson*).

278. (14.) **Teracolus Topha**, (Wallengren).

♀ *Anthopsyche Eucharis*, ? Var., Wallengr., K. Sv. Vet.-Akad. Handl., 1857; Lep. Rhop. Caffr., p. 15.[1]
♂ *Anthopsyche Topha*, Wallengr., Wien. Ent. Monatschr., 1860, p. 34, n. 5.
♂ ♀ *Anthocharis Keiskamma*, Trim., Rhop. Afr. Aust., i. p. 56, n. 37 (1862); and pl. 2, ff. 2 [♂], 3 [♀] (1866).
♂ ♀ *Callosune Topha*, Wallengr., K. Sv. Vet.-Akad. Förhandl., 1872, p. 46, n. 7.

Exp. al., (♂) 1 in. 6–8 lin.; (♀) 1 in. 6–8 lin.

♂ Bright sulphur-yellow, with orange apical patch, shot with pink, occupying *about half the area of fore-wing*. *Fore-wing:* apical patch with *no* internal blackish margin, and externally only *very narrowly* edged with faint-brownish, which extends but very little on either side of apex,—the orange reaching nearly to posterior angle, and slightly infringing on extremity of discoidal cell; base very narrowly marked with blackish; no disco-cellular spot; costal edge very faintly tinged with pale-brownish; fringe pale reddish-brownish as far as first median nervule, between which and posterior angle it is sulphur-yellow. *Hind-wing:* spotless; base very narrowly marked with blackish; fringe sulphur-yellow as far as second or first median nervule,—thence, along inner margin to base, pale russet-brownish. UNDER SIDE.—*Fore-wing:* ground-colour paler than on upper side; position of apical patch only indicated by a deeper shade of yellow; costa narrowly, apex more widely, tinged with very faint brownish,—the apex, in addition, mottled or irrorated with pale-brownish atoms; along hind-margin a row of minute black dots *between* nervules, and two similar dots on costa immediately before apex. *Hind-wing:* dull, pale, reddish-ochreous (*paler on apical half of hind-marginal portion*), *rather thickly irrorated with brownish atoms*; on costa, beyond middle, a faint blackish mark, which, with some much fainter marks between third median nervule and submedian nervure, indicate an angulated macular stripe across wing; along hind-margin a row of minute black dots between nervules.

[1] Wallengren in 1872 (*loc. cit. infra*) referred his ?*Eucharis* ♀ var. to *Topha*; and a typical example lent to me by Mr. C. Aurivillius quite confirmed this determination.

♀ *Ground-colour much paler, sometimes nearly white; orange apical patch paler and duller, with very slight pink gloss. Fore-wing:* a rather large, black, disco-cellular spot; beyond it a curved discal transverse row of five good-sized blackish spots crossing the orange,—the first and second respectively above and below first discoidal nervule,—the third and fourth respectively above and below second median nervule,—the fifth and largest between first median nervule and submedian nervure; base and costa greyish; apical patch margined outwardly with blackish, which slightly indents the orange on nervules, sometimes forming a large blackish mark at posterior angle, below and touching the orange. *Hind-wing:* along hind-margin a row of large, ill-defined, blackish spots at extremities of nervules, but not reaching farther than extremity of second median nervule. UNDER SIDE.—*Fore-wing:* white, or very pale-yellowish; disco-cellular spot and row of spots as on upper side, but not so strongly marked; costa and apex darker than in ♂, the former faintly irrorated with brown, the latter more widely so than in ♂; hind-marginal dots as in ♂. *Hind-wing: clearer in tint than in ♂, not so reddish, irrorations darker;* a conspicuous, shining-white, disco-cellular spot; a distinct angulated, brown, transverse discal stripe or shade, the edges of which are not clearly defined; hind-marginal dots as in ♂ *between* nervules. *Apex of fore-wing more rounded than in ♂.*

In some specimens of the ♀, the discal blackish spots on the upper side are considerably smaller and fainter than as above described, and occasionally they are all but obsolete.

LARVA.—When first hatched, bright-orange; afterwards brownish-green; finally, dull glaucous bluish-green, with a darker median dorsal stripe, and a pale-yellow (almost white) stripe on each side above the legs. Food-plant, *Cadaba Natalensis (Capparideæ).*

PUPA.—Bright-green, with a thin yellow lateral line.

Mr. J. P. Mansel Weale, to whom I owe the above note of the larva and pupa, mentioned in a letter to me (and has also recorded in *Trans. Ent. Soc. Lond.*, 1877, pp. 273-275) that the ♂ *Keiskamma* (= *Topha*, Wallengr.) laid her small, fluted orange-coloured eggs singly on the summit of the flower-buds of the food-plant, and that the young larva penetrates the bud, where it passes its first stage. The latest colouring of the caterpillar assimilates so nearly to that of the leaves that it is difficult to discover the insect, and Mr. Weale obtained most of his specimens by beating the shrub. Finding that the pupæ varied a good deal in colour when developed in confinement, Mr. Weale tried the effect of rearing some specimens " in glass test-tubes exposed on coloured cards, in which they were partially enveloped," with the result that on a vermilion card one pupa was pale-ochreous and another pale bluish-green; on a gamboge-yellow card, bright-green; on green card (cobalt and gamboge), ochreous; on cobalt-blue card, greenish-white. Exposed on the food-plant in nature, the pupa was bright-green; on dead leaves away from the light, dark-brown; on dry stem of food-plant, pale-ochreous; and on a glass tumbler, pale yellowish-green.

Topha is distinguished from *Auxo*, Lucas, by its considerably smaller size and somewhat acuter fore-wings; by the comparatively larger and more deeply-tinted orange patch in the ♂, always wholly free from any trace of inner dark edging, and with only an extremely faint and attenuated brownish outer edging; by the much reduced discal spotting of the ♀, which in the hind-wings disappears altogether, and by the deep dull-reddish colouring of the under side in both sexes. One ♂ example, from Weenen County, in Natal, exhibits considerable deviation from the last-named character, having the under side of the hind-wings yellow, with only a basal reddish stain and some very sparse brownish irroration; and another ♂ from the same district exhibits in the same wings an ill-defined dusky longitudinal streak traversing discoidal cell from base to extremity.

Mr. W. S. M. D'Urban, who in 1860 made known to me this interesting *Teracolus*, was so struck with its abundance in and apparent restriction to the neighbourhood of the Keiskamma in the then Colony of British Kaffraria, that he proposed for it the name of that river, which I accordingly adopted in my *Rhopalocera Africæ Australis*, not being aware that the butterfly had already been described by Wallengren under the name *Topha*. It has since been found to have a considerable range eastward and northward, but seems to occur nowhere in such profusion as in the locality where Mr. D'Urban discovered it.

In the paper already quoted, Mr. Mansel Weale records the interesting fact that while from January to May 1876 he for the first time met with *Keiskamma* (= *Topha*) numerously about a bush of what proved to be their food-plant, a few miles from King William's Town, and saw no examples of *Evarne* (= *Auxo*), yet in the next succeeding summer, from the end of 1876 to the end of April 1877, he found no *Topha* on or about the same *Cadaba* bush, but only *Auxo*, which he had not before noticed in this neighbourhood. He observed the ♀ *Auxo* laying her eggs in the same manner as the ♀ *Topha*; the eggs and the resulting larvæ and pupæ did not differ from those of *Auxo*, and the pupæ exhibited the same liability to vary in colour.

Mr. Weale wrote to me that he regarded these observations as proving the species-identity of the two butterflies, and certainly the evidence in that direction is strong. But a difficulty occurs in the circumstance that the two forms were not seasonal ones in the ordinary sense, but appeared in the corresponding (summer) season in both years of observation; and it is also to be noted that Mr. D'Urban (who did not meet with *Auxo* during his stay in British Kaffraria) expressly wrote that *Topha* was on the wing "all the year round." Mr. Weale mentions, however, that the summer when *Topha* prevailed was a wet one, and that the succeeding one, when only *Auxo* was seen, was unusually dry. Until one form has been shown by direct observation to result from ova laid by the other, I think it advisable, in view of the very marked differences exhibited in both sexes, to keep the two apart.

Localities of *Teracolus Topha*.

I. South Africa.
 B. Cape Colony.
  b. Eastern Districts.—"Kingscote to Chalumna and Line Drift," Keiskamma (*W. S. M. D'Urban*). "King William's Town."—J. P. Mansel Weale.

D. Kaffraria Proper.—Special locality not noted (*J. H. Bowker*).
E. Natal.
  b. Upper Districts.—Weenen County (*J. M. Hutchinson*).
F. Zululand.—St. Lucia Bay (the late *Colonel H. Tower*).
K. Transvaal.—Lydenburg District (*T. Ayres*).

279. (15.) Teracolus Evenina, (Wallengren).

♀ *Anthopsyche Evenina*, Wallengr., K. Sv. Vet.-Akad. Handl., 1857; Lep. Rhop. Caffr., p. 12, n. 3.
♂ *Anthopsyche Deidamia*, Wallengr., Wien. Ent. Monatschr., 1860, p. 35, n. 7.
♀ *Anthocharis Evenina*, Trim., Rhop. Afr. Aust., ii. p. 322, n. 216 (1866).
♀ *Callosune Evenina*, Trim., Trans. Ent. Soc. Lond., 1870, p. 380, pl. vi. f. 11.
♂ *Anthopsyche Deidamia*, Wallengr., K. Sv. Vet.-Akad. Förhandl., 1872, p. 44.
♂ ♀ *Callosune Evenina*, Wallengr., *op. cit.*, 1875, p. 90, n. 51.
♂ *Callosune casta*, Gerst., Gliederth.-Faun. Sansib.-Gebiet., p. 365, tab. xv. ff. 1, 1a (1873).
Var. ♂ *Callosune deidamioides*, Auriv., K. Sv. Vet.-Akad. Förhandl., 1879, p. 45, n. 22.
♂ (*nec* ♀) *Teracolus Sipylus*, Swinhoe, Proc. Zool. Soc. Lond., 1884, p. 444, pl. xl. f. 11.

Exp. al., (♂) 1 in. 9 lin.—2 in.; (♀) 1 in. 8–10½ lin.

♂ *Pure white, with rather large orange apical patch in fore-wing, and a blackish longitudinal stripe (greyish-clouded in fore-wing) in both wings.* Fore-wing: orange patch edged externally and superiorly by a narrow but very clearly defined black border (piercing the orange to some depth on nervules, especially on third and second median nervules), and internally by a diffused but usually rather broad and conspicuous blackish bar from costa, joining the hind-marginal border below second median nervule; costa before middle very thinly black-edged; along inner margin a more or less developed blackish longitudinal stripe (all the basal part of which is grey with a dense irroration of white scales), sometimes fading out before middle, but, when fully developed, ending abruptly beyond middle; this stripe is superiorly more or less diffused, sometimes so wide near base as to rise above median nervure; above its outer terminal portion a grey shade indicates position of the wider corresponding black band of the *under side*. Hind-wing: along costa a similar black stripe, greyish and diffused in that part of its wide base which lies between costal and submedian nervures, abruptly and excavately truncate beyond middle; along hind-margin a series of black nervular spots, very variable in size and acutely pointed inwardly, and usually a thin black bounding line. UNDER SIDE.—*White*. Fore-wing: apical patch pale-orange inwardly, pale-creamy outwardly, the inner blackish border of the upper side faintly indicated; nervules crossing patch white, but in some specimens with some faint diffused fuscous clouding where they cross the pale-orange; inner marginal blackish stripe not rising above median nervule, its outer extremity broader, blacker, more distinctly truncate than on upper side, bounded superiorly by first median nervule. *Hind-wing:* costa broadly edged with bright-orange from base to before middle; an indistinct grey shade marks position of costal stripe of upper side, and a small dusky mark (usually mixed with rufous) its extremity beyond middle; a hind-marginal series of minute nervular spots, usually wanting above third median nervule; rarely a faint dusky longitudinal ray along disco-cellular fold to beyond middle.

♀ *Blackish markings largely developed, especially in basal areas. Fore-wing:* apical patch duller than in ♂, inclining to yellow along its inner edge (which has no trace of a blackish border), and intersected by a well-marked, curved submacular blackish stripe from costa to second median nervule, or a little below it; apical hind-marginal border of patch brownish, broader than in ♂, more broadly and deeply penetrating the orange on nervules, but itself pierced almost or quite to hind-marginal edge by the prolonged acuminate points of the inter-nervular orange rays thus formed; basal blackish, more or less white-scaled, not reaching to middle of inner margin, but filling the whole of discoidal cell (leaving costal border white), concavely excised at extremity of cell, extending beyond and below cell, and ending in an abruptly truncate projection on median nervules; below first median nervule, beyond middle, a large quadrate blackish spot (representing termination of inner-marginal stripe in ♂), the upper inner angle of which touches, or almost touches, truncate extremity of basal blackish area. *Hind-wing:* basal area more or less deeply suffused with greyish-fuscous, having a white or whitish space on costa about middle, immediately followed by a large inferiorly acuminate blackish costal marking; extremity of discoidal cell usually blackish, enclosing a thin whitish terminal lunule; hind-marginal spots much enlarged and diffused, forming a broad border, enclosing externally a series of small ill-defined whitish inter-nervular spots; preceding the border, but partly touching it, a discal transverse blackish ray on median nervules. UNDER SIDE.—*Hind-wing and outer part of apical patch of fore-wing more or less tinged with creamy, rarely of a decided yellowish or reddish tint. Fore-wing:* blackish markings of upper side reproduced, but much fainter, the large basal one densely irrorated with yellowish except on its outer projections; intersecting ray of apical orange faintly indicated. *Hind-wing:* a very faint grey shade representing basal clouding; costal mark and discal ray ill-defined, more or less red; a rather conspicuous orange terminal disco-cellular spot centred with white; often a slight irroration of grey atoms over all the wing, except upper half beyond middle.

VARIETY A. ♂ and ♀ (*T. Deidamioides,* Auriv.).

♂ *Apical patch of fore-wing with much reduced, sometimes almost evanescent, inner blackish edging, usually broken at second median nervule; inner-marginal stripe only narrowly developed near base;* costal stripe of hind-wing fainter and narrower. UNDER SIDE.—*Hind-wing and* outer part of apical patch of fore-wing more or less tinged with pinkish-creamy, irrorated (and in parts shortly striolated) with grey; in hind-wing a more or less conspicuous fuscous or dusky longitudinal ray along disco-cellular fold to a little before hind-margin; in fore-wing the inner-marginal streak is usually represented by a greyish mark at base and a faint discal spot, but is sometimes wanting altogether.

♀ *Under-side colour duller than in ♂, and more closely and generally hatched and irrorated in hind-wing, so that disco-cellular dusky ray is less distinct or sometimes hardly distinguishable;* costal and discal markings of hind-wing also diffused, and not, or but little, red-tinged.

(*Hab.*—North of Cape Colony, Western Basutoland, Griqualand West, Transvaal, Eastern Tropical Interior, and Damaraland.)

I was enabled by the kindness of Mr. Chr. Aurivillius to examine a type specimen of his *Deidamioides*, a ♂ in which the disco-cellular ray of the under side of the hind-wings was not very marked. In the South-African Museum and in my own collection there are several Transvaal ♂ s in which the under-side colouring is intermediate between this variety and the typical *Evenina* ♂. *Casta* of Gerstäcker, founded on a single ♂ from Lake Jipè in East Africa, in some respects is very near *Deidamioides*, but the white under side of the hind-wings is quite like that of *Evenina* ♂, and I do not think that it can be separated from the latter. The South-African Museum possesses three unusually small ♂ s of the variety from Damaraland, expanding only 1 in. 7 lin. across the wings; and some still smaller ♂ s (*exp. al.* 1 in. 4–5 lin.) and a ♀ (*exp. al.* 1 in. 5 lin.) were captured in Namaqualand by Mr. L. Péringuey. The pinkish tint of the under side of the hind-wings is faint in these dwarfed examples, and its irroration sparse and limited.

Evenina ♂ is highly variable both in size and markings. The largest and most strongly marked specimens I have seen are two[1] taken by Mr. J. L. Fry at Tati on 8th January 1887,—the black borders, stripes, and spots being throughout larger and darker, especially the inner-marginal stripe of the fore-wings, and the hind-marginal spots of the hind-wings. The smallest and most faintly marked (excluding the Variety A.) are from Bechuanaland and Damaraland. The colour of the apical patch is very near that shown by the European *Anthocharis Cardamines* ♂, but in fine examples exhibits a slight pink gloss. *Evenina* ♀ varies very much less than the ♂, and chiefly in the extent and intensity of the basal and hind-marginal blackish of the hind-wings. Two ♀ s from Tati which accompanied the ♂ s just mentioned are, however, remarkable for the corresponding great development of its dark

---

[1] Another specimen is rather below the average size, and distinctly but not heavily marked; while two others hold an intermediate position, but are rather larger than the average. I have examined the types of Swinhoe's *Sipylus* from Zanzibar in the British Museum. The ♂ is inseparable from the larger darker specimens of ♂ *Evenina* just mentioned, though it is somewhat more heavily marked; while the ♀ is plainly referable to *Garisa*, Wallengr.

upper-side markings, and for the strong yellow tinge of its under-side colouring.

Wallengren (*loc. cit.*) in 1875 recorded the confirmation of his previous suggestion (in 1872), that his *Deidamia* would prove to be the ♂ of *Evenina*, by the capture in the Transvaal (on 27th March 1870) by Mr. N. Person of the paired sexes. Mr. W. Morant on 3d March 1872 took near Potchefstroom (as he wrote to me) a *Deidamia* "at the same spot as the *Evenina* now sent;" and I had for some time previously looked upon the two as in all probability sexes of the same species.

The ♂ of this beautiful *Teracolus*, as Wallengren pointed out in his original diagnosis, has some affinity to *T. Ephyia* of Klug (*Symb. Phys.*, t. vi. ff. 9, 10), but judging from the latter author's figures and brief description, it is a far larger butterfly, with a relatively much larger and brighter orange apical patch than *Ephyia* : the Dongolan species is, moreover, entirely devoid of any longitudinal blackish striping in both wings, has a distinct black terminal disco-cellular spot in the fore-wing, and the white under side of hind-wing absolutely without marking of any kind.

The ♀ *Evenina* appears to stand quite alone in the peculiar distribution and outline of the basal blackish clouding, especially in the fore-wing, where, as above described, it completely fills the discoidal cell up to the subcostal nervure, and forms a singular truncate projection below and beyond the extremity of the cell.

Evenina seems to be tolerably prevalent in the Transvaal and farther to the northward, spreading far into the South Tropical regions, but only penetrates sparingly the northern parts of the Cape Colony, and has not been recorded from any locality to the southward of the 30th parallel. There were thirty-eight examples in Mr. J. A. Bell's collection, made in Damaraland.

Localities of *Teracolus Evenina*.

I. South Africa.
  B. Cape Colony.
    a. Western Districts.—Springbokfontein (*G. A. Reynolds*), and Spectakel (*L. Péringuey*).—Var. A.
    b. Eastern Districts.—Hope Town (*J. H. Bowker*).—Var. A.
    c. Griqualand West.—Vaal River (*Mrs. Barber* and *J. H. Bowker*).—Typ. and Var. A.
    d. Basutoland.—Maseru (*J. H. Bowker*).—Typ.
    e. Orange Free State.—Hebron (*W. Morant*).
  J. Delagoa Bay.—Lourenço Marques (*Mrs. Monteiro*).
  K. Transvaal.—Limpopo River (*F. C. Selous*).—Typ. Potchefstroom and Lydenburg Districts and Origstadt Valley (*W. Morant* and *T. Ayres*).—Typ. and Var. A.
  L. Bechuanaland.—Motito (the late *Rev. J. Frédoux*).—Typ.

II. Other African Regions.
  A. South Tropical.
    a. Western Coast.—Damaraland (*J. A. Bell* and *W. C. Palgrave*).—Typ. and Var. A.
    b. Eastern Coast.—Zanzibar.—Coll. Brit. Mus. [Type of *Sipylus*, Swinhoe].
    *b*1. Eastern Interior.—Tati (*J. L. Fry*).—Typ. Mombasa: "Lake Jipé (*Kersten*)."—Gerstäcker [*Casta*, Gerst.]. "Victoria Nyanza."—Butler [*Casta*, Gerst.].

280. (16.) **Teracolus simplex**, Butler.

♂ *Teracolus simplex*, Butl., Proc. Zool. Soc. Lond., 1876, p. 148, n. 71.
♂ *Callosune Damarensis*, Auriv., K. Sv. Vet.-Akad. Förhandl., 1879, p. 46, n. 23.

Exp. al., (♂) 1 in. 9–10 lin.

♂ *White, with moderate-sized scarlet-orange apical patch; base of both wings with a very slight grey irroration.* Fore-wing: a minute terminal disco-cellular black spot; inner edge of apical patch very oblique, irregularly denticulated, *without any blackish or dusky scaling*,—costal and hind-marginal edge with an extremely narrow border of fuscous-grey, from apex downward emitting short nervular rays piercing the orange, and ending between second and first median nervules. Hind-wing: a hind-marginal series of terminal nervular black dots. UNDER SIDE.—Fore-wing: disco-cellular dot well marked; apical patch creamy-yellowish faintly irrorated and sparingly striolated with grey, its inner border tinged with very pale-orange. Hind-wing: creamy-whitish, very faintly tinged with pink, sparsely irrorated and striolated with grey (rather more closely so near base and on costa a little before apex); a terminal disco-cellular black dot, very faintly orange-bordered inwardly.

This is a form of which the ♀ is as yet unknown. The ♂ is to some extent a link between the *Evenina* and *Achine* groups of the genus, and may be recognised by the peculiar form and colour of the apical patch, with its total want of any inner dark edging, and extremely thin and dull costal and outer edging, as well as by the complete absence of any trace of fuscous bordering along either the inner margin of the fore-wings or the costa of the hind-wings.

The type of *Simplex* is a single injured ♂, ticketed "D'Urban," in the collection of the British Museum. By the kind aid of Mr. C. Aurivillius, I was able to compare with it a typical example of his *Damarensis*, and found the two to be unquestionably the same species. A third ♂ has reached me from Mashunaland near the Lower Zambesi, in a collection made by Mr. F. C. Selous. Mr. Aurivillius (*loc. cit.*) notes the near apparent alliance of this butterfly to *Anterippe*, Boisd., and *Zera*, Lucas, both from North Tropical Africa. These two forms (judging from specimens so named in the collection of the British Museum) belong to the *Achine* group, and do not appear to me to be separable as species.

Localities of *Teracolus simplex*.

I. South Africa.
  E. Natal.
    a. Coast Districts.—D'Urban (*Shelley*).—Coll. Brit. Mus.
II. Other African Regions.
  A. South Tropical.
    a. Western Coast.—Damaraland (*G. De Vylder*).—Coll. Stockholm Mus.
    b1. Eastern Interior.—Mashunaland (*F. C. Selous*).

281. (17.) Teracolus Achine, (Cramer).

♀ *Papilio Achine*, Cram., Pap. Exot., iv. pl. ccexxxviii. ff. E, F (1782).
♂ ♀ *Aphrodite Achine*, Hübn., Samml. Exot. Schmett., ii. t. 128, ff. 1, 2 [♂]. 3. 4 [♀] (? 1816).
♀ *Pieris Achine*, Godt., Enc. Meth., ix. p. 122, n. 13 (1819).
♂ ♀ *Anthocharis Achine*, Boisd., Sp. Gen. Lep., i. p. 574, n. 21 (1836).
♂ ♀ „ „ Trim., Rhop. Afr. Aust., i. p. 46, n. 29 (1862).
♀ *Anthopsyche Roxane*, Feld., Reise d. Novara, Lep., ii. p. 187, n. 182 (1865).
♀ *Teracolus Achine*, Butl., Proc. Zool. Soc. Lond., 1876, p. 149, n. 77.
♂ ♀ *Teracolus Trimeni*, Butl., loc. cit., n. 79.
Var. ♂ ♀ *Teracolus Hero*, Butl., loc. cit., n. 81, pl. vi. f. 11.
Var. ♂ ♀ *Teracolus Hyperides*, Butl., loc. cit., n. 76.

Exp. al., (♂) 1 in. 9–11 lin; (♀) 1 in. 7 lin.—2 in.

♂ *White; apical patch bright red, inclining to orange, but glossed with a rosy lustre.* Fore-wing: apical red divided into six by the dark nervules crossing it, margined outwardly by a narrow black stripe sharply indenting the red by radiating upon the nervules, inwardly by a thin, blackish, ill-defined streak, within which the nervules are usually black, piercing white ground for a little distance; costa edged with blackish, widening above apical red; a distinct round, black dot at extremity of discoidal cell; a broad, blackish band along inner margin to beyond middle, where it very slightly curves upwards and ends abruptly. Hind-wing: base greyish; a blackish band along costa adjoining that on inner margin of fore-wing, and abruptly ending immediately below where the latter ends; on hind-margin, at extremities of nervules, a row of black spots, variable in size, radiating on nervules, and inclining to form a continuous band, generally diminishing in size towards posterior angle. UNDER SIDE.—Fore-wing: apical marking soft creamy-yellow, inwardly pale-orange; disco-cellular spot distinct; inner-marginal band pale-greyish from base, ending in a darker spot. Hind-wing: varying from creamy-white to creamy-yellowish or creamy-reddish, in the more deeply tinted examples more or less irrorated with grey atoms; costa, from base to beyond middle, edged with bright orange-yellow; disco-cellular spot orange, outwardly black-dotted; a scarcely perceptible greyish tint along costa indicates the position of the costal band, the extremity of which is always shown by a greyish mark, sometimes mixed with reddish; rarely some indistinct traces on lower disc of an angulated band.

♀ *Very variable in size, frequently smaller than ♂; of a duller white, or sometimes decidedly yellowish.* Fore-wing: red at apex much narrower than in ♂, *of a more orange tint, and without rosy lustre*, its black borders very much wider, especially the inner one (which latter is, however, often intersected by the red, and occasionally with faint reddish colouring on its inner side); spot at extremity of discoidal cell larger than in ♂; *inner-marginal blackish band very much broader and darker*, its upper edge not clearly defined, but gradually shading off in discoidal cell to near costa, and its outer extremity usually united to inner black border of apical marking, on third median nervule, by a narrow, upward, blackish streak. *Hind-wing:* grey at base and costal band similar to the same in ♂, but broader and darker; from extremity of band a more or less marked blackish streak extends almost to middle of hind-margin, where it unites with a broad diffused submacular blackish border, and from whence it turns off at right angles to beyond middle of inner margin, this latter portion being very much broader than the rest. UNDER SIDE.—*Duller in colour than in ♂, and usually more or less tinged with yellow. Fore-wing:* a blackish streak bordering inner side of apical marking; inner-marginal band greyish, ending in a well-marked black spot, which is often united by a thin, faint, blackish streak, to the inner border of apical marking, as on upper side; base sometimes tinged with pale yellow. *Hind-wing:* greyer in tint than in ♂, but marked similarly, excepting more or less indistinct pale-brownish traces of the angulated band of upper side.

The apical red in the ♀ presents various gradations through orange to dull-yellowish, and is rarely wanting altogether, being replaced by whitish spots.

The ♂ varies considerably, not only in the tints of the under side above mentioned, but in the development of the black markings, more especially as regards the inner edging of the apical patch, the longitudinal blackish band in both fore and hind wings, and the hind-marginal spots of the hind-wings.[1] The ♀ also presents much variation in her considerably heavier black markings, which are sometimes so enlarged as to leave only isolated patches of the whitish or yellowish ground-colour.

The *T. Hero* of Butler (the types of which, a ♂ and a ♀, I have examined) is founded upon a ♀ which presents these very strongly developed blackish markings in conjunction with a dull-yellowish ground-colour and a scarcely brighter apical patch; and upon a ♂ in which the blackish markings of the fore-wings are reduced,—the inner-marginal band being dusky-grey and becoming obsolete beyond middle, and the inner edging of apical patch being also almost obsolete. Butler's *T. Trimeni* (of which I have also examined the types) is founded on specimens identified with Hübner's figures above cited (including some taken by myself in the Knysna District of Cape Colony), which cannot, in my opinion, be held distinct from typical *Achine*. Hübner's figures represent a ♂ and ♀ in which the upper-side black markings are, though very distinct, considerably narrowed; but it seems to me impossible to separate Hübner's ♀ from that originally figured by Cramer, which only differs in its duller white ground-colour and heavier black markings.

Hyperides, Butl., is founded on a ♂ and two ♀s collected by

---

[1] In two ♂s among those that I captured near Grahamstown in the Cape Colony, there are faint blackish traces of an angulated discal ray on the *upper side* of the hind-wings.

the late Mr. E. C. Buxton in Swaziland, of rather smaller size than usual (♂ 1 in. 9 lin., ♀ 1 in. 8 lin.). The ♂ has the longitudinal stripe of the fore-wing reduced and grey, while that of the hind-wing is obsolete; and the under side of the hind-wings is white. The ♀ s cannot, in my opinion, be separated from the smaller examples of *Achine*.

A dwarf ♂ in my collection is only 1 in. 6 lin. in expanse, while two still smaller ♀ s expand respectively 1 in. 5 lin. and 1 in. 3 lin. I captured these three examples at Knysna.

From Natal, Delagoa Bay, the Transvaal,[1] and the Eastern South-Tropical Interior, all the ♂ examples I have seen have the under side white or creamy-white, and several of those from the two tracts last-named (as well as two of the ♀ s) exhibit a tendency towards the very nearly allied *Garisa*, Wallengr., in the nervules on the under side being blackish close to the hind-margin.

This gaily-coloured *Teracolus* is numerous, and generally distributed over most of Eastern South Africa, but in the Cape Colony does not appear to extend westward beyond the Oudtshoorn District. I have met with it freely at Knysna, Port Elizabeth, Uitenhage, and near Grahamstown, and took the paired sexes at the first-named place on 18th November 1858. In Natal it seems on the coast to be replaced almost entirely by the doubtfully distinct but larger and more heavily marked *Garisa*, Wallengr. During the summer and early autumn it is constantly to be found coursing actively along the edges of woods, seldom penetrating for any distance among the trees or ranging far into open ground.

Localities of *Teracolus Achine*.

I. South Africa.
   B. Cape Colony.
      a. Western Districts.—Oudtshoorn (— *Adams*). Knysna and Plettenberg Bay.
      b. Eastern Districts.—Port Elizabeth. Uitenhage. Kowie River Mouth (*J. L. Fry*), Kleinemond River (*H. J. Atherstone*), and Tharfield (*Miss M. L. Bowker*). Grahamstown. King William's Town (*W. S. D'Urban*).
   D. Kaffraria Proper.—Butterworth and Bashee River (*J. H. Bowker*).
   E. Natal.
      a. Coast Districts.—Verulam.
      b. Upper Districts.—Weenen County (*J. M. Hutchinson*).
   G. "Swaziland."—The late E. C. Buxton.
   H. Delagoa Bay.—Lourenço Marques (*Mrs. Monteiro*).
   K. Transvaal.—Limpopo and Marico Rivers (*F. C. Selous*), and Junction of those rivers (*A. W. Eriksson*). Potchefstroom and Lydenburg Districts (*T. Ayres*).

---

[1] I have not seen the types of *T. fumidus*, Swinhoe, from Transvaal; but the figures and descriptions (*Proc. Zool. Soc. Lond.*, 1884, pl. xl. ff. 4, 5, p. 442) of both sexes induce a strong belief that the species in question has been founded on dwarf examples of *Achine*. The expanse of fore-wings noted is 1½ inch.

II. Other African Regions.
   A. South Tropical.
      a. Western Coast.—Damaraland (*J. A. Bell* and the late *C. J. Andersson*). "Angola (*Pogge*)."—Dewitz.
      b1. Eastern Interior.—Tauwani River (*F. C. Selous*). Between North-West Transvaal and Gubulewayo (*A. W. Eriksson*). Matabeleland (*H. Barber*). Zambesi: "Tette."—Hopffer. Mombasa: "Lake Jipè (*Kersten*)."—Gerstäcker.

282. (18.) Teracolus Gavisa, (Wallengren).

♂ ♀ *Anthocharis*, "allied to *A. Achine*," Angas, Kafirs. Illustr., pl. xxx. ff. 4 [♂], 5 [♀] (1849).
? ♀ *Anthocharis Eeole*, ♀, Reiche, Ferr. et Gal. Voy. Abyss., iii. pl. 31, ff. 5, 6 (1849).
♂ *Anthopsyche Gavisa*, Wallengr., K. Sv. Vet.-Akad. Handl., 1857; Lep. Rhop. Caffr., p. 13, n. 6.
♀ *Anthopsyche Omphale*, Wallengr., *loc. cit.*, p. 11, n. 2.
Anthocharis Achine, Trim. [part.], Rhop. Afr. Aust., i. p. 48, *obs*.
♂ *Anthocharis Gavisa*, Trim., *op. cit.*, ii. p. 324, n. 218 (1866).
♂ ♀ *Teracolus Gavisa*, Butl. [part.], Proc. Zool. Soc. Lond., 1876, p. 150, n. 80.
♀ [and ? ♂] *Teracolus subvenosus*, Butl., Ann. and Mag. Nat. Hist., 5th ser., vol. xii. p. 105 (1883).
♀ [nec ♂] *Teracolus Sipylus*, Swinhoe, Proc. Zool. Soc. Lond., 1884, p. 445. pl. xl. f. 12.
♂ *Callosune Gavisa*, Staud., Exot. Schmett., i. pl. 23 (1884).

Exp. al., (♂) 1 in. $9\frac{1}{2}$ lin.—2 in. $0\frac{1}{2}$ lin.; ♀ 1 in. $9\frac{1}{2}$–$11\frac{1}{2}$ lin.
Very intimately related to *T. Achine*, Cram.

♂ Markings and pattern as in *Achine*, but all the black borders and bands broader and darker. *Hind-wing*: a continuous moderately wide hind-marginal black border, formed of the enlarged and confluent nervular spots. UNDER SIDE.—*Neuration of fore-wing black beyond middle, that of hind-wing black throughout, except main nervures for a little distance from base*. *Hind-wing*: creamy-white; orange-yellow mark immediately preceding terminal disco-cellular black spot more or less prolonged in direction of base; from hind-margin to about middle, a black line on inter-nervular fold between first median nervule and submedian nervure.

♀ Like *Achine* ♀, but black markings always very broad and dark. *Fore-wing*: terminal disco-cellular spot sometimes merged in the dark clouding extending from base; spaces of whitish or yellowish ground-colour considerably reduced. *Hind-wing*: spaces of ground-colour reduced by enlargement of black markings to a moderate-sized central patch, and a series of submarginal spots (of which the middle two are more or less obliterated). UNDER SIDE.—Hind-wing and outer portion of apical patch of fore-wing strongly yellow-tinged (sometimes with a greenish cast); *black neuration as in ♂*. *Hind-wing*: angulated discal ray well marked in diffused dull orange-yellow; bright orange-yellow mark preceding terminal disco-cellular spot usually larger than in ♂, in one example much wider externally and prolonged almost to base.[1]

It is with much hesitation that I accord *Gavisa* specific rank, having in view the examples of both sexes of the variable *Achine*, which exhibit a partial blackening of the neuration as above mentioned. It seems, however, to stand out as a prominent race, and perhaps may breed apart from *Achine*, if we can judge by a single case of pairing observed by Colonel Bowker at D'Urban in December 1879. In this pair (sent to me by Colonel Bowker) the characteristic complete black neuration of the under side is equally well marked in both sexes.

A typical specimen (♂) of Wallengren's *Gavisa* was kindly lent to me by Mr. Aurivillius, and it should be noted that it agreed with the author's description (*loc. cit.*) in having the black neuration of the under side limited to near the hind-margins, and in that respect more resembled the intermediate examples which I have referred to *Achine*.

Gavisa seems limited to the eastern side of South Africa, the most western locality I have recorded for it being the right bank of the Kei River, whence Colonel Bowker sent one of each sex in 1873. These two specimens, though strongly black-veined beneath (the ♂ especially), are less heavily black-marked above, and the ♀ has the under side of the hind-wings tinged with ochre-yellow.

I met with this form of *Teracolus* not uncommonly in Natal, especially in the neighbourhood of D'Urban; it had the ordinary flight and habits of the genus, and I observed it on the wing at the end of January and throughout February and March.[2]

Localities of *Teracolus Gavisa*.

I. South Africa.
    B. Cape Colony.
        b. Eastern Districts.—Fort Warden, Kei River (*J. H. Bowker*).
    D. Kaffraria Proper.—Bashee River (*J. H. Bowker*).
    E. Natal.
        a. Coast Districts.—D'Urban. Pinetown (*J. H. Bowker*). Tongaati River. Little Noodsberg.
        b. Upper Districts.—Estcourt (*J. M. Hutchinson*).
    G. "Swaziland."—The late *E. C. Buxton*.
    K. Transvaal.—Junction of Limpopo and Marico Rivers (*A. W. Eriksson*). Lydenburg District (*T. Ayres*).

---

[1] As Gerstäcker (*Gliedl.-Fauna des Sansibar-Gebietes, Ins.*, p. 364) has noted, the ♀ figured by Reiche as that of his *Exole* (*loc. cit.*) undoubtedly belongs to the *Achine* group, and I concur with Mr. Butler in thinking that it should be regarded as identical with the ♀ of Wallengren's *Gavisa*.

Having examined the types of *Subvenosus*, Butl., from Victoria Nyanza, I find the ♀ inseparable from that of *Gavisa*, while the ♂, though very near the corresponding sex of the species named, differs in wanting the inner black edging of the apical patch, and in the feeble development of the inner-marginal blackish bar of the fore-wings and the costal one of the hind-wings.

[2] Five examples sent to me from D'Urban by Mr. A. D. Millar are dated 17th December 1887.

II. Other African Regions.
  A. South Tropical.
    b. Eastern Coast.—"Zanzibar."—Swinhoe [*Sipylus*, S., ♀].
    b1. Eastern Interior.—Between North-West Transvaal and Gubule-wayo (*A. W. Eriksson*).
  B. North Tropical.
    b1. Eastern Interior.—"Abyssinia."—Reiche [♀ *Evote*, R.].

283. (19.) Teracolus Antevippe, (Boisduval).

♂ ♀ *Anthocharis Antevippe*, Boisd., Sp. Gen. Lep., i. p. 572, pl. 18, f. 3 [♂] (1836).
♀ *Aphrodite Antevippe*, Geyer, Forts. Hübn. Zutr. Exot. Schmett., p. 37, ff. p. 949–950 (1837).
♂ *Anthocharis Antevippe*, Trim., Rhop. Afr. Aust., i. p. 48, n. 30; and (♀) p. 49, *note* (1862).
♂ ♀ *Teracolus Ithonus*, Butl., Proc. Zool. Soc. Lond., 1876, p. 146, n. 66, pl. vi. f. 8.
Teracolus Harmonides, Butl., *loc. cit.*, p. 146, n. 167.
Teracolus Hipporrene, Butl., *loc. cit.*, p. 147, n. 68.
Teracolus ignifer, Butl., *loc. cit.*, p. 147, n. 69.

Exp. al., (♂) 1 in. 8 lin.—2 in.; (♀) 1 in. 5½–10 lin.

♂ *White, with bright-red, slightly rosy-glossed apical patch, without* (or with extremely thin and faint) *internal blackish edging; no longitudinal blackish band in either wing*. *Fore-wing*: terminal disco-cellular spot very small and narrow; lower portion of base narrowly blackish; apical patch very similar in colour and form to that of *Achine* ♂, but with the outer black border narrower and radiating more thinly on the nervules, so that the inter-nervular rays of red are blunter terminally; costal black-edging as in *Achine*. *Hind-wing*: a narrow blackish suffusion at base, more extended between median and submedian nervures; nervular hind-marginal black spots small or very small, more or less linear and acuminate inwardly, but wider outwardly (rarely so wide that they almost touch). UNDER SIDE.—*Fore-wing*: terminal disco-cellular spot rather larger than on upper side; apical patch creamy-yellowish, more or less freckled with reddish-brown atoms and striolæ,—along its inner portion a pale-orange suffusion almost devoid of freckling. *Hind-wing*: varying from creamy-white to creamy-pinkish, always more or less freckled and striolated with brownish or reddish-brown,—these markings being densest along costa near base, and in, below, and beyond discoidal cell; a small but distinct terminal disco-cellular black dot, faintly edged or bordered with yellow inwardly.

♀ *Pattern of Achine* ♀, *but much less heavily marked with black, especially in hind-wing;* apical patch of fore-wing with the coloured portion varying from reddish-orange to pale-yellow, and with the traversing black macular ray near its inner portion usually of moderate width, but in some specimens broadened so as to form a border with no orange or yellow preceding it. *Fore-wing:* disco-cellular spot larger and rounder than in ♂; inner-marginal stripe moderately broad, varying from pale-grey to blackish,—its truncate extremity always blackish, and often emitting a short thin upward streak,—its basal portion often more or less thinly diffused over discoidal cell. *Hind-wing:* costal stripe rather ill defined and diffused,—its blackish extremity almost separate, elongate, forming the commencement of the ordinary angulated discal ray, which is usually obsolete or obsolescent between second subcostal and radial nervules, and sometimes so near inner margin also; basal dusky-grey suffusion usually filling greater part of discoidal cell, and extending below it to beyond middle; hind-marginal black spots very variable in development, sometimes scarcely larger than in ♂, but usually much more so,—especially the three superior ones, which occasionally coalesce into a short border. UNDER SIDE.—*Fore-wing:* apical patch darker than in ♂, being more densely freckled, the traversing ray of the upper side reproduced in dusky grey; base faintly tinged with yellowish; disco-cellular spot well defined; inner-marginal stripe fainter and narrower, except at extremity. *Hind-wing:* pinkish-creamy or yellowish-creamy, more closely striolated and freckled generally than in ♂; incomplete discal ray faintly indicated in brownish.

LARVA.—Dull reddish-sandy above, with a broad dull-grey median longitudinal stripe, interrupted on middle of each segment by a large quadrate purplish-black spot. Sides superiorly whitish, lower down greyish-sandy, and inferiorly (with legs, ventral surface, and head) dark-brown. General surface very sparsely set with very short white hairs. Dorsal surface minutely shagreened with raised whitish dots, but also with a number of considerably larger and more prominent deep gamboge-yellow dots, in transverse rows of six and two on each segment.

(Described from a single specimen, advanced towards pupation, received from Colonel Bowker in August 1887, which pupated on 25th.)

PUPA.—Very pale ochreous-yellow, with a slight reddish tinge over all the lower surface. On back, a median and two lateral grey longitudinal streaks,—each of the latter bounded interiorly by a yellowish-white line; between median and lateral streaks, on each side, two rows of more or less distinct small blackish spots; mixed with these some minute greyish speckles. Wing-covers and leg-covers here and there finely streaked with grey, the former with a row of minute black dots near hind-margin. Abdomen inferiorly with a median and two lateral dark-grey longitudinal streaks. Length, 9 lin.; depth (from back to bulge of wing-covers), 3 lin.

(Described from two specimens received from Colonel Bowker (D'Urban, Natal) in August 1887, resulting in ♂ imagines on the 23d and 25th respectively.)

The pupa formed by the larva above described produced a ♀ imago on 14th September. It was more brightly and deeply coloured than the two ♂ pupæ, pale-yellow on the back, the dorsal median streak purple, and most of the other streaks dull vinous-red.

After careful comparison of a large number of South-African specimens, and examination of the types of Mr. Butler's four species above named, I have found no characters serving to distinguish the latter from one another (except size), or the Southern *Antevippe* generally from the Senegal one constituting Boisduval's type. This author figures the upper side of a ♂, showing no trace of any blackish on the inner edge of the apical patch, but in which the hind-marginal spots of the hind-wing are sufficiently enlarged to form a narrow continuous edging. Geyer's figures (*op. cit.*) of a ♂ from the same country represent the same total absence of any inner blackish to the patch, but differ in the more reduced and almost macular state of the outer black edging; while the hind-wing spots are much smaller and not united, and the under side of the hind-wing has a decidedly pinkish-creamy tinge.

In Mr. Butler's series, *Ignifer* is the largest, and the ♂ has a slight blackish inner edge to the apical patch. It is to his " Var. ♂ " of this that are referable the specimens described by me as *Antevippe* (*op. cit.*) in 1862; in them the inner edging referred to is better expressed, and but for the entire want of the longitudinal blackish stripe in both wings, they closely resemble the ordinary ♂ *Achine*.[1] The specimens of *Ithonus*, Butl., seemed to me quite inseparable from *Ignifer* except by their smaller size in both sexes; *Hippocrene* again was represented by still smaller examples and a ♀ with dull pale-yellowish upper-side apical marking, and *Harmonides* by the smallest of the series (*exp. al.* 1 in. 5 lin.).[2]

In October 1885 I received from Colonel Bowker the paired sexes, then recently captured by him near D'Urban in Natal. The ♂ of this pair has no blackish whatever along the inner edge of the apical red; and the ♀ has the apical orange-red well developed, with the traversing macular blackish ray thin and faint, while the discal ray of the hind-wings is almost obsolete. The under-side irroration is well developed in both ♂ and ♀.

There is nothing special about the habits of this *Teracolus*, which on the wing resembles *Achine*. It was much scarcer than the latter when I was collecting in the Knysna district of the Cape Colony, and I only fell in with it during the month of November. On the coast of Natal it is evidently abundant in the dry (winter) season, Colonel Bowker having collected a large number, chiefly in the month of August;[3] of which the ♀ s exhibit every gradation of size, development of upper-side black markings, and tint of apical patch. A few specimens were sent from the Trans-Keian territory by Colonel Bowker as long ago as 1861-63, and are still in the collection of the South-

---

[1] Colonel Bowker has sent one example from D'Urban, Natal, which shows a still nearer approach to *Achine* by possessing a very faint diffused sparse blackish irroration along the inner margin of the fore-wings.

[2] Judging from the figure (*Ent. Schmett.* i. pl. 23, 1884), Staudinger's *Havernickii* from "Transvaal" is a small ♂ of *Antevippe*.

[3] One of two ♂ s collected at D'Urban by Mr. A. D. Millar is ticketed " 17th September 1887," and the other " 10th February 1888."

African Museum; the ♂s agree closely with the examples above referred to as belonging to the "Var. ♂" of Mr. Butler's *Ignifer*, but the two ♀s have the macular ray of the apical patch so expanded inwardly that no orange-red precedes it.

Localities of *Teracolus Antevippe*.

I. South Africa.
  B. Cape Colony.
    a. Western Districts.—Knysna.
    b. Eastern Districts.—Between Zwartkops and Coega Rivers, Uitenhage District (*J. H. Bowker*). King William's Town District (*W. S. M. D'Urban*).
  D. Kaffraria Proper.—Butterworth and Bashee River (*J. H. Bowker*).
  E. Natal.
    a. Coast Districts.—D'Urban (*J. H. Bowker* and the late *M. J. M'Ken*).
    b. Upper Districts.—Estcourt and Weenen County (*J. M. Hutchinson*).
  F. Zululand.—St. Lucia Bay (the late *Colonel H. Tower*).
  G. "Swaziland."—The late E. C. Buxton.
  H. Delagoa Bay.—Lourenço Marques (*Mrs. Monteiro*).
  K. Transvaal.—Lydenburg District (*T. Ayres*).

II. Other African Regions.
  A. South Tropical.
    a. Western Coast.—Damaraland (— *Hutchinson*). "Angola."—Hopffer.
    b1. Eastern Interior.—Mashunaland (*F. C. Selous*). "Tette."—Hopffer.
  B. North Tropical.
    a. Western Coast.—"Senegal."—Boisduval, Geyer, and Hopffer.
    b. Eastern Coast.—Red Sea: "Massowah (*Raffray*)."—Oberthür. White Nile (*Petherick*).—Coll. Brit. Mus.

284. (20.) Teracolus Halyattes, Butler.

♂ ♀ *Teracolus Halyattes*, Butl., Proc. Zool. Soc. Lond., 1876, p. 145, n. 65, pl. vi. f. 7.

Exp. al., (♂) 1 in. 2–5 lin.; (♀) 1 in. 5–8 lin.

♂ *White, with rather large orange apical patch*, divided into six by dark nervules crossing it,—its outer blackish border moderately wide, indenting orange a little on nervules,—its inner blackish border variable in development, rather diffused, more or less obsolescent superiorly. *Fore-wing:* no inner-marginal stripe; base slightly blackish; a terminal disco-cellular black spot, always small, but varying in size. *Hind-wing:* base slightly blackish; a hind-marginal series of small nervular spots. UNDER SIDE.—*Hind-wing and apical patch of fore-wing reddish-creamy, irregularly sprinkled with fuscous atoms. Fore-wing:* a terminal disco-cellular black dot. *Hind-wing:* on costa, not far from apex, a small cluster of fuscous atoms; a terminal disco-cellular black dot internally bordered with orange.

♀ Much resembling a small example of *Achine* ♀. White or yellowish; the orange of apical patch varying from dull-orange to dull-yellowish or whitish, and traversed near its inner border by a strongly marked black bar; blackish longitudinal stripes, and discal ray and hind-marginal spots of hind-wing, moderately developed. UNDER SIDE.—*Hind-wing and apical patch of fore-wing much darker than in ♂, closely hatched and irrorated with fuscous. Fore-wing:* base more or less strongly suffused with pale-yellow; inner-marginal stripe dusky-grey, darker at its extremity. *Hind-wing:* angulated discal ray moderately developed.

This appears to be a distinct little species, combining some of the characters of *Theogone*, Boisd., and *Achine*, Cram. I made the foregoing description from the types in the British Museum, which are the only specimens that I have seen, and are noted as from the north-east of Natal, collected by the late Mr. E. C. Buxton. The form and size of the apical patch and its bordering, and the under-side colouring, in the ♂ are very similar to the same characters in the ♂ *Theogone;* but the corresponding features in the ♀ approximate more closely to those of *Achine* ♀.

Locality of *Teracolus Halyattes*.

I. South Africa.
E. Natal.
a. Coast Districts.—"North-East" (the late *E. C. Buxton*).

285. (21.) Teracolus Evippe, (Linnæus).

♂ *Papilio Evippe*, Linn., Mus. Lud. Ulr. Reg., p. 239, n. 58 (1764); and Syst. Nat., i. 2, p. 762, n. 87 (1767).
♀ *Papilio Arethusa*, Dru., Ill. Nat. Hist., ii. pl. xix. ff. 5, 6 (1773).
♂ *Papilio Evippe*, Cram., Pap. Exot., i. pl. xci. ff. F, G (1779).
♀ *Papilio Arethusa*, Cram., Pap. Exot., iii. pl. ccx. ff. E, F (1782).
♂ *Pieris Evippe*, Godt., Enc. Meth., ix. p. 122, n. 10 (1819).
♀ *Pieris Amytis*, Godt., *loc. cit.*, p. 123, n. 14.
♂ *Anthocharis Evippe*, Boisd., Sp. Gen. Lep., i. p. 573, n. 20 (1836).
♀ *Anthocharis Arethusa*, ♂, Boisd., *op. cit.*, p. 582, n. 35.
♀ *Anthocharis Cebrene*, Boisd., *op. cit.*, p. 583, n. 36.
♂ ♀ *Teracolus Pseudevale*, Butl., Proc. Zool. Soc. Lond., 1876, p. 154, n. 93, pl. vi. f. 10.
♂ ♀ *Teracolus Angolensis*, Butl., *loc. cit.*, n. 94.

Exp. al., (♂) 1 in. 7–10 lin.; (♀) 1 in. 5–8½ lin.

♂ *White, with apical patch of orange-red narrow, curved, very broadly black-bordered, especially on its inner edge. Fore-wing:* base sparsely but usually widely irrorated with blackish; costa narrowly black-edged to about middle; a distinct but thin terminal disco-cellular black dot; apical red divided into six unequal portions by crossing black nervules,—the second portion linear, very small, and the lowest often smaller than the remainder; outer black border of red broad and even, and prolonged (usually more narrowly) to posterior angle,—inner black border extremely broad, rather diffused and irregular along its inner edge, joining outer border on second and first median nervules. *Hind-wing*: basal fuscous irroration less extended than in fore-wing; a series of large hind-marginal nervular black spots, usually uniting to form a border of moderate width, sharply dentating the white on nervules; on costa, beyond middle, a moderate-sized blackish marking, between costal nervure and first subcostal nervure. UNDER SIDE.—*White*. *Fore-wing*: base rather widely suffused with pale sulphur-yellow; disco-cellular spot usually better marked and rounder than on upper side; apical orange much paler and less defined, but considerably broader than on upper side, its borders only very faintly indicated by the palest grey and the outer one tinged with sulphur-yellow; beyond middle, between first median nervule and submedian nervure, a diffused or faint blackish spot, sometimes obsolete. *Hind-wing*: costa for a little distance from base tinged narrowly with yellow,—sometimes edged with orange yellow; a distinct terminal disco-cellular black dot, with a conspicuous (sometimes large) bright-orange spot immediately preceding and touching it; rarely on costa, and still more rarely on median nervules, some exceedingly faint traces of a discal ray. Minute but distinct terminal nervular black dots in both wings.

♀ Yellowish-white or yellowish; apical patch brownish-black, usually uniform, but sometimes *traversed by a series of three or four small dull pale-orange or yellowish spots (almost obsolete in some specimens) and more rarely by a well-developed six-partite pale-orange band, much narrower and more curved than in ♂*. *Fore-wing*: basal suffusion closer and wider than in ♂, and on inner margin about middle merged in an ill-defined fuscous stripe, of which the extremity beyond middle is blackish and truncate, and usually emits an upward projection; disco-cellular terminal dot rather larger. *Hind-wing*: basal suffusion very much wider and closer than in ♂, extending inferiorly to beyond middle, and more narrowly along costa; discal ray strongly blackish-marked at costal origin, usually continuous, and also broadly and strongly marked in its inferior portion; hind-marginal blackish spots similar in size and form to those of the more strongly marked ♂ s. UNDER SIDE.—*Fore-wing sulphur-yellow, with broad dull pale-orange apical patch; hind-wing varying from pale-yellow to deep ochrey-yellow and pale ochrey-reddish*. *Fore-wing*: blackish spot indicating termination of inner-marginal stripe well marked. *Hind-wing*: discal ray reproduced in dull pale brownish-ferruginous.

The development of the apical black is very strong in this species; and I have a ♂ from Cape Coast Castle in which the orange-red patch is quite obliterated by black below the lower radial nervule. Drury (whose figures are exceedingly inferior to most of his illustrations) and Cramer both figure a ♀ in which the apical patch is wholly black except for a slight reddish scaling along its inner edge, but Cramer's ♀ is more heavily marked, and with a much deeper tinted under side. The

♂ represented by Cramer exhibits a very strong orange-red edging all along the costa of the hind-wings on the under side, but I have not observed anything approaching this in the specimens I have examined.

After a very careful inspection and comparison of Mr. Butler's types of *T. Pseudocale* and *T. Angolensis* (in 1881 and again in 1886), I could find nothing (except the smaller size of the former) to distinguish the two from each other, nor was I able to regard them as distinct from the variable *Erippe*. The ♀ in both cases has in the apical patch a narrow pale-orange five-partite band.[1]

Erippe is the oldest recorded species of the genus, the ♂ having been described by Linnæus from Angola in 1764; and it is one of the comparatively few *Teracoli* which are known from the West-African Coast north of the Equator, appearing to be very numerous on the Gold Coast and at Sierra Leone. The only South-African examples that I have seen are those taken by the late Mr. E. C. Buxton in Swaziland, and presented by him to the British Museum. They consist of a ♂, referred by Mr. Butler to *Arethusa*, Drury, and three ♂s (two small) and two ♀s (one small), which constitute the types of his *Pseudocale*. In the coloured photographs sent to me by Mr. Buxton there can be distinctly recognised four ♂s (two small) and a ♀ of *Erippe*, and in his numbered list accompanying them all these examples are noted as from Swaziland. I have no record of the haunts or habits of this butterfly.

Localities of *Teracolus Erippe*.

I. South Africa.
  G. Swaziland.—(The late *E. C. Buxton*).
II. Other African Regions.
  A. South Tropical.
    a. Western Coast.—Angola (*E. C. Buxton*).—Coll. Brit. Mus. Congo (*Curror*).—Coll. Brit. Mus. "Chinchoxo (*Falkenstein*)."—Dewitz.
    b1. Eastern Interior.—"Tati, and between Inyati and Gubulewayo (*Oates*)."—Westwood.
  B. North Tropical.—Gold Coast; Cape Coast Castle (*J. M. Pask* and *A. N. Innes*). Accra and Ashanti.—Coll. Brit. Mus. "Lower Niger (the late *W. A. Forbes*)."—Godman and Salvin. Sierra Leone (*Foxcroft*, &c.). "River Gambia (*Moloney*)."—G. E. Shelley.

286. (22.) Teracolus Omphale, (Godart).

♂ *Pieris Omphale*, Godt., Enc. Meth., ix. p. 122, n. 12 (1819).
Anthocharis Omphale, Boisd., Sp. Gen. Lep., i. p. 574, n. 22 (1836).
♂ *Anthocharis Exole*, Reiche, Ferr. et Gal. Voy. Abyss., pl. 31, f. 4 (1849).
♂ *Anthopsyche Achine*, Wallengr., K. Sv. Vet.-Akad. Handl., 1857; Lep. Rhop. Caffr., p. 11, n. 1.

---

[1] I think that *Ocale*, Boisd. (*op. cit.*, p. 584, n. 37), of which only the ♀ is described, is referable to *Erippe*; it is described as possessing a curved orange band of five or six divisions in apical patch, but has the discal ray of the hind-wings more or less merged with the hind-marginal black border. The specimens of both sexes referred to *Ocale* in the British Museum collection (October 1886) seemed identical with *Angolensis*, Butl.

♀ *Anthopsyche Acte*, Feld., Reise Novara, Lep., ii. p. 187, n. 181 (1867).
♂ ♀ *Teracolus Omphale*, Butl., Proc. Zool. Soc. Lond., 1876, p. 151, n. 83.
♂ *Callosune Achine*, Staud., Exot. Schmett., i. pl. 23 (1884).
VAR. ♂ ♀ *Teracolus Omphaloides* and *T. hybridus*, Butl., *loc. cit.*, pp. 151, 152, nn. 82, 86.
VAR. ♂ ♀ *Anthocharis Omphale*, Trim., Rhop. Afr. Aust., i. p. 50, n. 31 (1862).

Exp. al., (♂) 1 in. 7–11 lin.; (♀) 1 in. 7–10 lin.

♂ *White, with broadly black-bordered orange-red (not rosy-glossed) apical patch; longitudinal black stripes dark and strongly developed, as well as lower portion of black discal ray in hind-wing.* Fore-wing: terminal disco-cellular spot wanting, or extremely faint and minute (except in one very large and blackly-marked example, where it is distinct); apical orange-red patch not subtriangular, but a broad curved bar, variable in width, divided into six unequal portions by crossing black nervules,—its outer black border (prolonged to posterior angle) moderately broad, even, only slightly indenting red on nervules,—its inner border broad or very broad (except in some specimens near costa), with a rather diffused inner edge slightly radiating on nervules; costa black-edged almost to base; inner-marginal black bar very broad and black, superiorly infringing a little on discoidal cell, usually ending slightly diffusedly beyond middle, but sometimes united by blackish scaling to black border at posterior angle. Hind-wing: costal black border strongly marked, slightly diffuse and irregular along its lower edge,—at its extremity beyond middle a slight downward projection indicating beginning of a discal ray; beyond middle, inferior portion of discal ray (between radial nervule and inner margin) forming a somewhat curved conspicuous broad bar; a hind-marginal submacular black border of moderate width, emitting short nervular rays, and more or less completely united with discal bar on radial and third median nervules. UNDER SIDE.—*White; all the black markings of the upper side faintly represented in a very pale grey shade.* Fore-wing: costal and hind-marginal border of apical patch (which is pale-orange) tinged with pale-yellow so as to present a greenish tint; terminal disco-cellular dot present; termination of greyish inner-marginal bar blackish. Hind-wing: a very faint tinge of yellow along costal and hind-marginal border gives a slight greenish cast like that shown near apex of fore-wing; disco-cellular terminal dot small, black, immediately preceded by a rather conspicuous mark of bright orange; discal ray represented by a pale ferruginous mark on costa beyond middle, and a narrow curved streak of the same colour between radial nervule and inner margin; a hind-marginal series of nervular minute black dots.

♀ *Duller-white, yellowish-white, or yellowish; apical patch considerably narrower, orange or dull orange; black markings generally broader, but not so dark as in ♂.* Fore-wing: disco-cellular dot usually well marked; inner-marginal black exceedingly broad, rising over more than lower half of discoidal cell, touching, or almost touching, terminal dot, and by a short upward diffused projection meeting very broad inner border of apical patch on second median nervule; in specimens where inner-marginal black extends to posterior angle a white or yellowish spot is isolated just below lower extremity of apical orange. *Hind-wing:* upper part of discal ray prolonged to angulation on radial nervule, lower part very much broader than in ♂,—the whole ray more or less confluent broadly on nervules with hind-marginal macular black border. UNDER SIDE.—*Hind-wing and apical border of fore-wing more or less strongly tinged with creamy yellow;* indication of dark upper-side markings not so pronounced as in ♂, except termination of inner-marginal band of fore-wing, which is blackish. *Fore-wing:* basal area widely suffused with pale-yellow. *Hind-wing:* ferruginous discal ray more strongly marked.

VARIETY A. ♂ and ♀ (*Omphaloïdes*, Butler).

Exp. al., (♂) 1 in. 5½–8 lin.; (♀) 1 in. 5–8 lin.

♂ *All the black markings more or less reduced,* but especially the lower portion of discal ray of hind-wing, which is never broad, often diffused, and sometimes obsolescent.[1] UNDER SIDE.—*Hind-wing and apical border of fore-wing more or less tinged with creamy-pink,*—in the darker specimens sparsely irrorated with brown atoms.

♀ Black markings similarly reduced, as a rule, but variable; in some examples as strongly developed as in type-form. UNDER SIDE.—Pinkish tinge usually deeper but duller than in ♂; in some examples replaced by pale creamy-brownish; hind-wing faintly striolated with brown.

(*Hab.*—Cape Colony and Delagoa Bay.)

I am unable to separate from this variety a series of still smaller specimens which I captured at Robertson, Western Cape Colony, in January of 1876 and 1877. The ♂s expand only 1 in. 3½–6 lin., and the ♀s 1 in. 5 lin. The former exhibit a complete gradation, from one with black and well-developed longitudinal bands and a thin discal ray in the hind-wing, to an individual in which all these markings are wholly absent. Two similarly unbanded ♂s have reached me from Namaqualand District and one from Carnarvon District. The ♀s do not differ in marking or colouring from the smaller ordinary ones of the variety.[2]

The *Exole* ♂ of Reiche (*op. cit.*) seems to me identical with the larger and darker examples of the ♂ *Omphale*. My conclusion from Wallengren's description (*loc. cit.*), that his *Achine* ♂ was referable to the ♂ *Omphale*, has been confirmed by the loan of a typical example from the Stockholm Museum. After a careful comparison of the types of Mr. Butler's *Omphaloïdes* and *Hybridus*—the ♂ of the latter being one of my own captures at Plettenberg Bay—I was unable to keep

---

[1] In a single ♂ which I captured at Plettenberg Bay all the black markings but the borders of apical patch are wanting.

[2] I believe that the only *Teracolus* I ever saw on the wing at Cape Town was a ♂ of this variety. It was flying in the Cathedral enclosure on the 14th December 1878; but my attempts to capture it with a pith helmet were without success.

them apart, or to find any distinct character to separate them from *Omphale*; and, indeed, Mr. Butler himself (*loc. cit.*) expresses some doubt on the point.

I met with the typical form very numerously at Port Natal, and with the variety abundantly at Plettenberg Bay and near Grahamstown in the Cape Colony. It keeps very closely to wooded spots, especially in country where woods are sparse and stunted, and usually confined to small valleys or kloofs. It is less swift on the wing than *Achine*, Cram., and settles more frequently, never rising far from the ground. The largest and finest South-African specimens are from Natal and the Eastern Transvaal, but I have taken at D'Urban a ♂ only 1½ inches in expanse. This typical form of the species is not known to me to occur to the westward of the Bashee River in Kaffraria. It is remarkable that both the typical form and the Variety A. occur at Delagoa Bay.

Localities of *Teracolus Omphale*.

I. South Africa.
  B. Cape Colony.
    a. Western Districts.—[Var. A.] Spectakel (*L. Péringuey*) and Springbokfontein (*G. A. Reynolds*), Namaqualand District. Van Wyk's Vley, Carnarvon District (*E. G. Alston*). Robertson. Knysna (*Miss Wentworth*). Plettenberg Bay.
    b. Eastern Districts.—[Var. A.] Port Elizabeth. Uitenhage. Grahamstown. Kleinemond River (*H. J. Atherstone*) and Tharfield (*Miss M. Bowker*), Bathurst District. East London (*P. Borcherds*).
  D. Kaffraria Proper.—Bashee River (*J. H. Bowker*).
  E. Natal.
    a. Coast Districts.—D'Urban. Pinetown (*J. H. Bowker*).
    b. Upper Districts.—Maritzburg (*Miss Colenso*). Estcourt (*J. M. Hutchinson*).
  G. Swaziland.—(The late *E. C. Buxton*.)
  H. Delagoa Bay.—Lourenço Marques (*Mrs. Monteiro*).—[Typ. and Var. A.]
  K. Transvaal.—Limpopo and Marico Rivers (*F. C. Selous*). Potchefstroom and Lydenburg District (*T. Ayres*).

II. Other African Regions.
  A. South Tropical.
    b. Eastern Coast. — "Mozambique."—Hopffer. [*Evole*, Reiche.] "Zanzibar (*Raffray*)."—Oberthür.
      b1. Eastern Interior.—Zambesi, "near Victoria Falls (*Oates*)."—Westwood.
  B. North Tropical.
    a. Western Coast.—Gambia.—Coll. Hope, Oxon.
    b1. Eastern Interior.—"Abyssinia."—Reiche.

287. (23.) Teracolus Theogone, (Boisduval).

♂ ♀ *Anthocharis Theogone*, Boisd., Sp. Gen. Lep., i. p. 575, n. 23 (1836) [excl. "Var. A."].
♂ ♀ *Teracolus Theogone*, Butl., Proc. Zool. Soc. Lond., 1876, p. 152, n. 87.
♀ *Anthopsyche Proene*, Wallengr., K. Sv. Vet.-Akad. Handl., 1857; Lep. Rhop. Caffr., p. 12, n. 4.

♀ *Anthocharis Proene*, Trim., Rhop. Afr. Aust., ii. p. 323, n. 217 (1866).
♂ ♀ *Teracolus Proene*, Butl., Proc. Zool. Soc. Lond., 1876, p. 153, n. 89.
♂ ♀ *Teracolus Loandicus*, Butl., Proc. Zool. Soc. Lond., 1871, p. 724 [♂ only]; Lep. Exot., p. 91, pl. xxxiv. f. 10 [♂]; and Proc. Zool. Soc. Lond., 1876, p. 153, n. 91 [♂ and ♀].

Exp. al., (♂) 1 in. 3–9½ lin.; (♀) 1 in. 4–8½ lin.
Allied closely to *T. Omphale*, (Godt.), Var. A.

♂ *White, with orange-red of apical patch broad, but narrowly bordered with blackish, especially on its inner edge. Fore-wing:* base blackish, the suffusion chiefly inferior, usually extending for a little distance inner-marginally; rarely a narrow, somewhat diffused, blackish inner-marginal stripe to beyond middle; no terminal disco-cellular dot; *apical red considerably wider in its upper part than Omphale's, the nervules crossing it only black near the blackish borders,—the inner border very much narrower than in Omphale, sometimes diffused, and occasionally obsolescent, except in its middle. Hind-wing:* base narrowly or very narrowly suffused with blackish; in those examples which possess an inner-marginal stripe in fore-wing, a more or less imperfect similar costal stripe; hind-marginal nervular black spots small or very small,—most developed in the blackish-striped examples. UNDER SIDE.—Very closely resembling that of *Omphale*, Var. A., *but in hind-wing the creamy-pinkish tinge is sometimes very faint or wanting, while the irroration and minute striolation with dusky-brown is much stronger, especially along costa and over lower half of discoidal cell.*

♀ *Apical orange paler, but often brighter and usually broader than in* ♀ *Omphale*, Var. A.,—*the nervules crossing it scarcely ever black beyond the short external indentations,—its inner black edging usually well developed, but ending abruptly about third median nervule, leaving broad termination of orange band in contact with the white field on and below second median nervule. Fore-wing:* no terminal disco-cellular dot; inner-marginal blackish stripe very variable in development, but never nearly so broad as in *Omphale*, Var. A., and often very narrow and sub-obsolescent except at base and extremity. *Hind-wing:* costal blackish usually faint and diffused except at base and extremity; discal ray interrupted widely below costal commencement,—its lower portion not more developed than in ♂ *Omphale*, Var. A.,—usually narrow, and frequently obsolescent (rarely quite obsolete) below point of angulation on radial nervule; hind-marginal spots always small or of moderate size, and (except rarely the three upper ones) quite separate. UNDER SIDE.—As in ♂, but fore-wing with a dusky indication of both dark inner border of apical patch and inner-marginal stripe, and hind-wing with closer and darker striolation, and usually more strongly marked discal ray.

ABERRATION.—♂ *Apical patch of fore-wing bright golden-yellow;* its inner black edging strongly developed; inner-marginal stripe slightly developed.

This beautiful "sport" resulted from a pupa forwarded in August 1887 from D'Urban, Natal, by Colonel Bowker. The pupa was attached to the base of the lanceolate leaf of a "small tree;" it very closely resembled that of *T. Antevippe*, (Boisd.), received at the same time (see above, p. 137), but was of a deeper colour, and had a pale apical space on the wing-covers.

Mr. Aurivillius, of the State Museum in Stockholm, kindly obtained for me a carefully executed coloured figure of Wallengren's type (♀) of *Proene;* it agrees very closely with those specimens from Natal and Zululand in which the inner black border of the apical patch is most developed.

The united sexes have twice been taken by Colonel Bowker, viz., at D'Urban, Natal, on 15th August 1878, and at Coega Railway Station in the Uitenhage District of Cape Colony in October 1879. The ♂ s of these pairs are smaller than usual, and closely alike, except that the D'Urban individual has a little more blackish inner-marginally and less on inner edge of apical patch than the Coega one; while in the ♀ s just the reverse is the case, besides which the Coega ♀ has the lower part of the discal ray of the hind-wings more complete. Small ♂ s appear to occur more commonly than small ♀ s. I have received about three times as many of the former as of the latter. The smallest ♂ is from the Bashee River in Kaffraria,—the smallest ♀ from Delagoa Bay. I have not found any trace of the longitudinal blackish bars in any of these under-sized ♂ s. The largest examples of both sexes, and those in which the black markings—especially the inner border of the apical patch in the ♂—are best developed, are from Natal, but very small ♂ s have also been received from that Colony.

The characters emphasised in my description seem to warrant our holding *Theogone* as distinct from its congeners. The width of the apical red or orange, with the absence of black neuration crossing it, and (in the ♀) the way in which the broad lower extremity of the curved orange band meets the white field, are notable features.

The butterfly has a wide range over Eastern South Africa, and seems particularly prevalent on the Natal Coast in the winter (dry) season. I took the ♂ at D'Urban as early as the 8th April, and at Grahamstown, Cape Colony, on the 5th June.

Localities of *Teracolus Theogone*.

I. South Africa.
   B. Cape Colony.
      b. Eastern Districts.—Uitenhage (*S. D. Bairstow*) and Coega (*J. H. Bowker*), Uitenhage District. Grahamstown.
      c. Griqualand West.—Vaal River (*J. H. Bowker*).
   D. Kaffraria Proper.—Bashee River (*J. H. Bowker*).
   E. Natal.
      a. Coast Districts.—D'Urban. Umgeni (*J. H. Bowker*).
      b. Upper Districts.—Estcourt and Weenen County (*J. M. Hutchinson*).

F. Zululand.—St. Lucia Bay (*H. Tower*).
H. Delagoa Bay.—Lourenço Marques (*Mrs. Monteiro*).
K. Transvaal.—Potchefstroom and Lydenburg Districts (*T. Ayres*). Barberton (*J. P. Cloete*).

II. Other African Regions.
A. South Tropical.
    a. West Coast.—Angola: Quanza (*J. J. Monteiro*).—Coll. Brit. Mus.; "Loanda and Ambriz (*Monteiro*)."—Butler [*Loandicus*, Butl.]. Congo: "Kinsembo (*Ansell*)."—Butler [*Loandicus*].
    b. East Coast.—"Querimba."—Hopffer.

288. (24.) Teracolus Antigone, (Boisduval).

♂ *Anthocharis Antigone*, Boisd., Sp. Gen. Lep., i. p. 572, n. 19 (1836).
♂ ♀ *Anthocharis Delphine*, Boisd., *op. cit.*, p. 577, n. 28.
♂ *Anthopsyche Eucharis*, Wallengr., K. Sv. Vet.-Akad. Handl., 1857; Lep. Rhop. Caffr., p. 15, n. 10 [*nec* ♀].
♀ *Anthocharis Antigone*, Trim., Rhop. Afr. Aust., i. p. 52, n. 33 (1862).
♂ ♀ *Teracolus subfumosus*, Butl., Proc. Zool. Soc. Lond., 1876, p. 139, n. 44, pl. vi. f. 3.
♂ ♀ *Teracolus Flaminia*, Butl., *loc. cit.*, p. 140, n. 46.
♂ ♀ *Teracolus Lycoris*, Butl., *loc. cit.*, p. 140, n. 45, pl. vi. f. 1.
♂ ♀ *Teracolus Friga*, Butl., *loc. cit.*, p. 142, n. 48, pl. vi. f. 5.
♂ ♀ *Teracolus Galathinus*, Butl., *loc. cit.*, p. 142, n. 49.
Var. ♂ (! ♀) *Teracolus Lyceus*, Butl., *loc. cit.*, p. 141, n. 47, pl. vi. f. 2.

Exp. al., (♂) 1 in. 3–6 lin.; (♀) 1 in. $0\frac{1}{2}$–$4\frac{1}{2}$ lin.

♂ *White, with orange slightly pink-glossed apical patch (marked inferiorly on its inner side by a small black spot,—reduced in some examples to two faint blackish dots or one dot, and rarely altogether wanting); black longitudinal stripes in some examples well marked and rather broad, in others narrower and fainter, in others reduced to a very faint diffused state, slightly darker at extremity, and sometimes obsolete altogether. Fore-wing:* basal blackish diffuse, extending chiefly along inner margin; costal, apical, and hind-marginal black border of apical orange narrow or very narrow, but widening costally at beginning of orange, and externally rather deeply and sharply piercing the orange on five or six nervules; hind-marginal border in the darker-banded examples emitting on second median nervule a broader ray than the rest (which inferiorly bounds the orange and unites with black spot on inner edge of latter), but not continued to posterior angle except by separate black terminations of first median nervule and submedian nervure, with occasionally a smaller mark midway between them; inner-marginal stripe at its widest not encroaching on discoidal cell, its termination broad and irregularly truncate; in some of the more faintly marked specimens all but the termination of this marking is next to obsolete, in others there is but the faintest indication of the entire stripe, while in others only the basal commencement exists. *Hind-wing:* costal stripe as variable as inner-marginal one, and wholly wanting more frequently; basal blackish central, diffused, usually very limited; nervular hind-marginal black spots (narrowed and acuminate inwardly) usually small (sometimes very small) and separate, but in the darker specimens larger, rather diffused, and the upper three touching each other; rarely a faint diffused trace of lower part of discal blackish ray between radial nervule and submedian nervure. UNDER SIDE.—*Fore-wing*: white, occasionally with a faint sulphur-yellow tinge at base; a black disco-cellular terminal dot; apical patch varying from yellowish to reddish-creamy, more or less irrorated finely with dark-grey; its inner border rather narrowly dull pale-orange, marked out from the broad outer portion of patch by an ill-defined (sometimes scarcely traceable) greyish submacular ray. *Hind-wing*: whitish, sandy-whitish, yellowish, or reddish-creamy, more or less irrorated (and occasionally in parts also slightly striolated) unequally with dark-grey; discal ray very feebly expressed by a dusky mark on costa beyond middle, and another on disc between third and second (or sometimes as far as first) median nervules; costa from base to beyond middle edged with orange or pale orange, sometimes rather faint; a terminal disco-cellular black dot, bounded internally by a small (sometimes very small and faint) orange mark.

♀ *White or yellowish-white (rarely pale-yellow); apical patch with the orange paler and narrower than in ♂; traversed superiorly (near its inner edge) and bounded inferiorly by a black curved ray of variable development, and bordered more broadly by black outwardly, or with the orange replaced by a much-reduced (sometimes almost obsolete) series of dull-yellowish or dull-creamy separate marks. Fore-wing*: inner-marginal blackish stripe more or less mixed with grey except at extremity, diffuse superiorly, in the most strongly-marked examples very slightly encroaching on discoidal cell near base, sometimes very narrow and faintly marked, rarely obsolete from a little before middle; a well-marked usually conspicuous, rounded, terminal disco-cellular black spot. *Hind-wing*: costal stripe as variable in darkness and development as inner-marginal stripe of fore-wing (in one example obsolete); discal ray also variable, but its lower portion usually well-expressed and sometimes rather broad; hind-marginal black almost always larger than in ♂, and usually united to form a border (broader superiorly), emitting strong nervular dentations, of which the middle two or three sometimes meet discal ray. UNDER SIDE.—*Hind-wing and apical patch of fore-wing always yellowish or reddish-tinged, duller than in ♂, being usually more closely and evenly irrorated with dark-grey. Fore-wing*: basal sulphur-yellow always more developed, sometimes very wide; pale-orange on inner border of apical patch traversed (and in some specimens more or less obliterated) by a dusky-blackish ray; inner-marginal stripe always fainter than on upper side. *Hind-wing*: discal ray exceedingly variable in development, sometimes barely traceable, or presenting only the portions shown in ♂, but often dark, and wanting only the thin part between first subcostal and radial nervules, and occasionally complete but macular.

ABERRATION.—♂ *Apical patch not orange, but sericeous pinkish-grey, with a thin line of dull-orange along its inner edge, and also on costa and third subcostal nervule;* inner black spot of apical patch very small; no longitudinal blackish stripes; under-side tint creamy, rather densely irrorated with dark-grey.

Hab.—Estcourt, Natal (*J. M. Hutchinson*).

I think that there can be no doubt, on carefully comparing the descriptions given by Boisduval, of the species-identity of his *Antigone* and *Delphine*, the former being a West-African ♂, with the longitudinal dark stripes very faintly marked and the under side of the hind-wings whitish, and the latter the more prevalent South-African ♂ and ♀, with reddish-creamy under side of hind-wings,—the ♂ without, the ♀ sometimes with, the longitudinal dark stripes.

I have been enabled to identify the *Eucharis* ♂ of Wallengren with this species by means of a coloured drawing of the type-specimen kindly obtained for me by Mr. Chr. Aurivillius. It is a ♂ in which all the blackish markings of the upper side are very small and attenuated, but there exists an extremely faint inner-marginal spot indicating the termination of the longitudinal stripe of the fore-wings, while the under side of the hind-wings is creamy with partial dark irroration and a costal spot indicating the commencement of the discal ray. I possess an almost identically marked ♂ from Estcourt in Natal, only wanting the inner-marginal spot in the fore-wings.

The most painstaking examination—in 1881, and renewed in 1886—of the types of Mr. Butler's six species above named has brought me to the conclusion that they are really inseparable from each other and from *Antigone*, Boisd., being founded on individual variations prevailing in both sexes in respect alike of size, marking, and colouring. As regards the ♂ s, there is a regular gradation from *Subfumosus*, which has no trace of dark longitudinal stripes, through *Flaminia*, *Lycoris*, and *Friga*, to *Galathinus*, in which those markings are well developed; and *Lyons*, though larger and with fainter stripes, is quite of the *Galathinus* character. The specimens of ♀ s respectively assigned to the ♂ s of this series cannot be clearly discriminated, varying too much to be with any consistency grouped in any order corresponding with that of the ♂ s. All the specimens on which these six presumed species are founded were collected by the late Mr. E. C. Buxton in Swaziland.

The specimens in which the blackish stripes are best developed certainly exhibit close resemblance to *Phlegetonia*, Boisd., and it seems not improbable that eventually the latter will be ascertained to be conspecific with the series included under *Antigone*.

I have only once met with this widely-distributed *Teracolus*, which appears to be less numerous than *Phlegetonia* in the eastern tracts of South Africa, but a good many examples have reached me from correspondents in various quarters. At Grahamstown and Uitenhage in the Cape Colony, D'Urban in Natal, and Lourenço Marques (Delagoa Bay), as well as in Swaziland, and in the Lydenburg District of Transvaal, varying grades of it occur, together with the darker *Phlegetonia*.

Localities of *Teracolus Antigone*.

I. South Africa.
  B. Cape Colony.
    b. Eastern Districts.—Uitenhage (*S. D. Bairstow* and *J. H. Bowker*).
    Grahamstown. Fish River, Albany District (*Mrs. Barber*).
    King William's Town (*W. S. M. D'Urban*). Bedford (*J. P. Mansel Weale*).
  E. Natal.
    a. Coast Districts.—D'Urban (*J. H. Bowker*).
    b. Upper Districts.—Estcourt and Bushman's River, Weenen County (*J. M. Hutchinson*).
  C. Swaziland.—The late *F. C. Buxton*.
  H. Delagoa Bay.—Lourenço Marques (*Mrs. Monteiro*).
  K. Transvaal.—Lydenburg District (*T. Ayres*). Limpopo River (*F. C. Selous*).

II. Other African Regions.
  B. North Tropical.
    a. Western Coast.—" Guinea."—Boisduval.
    b1. Eastern Interior.—Soudan : " Atbara River."—Butler.]

289. (25.) **Teracolus Phlegetonia,** (Boisduval).

♂ *Anthocharis Phlegetonia*, Boisd., Sp. Gen. Lep., i. 576, n. 25 (1836).
Var. ♂ *Anthopsyche Phlegetonia*, Wallengr., K. Sv. Vet.-Akad. Handl., 1857; Lep. Rhop. Caffr., p. 13, n. 5.
Anthocharis Phlegetonia, Hopff., Peters' Reise Mossamb., Ins., p. 360 (1862).
Var. ♂ ♀ *Anthocharis Eione*, Boisd., *loc. cit.*, p. 578, n. 29.
Anthocharis Delphine, Trim., Rhop. Afr. Aust., i. p. 53, obs. and note (1862).
Teracolus Eione, Butl., Proc. Zool. Soc. Lond., 1876, p. 144, n. 54.

Exp. al., (♂) 1 in. 3–6 lin. ; (♀) 1 in.—1 in. 6 lin.

♂ *White, with bright-orange slightly pink-glossed apical patch (marked inferiorly on its inner side by a black spot); black longitudinal stripes (especially that of fore-wing) very broad. Fore-wing:* usually a more or less distinct terminal disco-cellular black dot; orange apical patch wide superiorly and touching white field, but abruptly narrowed inferiorly by inner rather large outwardly-excavate black spot; costal black border of orange rather wide at beginning, then narrow, but widening again before apex; hind-marginal black border rather wide, radiating on all nervules, deeply and sharply indenting the orange on five of them, and on and below second median nervule bounding inferior extremity of orange by a broad ray that unites with black internal spot of patch, and sometimes diffusedly touches upper corner of extremity of inner-marginal black stripe; this stripe very broad, its upper edge rather diffused, and near base extending a little into discoidal cell, its extremity broad and more or less truncate. *Hind-wing*: base rather widely blackish except on inner-marginal border; a wide costal black stripe, a small downward projection of its extremity marking commencement of discal ray; this ray linear or obsolete as far as radial nervule, but inferiorly broad, black, diffused internally, but externally more or less completely confluent with a wide hind-marginal black border radiating deeply on nervules, so as to leave nothing more than four narrow white rays or white lines on inter-nervular folds. UNDER SIDE.— *Fore-wing*: base sometimes slightly tinged with sulphur-yellow; apical orange well developed, rarely with the faintest trace of inner dark spot, externally bordered (very broadly apically) with pale sulphur-yellowish; disco-cellular terminal dot always distinct; inner-marginal stripe grey, with extremity blackish. *Hind-wing*: white, yellowish-white, or dull-creamy; costa conspicuously orange-edged from base to beyond middle; a disco-cellular terminal black dot, in contact anteriorly with a much larger pointed orange spot; a very faint shade indicating position of costal stripe of upper side; discal ray and hind-marginal border also usually scarcely indicated, but the former sometimes better expressed (especially at its costal origin), and in a few specimens diffusedly marked in brownish-red or pale-ferruginous. In both wings a terminal nervular series of minute but very distinct black dots.

♀ *White, or yellowish-white; orange of apical patch exceedingly variable in extent,—in some examples almost as in ♂, except for a very faint and narrow traversing series of dusky nervular marks,—more often reduced to a narrow curved band only partly invading a broad inner blackish border,—still more diminished to a central series of three to five separate very dull pale-orange, yellowish, or even whitish, spots in a black apical patch,—rarely wholly obsolete, leaving the patch dusky-blackish only. Fore-wing*: disco-cellular terminal spot much larger than in ♂, and almost always touching inner-marginal black stripe, which is broader, and almost fills discoidal cell,—its superior terminal portion projecting so as to join inner border of apical patch on second median nervule. *Hind-wing*: basal blackish extending thinly beyond cell as far as lower part of discal ray; costal stripe as in ♂; a small, but usually quite distinct, terminal disco-cellular black dot; discal ray complete from its costal origin, and usually more distinguishable from hind-marginal border than in ♂, the white inter-nervular rays being broader, and sometimes themselves confluent. UNDER SIDE.—*Much duller and deeper-tinted than in ♂; hind-wing and apex of fore-wing varying from dull greyish-creamy or yellowish-creamy to reddish-creamy, sparsely irrorated with dark-grey;* markings similar to those of ♂, but stronger. *Fore-wing*: apical orange very variable, usually much narrower and fainter than in ♂, never wholly absent; a dusky streak traversing it near its inner border.

A typical specimen (♂) of Wallengren's *Phlegetonia*, kindly lent from the Stockholm Museum by Mr. Chr. Aurivillius, agrees with Wallengren's description (*loc. cit.*), and with a ♂ from Delagoa Bay and another from Transvaal, in wanting altogether the white inter-nervular streaks usually marking the confluent discal ray and hind-marginal border of the hind-wings, so that a very broad uniform black border covers the lower two-thirds of the outer part of those wings. Mr. J. L. Fry has sent me a ♀ similarly marked from Tati.

I am unable to distinguish Boisduval's *Eione* from his *Phlegetonia*, the only characters noted in his descriptions being the smaller size and greyish-yellowish under side of the former,—features highly variable, and presenting every gradation.

Mr. Butler (*loc. cit.*, pp. 143–144) identifies *Phlegetonia*, Boisd., with *Antigone*, Boisd. (*Sp. Gen.*, p. 572), but the description of the latter (only the ♂ is described) as having on the upper side "*la base et la commissure légèrement saupoudrées de gris*," and as regards the hind-wings, simply "*une bordure noire crenelée*," indicates a different insect from the heavily black banded and bordered *Phlegetonia*; although it must be admitted (as stated under *T. Antigone*) that the darkest examples do nearly approach *Phlegetonia* in marking.

This is one of the most striking of the smaller species of *Teracolus*, the relatively large size of the orange patch and the depth and extent of the black markings rendering it very conspicuous. I found it very seldom during my stay in Natal, but abundantly in the scrub bush at Uitenhage, Cape Colony. It is an active insect, but not swift, and flies close to the ground, appearing in the height of summer (January to March). Near Grahamstown I also met with the species, and on the 12th February 1870 captured the united sexes. In this pair neither sex had the black markings at their highest development, and the well-marked apical orange of the ♀ partly penetrated the upper part of its broad inner blackish border. In another pair, taken *in copula* by Colonel Bowker near the Upper Tugela on 2d April 1880, the ♂ has heavy dark markings, but the ♀ is scarcely more heavily marked than the darkest ♀ s of *Antigone*, and has the apical orange superiorly almost as in the ♂, the usual dark inner border being reduced to a series of dusky nervular marks well within the orange.[1]

Localities of *Teracolus Phlegetonia*.

I. South Africa.
   B. Cape Colony.
      b. Eastern Districts.—Uitenhage. Grahamstown.
   D. Kaffraria Proper.—Bashee River (*J. H. Bowker*).
   E. Natal.
      a. Coast Districts.—D'Urban. Verulam.
      b. Upper Districts.—Greytown. Valley of Tugela and Mooi Rivers (*J. H. Bowker*).
   G. Swaziland.—The late *E. C. Buxton*.
   H. Delagoa Bay.—Lourenço Marques (*Mrs. Monteiro*).
   K. Transvaal.—Lydenburg District (*T. Ayres*). Marico and Limpopo Rivers (*F. C. Selous*).

---

[1] A very fine ♀, sent from Weenen County in Natal by Mr. J. M. Hutchinson, considerably surpasses that here noticed as regards the width of the apical orange and the faintness of the traces of the inner dark border of the orange, but presents, on the contrary, all the great black markings in their highest development on both wings.

II. Other African Regions.
  A. South Tropical.
    a. Western Coast.—Damaraland (*J. A. Bell* and the late *C. J. Andersson*).
    b1. Eastern Interior.—Tauwani River (*F. C. Selous*). Between Transvaal and Gubulewayo (*A. W. Eriksson*). Tati (*J. L. Fry*). Zambesi: "Tette."—Hopffer.
  B. North Tropical.
    a. Western Coast.—"Accra (*Carter*)."—Butler. "Senegal."—Boisduval.
    b1. Eastern Interior.—Abyssinia: "Shoa (*Antinori*) and Bogos (*Beccari*)."—Oberthür.

290. (26.) Teracolus Microcale, Butler.

♂ ♀ *Teracolus Microcale*, Butl., Ann. and Mag. Nat. Hist., 4th Ser., xviii. p. 487 (1876).

Exp. al., (♂) 1 in. 6 lin.; (♀) 1 in. 3½ lin.

♂ *White, with rather large orange apical patch bordered rather widely throughout with blackish. Fore-wing:* base rather broadly irrorated with blackish; apical orange of six divisions, being outwardly deeply pierced by black nervules. *Hind-wing:* a series of rather large blackish nervular hind-marginal spots, of which the three upper ones are confluent. UNDER SIDE.—*White. Fore-wing:* a terminal disco-cellular black dot; apical orange of upper side showing faintly through. *Hind-wing:* a terminal disco-cellular black dot attached to a preceding small orange spot.

♀ *Similar, smaller; with apical orange narrow, curved, open at its lower extremity so as to touch white field beneath third median nervule. Hind-wing:* hind-marginal spots smaller, more separate. UNDER SIDE.—*Hind-wing and apical patch of fore-wing creamy tinged with yellowish, very thinly irrorated with brownish. Fore-wing:* base tinged with pale sulphur-yellow. *Hind-wing:* a faint trace of a brownish discal elbowed bar across median nervules.

This little species comes perhaps closer to the *Eeippe* group than any other, and is noticeable for the wide even dark bordering of the apical patch on all sides. Neither sex has any longitudinal black markings in either wing. Mr. Butler founded the species on a ♂ and a ♀ from the Atbara (Soudan), and my description is made from those type specimens and from a second ♀, ticketed "Orange River," placed with the others in the British Museum collection. I did not ascertain who had signified the locality of the last-named example, but I could detect no difference of any importance between it and the type ♀.

Localities of *Teracolus Microcale*.

I. South Africa.
  B. Cape Colony.—Orange River.—Coll. Brit. Mus.

II. Other African Regions.
  B. North Tropical.
    b1. Eastern Interior.—Soudan: Atbara River.—Coll. Brit. Mus.

291. (27.) Teracolus Lais, Butler.

♂ *Teracolus Lais*, Butl., Proc. Zool. Soc. Lond., 1876, p. 145, n. 64.

Exp. al., (♂) 1 in. 4–4½ lin.; (♀) 1 in. 4–7 lin.

♂ *White, with rather small dull orange-yellow apical patch blackish-bordered both internally and externally. Fore-wing:* costa with a very fine linear black edging; base narrowly and very sparsely powdered with blackish; a well-defined terminal disco-cellular black dot; apical orange bordered internally rather widely with blackish (which in type is even throughout, but in a second specimen becomes rather thin and diffused between third and second median nervules), and externally by narrower border, piercing the orange to some depth on nervules, and itself pierced by four or five inter-nervular acuminate rays of the orange, and ending about second median nervule. *Hind-wing:* sparse basal powdering rather more extended than in fore-wing; a hind-marginal series of six moderate-sized or small separate nervular blackish spots. UNDER SIDE.—*Hind-wing and apical patch of fore-wing creamy-white, finely irrorated with fuscous. Fore-wing:* terminal disco-cellular dot larger, well defined; a very faint indication of apical orange and its dark inner border. *Hind-wing:* costa edged with orange-yellow for a little distance from base; a terminal disco-cellular black dot, bounded internally by a small bright orange spot; nervular hind-marginal black dots very minute.

♀ *Apical orange-yellow patch narrow, much curved, six-partite (unequally) by very fine black nervules, much more broadly blackish-bordered internally and externally than in ♂; in each wing an imperfect longitudinal blackish stripe, and in hind-wing also an irregular elbowed discal blackish ray. Fore-wing:* basal area finely irrorated with blackish (in one example also tinged with sulphur-yellow); inner-marginal blackish broad, somewhat greyish, and diffused, except at its extremity not far beyond middle (where it emits a faint thin upward ray) joining inner border of apical patch; terminal disco-cellular spot large, sub-ovate; outer border of apical patch broad, not deeply indenting the orange, and not pierced by the latter, ending abruptly on or just below first median nervule; the two borders unite rather broadly on second median nervule. *Hind-wing:* basal irroration sparse, but widely extended inferiorly; costal stripe broad, rather faint, except at its abrupt termination; a small but distinct terminal disco-cellular dot; discal elbowed ray narrow, irregularly angulated, fainter superiorly, and becoming obsolete towards inner margin; a rather wide hind-marginal blackish border, continuous from apex to second subcostal nervule (where it is connected with elbow of discal ray), but thence with inter-nervular interruptions and becoming obsolete not far from anal angle. UNDER SIDE.—*Hind-wing and apical patch of fore-wing very pale creamy-yellowish or dull creamy, very finely irrorated with fuscous. Fore-wing:* basal area faintly suffused with sulphur-yellow, but without fuscous irroration above median nervure; inner-marginal stripe fainter than on upper side, but its terminal upward ray well marked, and (more faintly) continued to near costa by diffused macular inner border of apical patch; disco-cellular spot well developed; orange of apical patch very faintly indicated. *Hind-wing:* only a very faint indication of elbowed discal ray.

VARIETY A. (♂ and ♀).

♂ *Larger (exp. al.,* 1 in. 7–7½ lin.), *with orange-yellow apical patch somewhat redder but duller in tint, and its inner blackish border much narrower and more diffused. Fore-wing:* disco-cellular dot smaller. *Hind-wing:* hind-marginal nervular spots very small. UNDER SIDE.—*Hind-wing and apical patch of fore-wing very pale reddish-creamy, irrorated more closely than in type-form.*

♀ *Exp. al.,* 1 in. 6–7 lin. Pattern as in type-form; but apical orange-yellow in one example represented only by a few scales, and in the other by a diffused *ill-defined ray of three spots in the middle of a dusky blackish patch.* UNDER SIDE.—Hind-wing and apical patch of fore-wing paler and duller than in typical ♀.

(*Hab.*—Vaal River, Griqualand West.)

The ♂ of this very distinct species has some resemblance to Klug's figures of his *Ephyia*, from Ambukohl (*Symb. Phys.*, pl. vi. ff. 9, 10), but presents a smaller apical patch and an irrorated under side. In both these characters it approaches *T. Bowkeri*, Trim. (see p. 100), but differs in the much deeper, warmer colour of its apical patch with its narrower and fainter inner blackish border. In this last feature it is not unlike *Agoye*, Wallengr., ♂, but wants the conspicuous upper-side black neuration and irroration of the latter. As regards the ♀, *Lais* is altogether different from the two South-African congeners just named, that sex having the upper-side pattern and facies partly of *Anterippe*, Boisd., ♀, and partly of *Antigone*, Boisd., ♀, whereas the ♀s of *Agoye* and *Bowkeri* are wholly, or almost wholly, devoid of the characteristic dark stripes and rays.

The variety above described co-exists with the type-form on the Vaal River, and may possibly be a seasonal form of the latter.

Very few examples of this butterfly have come under my notice. The type (a ♂) in the British Museum is ticketed "Orange River," and registered as collected by "C. H. Pilcher, 1872." A ♀ from Damaraland was sent to me for determination by Mr. Aurivillius in 1881, and a ♂ and ♀ taken on the Vaal River were presented to the South-African Museum by Mr. H. L. Feltham during the current year (1888).

Of the Variety A., four examples (two of each sex) were taken by Colonel Bowker on the Vaal River in 1871, but no note accompanied them except the general one (since confirmed by Mr. Feltham), that in that tract of country all the *Teracoli* were almost confined to the immediate vicinity of the river.

Localities of *Teracolus Luis*.

I. South Africa.
  B. Cape Colony.
    ? b. Eastern Districts.—Orange River (*C. H. Pilcher*).—Coll. Brit. Mus.
    c. Griqualand West.—Vaal River (*J. H. Bowker*—Var. A.; and *H. L. Felthan*).
II. Other African Regions.
  A. South Tropical.
    a. Western Coast.—Damaraland (*De Vylder*).

292. (28.) Teracolus Celimene, (Lucas).

♂ ♀ *Anthocharis Celimene*, Lucas, Rev. et Mag. Zool., 2nd Ser., iv. p. 426 (1852).
♂ ♀ *Anthocharis Amina*, Hewits., Exot. Butt., iii. pl. 5, ff. 1–3 (1866).
♂ *Callosune Amina*, Staud., Exot. Schmett., i. pl. 23 (1884).

Exp. al., (♂) 1 in. 7–9½ lin.; (♀) 1 in. 10 lin.

♂ *White, with very large dull purplish-lake, violet-glossed, black-bordered apical patch, intersected by a black streak.* Fore-wing: base and costal margin before middle narrowly and sparsely powdered with blackish; a very faint indication of the terminal disco-cellular black dot of the under side; costa with a linear black edging as far as beginning of apical patch; *nervules crossing apical patch heavily clouded with black, especially externally, dividing the purple into seven (rarely eight) unequal rays, the hind-marginal tips of six of which (between apex and first median nervule) are acuminate and conspicuously pale sulphur-yellow;* costal and hind-marginal black borders of apical patch very narrow, but the latter suddenly broadening at (or rarely below) first median nervule, where it unites with the narrow rather irregular and (internally) diffused inner black border of patch, and whence the united borders continue to posterior angle; black streak traversing patch nearer inner than outer edge, running a little obliquely outward from costa as far as third median nervule, but thence deflected inwardly and joining inner black border on second median nervule. Hind-wing: a hind-marginal black border of moderate width, emitting inwardly nervular dentations, gradually narrowing from apex to anal angle, and enclosing near its outer edge a series of five (rarely six) small sub-rhomboidal inter-nervular white spots, lessening in size as the border narrows; base very sparsely and narrowly powdered with blackish, which extends a little along both sides of median and submedian nervures; before and beyond middle respectively, the two dark oblique sub-transverse streaks of the under side are faintly visible. UNDER SIDE. —*Hind-wing and outer apical hind-marginal area of fore-wing varying from cream-colour to pale sulphur-yellow; neuration across apical patch* of fore-wing, and in hind-wing from before middle, pale ferruginous-reddish; hind-margins edged throughout by a fine line of black. *Fore-wing*: intersecting streak of apical patch more macular and conspicuous than on upper side, but sometimes brownish superiorly; *immediately preceding it a short broad bar of pale vermilion-red with a tinge of pink*; parallel to hind-margin a reddish-brown streak (sometimes indistinct on radial nervules) extending from costa to first median nervule, where it becomes broader and blackish, and ends in a blackish space, externally marked by two small yellowish spots immediately above posterior angle; *base with a broad suffusion of orange-yellow*, filling basal half of cell and extending to costal nervure, but not (or very faintly) below median nervure; a minute terminal disco-cellular blackish dot. *Hind-wing*: nervures faintly clouded with blackish for a little distance from base; a conspicuous but rather diffused slightly sinuated blackish streak running obliquely from costa considerably before middle to between first median nervule and submedian nervure rather beyond middle; another highly irregular, often fainter or browner streak, from costa scarcely beyond middle to third median nervule far beyond middle,—there sharply broken, but resumed indistinctly nearer origin of same nervule, and prolonged obliquely to inner margin just before anal angle; a submarginal brown streak as in fore-wing, but much broader, and so radiating outwardly on the reddish nervules as to shut off five large internally-rounded hind-marginal spots of the ground-colour; a wide edging along costa from base to middle, a mark between costal and subcostal nervures, a short longitudinal disco-cellular streak (with a small ill-defined terminal spot), a long streak from base to anal angle above submedian nervure, and a short (sometimes indistinct) one below it,—all bright orange-yellow.

♀ *Apical patch inferiorly broader than in* ♂, *black, with only the very faintest trace of purple-lake on upper part of its inner side, but traversed centrally by a curved row of five yellowish-white spots (of which the two lower are in one example very small and indistinct)*. Fore-wing: basal irroration wider than in ♂; a distinct terminal disco-cellular dot; hind-marginal series of inter-nervular acuminate pale sulphur-yellow spots larger. *Hind-wing*: hind-marginal border broader. UNDER SIDE. —Quite as in ♂, but slightly paler.

In the ♂ the apical portion of the fore-wings is much produced, and the hind-margin thus slightly convex from the apex as far as third median nervule, but thence slightly hollowed as far as first median nervule; and the anal-angular portion of the hind-wings is also, but less markedly, produced. In the ♀ these peculiarities of outline are not nearly so pronounced.

Lucas's description of the Abyssinian type agrees too closely with Hewitson's *Amina*, from the Zambesi, to admit of the recognition of the latter as a distinct species. Wallengren's *Pholoë* (*Wien. Ent. Monatschr.*, 1860, p. 35), from Lake Ngami—with which my *Phænon* (*Trans.*

Ent. Soc. Lond., 3rd Ser., i. p. 522, 1863), from Damaraland, is identical—seems entitled to species rank. Only the ♂ of this butterfly is known,[1] and I have not seen any example except the two on which (in ignorance of Wallengren's diagnosis) I proposed the species *Phœnon*. Compared with *Celimene*, these examples of *Pholoë* are much smaller (*exp. al.* 1 in. 6 lin.); the apical patch is duller, with less crimson in its tint, with all its black borders and neuration, as well as the intersecting streak, duller and narrower; the disco-cellular dot is distinct on the upper side of the fore-wing, while the hind-wing has no black border, but merely a thin blackish line on hind-marginal edge as far as second subcostal nervule and the extremities of four nervules clouded with black (the first median nervule and the submedian nervure being unmarked). On the under side the ground-colour is paler, nearly white; the neuration clouding yellower; in the fore-wings, the basal orange-yellow is more extended and deeper; the disco-cellular dot enlarged; and the red on inner side of apical patch duller, paler, without tinge of pink, and very much reduced, being almost obsolete above lower radial nervule; while in the hind-wings the two subtransverse striæ are duller and yellower and less irregular—the outer one especially being not nearly so widely disjoined on third median nervule.

As far as at present known, *Pholoë* may be regarded as the Western representative of the East-African *Celimene*.

The latter appears to be by no means frequent in collections. I examined a pair from the Zambesi (the types of *Amina*) in the Hewitson Collection, and the British Museum also possesses a ♂ taken by the late Mr. E. C. Buxton in Swaziland. In the North-West Transvaal, on the Limpopo and Marico, the butterfly seems to be not very rare, Mr. F. C. Selous having forwarded four ♂ s and a ♀ from thence, and also a ♂ from the Makloutze River within the Tropic. Mr. T. Ayres' collection contained a ♂ (now in the South-African Museum), captured "between the Limpopo and Zambesi."

Celimene and *Pholoë* constitute a singular section of *Teracolus*, which, as Mr. Butler (*loc. cit.*) has pointed out, is less isolated than it appeared to be since the discovery in Somaliland of the beautiful *T. præclarus*, Butl. (*loc. cit.*, p. 769, pl. xlvii. f. 7, ♀). The ♂ of this species is on the upper side very like the ♂ *Celimene*, but has the purple-red patch even larger, while the ♀ presents in the equally large black apical patch a central lake-red macular band and six small hind-marginal spots of the same colour. On the under side, however, though the pattern of the fore-wings is not widely different from that shown by *Celimene*, in the hind-wings the absence of red neuration and of the striæ, and the presence of a brownish discal band and of reddish-ochreous colouring beyond it, exhibit more resemblance to the splendid *T. Zoë*, Grandidier, of Madagascar. The under-side colouring and pattern in *Celimene* and *Pholoë* combine to some extent the features of the very different *T. Eulimene*, Klug, from Dongola, and *T. Vesta*, Reiche, while the upper side presents no likeness to that of either of those species.

---

[1] Mr. Butler (*Proc. Zool. Soc. Lond.*, 1885, p. 770) regards the diagnosis of Wallengren's *Pholoë* as relating to the ♀; but it is clear to me that the description of the apical patch, "maculis biseriatis rubris violaceo-micantibus, lineolisque marginalibus flavis," can only apply to the ♂.

Localities of *Teracolus Calimene.*

I. South Africa.
  G. Swaziland.—The late F. C. Buxton.
  K. Transvaal.—Limpopo and Marico Rivers (F. C. Selous).

II. Other African Regions.
  A. South Tropical.
    b1. Eastern Interior.—Makloutze River (F. C. Selous). Between Limpopo and Zambesi Rivers (T. Ayres). Zambesi.—Coll. Hewitson.
  B. North Tropical.
    b1. Eastern Interior.—"Abyssinia."—Lucas.

293. (29.) **Teracolus Vesta**, Reiche.

Idmais Vesta, Reiche, Ferr. et Gal., Voy. Abyss., iii. p. 463, pl. xxxi. ff. 7, 8 (1849).
  „   „   Hopff., Peters' Reise nach Mossamb., Ins., p. 361 (1862).
Var. *Teracolus argillaceus*, Butl., Ann. and Mag. Nat. Hist., 4th Ser., xix. p. 459, n. 3 (1877).
Idmais Vesta, Staud., Exot. Schmett., i. pl. 23 (1884).

Exp. al., ($\male$) 1 in. $7-9\frac{1}{2}$ lin.; ($\female$) 1 in. $7\frac{1}{2}-10\frac{1}{2}$ lin.

$\male$ *Deep cream colour, inclining sometimes to a slightly pink, sometimes to an ochrey-yellow tint; bordered and transversely banded with black; basal areas white, near thorax more or less clouded with bluish-grey; neuration beyond middle black-clouded in both wings.* Fore-wing: basal bluish-grey variable in extent, covering from one-third to two-thirds of discoidal cell, so that white area beyond (extending to extremity of cell and to a little before middle inner-marginally) is wider or narrower accordingly; costa blackish-edged, but covered as far as end of cell by basal bluish-grey; a broad large elongated slightly-curved terminal disco-cellular black marking, superiorly touching costal blackish; black nervular clouding only failing on lower radial nervule; a highly irregular discal black band, strongly sinuated both superiorly and mesially, broad superiorly and inferiorly, but rather thin in its middle sinuation (on median nervules), runs from costa to inner margin; a broad black hind-marginal border, marked externally by six short inter-nervular linear marks of the ground-colour, and internally more or less intimately united with discal band by the black-clouded nervules, so as to enclose eight unequal spots of the ground-colour (the four upper smaller than the four lower ones). *Hind-wing:* basal bluish-grey variable in extent, sometimes so wide as to obliterate the white usually present beyond it; about middle of costa commences a transverse black band of variable width, both internally and externally rather irregularly dentate on nervules, extending almost to submedian nervure, not far before anal angle; a broad hind-marginal black border as in fore-wing, but with seven external spots (not streaks) of the ground-colour, and so united with central black band by clouded nervules as to enclose a regular series of eight elongate ground-colour spots (the first and the last two much smaller than the rest). UNDER SIDE.—*Hind-wing and apical hind-marginal area of fore-wing sulphur-yellow, with ferruginous-red stripes and nervular clouding.* Fore-wing: warm ochreous-yellow with a tinge of orange; no basal grey or white or costal blackish, but costa near base bordered with sulphur-yellow; terminal disco-cellular black mark much narrower than on upper side; discal band reduced, sub-macular or even macular, near costa ferruginous-red; hind-marginal border reduced to a thin ferruginous streak, and ferruginous nervules marking off a hind-marginal series of large sulphur-yellow spots, except towards its lower extremity, where it is blackish with only a minute external ochre-yellow spot; upper four of spots between hind-marginal border and discal band sulphur-yellow. *Hind-wing:* central band and inner side of hind-marginal border ferruginous-red, the nervules clouded with the same colour, uniting them and serving to separate seven hind-marginal sulphur-yellow spots, all very large and rounded except a linear one at apex; before middle, from costal nervure, a narrow irregularly-dentate ferruginous-red transverse streak, bounded outwardly by disco-cellular nervule, and becoming obsolete a little below it; costa rather widely edged with orange-yellow to middle; from base four longitudinal streaks of orange-yellow, viz., a very short one below costal nervure, a short one to extremity of cell, a very long one between median and submedian nervures (extending to end of central band), and a rather short one just below submedian nervure.

♀ *Like ♂, but ground-colour duller and paler (especially in hind-wing), and sometimes creamy-whitish or almost white; black markings usually not so dark; hind-marginal fore-wing streaks and hind-wing spots ochrey-yellow, usually larger (especially in hind-wing).* Fore-wing: basal bluish-grey wider and duller, rarely leaving any white beyond it; terminal disco-cellular mark broader (in some specimens superiorly radiating towards base). UNDER SIDE.—As in ♂.

VARIETY A. ♂ and ♀ (*T. argillaceus*, Butl.).

Exp. al., (♂) 1 in. 6–8½ lin.; (♀) 1 in. 8½ lin.

♂ Smaller; all the black markings much narrower; hind-marginal spots of both wings larger, and those of fore-wing much less elongate. UNDER SIDE.—*Hind-wing and apical hind-marginal area of fore-wing completely suffused with dull creamy-reddish,* the usual sulphur-yellow spots being obliterated and the ferruginous stripes represented by dull ashy-grey.

♀ As in ♂, but the under-side suffusion much paler and tinged with pale-greyish.

(*Hab.*—Natal and Transvaal.)

A dwarf ♂ of the type-form, from the interior of Natal, expands only 1 in. 4 lin., and a dwarf ♀, from the Upper Limpopo, the same.

Boisduval (*App. to Voy. de Deleg. dans l'Afrique Aust.*, p. 588,

1847) mentioned "*Idmais Vesta*, Boisd., ined.," as having been figured in Doubleday and Westwood's "Genera" (pl. 7, f. 5) as the ♀ of *Chrysonome*, Klug (*Symb. Phys.*, t. vii. ff. 9–11). I accordingly described (*Rhop. Afr. Aust.*, i. p. 62, n. 41, 1862) the British Museum specimens (from Congo), from which the "Genera" figure was taken, as probably *Vesta*, Boisd. Mr. W. F. Kirby, in 1869, pointed out to me that Doubleday's butterfly was quite distinct both from *Chrysonome*, Klug, and also from the Abyssinian *Vesta*, Reiche (*op. cit., supra*), and in his *Synonymic Catalogue* (1871) he gave the species the name of *Idmais Hewitsoni*. I subsequently found, however, that Hopffer had previously (*Peters' Reise n. Mossambique*, pp. 362–363, 1862) done the same, but had named Doubleday's species *I. Doubledayi*.

Vesta, Reiche, described from Abyssinian specimens, is unquestionably identical with the South-African examples here described. It may at once be distinguished from the Congo ally (*Doubledayi*, Hopffer) by its conspicuous black central band across the upper side of the hind-wings, but the other black markings are very similar, *Doubledayi* holding an intermediate position between *Vesta* and the much smaller and duller *Chrysonome*, Klug, with dusky almost unmarked upper side of hind-wings, from Dongola.[1]

I find that the basal white of the upper side (in conjunction with a yellower general ground-colour) is usually more developed in ♂ specimens from tracts within or near the Southern Tropic than in those inhabiting Natal; the palest ♀ s also (some nearly white) come from the same tracts.

The *Variety* A. is linked to the type-form by ♂ s and a ♀ from the Tugela River and Delagoa Bay, in which the under-side colouration is dull and suffused, and the crossing bands grey, but the reddish tint not decided. Colonel Bowker, on 30th July 1878, took at the mouth of the Tugela a very pronounced ♂ of the variety in company with the ♀ of the intermediate form. This dull creamy-reddish suffusion of the under side appears to be a tendency widely prevalent throughout the greater part of the genus *Teracolus*.

I only once fell in with this species during my stay in Natal, on the 12th March 1867, at a spot of limited extent in the rough thorny country called the "Doorns," near Greytown, where I took several specimens about flowers at a steep bank by the roadside. It was very conspicuous on the wing, and but for the roughness of the rocky ground would have been easy to capture, not flying at all swiftly At the end of March 1880, Colonel Bowker took several examples near the junction of the Mooi and Tugela Rivers, and, on 2d April, the sexes *in copulâ*. All these examples were of the type-form, and the ♂ and ♀ taken paired were both noticeable for the heaviness of their black markings on the upper side. Mr. John L. Fry brought me specimens taken in the interior at the Makloutse River on the 20th April, and at Tati on the 20th May 1887.

---

[1] Mr. Butler records this species as also inhabiting Somaliland. See *Proc. Zool. Soc. Lond.*, 1885, p. 708.

Localities of *Teracolus Vesta*.

I. South Africa.
   E. Natal.
      a. Coast Districts.—Mouth of Tugela River (*J. H. Bowker.*—Subtyp. and Var. A.). Greytown.—Typ. Weenen County (*J. M. Hutchinson.*—Typ. and Var. A.). Between Tugela and Mooi Rivers (*J. H. Bowker.*—Typ.).
   G. Swaziland.—(The late *E. C. Buxton.*—Typ. and Var. A.).
   H. Delagoa Bay.—Lourenço Marques (*Mrs. Monteiro.*—Typ. and Subtyp.).
   K. Transvaal.—Lydenburg District (*T. Ayres.*—Typ. and Var. A.). Limpopo River (*F. C. Selous*).

II. Other African Regions.
   A. South Tropical.
      a. Western Coast.—Damaraland (*J. A. Bell.*—Typ.).
      b. Eastern Coast.—"Querimba."—Hopffer.—Sub-typ. and Var. A. "Mombasa."—Staudinger.
      b1. Eastern Interior.—Lotsani River, Tchakani Vley, and Tati (*F. C. Selous.*—Typ.). Makloutse River (*J. L. Fry.*—Typ.). Zambesi (*Rev. H. Rowley.*—Typ.);—"Tette."—Hopffer.
   B. North Tropical.
      b1. Eastern Interior.—"Abyssinia."—Reiche. "Shoa (*Antinori*)."—Oberthür.

Genus COLIAS.

Colias, Fabricius, "Illiger's Mag., vi. p. 284 (1807)."
  ,,   Latreille, Enc. Meth., ix. p. 10 (1819).
Eurymus, Swainson, Zool. Illustr., 2nd Ser., ii. pl. 60 (1831-32).
Colias, Boisduval, Sp. Gen. Lep., i. p. 633 (1836).
  ,,   Doubleday, Gen. D. Lep., i. p. 72 (1847).
  ,,   Trimen, Rhop. Afr. Aust., i. p. 70 (1862).
  ,,   (revision), Elwes, Trans. Ent. Soc. Lond., 1880, p. 133, and 1884, p. 1.

IMAGO.—*Head* of moderate size, scaly, densely clothed with hair, especially frontally, where there is a large projecting tuft; *palpi* rather short, compressed, densely scaled, clothed beneath with moderately long fine hairs; basal and middle joints about equal in length, terminal joint extremely small and short, not acute; *antennæ* short, stout, straight, with a very gradually formed cylindrical club, truncate at tip.

Thorax rather stout, densely clothed with hair, which is long and silky above (especially posteriorly). *Fore-wings* rather truncate; costa almost straight beyond basal curve; apex marked, but not acute; costal nervure very stout, ending considerably beyond middle; subcostal nervure four-branched,—first nervule long, given off not far beyond middle of discoidal cell,—second not so long, given off at, or a little (or, rarely, a good way) beyond, extremity of discoidal cell,—third short, ending on costa just before apex,—fourth short, ending on hind-margin a little below apex; upper radial united to subcostal nervure, usually about midway between forking of third and fourth nervules and extremity of cell, but sometimes much nearer the latter; upper disco-cellular nervule very short, slightly oblique, sometimes a little curved,—lower one long, arched, occasionally angulated; discoidal cell short, less than half length of wing, moderately wide, truncate at extremity. *Hind-wings* rather bluntly sub-ovate; costa moderately arched; anal angle rather pronounced; convex inner-margins forming a groove covering three-fourths of abdomen; first subcostal nervule arched, given off some distance before extremity of cell; upper disco-cellular nervule of moderate length, straight, oblique,—lower one considerably longer, angulated above its middle point; internal nervure rather long, ending at some distance beyond middle; ♂ (in the *Edusa* group) with an elongate-ovate patch or badge of small, closely-set, elevated scales near base, just above subcostal nervure. *Legs* rather short and stout, scaly; femora with long and rather sparse hair beneath; middle and hind tibiæ finely and sparsely spinulose beneath, their terminal spurs of moderate size; tarsi finely spinulose generally, their claws very deeply bifid, without paronychia.

Abdomen rather short, compressed but not very slender.

This very compact and well-defined genus (of which the well-known "Clouded Yellows" of Europe are typical) consists of a limited number—variously estimated at from twenty-six to forty-eight—of species, very closely allied, and mainly characteristic firstly of the Palæarctic, and secondly of the Nearctic Regions. Several species are circumpolar in the Northern Hemisphere, and the majority is in temperate regions limited to high alpine tracts. It seems most probable that *Colias* is one of the groups which was of very wide and general prevalence during the last glacial period, but has since in tropical regions been compelled to retreat to the mountains. Though extending all through America, from the extreme north (Grinnel Land) to Cape Horn, it is confined to the Andes in the great tropical belt; and in the Oriental Region only appears on the Himalayan boundary, with the exception of an isolated species on the Nilghiris in Southern India. Tropical Western Africa has yielded no representative of the genus, but South Africa has one exceedingly common and generally distributed species (*Electra*, Linn.), a very close ally of the European *Edusa*. The latter is abundant on the African shores of the Mediterranean, and *Electra* is recorded by M. Oberthür as having occurred in Shoa on the Abyssinian plateau, while Mr. Godman has noted the occurrence of a *Colias* (referred by him to *Edusa*) among Mr. Johnston's captures on Kilima-njaro.

These butterflies are orange, orange-yellow, sulphur-yellow, or yellowish-white above, with a more or less developed blackish border, a black terminal disco-cellular spot in the fore-wings, and a larger orange one in the hind-wings; beneath, they are of a paler yellow, always more or less tinged with green; their antennæ and legs and wing-cilia are pinkish-red. There is in many of the deeper-orange species a beautiful rosy and violaceous gloss, better shown in the ♂. The ♀ is constantly paler and duller than the ♂, and in the largest (*Edusa*) group alone possesses yellow spots in the border of the forewings; she also, in the same group, usually presents a second white or very pale form, with in some instances individuals intermediate between this and her normal colouring. In some of the Arctic forms greenish colouring pervades the upper as well as the under surface of the wings; and the blackish border is very faint, and in the ♀ evanescent.[1]

In *Rhopalocera Africæ Australis* I included, on what appeared to be good authority, *C. Hyale* (Linn.), in the South-African list; but, as no properly authenticated specimen has ever since been brought to notice in any part of the country, I feel convinced that some error (perhaps the old one of confusing *Hyale* with the pale form of the ♀ *Edusa* or *Electra*) must have led to Mr. Layard's belief that he and other collectors had taken *Hyale* in South Africa. It is at the same time to be noticed that M. C. Oberthür records this species as among the late Marquis Antinori's captures in 1879-81 in Shoa, Abyssinia.

294. (1.) Colias Electra, (Linnæus).

Papilio Electra, Linn., Syst. Nat., i. 2, p. 764, n. 101 (1767).
♂ ♀ *Papilio Hyale*, Cram., Pap. Exot., iv. pl. cccli. ff. E, F [♂], G, H [♀], (1782).
VAR. ♀ *Papilio Palæno*, Cram., *op. cit.*, pl. cccxl. ff. A. B.
♂ ♀ ♀ VAR. *Colias Electra*, Godt., Enc. Meth., ix. p. 102, n. 39 (1819).
♂ ♀ ♀ VAR. *Colias Electra*, Boisd., Sp. Gen. Lep., i. p. 637, n. 3 (1836).
♂ *Colias Edusina*, Feld., "Wien. Ent. Monatschr., iv. p. 100, n. 55 (1860)."
♂ ♀ ♀ VAR. *Colias Electra*, Trim., Rhop. Afr. Aust., i. p. 71, n. 47 (1862).
LARVA and PUPA, Trim., *op. cit.*, ii. p. 332, pl. 1, ff. 2, 2*a* (1866).

Exp. al., (♂) 1 in. 9 lin.—2 in. 2 lin.; (♀) 1 in. $8\frac{1}{2}$ lin.—2 in. 3 lin.

♂ Deep chrome-yellow, inclining to orange, glossed with a pink lustre in certain lights; with broad black bands on hind-margins. *Fore-wing*: base very narrowly blackish; an ovate, usually rather narrow, black spot at extremity of discoidal cell; hind-marginal band broadest at apex, irregularly dentate on nervules on its internal edge, dusted with pale-yellow scales, some of the nervules crossing it near apex being also pale-yellow. *Hind-wing*: base blackish, extending along median and submedian nervures; costa and inner-margin pale yellowish-green; a large, rounded, deep orange disco-cellular spot; hind-marginal black

---

[1] The opposite extreme in this last character is reached in the very remarkable species *C. Wiskotti*, Staudr., from Samarkand, the ♂ of which has the black border in both wings occupying all the outer area from a little beyond the end of the discoidal cell. (See *Berl. Entom. Zeitschr.*, 1882, p. 165, t. ii. ff. 9 ♂, 10 ♀.)

band *narrower than in fore-wing*, irregularly dentate on nervules on its inner edge, gradually diminishing as far as first median nervule, where it ends; a clothing of pale-yellow hairs on the dusky portion of wing near base and inner margin; yellow field generally more or less finely irrorated with blackish. UNDER SIDE.—*Fore-wing*: pale ochreous-yellow, almost cream-colour on inner margin; costa rather narrowly from base, and broad hind-marginal border, of a lighter or darker *greenish-yellow*; disco-cellular spot black and conspicuous; along inner edge of hind-marginal band, and marking its separation from the yellow ground-colour, is a row of sub-triangular dusky-blackish spots, commencing on costa (*in* the hind-marginal band), with some ferruginous-blackish, often half-obsolete spots, but below third median nervule, where the band narrows, marking its internal edge. *Hind-wing*: wholly of the same greenish-yellow as border of *fore-wing*: disco-cellular spot bright silvery in a ferruginous ring, usually with a smaller, similar, but duller spot touching its upper edge; near hind-margin, and parallel to it (as if continuous of the transverse row in fore-wing), a row of seven more or less conspicuous ferruginous spots, commencing with a rather larger spot on costa, and extending to immediately above submedian nervure.

♀ *Ground-colour much the same as in* ♂, *sometimes rather darker, the pink lustre much fainter; hind-marginal bands spotted with yellow.* *Fore-wing*: basal blackish much broader than in ♂, shading off into dusky-greenish, which extends along costa to hind-marginal band and to before middle of inner margin; disco-cellular spot rounder and usually rather larger; hind-marginal band usually rather broader and blacker, containing six or seven *sulphur-yellow spots*, five of which form a curved row near apex, from costa to third median nervule (the first three of these are elongate, the two others rounded); another, always present, between second and first median nervules; and the seventh spot, generally indistinct, and often wanting, immediately below first median nervule. *Hind-wing*: wholly suffused with dusky-greenish; disco-cellular orange spot more conspicuous than in ♂; hind-marginal black band not so dark as in fore-wing, not well-defined on its inner edge, in some specimens extending almost to submedian nervure, in others only to second median nervule, often spotless, but sometimes containing on its inner side from three to five pale-yellow spots (those nearer anal angle becoming merged with the ground-colour). UNDER SIDE.—Usually *somewhat deeper in colour*. *Fore-wing*: ochreous-yellow ground-colour darker, not paler, on inner margin. *Hind-wing*: transverse row of ferruginous spots not so conspicuous. In both sexes, down on forehead and front of thorax, antennæ, legs, and costal edges of a *reddish-pink* colour; fringes varied with pink above, wholly pink beneath.

Dimorphic form of ♀.—*Orange-yellow ground-colour replaced by greyish-white.* *Fore-wing*: dusky-grey suffusion from base forms a conspicuous basal patch, abruptly ending a little before extremity of discoidal cell; disco-cellular spot often much larger than in ordinary specimens; spots in hind-marginal band conspicuous for their *whiteness*. *Hind-wing*: the whole or greater part suffused with grey; disco-cellular spot whitish, centred with yellowish; hind-marginal spots in band whitish, variable in distinctness. UNDER SIDE.—*Fore-wing*: basal portion, costa to beyond middle, and all but apical portion of hind-marginal band, *bluish grey*; apical portion of band dull greenish-yellow or yellowish-green; black spots as in ordinary specimens. *Hind-wing*: marked as usual, but of the same dull mixture of yellow and green as apical colouring of fore-wing. Antennæ, legs, margins, and fringes *pink*, as in the common examples.[1]

LARVA.—Attenuated posteriorly; slightly and thinly pubescent. Yellowish-green, closely irrorated with darker atoms. Along centre of back, from head to tail, a dark-green streak; on each side of back an ill-defined yellow streak, shading above into the pale-green. On each side, touching yellow streak, a broad dark-green band, edged inferiorly by a conspicuous pure-white or yellowish-white narrow stripe above spiracles. Head granulated, clothed with very short pale hairs.

PUPA.—Attached invariably head uppermost. Pale-green, semi-transparent, darker anteriorly. Prominences of head and back of thorax rather obtuse; outline of wings projecting in a convex ridge below breast. A dark-green dorsal line. Median lateral line of frontal prominences marked with a black streak inferiorly edged with greenish-yellow; inner-marginal edge of wing-cases also defined with blackish. On each side of abdomen a yellowish-white stripe; below this, touching hind-margin of wing-covers, a short abruptly-ending black streak.

The larvæ were found in May, feeding on *Vicia sativa* (lucerne). The pupa state continued from eighteen to twenty-three days, but this was in the beginning of winter.

Except in size and in the width of the hind-marginal border (especially in the fore-wing), the ♂ *Electra* varies very little; but the ♀ exhibits many variations in tint of ground-colour and development of the basal blackish space in the fore-wing, leading in the direction of the white form above described, which is figured by Cramer as *Palæno*.

Very small examples of both sexes are often met with; a ♂ in my own collection expands only 1 in. 5½ lin., and a ♀ in the South-African Museum only 1 in. 4 lin.

The pale form of ♀, though not rare, is very much scarcer than the orange-yellow one.

Electra is nearly related to the well-known European *C. Edusa*,

[1] Mr. Möschler (*Verh. K.K. Zool.-Bot. Gesellsch. Wien*, 1883, p. 279) states that I in error described this form of the ♀ as *Hyale*, L. That I did not do so will be evident on reference to my *Rhop. Afr. Aust.*, i. p. 72, where the pale "Var. ♀" of *Electra* is described, and p. 74, where the true *Hyale* is fully described in reference to some reputed South-African specimens.

Fab., but is distinguished by its deeper ground-colour and strong rosy gloss, as well as by the longer and sharper inward nervular dentations of the hind-marginal border of the fore-wings; the ♀, too, is almost always more marked with dusky suffusion at the base of the fore-wings, and has the hind-marginal border of the hind-wings blacker and better defined. In all these characters, except the more sharply-dentated inner edge of the hind-marginal border, *Electra* is more allied to the Himalayan *C. Fieldii*, Ménét., a species remarkable for the large size of the black disco-cellular spot of the fore-wings, which is on the under side (and in one ♀ more faintly on the upper side also) centred with silvery-white. The larva of *Electra* differs from the descriptions and figures of that of *Edusa* known to me by possessing more conspicuous and defined longitudinal stripes of dark and light green, and by wanting altogether the conspicuous orange spots on the lateral white or yellowish-white streak.

The ♀ deposits her eggs singly, each on a separate leaflet of lucerne, clover, or trefoil.[1] The egg is very pale-yellow, very elongate, sub-cylindrical, slightly fluted longitudinally; it is attached by the broader extremity only, so as to stand erect, the smaller free end being subconical.

This *Colias* seems distributed throughout South Africa; it is almost everywhere abundant, and flies throughout the year, but is more prevalent in the summer months. On the wing it is less swift than *Edusa*, and much less so than *Hyale*. Some of the duller-tinted ♀s, especially if somewhat worn, look very dingy in contrast to their brilliant mates. I have frequently taken the paired sexes, and Colonel Bowker took them at Isipingo in Natal, but in no case that has come under my notice was the ♀ of the white form. The most northern localities for the species known to me are Damaraland on the west and the Upper Limpopo on the north-west border of the Transvaal; but according to Oberthür, it was met with by Antinori in Abyssinia, and I think it not improbable that the specimens taken by Mr. H. H. Johnston in Kilima-njaro—referred by Mr. Godman (*P. Z. S. L.*, 1885, p. 540) to *C. Edusa* —were actually *Electra*.

Localities of *Colias Electra*.

I. South Africa.
  B. Cape Colony.
    a. Western Districts.—Cape Town. Ookiep, Namaqualand District (*L. Péringuey*). Stellenbosch. Wellington and Paarl. Vogel Vley and Ceres, Tulbagh District. Robertson. Swellendam (*L. Taals*). Van Wyk's Vley, Carnarvon District (*E. G. Alston*). Knysna and Plettenberg Bay. Oudtshoorn (—*Adams*).
    b. Eastern Districts.—Uitenhage (*S. D. Bairstow*). "Port Elizabeth."—W. S. M. D'Urban. Grahamstown. Kowie Rivermouth (*J. L. Fry*). King William's Town (*W. D'Urban*). Windvogelberg, Queenstown District (*Dr. Batho*). Murraysburg (*J. J. Muskett*). Colesberg (*A. F. Ortlepp*).

---

[1] Mrs. Barber informs me that the food plants of the larva at Highlands, near Grahamstown, are a species of *Indigofera*, *Trifolium Burchellianum*, and *T. Africanum*.

c. Griqualand West.—"Kimberley."—H. L. Feltham. Hebron (W. Morant).
d. Basutoland.—Maseru (J. H. Bowker).
C. Orange Free State.—Bloemfontein (Dr. H. Exton).
D. Kaffraria Proper.—Butterworth and Bashee River (J. H. Bowker). St. John's River-mouth (Sir H. Barkly).
E. Natal.
  a. Coast Districts.—"Lower Umkomazi."—J. H. Bowker. Isipingo (J. H. Bowker). Mapumulo.
  b. Upper Districts.—Udland's Mission Station Hermansburg. Great Noodsberg. Greytown. Pietermaritzburg. Estcourt (J. M. Hutchinson). Karkloof (W. Morant). Colenso (W. Morant).
F. Zululand.—Isandlhwana (J. H. Bowker). Etshowe (A. M. Goodrich).
G. "Swaziland."—The late E. C. Buxton.
K. Transvaal.—Potchefstroom District (T. Ayres). Upper Limpopo River (F. C. Selous).
L. Bechuanaland.—Motito (the late Rev. J. Frédoux).

II. Other African Regions.
A. South Tropical.
  a. Western Coast.—Damaraland (J. A. Bell).
B. North Tropical.
  b1. Eastern Interior.—"Abyssinia: Shoa (Antinori)."—Oberthür.

GENUS ERONIA.

Eronia (Hübner, *in icon.*, 1816-36), Boisd., Sp. Gen. Lep., i. p. 604 (1836).
Dryas, Boisd., App. Voy. de Deleg., ii. p. 588 (1847).
Eronia, Doubl. [part], Gen. D. Lep., i. p. 64 (1847).
Nepheronia, Butl. [part]. Cist. Ent., i. pp. 38 and 53 (1870).

IMAGO.—*Head* rather broad, clothed with dense, rather coarse, short hair (longer in front); *palpi* very short, with long rough hair beneath,—basal joint long and curved, middle one not half as long, terminal joint minute and rounded; *antennæ* rather short and stout, with a very gradually-formed almost cylindrical club, blunt and rounded at tip.

Thorax rather stout, densely clothed with fine hair, which is longer and silky on back; prothorax well marked, forming a short neck. *Fore-wings* with costa well marked; apex well defined, sometimes rather produced and subacute; hind-margin slightly concave below apex, but prominent inferiorly; subcostal nervure five-branched,—first and second nervules given off near each other at two-thirds the length of discoidal cell, the third short, given off either about midway between end of cell and apex, or nearer the latter; fourth and fifth also short, the fourth ending at apex or just before it, the fifth a little below it; upper radial nervule originating at extremity of cell; upper discocellular nervule somewhat variable in length, curved or sub-angulated, oblique inferiorly,—lower one much longer (twice, or nearly thrice), also curved and slightly oblique; submedian nervure sinuated before middle; discoidal cell about half, or a little more than half, the length of wing, rather broad, truncate at extremity. *Hind-wings* large, rounded, somewhat prolonged inferiorly; costa variable in convexity; hind-margin entire, or (in *Cleodora*) moderately dentate; costal nervure usually extending to apex, but in *Leda* ending at some distance before it; first subcostal nervule much arched, given off far before end of cell; discocellular nervules very oblique,—lower one more than twice, or nearly thrice, as long as upper, and strongly sinuated; discoidal cell more than half as long as wing, more or less acuminate inferiorly at extremity; abdominal channel formed by inner margins complete, but not deep. *Legs* short, slender; femora hairy inferiorly for more than half their length from base; tibiae set with short appressed silky bristles, and sparsely and finely spinulose beneath,—their terminal spurs very fine; tarsi rather thickly spinulose.

Abdomen rather long, deep, arched, clothed with long silky hair at base.

I do not find ground for following Mr. Butler in limiting *Eronia* to "*E. Cleodora* and the *Leda* group," but would retain in it the other African species, only separating from it the Asiatic and Austro-Malayan species (*Valeria*, Cram., *Hippia*, Fab., &c.), whose much longer abdomen, antennae, and fore-wings, as well as their totally different pattern and facies, amply distinguish them.[1]

Eronia in general structure and neuration is near *Teracolus*, differing, however, in having the subcostal nervure of the fore-wings five- (instead of four) branched, much smaller and blunter terminal joint of palpi, and gradually formed, not flattened, club of antennae. In its robuster body and gradually clavate antennae, and in outline of wings, it exhibits some approach to *Callidryas* (one species, *E. Buquetii*, in tint and marking has quite the aspect of the pale species of that genus).

The seven or eight species which *Eronia* contains are good-sized butterflies, and present remarkable differences in pattern and colouring, the ♀ being in some instances highly variable and quite unlike the ♂. The type of the genus is *E. Cleodora*, in which the sexes are alike white or yellowish, with a sharply-defined black hind-marginal border, and have the under side of the hind-wings creamy-yellow bordered and spotted with mixed brown and silvery-grey. *E. Leda* (on which Boisduval founded his genus *Dryas*) is vivid-yellow, with a brilliant-orange apical patch in the fore-wings, and looks like a magnified *Teracolus Auxo*; and the Madagascar *E. Lucasi*, Grandidier, is white with a lemon-yellow apical patch. The males in *E. Argia* and *Thalassina* are greenish-white and greenish respectively, with black borders, but their females are coloured with ochre-yellow and orange-red in imitation of certain species of *Pieris* and *Mylothris*. Both sexes of *E. Buquetii* are alike of a plain greenish-white, with a blackish apical hind-marginal

---

[1] The ♂s of this group are bluish or bluish-white, with black border and neuration, and the ♀s, by a modification of these markings, closely mimic various species of *Danais*. (See Wallace, *Trans. Ent. Soc. Lond.*, 1867 (vol. iv.), p. 309.)

border of exceedingly variable development, and their under side has a shining dull-greenish tint, with a terminal disco-cellular spot in the hind-wings, so that specimens with scarcely any blackish border look very like *Callidryas Florella*. *E. Pharis* (which I have not had the opportunity of examining) is described by Doubleday (*op. cit.*, p. 65) as having "the wings nearly as round as the genus *Pontia*, and of as delicate a texture; the apex of the anterior just touched above with black, below varied with brown; the posterior wings above immaculate, below sometimes nearly immaculate, at others varied with large clouds of satiny-brown and silvery-white."

The four species found in South Africa all have a tropical range, *Argia* and *Buquetii* both inhabiting alike the East and West Coasts, while *Leda* and *Cleodora*, with an extensive East-African distribution, do not appear to have been hitherto recorded from West Africa. They are all found on the south-east, *Leda* and *Argia* not spreading farther south-westward than the Bashee River, but *Cleodora* extending to Van Staden's River, and *Buquetii* to Knysna, and even straggling to Cape Town. On the wing they are decidedly swift, but constantly check their onward flight to visit flowers.

The larva of *E. Cleodora* has been found on *Capparis Zeyheri*, but I have not seen either specimens or drawings of it.

295. (1.) Eronia Cleodora, Hübner.

♂ *Eronia Cleodora*, Hübn., Samml. Exot. Schmett., ii. pl. 130 (1806?).
    ,,   ,,   Boisd., Sp. Gen. Lep., i. p. 605. n. 1 (1836).
♂ ,,   ,,   Doubl., Gen. D. Lep., i. pl. ix. f. 1 (1847).
♂ ♀ ,,   ,,   Trim., Rhop. Afr. Aust., i. p. 64, n. 43 (1862).
♂ ,,   ,,   Hewits., Exot. Butt., iv pl. 5, f. 7 (1867).
♂ *Eronia Excia*, Hewits., *op. cit.*, p. 8
♂ *Eronia Cleodora*, Staud., Exot. Schmett., i. pl. 21 (1884).

Exp. al., (♂) 2 in. 2–7½ lin.; (♀) 2 in. 6–9 lin.

♂ *White, yellowish-white, or pale-yellowish, with a black hind-marginal border, variable in width and irregular on its inner edge. Fore-wing:* base usually marked very faintly and narrowly with blackish; hind-marginal border varying from very narrow to broad,—wide apically, rather abruptly narrowed on third median nervule, and thence either tapering to a point or to a narrower or broader termination (in accordance with the less or greater width of the border) to posterior angle; inner edge very variable and irregular in outline, generally more or less denticulate on nervules and excavate between them,—always a very marked and conspicuous prominence or projection (rounded, subacute, or truncate) immediately above third median nervule; in apical extremity of border two spots of the ground-colour,—a small one between second and third subcostal nervules, and a larger one (nearer apex) between fifth subcostal and upper radial nervules. *Hind-wing:*

hind-marginal border varying in width in accordance with that of fore-wing, attenuated to a point at apex and anal angle, its inner edge irregularly dentate on nervules, and almost always exhibiting a considerable projection (extremely variable in size and shape) between second subcostal and radial nervules. *Cilia* mixed brownish and whitish in fore-wing; whitish or yellowish in hind-wing. UNDER SIDE.—*Fore-wing: hind-marginal border internally blackish, but externally silvery-grey stippled and clouded with ferruginous,* which constitutes an outer border wide superiorly, and tapering to a point about first median nervule, and is bounded superiorly by a costal-apical creamy-yellow or ochrey-yellow irregular marking, the inner edge of which corresponds with the position of the two white or yellowish spots of the upper side. *Hind-wing: creamy or ochreous yellow; hind-marginal border silvery-grey, more or less densely stippled and clouded with ferruginous or ferruginous-brown;* inner edge of border more irregular than on upper side, and sharply defined by a ferruginous-brown line,—the inward projection between second subcostal and radial nervules always narrower and longer, almost always extending as far as third median nervule a little beyond extremity of discoidal cell, and thence often prolonged by some ferruginous scaling so as to join an irregular compound double spot of the same colouring between first median nervule and submedian nervure; close to inner edge of border (and often so united with it as to enclose a small spot of the ground-colour), between first and second median nervules, an elongate grey-and-ferruginous spot;[1] a larger irregular elongate mark, of the same colouring, on costa a little before middle, ending on first subcostal nervule; two small rounded spots near base, one in cell and the other below it; and a still smaller terminal disco-cellular spot.

♀ *Like ♂, but hind-marginal border broader (especially about posterior angles), of a rather duller black, and its inner edge more irregular, emitting (especially on second and first median nervules of fore-wing and third median nervule of hind-wing) much larger and deeper dentations;* the principal projection in fore-wing less prominent (owing to width of lower half of border) and excavated at extremity, and that in hind-wing very much broader and more truncate. UNDER SIDE.—As in ♂, but the grey of the markings not silvery, the ferruginous-brown darker, the lesser markings of the hind-wing often smaller, and the great projection of the hind-wing border not united to the spot below first median nervule by ferruginous scaling.

The ♂ varies very greatly in the development of the black border. Specimens like that figured by Hübner, with a rather narrow border, are frequently met with; while not only are numerous gradations found up to and beyond as broad a border as that figured by Doubleday (which Hewitson, *op. cit.*, proposed to separate as *E. Erxia*), but instances are not rare of ♂s with very much narrower borders than Hübner

---

[1] This marking is rarely (two examples) reproduced in black on the upper side.

depicts. In these examples the subapical spots of the ground-colour in the fore-wings are almost invariably larger than usual, and are often so extended as to become confluent, thus forming as large a marking as that shown in yellow on the under side. The extreme in this direction is reached in an individual sent from King William's Town by Miss F. Bowker, where the confluent spots make a small apical patch, broadly bounded by black inwardly, but inferiorly completely united with the ground-colour, and the hind-margin bears only a narrow blackish edging; while in the hind-wing the border is reduced to a series of quite separate ill-defined thin nervular fuscous marks with a small separate spot indicating the customary marked projection.

The ♀ is not nearly so frequently met with as the ♂; among fifty-five specimens, from all parts of South Africa, now before me, only eight are ♀s. In a pair taken *in copulâ* by Colonel Bowker in March 1879, the sexes do not differ much either in size or pattern, the white ♂ being the broadest-bordered that I have seen, while the yellowish ♀ is of medium character in that respect.

A dwarf very narrowly-bordered ♂, taken at D'Urban, Natal, by the late Mr. M. J. M°Ken, expands less than 1 in. 10 lin. In this example the subapical spots of the fore-wing border are normal, but the hind-wing border has its inner edge quite even.

Hopffer (*Peters' Reise n. Mossamb., Ins.,* p. 363) and Oberthür (*Études d'Ent.,* iii. p. 21) independently note the large size and broad border of East-African examples (respectively from Querimba and Zanzibar) in comparison with those from South Africa; it is probable that they both refer to specimens of the ♀.

Mrs. Barber wrote to me that the larva of this species fed on a *Capparis*, and subsequently Mr. J. P. Mansel Weale informed me that *Capparis Zeyheri* was the food-plant. Mr. Weale added that the larva much resembled that of *Teracolus Auxo* (Lucas); it was difficult to find, its reddish-yellow lateral stripe matching in tint the edge of the leaves.

This strikingly marked *Eronia*, which has peculiarly soft rich under-side colouring, is widely distributed, and apparently numerous in most parts of South-Eastern Africa. It has been recorded from various places on the Tropical East Coast, and from as far north as Shoa in Abyssinia. It is swift on the wing, but has the family habit of pitching frequently on flowers; and at D'Urban, Natal, I took a fine series at flowers of the introduced *Vinca rosea* during the month of February. These summer specimens (including a few ♀s) were all of the larger broad-bordered description; but two ♂s which I captured, one in June and the other in August 1865, were small and narrow-bordered. As two ♂s taken by Colonel Bowker (at King William's Town in May and at D'Urban in August respectively) are also small and narrow-bordered, I am disposed to think that this form of the butterfly may possibly be the winter brood, but hitherto there have been no correspondingly narrow-bordered ♀s recorded as occurring in the winter months.[1]

---

[1] I have lately (1888) discovered in the South-African Museum a pair marked as taken *in copulâ* by Colonel Bowker at D'Urban on 11th May 1879. These specimens are much smaller (♂, 2 in. 2 lin., ♀, 2 in. 4 lin. across fore-wings) than the pair above mentioned as taken in March of the same year, and both have the black border considerably narrower.

Mr. Alfred D. Millar informs me that the butterfly is very common all the year round at

Localities of *Eronia Cleodora*.

I. South Africa.
   B. Cape Colony.
   b. Eastern Districts.—"Near Port Elizabeth."—W. S. D'Urban. Van Staden's River.—Burchell Collection in Mus. Oxon. Grahamstown (*M. E. Barber*). Bathurst District: Mouth of Kowie River (*J. L. Fry*), Kleinemond River (*H. J. Atherstone*), and Tharfield (*Miss M. L. Bowker*). King William's Town (*J. H. Bowker*). Kingscote, near Bodiam (*W. S. D'Urban*). East London (*P. Borcherds*).
   D. Kaffraria Proper.—Bashee River (*J. H. Bowker*).
   E. Natal.
   a. Coast Districts.—"Lower Umkomazi."—J. H. Bowker. D'Urban.
   b. Upper Districts.—Maritzburg (*S. Windham*). Estcourt (*J. M. Hutchinson*).
   F. Zululand.—Etshowe (*A. M. Goodrich*).

II. Other African Regions.
   A. South Tropical.
   b. Eastern Coast.—"Querimba (*Peters*)."—Hopffer. "Bagamoyo and Zanzibar (*Raffray*)."—Oberthür.
   b1. Eastern Interior.—"Taveta (*H. H. Johnston*)."—Godman.[1]
   B. North Tropical.
   b1. Eastern Interior.—Abyssinia: "Shoa (*Antinori*)."—Oberthür.

296. (2.) **Eronia Leda,** (Boisduval).

♂ ♀ *Dryas Leda*, Boisd., App. Voy. Deleg. Afr. Aust., p. 588, n. 30 (1847).
♀ *Eronia Leda*, Doubl., Gen. D. Lep., i. p. 65 (1847), and ii. p. 530 (1852).
♀ *Callidryas*, sp. inedit., Angas, Kafirs. Illustr., pl. xxx. f. 2 (1849).
♂ ♀ *Dryas Wahlbergi*, Wallengr., K. Sv. Vet.-Akad. Handl., 1857; Lep. Rhop. Caffr., p. 17.
♂ *Eronia Leda*, Hopff., Peters' Reise Mossamb., Ins., p. 364 (1862).
♂ ♀ *Eronia Leda*, Trim., Rhop. Afr. Aust., i. p. 63, n. 42 (1862), and ii. pl. ii. f. 5 [♀].
♀ *Eronia Trimenii*, Oberth., Étud. d'Ent., iii. p. 20 (1878).
♂ *Eronia Leda*, Staud., Exot. Schmett., i. pl. 21 (1884).

Exp. al., (♂) 2 in. 2-7 lin.; (♀) 2 in. 2-7½ lin.

♂ Bright-yellow; fore-wing with a broad orange apical patch. Fore-wing: apical orange commencing on costa a little before extremity of discoidal cell, and its outer edge occupying hind-margin as far as

D'Urban, and that "during the winter months it is much smaller, and varies considerably in the markings." Of the ten specimens he has sent me, five, ticketed "December 1887," are large and broad-bordered, while the remaining five, ticketed "August 1887," are small and narrow-bordered—three of them strikingly so.

[1] Mr. Butler (*Proc. Zool. Soc. Lond.*, 1888, p. 96) has described these specimens of Mr. Johnston's, with others from the same locality taken by the late Bishop Hannington, and at Kilima-njaro by Mr. F. J. Jackson, under the name of *Eronia dilatata*; but it seems perfectly clear to me that the paler yellow and broader border of the hind-wings on the under side, given by Mr. Butler as "the only satisfactory distinguishing characters," are altogether insufficient to warrant the separation of these examples from so highly variable a species as *E. Cleodora* has been shown to be.

first median nervule; apex edged with ferruginous-brown from first subcostal nervule on costa to third or second median nervule on hind-margin; a hind-marginal series of minute inter-nervular black dots. *Hind-wing*: slightly paler on inner margin; a hind-marginal series of dots as in fore-wing. UNDER SIDE.—*Hind-wing and apical area of fore-wing deep chrome-yellow with ferruginous marks. Fore-wing*: paler than on upper side, whitish on inner margin; close to apex, two or three small ferruginous spots with large silvery centres; a very small blackish terminal disco-cellular dot. *Hind-wing*: sometimes rather deeper in tint than apex of fore-wing, thickly or sparsely flecked with ferruginous or ferruginous-brown; a row of rather darker spots along costal edge (which is fringed with pinkish and brown hairs), the largest and most conspicuous spot elongate, variable in shape, a little beyond middle, touching first subcostal nervule; a curved discal row of six (as many as four sometimes obsolete) small ferruginous spots, of which the last, between first median nervule and submedian nervure, is much the largest and always well marked. Hind-marginal minute spots a little larger than on upper side in both wings, ferruginous.

♀ *Paler. Fore-wing*: orange apical patch duller, sometimes almost obsolete, rarely wanting wholly; apical ferruginous-brown edging wider, somewhat diffuse; a submarginal series of five inter-nervular ferruginous-brown spots, of which the lowest, between second and first median nervules, is indistinct. UNDER SIDE.—*Fore-wing*: pale lemon-yellow; beyond discoidal cell orange, deepening into ferruginous-red, tinged with pink at apex; spots at apex larger; other submarginal spots of upper side sometimes faintly indicated. *Hind-wing*: more thickly flecked than in ♂, the spots and atoms *pinker;* the largest spot on costa silvery-centred; *spots of discal row enlarged, rounded, silvery-centred, especially a greatly-enlarged spot lying between radial and third median nervules;* beyond this conspicuous spot and the smaller one just above it, a patch or cloud of ferruginous-red tinged with pink extending to hind-marginal edge; usually a diffused pink stain from base between median and submedian nervures to beyond middle.

In both sexes, but usually more prominently in the ♀, the under side of the hind-wing also bears an obliquely-transverse ill-defined ferruginous streak running from the extremity of the discoidal cell to submedian nervure before middle.

There is much variation in size as regards both sexes.[1] In colouring and pattern the ♂ is much more constant than the ♀; the upper side offering merely some modifications in the tint and development of the apical ferruginous-brown or dark-brown edging in the fore-wings, which when best expressed radiates slightly on the nervules, and the under side exhibiting a deeper or lighter chrome-yellow and

---

[1] Boisduval (*op. cit.*) observed that Abyssinian specimens were a third smaller than those which he had received from Natal; but I have received very small examples as well as very large ones from the latter country.

a denser or sparser flecking of ferruginous. Out of eleven ♀ specimens (all South-African) now before me, only four have the apical orange well developed; four present only irregularly-disposed diffused orange-streaking; one has scarcely any trace of that colour; and the remaining two are altogether without it. Usually the largest silvery spot of the under side of the hind-wings and the two small ones adjacent to it are perceptible on the upper side. The development and brilliancy of this large spot is very variable,—in some cases it is obscured greatly by ferruginous clouding, and the other spots of the discal row are very inconstant in size and shape, and in the brightness of their silvery centre.

M. Oberthür's separation (*op. cit.*), as *E. Trimenii*, of the small Kaffrarian ♀ figured in my *Rhopalocera Africæ Australis*, cannot be sustained.[1] It is noticeable, however, that the under side of the small individuals of both sexes is of a deeper yellow, with the flecking of the hind-wings denser and pinker. I think it possible that this may be the winter (dry-season) brood. I did not meet with any small specimens during my summer visit, and the only similar example whose month of capture is recorded is a ♀ taken in August 1878 by Colonel Bowker.

In upper-side colouring and pattern this brilliant butterfly has the aspect of a magnified *Teracolus Auro* (Lucas), but is of much more vivid and intense yellow and orange in the ♂. I found it pretty numerous on the coast of Natal in January, February, and April 1867, and had previously noticed a single specimen on the wing when calling at D'Urban on 23d June 1865. It is exceedingly conspicuous in flight, but so swift as to be difficult to secure. Fortunately it has the family taste for frequent sips of nectar, and nearly all my specimens were made captive while settling on the flowers of *Vinca rosea*. Colonel Bowker sent me the paired sexes taken by him near D'Urban in March 1879; the large ♀ of this pair has no apical orange. The same observer found the species common in Kaffraria Proper; it was noted as unusually numerous all through the summer and autumn of 1862.

Though well known to range widely through Eastern Africa, I have met with no record of the occurrence of *Leda* on the western side, except in Mr. Kirby's Catalogue of the Hewitson Collection, where Angola is given as a locality in addition to Natal.

Localities of *Eronia Leda*.

I. South Africa.
    D. Kaffraria Proper.—Bashee, Colossa, and Nabosa Rivers (*J. H. Bowker*).
    E. Natal.
        a. Coast Districts.—"Lower Umkomazi."—J. H. Bowker. D'Urban. Verulam.
    F. Zululand.—St. Lucia Bay (the late *Colonel H. Tower*). Etshowe (*A. M. Goodrich*).

---

[1] My friend's remark, "Jamais l'*Eronia Leda* n'a de taches dans sa macule apicale aurore," renders it clear that he was at the time unacquainted with the female of this butterfly.

II. Other African Regions.
   A. South Tropical.
      a. Western Coast.— "Angola [Hewitson Coll.]."—W. F. Kirby.
      b. Eastern Coast.—"Querimba (*Peters*)."—Hopffer.
   B. North Tropical.
      b. Eastern Coast.—"Massowah (*Raffray*)."—Oberthür.
      b1. Eastern Interior.—Abyssinia ; "Shoa (*Antinori*)."—Oberthür.

297. (3.) **Eronia Buquetii,** (Boisduval).

♂ *Callidryas Buquetii*, Boisd., Sp. Gen. Lep., i. p. 607, n. 1 (1836).
Eronia Buquetii, Boisd., App. Voy. de Deleg. Afr. Aust., p. 588 (1847).
♂ ♀ *Eronia Buquetii*, Trim., Rhop. Afr. Aust., i. p. 66, n. 44 (1862).
Eronia Buquetii, Hopff., Peters' Reise Mossamb., Ins., p. 363 (1862).

Exp. al., (♂) 2 in. 2½–6 lin.; (♀) 2 in. 2 6½ lin.

♂ *Greenish-white; fore-wing with an apical-hindmarginal fuscous border, varying in development from an almost linear edging to a band very broad at apex, and tapering gradually to posterior angle.* Fore-wing : costa with a very fine linear fuscous edging ; costal margin narrowly irrorated with reddish-brown and fuscous,—the irroration in specimens with very reduced apical-hindmarginal edging not extending beyond middle ; inner edge of apical-hindmarginal border (of whatever width), from its origin on costa beyond middle to its termination on hind-margin or at posterior angle, dentated on nervules and excavated between them ; in its most reduced condition this border is better expressed on costa than on hind-margin, where it is very attenuated and scarcely extends beyond third median nervule. Hind-wing : without markings. UNDER SIDE.—*Hind-wing and moderate-sized apical space of fore-wing dull greenish or yellowish-white (sometimes with a faint tinge of ochreous) shot with a pale bluish-violaceous gloss, and more or less indistinctly irrorated and partly striolated with brownish-grey ; on hind-marginal edge a series of minute nervular black spots, scarcely perceptible in fore-wing.* Hind-wing : costa tinged with pale-green on basal lobe ; at extremity of discoidal cell a well-marked elongate-ovate ferruginous-brown whitish-centred spot, bounded externally by an ill-defined small whitish mark ; sometimes a faint indication of a discal series of very small brownish-grey spots.

♀ Like ♂, but rather less greenish in tint. UNDER SIDE.—Hind-wing and apex of fore-wing usually deeper and more yellowish in tint ; terminal disco-cellular spot rather larger and rounder, with a better developed whitish centre.

Hopffer (*op. cit.*) has noted three varieties of this butterfly, in addition to the conspicuously black-bordered type-form from the West Coast of North Tropical Africa, viz., Var. β, *Arabica*, Var. γ, *Capensis*, and Var. δ, *Mossambicensis*, respectively inhabiting the regions indicated ; and Mr. Butler (*Proc. Zool. Soc. Lond.*, 1884, p. 493) observes that these local forms are permanent and should be kept distinct. Were the Cape Colony only in question here, I should be disposed to treat the Var. *Capensis* as a separate form, all the specimens that I have seen from that territory presenting only the almost linear rudiment of the dark border of the fore-wing; but in Natal, Transvaal, and Delagoa Bay the insect has the border more or less developed, being in the last-named locality identical with the Var. *Mossambicensis*. Two Transvaal specimens in the British Museum and five Natal examples in the South-African Museum present a complete gradation between the Vars. *Capensis* and *Mossambicensis*, which latter is scarcely separable from Congo *Buquetii*. The only examples that I have seen from Damaraland, viz., a ♂ taken by the late Mr. Andersson and a small yellowish ♀ in the Hewitson Collection, both have the border moderately broad and ending about first median nervule. From the Zambesi, two ♂s have the border black and broad, while in a ♀ it is dusky and narrower; and Oberthür notes that Zanzibar examples are like the type from Senegal. It is noticeable that the broad-bordered typical *Buquetii* has the under side greener in tint, more glossy, and less distinctly irrorated and striolated, with the terminal disco-cellular spot of the hind-wings narrower and less rounded.

I have not seen the Var. *Arabica*, but it is clear from Hopffer's account that only the apex of the fore-wings is dusky-blackish, while Butler's description of the under side applies very nearly to that of the type-form. A small ♀ from Madagascar in the South-African Museum is intermediate between this variety and *Mossambicensis*, the narrow border extending on hind-margin only a little below second median nervule.

As Doubleday (*Gen. D. Lep.*, i. p. 65) has remarked, this plainly-tinted *Eronia* both in colour and marking bears a strong resemblance to the white species of *Callidryas*; and Cape specimens are so like *C. Florella*, (Fab.), both in appearance and flight, that it is very difficult to distinguish the two before capture.

Wooded spots and gardens in their near neighbourhood are the favoured haunts of *Buquetii*, which appears at the end of January and remains out until the middle of April. At Plettenberg Bay and near Grahamstown I met with the ♂ abundantly, but ♀s were scarce; the flowers of *Plumbago* were their constant resort. They are remarkably swift on the wing, and their halt on a flower is exceedingly short. At Port Elizabeth I took one specimen and saw another at flowers of *Agapanthus*. Very rarely I have met with a straggler in the neighbourhood of Cape Town during the later summer months. I did not notice the species during my stay in Natal; and the few individuals I have received from that Colony were taken near Maritzburg.[1]

Localities of *Eronia Buquetii*.

I. South Africa.
  B. Cape Colony.
    a. Western Districts.—Cape Town [occasional visitor]. Knysna (*Mrs. Maskett*) and Plettenberg Bay.

---

[1] In the first week of February 1889, Mr. A. D. Millar met with about a dozen specimens near D'Urban, and captured seven of them. These examples have the blackish border moderately developed.

  b. Eastern Districts.—Port Elizabeth. Uitenhage (*S. D. Bairstow*). Kowie River Mouth (*J. L. Fry*). Grahamstown, King William's Town (*W. S. M. D'Urban*).
 D. Kaffraria Proper.—Bashee River (*J. M. Bowker*).
 E. Natal.
  b. Upper Districts.—Maritzburg (— *Quickett* and *S. Windham*—Sub-Typ.).
 G. "Swaziland."—The late E. C. Buxton.
 H. Delagoa Bay.—Lourenço Marques (Mrs. Monteiro.—Var. *Mossambicensis.*)
 K. Transvaal.—Coll. Brit. Mus.—Var. *Capensis* and Sub-Typ.

 II. Other African Regions.
  A. South Tropical.
   a. Western Coast.—Damaraland (the late *C. J. Andersson*—Sub-Typ.). "Angola (*J. J. Monteiro*)."—Druce. Congo.—Coll. Brit. Mus.—Typ.
   b. Eastern Coast.—Zambesi (*Rev. M. Rowley*.—Typ.). "Querimba (*Peters*)."—Hopffer.—Var. *Mossambicensis.* "Zanzibar (*Raffray*)."—Oberthür.—Typ.
   *b*1. Eastern Interior.—"Lake Nyassa."—W. F. Kirby, Cat. Hewits. Coll.
   bb. Eastern Islands.—Madagascar : Murundava (— *Grévé*.—Sub-Typ.).
  B. North Tropical.
   a. Western Coast.—Sierra Leone.—Coll. Brit. Mus.—Typ. "Senegal."—Boisd.—Typ.
   b. Eastern Coast. — Somaliland : "Eighty miles S. of Berbera (*Thrupp*)."—Butler.—Var. *Arabica.*
   *b*1. Eastern Interior.—Abyssinia : "Shoa (*Antinori*)."—Oberthür.
 IV. Asia.—Arabia: "Lahej and Haithalkim, near Aden (*Yerbury*)."—Butler.—Var. *Arabica.*

298. (4.) Eronia Argia, (Fabricius).

♂ *Papilio Argia,* Fab., Syst. Ent., p. 470, n. 118 (1775).
♂ *Papilio Cassiopea,* Cram., Pap. Exot., iii. pl. cci. f. A (1782).
♂ *Pieris Argia,* Godt., Enc. Meth., ix. p. 140, n. 77 (1819).
♂   ,,   ,,   Boisd., Sp. Gen. Lep., i. p. 442, n. 6 (1836).
♀ *Eronia varia,* Trim., ♂, Trans. Ent. Soc. Lond., 3rd Ser., ii. p. 175 (1864); and Rhop. Afr. Aust., ii. p. 327, n. 221 (1866).
Var. ♀, *Eronia varia,* Trim., ♀, *loc. et op. cit.*
♂ *Eronia Argia,* Staud., Exot. Schmett., i. pl. 21 (1884).

 Exp. al., (♂) 2 in. 8 lin.—3 in.; (♀) 2 in. 9 lin.—3 in. 1 lin.

 ♂ *Greenish-white; fore-wing with a black apical border, continued taperingly along hind-margin. Fore-wing:* costa narrowly edged with black; costal margin in basal third closely irrorated with blackish; black border very broad on costal edge, but below upper radial nervule narrowing rapidly (and below third median nervule sometimes becoming macular) to its terminating point at extremity of first or second median nervule; inner edge of black border more or less dentating (in one example deeply piercing) the ground-colour on nervules, and irregularly excavate between them. *Hind-wing:* along edge of outer half of inner margin, anal angle, and of hind-margin to midway between first and second median nervules, an ill-defined tinge of lemon-yellow. UNDER SIDE.—*Hind-wing and narrow costal edging and hind-marginal border of fore-wing pale shining greyish-yellow;* on hind-marginal edge an inter-nervular series of minute black spots. *Fore-wing:* apical broad portion of hind-marginal border of upper side represented by a large subquadrate dark ferruginous-brown patch, more or less clouded with violaceous, reaching costal edge superiorly, but scarcely extending to apex itself, and so leaving hind-marginal border beyond it narrowly yellowish; below patch some vague hoary-violaceous clouding marks hind-marginal border. *Hind-wing:* costal border broadly but faintly shot with a tint of hoary-violaceous; hind-marginal border similarly but more narrowly and more faintly shot; between second subcostal and radial nervules, near hind-margin, a faint, diffused, ill-defined ferruginous-brown mark (sometimes obsolete); costa narrowly edged with orange-yellow for a little distance from base.

♀ *Fore-wing with orange-red basal suffusion; hind-wing often pale creamy-yellow.* Fore-wing: apical hind-marginal border reduced to blackish nervular rays forming elongate spots, broad externally but acuminate inwardly, at apex very prolonged and diffusedly confluent, but below that quite separate, the lowest on first median nervule; inner edge of border faintly indicated near costa (and sometimes also on median nervules) by some indistinct fuscous scaling; basal orange-red occupies discoidal cell except near extremity, and extends faintly and diffusedly below it; in specimens with yellow hind-wings a faint yellowish edging along costa and hind-margin. *Hind-wing:* white or creamy-yellow—in the latter case the neuration is whitish; a nervular hind-marginal series of six rather ill-defined blackish spots, of which the first and sixth are small, the second prolonged ray-fashion along second subcostal nervule, and the third (on radial nervule) very small, minute, or even obsolete. UNDER SIDE.—*Hind-wing and narrow costal and hind-marginal border of fore-wing chrome-yellow with a tinge of ochreous; hind-margin of hind-wing with two violaceous-whitish, pale ferruginous-clouded blotches.* Fore-wing: basal orange-red deeper than on upper side, especially below cell,—its outer edge clearly defined; apical blotch coloured like the two in hind-wing, and enlarged inferiorly so as to reach lower radial nervule; usually a smaller similar but less defined blotch on hind-margin between third and second median nervules. *Hind-wing:* upper blotch the larger, between first subcostal and radial nervules, bifid internally; lower blotch prominent internally, rather hollowed outwardly, situated mainly between third and second median nervules, but a little overlapping both.

South-African specimens of the ♂ differ from the West-African type-form in the less-developed black border of the fore-wings on the upper side, and in a yellower tint on the under side of the hind-wings. The former character is, however, variable, and its reduced condition is not very marked except in two examples, taken at Etshowe in Zululand by Mr. A. M. Goodrich, which present a much narrowed apical patch, continued hind-marginally below lower radial nervule only by a small separate spot on third median nervule. One Zululand ♂ has the border very nearly as well developed as in West-African specimens.

An *aberration* of the ♀, captured by Colonel Bowker in Kaffraria Proper, has all the dark markings very faintly shown, the basal red of the fore-wings almost obsolete on the upper side, and replaced on the under side by ochreous-orange, the hind-wings on both surfaces yellow-ochreous, and the under-side blotches rather indistinct.[1]

I have not seen an ascertained ♀ of the typical *Argia* from West Africa, nor have I found any published description purporting to be one of that sex,[2] but I have examined twelve South-African examples. Of seven now before me, three have the hind-wings white on the upper side. The rich red and yellow colours borne by the ♀ are in singular contrast to the plain tints of the ♂, and there can be little doubt that they simulate those of the abundant and slow-flying *Mylothris Agathina*, (Cram.), a species which (as stated above, p. 45) appears to be also the object of direct mimicry by both sexes of *Pieris Thysa*, Hopff.

This fine *Eronia* seems to be more prevalent in Zululand than elsewhere in South Africa, only scattered specimens having occurred in Kaffraria and Natal. The two ♀ s noticed in my former work were noted by Colonel Bowker as taken near the Bashee River in February 1863; they settled on *Plumbago* flowers. The late Colonel Tower's Zululand examples were captured in the winter months, and the two ♂ s above specially noticed by Mr. A. M. Goodrich in October and November 1886 respectively.

Localities of *Eronia Argia*.

I. South Africa.
 D. Kaffraria Proper.—Bashee River (*J. H. Bowker*).
 E. Natal.—D'Urban (*J. H. Bowker*). Pinetown (*H. C. Harford*).

---

[1] In *Rhopalocera Africæ Australis*—having at the time only this specimen and one ordinary ♀—I mistakenly treated them as sexes of a presumed new species, which I named *E. varia*. Individuals of both sexes (from Zululand) did not reach me until 1867, and I then recognised the ♂ as *Argia*, Fab.

[2] Mr. Butler (*Proc. Zool. Soc. Lond.*, 1888, p. 96) identifies the ♀ with both *E. Poppea*, (Donov.), and (yellow form) *Idotea*, Boisd., and remarks that the females of *Argia* are known to be extremely variable. I inspected both these forms in the late Mr. Hewitson's collection. Both have the black border of the wings very much more developed than in the Southern ♀ (especially in the hind-wings) on both surfaces; the former sometimes has the fore-wing ochre-yellow, while *Idotea* is bright lemon-yellow in both wings, simulating the appearance of the West-African *Pieris Ianthe*, Doubl. Mr. Butler (*loc. cit.*) states that the ♂ *Argia* from Kilima-njaro does not differ from Sierra Leone specimens; his description of the single ♀ from that locality agrees with the characters of the Southern ♀, except that the border of the wings is apparently better developed.

F. Zululand.—Etshowe (*A. M. Goodrich*). St. Lucia Bay (the late Colonel *H. Tower*).
H. Delagoa Bay.—Lourenço Marques (*Mrs. Monteiro*).

II. Other African Regions.
  A. South Tropical.
    a. Western Coast.—"Angola (*J. J. Monteiro*)."—Druce. Congo: "Kinsembo (*Ansell*)."—Butler.
    b. Eastern Coast.—"Querimba."—Hopffer.
    b1. Eastern Interior.—"Kilima-njaro (*F. J. Jackson*)."—Butler.
  B. North Tropical.
    a. Western Coast.—Cameroon Mountains (*D. Rutherford*). Calabar.—Coll. Hewitson. Cape Coast Castle (*J. M. Pask*). "Accra: Aburi (*Weigle*)."—Möschler.

GENUS CALLIDRYAS.

Callidryas, Boisduval "Lep. Amer. Sept., p. 73 (1833);" Sp. Gen. Lep., i. p. 605 (1836).
Colias, Latr. (part), Enc. Meth., ix. p. 10 (1819).
    ,,   Swains., Zool. Illustr., i. pl. 5 (1820-21).
Callidryas, Doubl., Gen. D. Lep., i. p. 66 (1847).
    ,,   Trim., Rhop. Afr. Aust., i. p. 67 (1862).
    ,,   Butl., Cist. Ent., i. p. 46 (1870); and (Monograph) Lep. Exot., (1869-74).
Catopsilia, *Phœbis* (Hübner, 1816, part), *Callidryas*, and *Aphrissa*, Butl., Lep. Exot., pp. 154-155 (1874).
Catopsilia, Moore, Lep. Ceylon, i. p. 121 (1881).
    ,,   Distant, Rhop. Malay., p. 295 (1882-86).

IMAGO.—*Head* broad, clothed with short, compact, very dense hair, projecting frontally in an acute ridge; *palpi* very short, not rising nearly to level of top of head, strongly compressed, densely clothed with closely-appressed scales, which are longer and partly hair-like inferiorly,—basal joint long and curved,—middle one much shorter,—terminal one very small and short (except in the female of some species, in which it is considerably elongated and slender); *antennæ* short, stout, straight, very gradually thickening into an elongate, cylindrical, truncate club.

Thorax robust, clothed with close silky hair (very long posteriorly) above, and with densely-packed furry scales beneath. *Wings* thick, with strong projecting nervures. *Fore-wings* truncate, rather prominent and sub-acuminate at apex; costa usually finely serrated (in ♂ only),—beyond basal curve scarcely convex till towards apex; hind-margin very slightly concave in the middle; inner margin more or less convex near base and very slightly concave in middle; costal nervure two-thirds length of costa, or rather more; subcostal nervure four-branched,—first nervule given off usually before (but sometimes at or a little beyond) middle of discoidal cell; second a little before (sometimes only just before) extremity of cell; third and fourth either about midway between end of cell and apex, or considerably nearer the latter (rarely nearer the former), the third ending at or just before apex; upper radial nervule usually united to subcostal nervure midway between end of cell and bifurcation of its third and fourth nervules, but sometimes at end of cell; disco-cellular nervules not or scarcely oblique, both arched, the lower from once and a half to twice as long as the upper; discoidal cell not wide (there being unusual space between costal and subcostal nervures), truncate terminally; submedian nervure sinuated parallel to inner margin; internal nervure very short and slender, running into submedian nervure not far from base; ♂ almost always with a more or less extended space (varying from a narrow hind-marginal border to outer two-thirds or more of wing) bearing partly raised or tilted scales, and having a duller, almost chalky aspect; same sex in some species with inner-marginal edge near base recurved, and bearing a long tuft or brush of radiating silky hairs. *Hind-wings* prominent apically and inferiorly (rarely sub-caudate at anal angle); costa well arched, and with pronounced basal projection; hind-margin slightly dentate (sometimes more so inferiorly); inner margins forming a deep and complete groove entirely covering abdomen beneath, but markedly divergent beyond this; costal nervure extending to considerably beyond middle, and sometimes nearly to apex; first subcostal nervule arched, given off about middle or nearer extremity (upper) of discoidal cell; disco-cellular nervules both very oblique,— the lower one about or less than twice as long as the upper, curved or sub-angulated; internal nervure bent near base, running to end of inner-marginal groove; ♂ with space of raised scales much more limited than in fore-wing, and rarely extending as far as anal angle; same sex usually with a more or less ovate spot or short stripe of thickened scales near base, above subcostal nervures, and in some species with a long tuft of silky hairs below that nervure. *Legs* short, moderately stout; femora with only a very little extremely short hair beneath basally; fore-tibia much shorter, middle tibia rather shorter, hind-tibia much longer than femur,—all sparsely and finely spinulose; terminal spurs of middle and hind tibiæ small and slender, but not very short; tarsi nearly twice as long as tibiæ, rather strongly and thickly spinulose.

Abdomen arched, compressed, short, acuminate, dorsally tufted with silky hair at base.

LARVA.—Elongate, slightly tapering anteriorly and posteriorly; surface finely granulated.

PUPA.—Variable in shape, especially as regards convexity of wing-covers in front of breast (which, moderate in *Crocale*, *Gnoma*, *Florella*, &c., is very strongly arched and ridged in *Minuscula* and ? *Philea*), and length of acute projection on head (which is much prolonged in ? *Philea*).

These characters of larva and pupa are from Mr. Moore's and Mr. Butler's published figures, and Mrs. Barber's drawings of *Florella*.

Mr. Butler (*op. cit.*) has proposed to reduce *Callidryas* to the American *Eubule*, Linn., and eight allies, in which the ♂ has no brush of hairs in either wing. The other New World species he assigns to two genera, viz., *Aphrissa*, represented by *Statira*, Cram., and seven allies, in which also the ♂ has no tuft, but the ♀ has the terminal joint of the palpi lengthened,—and *Phœbis*, including *Cipris*, Cram., and eleven others, in which the ♂ has the tuft near the base of the hind-wings. The Old World species, *Crocale*, Cram., and thirteen others, characterised by the ♂ having the tuft near the base of the fore-wings, he places in the genus *Catopsilia*. While recognising these divisions as convenient and natural sections of the genus *Callidryas*, I cannot find that they are structurally entitled to generic rank. In the important feature of neuration there is but little variation, and the slight differences (above mentioned) that do occur are not characteristic of, or confined to any one of the divisions.[1]

In the ♂s of this genus, the inner edge of the space of raised scales is mostly well marked in both wings; it usually follows an irregular course, and when (as often occurs) the colouring is different—yellow and white or yellowish-white, or orange and yellow—on each side of it, the effect is very remarkable. In most of the South-American species the space forms a narrow hind-marginal border, with the inner edge regularly festooned. The tuft or brush of long silky hairs is often not apparent in ♂ specimens of the Old World group, being folded back on that part of the inner margin which is hidden by the costal convexity of the hind-wings.

Callidryas is in neuration and most other structural points rather nearly allied to *Gonepteryx*, but is well distinguished by its longer, much thinner, less stoutly-clubbed, not curved antennæ and shorter palpi, its less hairy thorax, and entire (instead of falcate and angulated) wings.

In addition to the special sexual badges above mentioned, wide distinctions in colour also mark most of the ♂s of *Callidryas*. This is very strikingly shown in some West-Indian species, of which *C. Avellaneda*, Herr.-Schæff.—the most splendid species in the genus—presents in the fore-wings an irregular median patch of deep crimson and an outer border of orange on a ground of lemon-yellow; and *C. Orbis*, Poey, a perfect circle of orange on the lemon-yellow basal half of the fore-wings, which outwardly are white. There is much less sexual disparity in the Old World species; several of the Indian forms (*Pyranthe*, Linn., and allies) present plain greenish-white black-bordered ♂s and ♀s; and the African *C. florella* includes an almost

---

[1] Mr. Distant (*op. cit.*) observes, in adopting *Catopsilia*, that the species of the Old and New Worlds "are generically quite distinct, the peculiarity in neuration of the wings being sufficient to easily separate them;" but he does not specify in what the peculiarity consists. Mr. Butler mentions no distinguishing characters of neuration in his proposed genera.

spotless greenish-white ♂ with a scarcely distinguishable ♀, as well as a pale sulphur-yellow ♀ with rusty-reddish marginal spots.

These robust and strongly-made butterflies are very swift on the wing, but their rapid and irregular onward flight is constantly interrupted by settling on flowers. They are also found settled in numerous drinking parties on the edges of streams and pools. As mentioned in my introductory remarks (vol. i. p. 31), the genus is famous for the amazing assemblages of some of its members in onward migration; records of their steadily advancing hosts are numerous from both Old and New Worlds, and the solitary African species (which ranges throughout the Ethiopian Region) is no exception to this remarkable tendency. Like *Terias*, it inhabits all tropical and sub-tropical countries, extending also northward as far as Ohio in the United States and Syria in the Mediterranean sub-region, and southward to Chili and the Cape.

The food-plants of the caterpillars in both hemispheres are *Leguminosæ*, and nearly all of those known and recorded are species of *Cassia*.

299. (1.) Callidryas Florella, (Fabricius).

♀ *Papilio Florella*, Fab., Syst. Ent., p. 479, n. 159 (1775); Ent. Syst., iii. 1, p. 213, n. 666 (1793).
♂ and white ♀. *Colias Pyrene*, Swains., Zool. Illustr., 1st Ser., i. pl. 51 (1820-21).
♂ and ? ♀. *Callidryas Florella*, Boisd., Sp. Gen. Lep., i. p. 608, n. 2 (1836).
♀ *Callidryas Hyblæa*, Boisd., op. cit., p. 612, n. 7.
♀ *Callidryas Rhadia*, Boisd., op. cit., p. 617, n. 11.
♂ ♀ (white). *Callidryas Florella*, Trim., Rhop. Afr. Aust., i. p. 68, n. 45, and Note (1862).
♀ *Callidryas Rhadia*, Trim., op. cit., p. 69. n. 46.
♂ ♀ *Callidryas Florella*, Hopff., Peters' Reise Mossamb., Ins., p. 365 (1862).
♀ Var. *Callidryas Florella*, Guen., Maillard, Notes Réunion (Bourbon), Lep., pl. 21, ff. 1, 2 (1862).
♂ ♀ *Callidryas Florella*, Trim., Trans. Ent. Soc. Lond., 1870, p. 382.
♂ ♀ (white). *Callidryas Pyrene*, Butl., Lep. Exot., p. 44, pl. xvi. ff. 8, 10 [♂], 9 [♀] (1870).
♀ *Callidryas Florella*, Butl., op. cit., p. 56, pl. xxii. ff. 1, 2.
♀ Var. *Catopsilia Aleurona*, Butl., Ann. and Mag. Nat. Hist., 4th Ser., xviii. p. 489 (1876).
♀ ? Var. *Catopsilia rufosparsa*, Butl., Ann. and Mag. Nat. Hist., May 1880, p. 395.
Callidryas Swainsonii, Westw., App. Oates' Matabeleland, p. 335 (1881).
PUPA (*West-African*), Butl., op. cit., pl. xxii. f. 2a.

Exp. al., (♂) 2 in. 3-9 lin.; (♀) 2 in. 2-9½ lin.

♂ Greenish-white. *Fore-wing*: a small narrow terminal discocellular spot, variable in development (especially in length), but usually well defined, and very rarely sub-obsolete; an exceedingly narrow faint reddish-brown edging on costa from beyond middle to apex, and thence (broken into ill-defined nervular spots) along hind-margin as far as third or second median nervule; a hind-marginal series of minute ferruginous spots at extremities of nervules,—sometimes barely perceptible or wanting altogether. *Hind-wing:* a similar hind-marginal series of minute spots,—often, however, wanting. UNDER SIDE.—*Hind-wing, and wide costal and very wide apical area of fore-wing, glossy pale greyish-yellow or greyish-green, more or less indistinctly freckled and hatched with brownish or reddish-grey;* cilia usually pale-reddish, except near posterior angle of fore-wing; minute ferruginous hind-marginal spots better marked; rarely altogether wanting. *Fore-wing:* costa at base with a thin pinkish-red edging; terminal disco-cellular spot larger, rounder, dull-red, with a large, paler, shining centre; between costa and second median nervule, not far from hind-margin, usually the more or less indistinct traces of a transverse series of five diffused reddish-grey spots, elbowed or angulated obliquely inwardly on upper radial nervule. *Hind-wing:* a terminal disco-cellular red spot, smaller than that in fore-wing, more circular, but with its pale centre usually whiter; in some specimens traces of a discal series of six or seven reddish-grey spots; at basal origin of nervures a small pinkish-red stain.

♀ *Like the ♂; but fore-wing with costa throughout usually bordered with brownish above costal nervure, and with apical and hind-marginal brownish edging rather more pronounced.* UNDER SIDE.—Like the less distinctly marked ♂ s, all the markings (including the terminal disco-cellular spots) more or less faint, and the discal spots very small or obsolete in both wings.

Second form of ♀ (*Florella*, Fab., = *Rhadia*, Boisd.).—*Sulphur-yellow; hind-marginal ferruginous spots enlarged. Fore-wing:* pale-brownish costal border tinged at base with pinkish-red, and becoming wider and ferruginous apically; six well-marked (though inwardly rather diffused), good-sized hind-marginal ferruginous spots, the lowest on first median nervule,—and a seventh minute spot on submedian nervure; terminal disco-cellular spot enlarged, black, rounded, conspicuous; more or less indistinct traces of the subapical elbowed transverse row of reddish spots which both sexes present on the under side. *Hind-wing:* costa and inner margin rather broadly bordered with somewhat iridescent pinkish-white; hind-marginal nervular ferruginous spots always much smaller than those of fore-wing, very variable in size, shape, and distinctness, nearly always a good deal diffused inwardly. UNDER SIDE.—*Hind-wing, and all fore-wing except yellowish or yellowish-white inner-marginal and sub-central area, rich ochreous-yellow inclining to orange, rather sparsely freckled and hatched with ferruginous. Fore-wing:* terminal disco-cellular spot usually larger than in ♂,—rather frequently with a second similar but usually smaller spot attached to it superiorly; elbowed submarginal series of spots dull-reddish, rather blurred and indistinct,—the spots below the elbow often united into a streak. *Hind-wing:* terminal disco-cellular spot rather larger than in ♂, with a silvery-white centre, and almost invariably accompanied by a similar spot immediately surmounting it, and by a third (commonly larger) one immediately preceding it, and so within discoidal cell; at base, immediately below origin of subcostal nervure, a ferruginous-red mark; eight spots of irregular discal row varying from dull-reddish to ferruginous, some or all of them often indistinct or almost obsolete. In both wings the minute spots of the hind-marginal nervular series are dark and distinct and bordered inwardly by some pink scaling.

As indicated in the foregoing descriptions, there is some diversity in the markings of the ♂, chiefly as relates to the development of those of the under side; but the two forms of the ♀ present a far greater discrepancy, one scarcely differing from the ♂, while the other is sulphur-yellow above with conspicuous disco-cellular and hind-marginal spots, and deep ochreous-yellow beneath, with one or two additional disco-cellular spots. There exist, however, several linking variations between these two forms of ♀, but almost all that I have seen[1] are nearer to the yellow than to the white form. These ♀s are either simply of a paler yellow, or exhibit a more or less extensive suffusion of white over the discal area of one or both wings. The *Hyblaea* of Boisduval, from Senegal, is one of the former category, and the *Aleurona* of Butler, from Abyssinia, one of the latter; and it is noteworthy that in both cases the describers mention only one terminal disco-cellular spot on the under side of hind-wing,—a feature characteristic of the white ♀ and of the ♂. A fine example of the full yellow coloration, taken near D'Urban by Colonel Bowker, also presents this solitary spot.

In South Africa, the yellow or *Rhadia* form of ♀ appears to predominate; out of 47 examples from all quarters, I find 26 yellow, 7 intermediate, and 14 white. These different ♀s are not local forms, but occur together (as Mr. Butler, *Proc. Zool. Soc. Lond.*, 1884, p. 486, has shown to be the case at Aden) in several places. For instance, I have white, yellow, and intermediate ♀s from the Bashee River and Potchefstroom in the Transvaal; and Mrs. Barber informs me that all three flew in the neighbourhood of Kimberley in the autumn of 1881.[2] I have met with both white and yellow ♀s among the stragglers that occasionally find their way to Cape Town, and have received the two forms together from Springbokfontein in Namaqualand. As respects Tropical Africa, I have noted the two in collections

---

[1] The exceptions are two very slightly yellowish-tinged ♀s from Aden, and a dwarf (exp. 2 in. 1 lin.) ♀, taken by Colonel Bowker in some South-African locality not specially noted, which is somewhat more yellow-tinged along hind-margins, and has in the fore-wings the disco-cellular spot very small and faint, and the reddish hind-marginal spots minute and indistinct.

[2] Mrs. Barber adds (12th September 1882): "I have seen both white and yellow ♀s laying their eggs on the same plant, and have reared the caterpillars, which produced butterflies of all the different tints that exist in this species."

from Sierra Leone, and also from Kilima-njaro, and have captured them flying together in Mauritius.[1]

Notwithstanding the Aden evidence, Mr. Butler (*loc. cit.*) keeps apart *Pyrene*, Swains., *Hyblaea*, Boisd., *Aleurona*, Butl.,[2] and *Florella*, Fab., but he admits the great difficulty of assigning ♂s to the several ♀s so separated; and I must say that the copious material arranged by him in the British Museum collection seemed to me, upon thorough examination last year (1886), to show very satisfactorily that there is not more than one variable species concerned.

Swainson (*op. cit.*) figures as *Pyrene* an ordinary ♂, with the under side rather dull yellowish and its hatching moderately distinct, and a small white ♀ (upper side only) nearly resembling the ♂. He notes that the ♂ was one of about twenty brought by Burchell from the interior of the Cape, but that the ♀ was discovered in Haworth's collection. Mr. A. G. Butler first called attention (*Cat. Fab. D. Lep.*, p. 224, 1868) to the resemblance borne by Fabricius' type of *Florella* in the Banksian collection to the *Rhadia* of Boisduval, and upon inspection of this example in 1881, I found it to be unquestionably (though very worn and faded) identical with Boisduval's insect. The lack of any ♂ approaching the *Florella* (= *Rhadia*) colouring led me early to the conjecture that these yellow ♀s could only be associated with *Pyrene*; and the subsequent capture (by myself at D'Urban[3] and by Colonel Bowker at King William's Town[4] respectively) of two pairs *in copulâ*, served to confirm that view, the ♂ in each case being of the ordinary *Pyrene* pattern and the ♀ a yellow *Florella*.[5] Mr. H. L. L. Feltham writes to me that in 1886, at the junction of the Modder and Vaal Rivers in Griqualand West, he found a large number of pairs at rest, and that in all the cases examined (from twenty to thirty pairs) the ♀ was yellow.

LARVA.—Yellowish-green dorsally, minutely granulated with black; pale glaucous-greenish laterally; the two colours separated by a rather wide, conspicuous yellow stripe. Head coloured like the back. Legs pale glaucous-greenish. Feeds on *Cassia arachoïdes*.[6]

---

[1] Mauritian specimens are in both sexes smaller than the ordinary Continental ones, and some from the Comoro Islands in the British Museum are still smaller.

[2] From Mr. Butler's description (*loc. cit.* in synom.), and his remarks in the paper just quoted on the Aden forms, I think that his *C. rufosparsa*, founded on a ♀ from Madagascar, cannot be separated from *Aleurona*.

[3] 26th March 1867.   [4] 12th September 1870.

[5] The capture *in copulâ* at Aden, by Major Yerbury, of similar sexes of the *Pyrene* pattern is recorded by Mr. Butler (*Proc. Zool. Soc. Lond.*, 1884, p. 487).

Mr. A. D. Millar has sent me the paired sexes captured by him at D'Urban on 10th February 1888. The ♀ in this case is worn and rather small; it appears to have had a faint tinge of yellow (with slight traces of spots) along the hind-margins, but is otherwise all greenish-white; the disco-cellular spot of the fore-wings is rounded much as in the yellow ♀.

[6] Mr. A. D. Millar has forwarded to me a specimen of the *Cassia* upon which he has observed *Florella* laying eggs at D'Urban, Natal. It has been kindly determined by Mr. MacOwan as *Cassia corymbosa*, an introduced South American species.

PUPA.—Pattern and colouring very like that of larva, but the green apparently more uniform and (except on wing-covers) inclining to glaucous; yellow lateral stripe paler. Acute cephalic projection tipped with reddish-brown. Attached to various bushes, grass, &c.

These descriptions of larva and pupa are made from a coloured drawing sent to me by Mrs. Barber in 1882. That of the pupa must be qualified by the following note, made by Mrs. Barber during the great abundance of the insect near Kimberley in 1881, viz., "The larvæ suspending themselves to various plants resulted in pupæ wonderfully adapted in colour to the particular plants occupied. Those upon dry grass were straw-coloured; one in a bunch of grass only half dry and half green was green on the under side and straw-coloured on the upper side. A number of caterpillars that I put into a tin-box suspended themselves on its sides and became pupæ of a leaden colour. I think, however, that when at liberty the larvæ preferred to pupate on the bluish-green upright stems of a small species of *Cyphonema*, common among the grass, for the bluish-green pupæ were crowded together upon it in great numbers. The chrysalis state seldom lasted more than ten days." As regards the larva, Mrs. Barber further notes: "I observed them literally in thousands on the *Cassia* plants; they cleared off every leaf, and then devoured the young shoots, and even the bark of the stems. I noticed no variation whatever in these caterpillars."

This butterfly is very closely allied to *C. Gnoma*, Fab., from India and China; but the latter has in both sexes a rather wider, darker (and in ♀ continuous) brown hind-marginal edging in the fore-wings, and the under-side tint yellower than in the ♂ *Florella*, but not nearly so deep as in the *Rhodia* form of ♀. The ♀ *Gnoma*, too, appears never to show more sulphur-yellow on the upper side than a rather narrow suffusion along the hind-margin of both fore and hind wings. From Captain De la Chaumette's description (quoted by Mr. Butler, *Lep. Exot.*, p. 43), and the figures given in Moore's *Lep. Ceylon.*, pl. 48, 2a), the earlier stages of *Gnoma* seem scarcely, if at all, to differ from those of *Florella*.

It is in the late summer and in autumn, from about the middle of February to the middle of May, that *Florella* is most prevalent in South Africa; but Mr. Feltham notes the occurrence of a few on the wing at Kimberley as early as 2d October, and I met with two flying at D'Urban as late as 23d June. During the time of the extreme abundance of the insect at Kimberley in 1881, Mrs. Barber noted that the yellow form of ♀ was at first very scarce, but later on became very numerous.[1] During my visit to Natal in the summer of 1867, the butterfly was by no means common. I took a good many ♂ s

---

[1] Mr. H. L. L. Feltham informs me that in the beginning of the season at Kimberley white ♀ s only are to be seen, and also notes the interesting circumstance that, as far as he has been able to observe, the ♂ s are then all less boldly marked on the under side than those ♂ s which fly later in the season, when the yellow ♀ s prevail.

and a few ♀s (all yellow)[1] about D'Urban, but did not meet with the species elsewhere. Like the *Eronia*, these swift flyers were fond of the flowers of *Vinca rosea*, and repeatedly stopped in their headlong flight to settle on them. The stragglers seen by me at Cape Town usually pitched on the scarlet flowers of a Pelargonium.

I have already referred (vol. i. p. 31) to Colonel Bowker's observation of an immense migrating host of this butterfly in Basutoland, and reproduce here his interesting account[2] of what he witnessed:—" During my trip to No-Man's-Land in March 1869, I crossed the Maluti Mountains at two different points, going and returning; and throughout the journey, whenever there was a gleam of sunshine between the prevalent showers, the exodus of *Florella* and *Rhodia* continued in one uninterrupted stream. These butterflies were to be seen in countless numbers, from the deepest and darkest valleys through which the Orange River forced its way, up to the highest peaks, 10,000 feet above the sea, *and all were steadily moving on eastward*. Sometimes one of them would stop to take a sip from a tempting gladiolus, or even turn back a few yards for that purpose; but it would be only for a minute, and then off he would hurry again, as if fearful of being left behind by his comrades. I have noticed the same swarms in the Trans-Keian country, and also in the Cape Colony. In the latter, I believe, other members of the *Pieridæ* were concerned."[3]

Judging from its known localities, this powerful insect ranges over all the Ethiopian Region (except the north-west extra-tropical tracts), including most of its islands, and penetrates as far northward as Syria. In South Africa it is generally distributed, but appears to be nowhere permanently abundant, though more prevalent in the northern tracts towards the tropic.

Localities of *Callidryas Florella*.

I. South Africa.
  B. Cape Colony.
    a. Western Districts.—Cape Town (occasional visitor). Van Wyk's Vley, Carnarvon District (*E. G. Alston*). Springbokfontein (*G. A. Reynolds*) and Ookiep (*G. Warden*), Namaqualand District. Knysna. Plettenberg Bay (the late *W. H. Newdigate*).
    b. Eastern Districts.—Grahamstown (*M. E. Barber*). Kleinemond River, Bathurst District (*H. J. Atherstone*). King William's Town (*W. S. M. D'Urban* and *J. H. Bowker*). Colesberg (*A. F. Ortlepp*). Burghersdorp (*D. R. Kannemeyer*).
    c. Griqualand West.—Kimberley (*M. E. Barber* and *H. L. Feltham*). Barkly, Vaal River (*M. E. Barber* and *J. H. Bowker*).
    d. Basutoland.—Maseru (*J. H. Bowker*). "Maluti Mountains."—J. H. Bowker.
  C. Orange Free State.—Bloemfontein (*Dr. H. Exton*).
  D. Kaffraria Proper.—Butterworth and Bashee River (*J. H. Bowker*).
  E. Natal.
    a. Coast Districts.—D'Urban. "Lower Umkomazi."—J. H. Bowker.
    b. Upper Districts.—Maritzburg (*Mrs. Francis*). Estcourt (*J. M. Hutchinson*).

---

[1] Mr. A. D. Millar informs me that, as a rule, the yellow ♀ is not so numerous about D'Urban as the white one. He notes that, "if any females are about, the males are blind to fear and easily caught."

[2] Published by me in *Trans. Ent. Soc. Lond.*, 1870, p. 383.

[3] As regards the Trans-Kei (Kaffraria Proper), Colonel Bowker noted in March 1863 that *Florella* suddenly appeared in thousands, but became rare by the middle of April.

F. Zululand.—Napoleon Valley (*J. H. Bowker*). Etshowe (*A. M. Goodrich*).
G. "Swaziland."—The late F. C. Buxton.
H. Delagoa Bay.—Lourenço Marques (*Mrs. Monteiro*).
K. Transvaal.—Potchefstroom and Lydenburg Districts (*T. Ayres*). Lydenburg (*A. F. Ortlepp*). Eureka, near Barberton (*C. F. Palmer*). Upper Limpopo (*F. C. Selous*).
L. Bechuanaland.—Motito (the late *Rev. J. Frédoux*).

II. Other African Regions.
A. South Tropical.
  a. Western Coast.—Damaraland (*J. A. Bell* and the late *C. J. Andersson*). "Angola (*Pogge*)."—Dewitz.
  b. Eastern Coast.—"Querimba" (*Peters*)."—Hopffer.
  *b*1. Eastern Interior.—Tauwani (*F. C. Selous*). "Motloutsi River (*Oates*)."—Westwood. Zambesi River: Zumbo (*F. C. Selous*). "Tette (*Peters*)."—Hopffer. Kilima-njaro.—Coll. Brit. Mus.
  bb. Eastern Islands.—Comoro Islands: Johanna.—Coll. Brit. Mus. "Madagascar (Coll. T. De Grey)."—A. G. Butler. Bourbon.—Coll. Brit. Mus. Mauritius.
B. North Tropical.
  a. Western Coast.—"Lower Niger (*Forbes*)."—Godman and Salvin. "Abomey (Coll. Saunders)."—Butler. Accra: "Aburi (*Weigle*)."—Möschler. Cape Coast Castle and Sierra Leone (*J. Burke*). "Senegal (Coll. Druce)."—Butler. "Gambia River (*Moloney*)."—G. E. Shelley.
  b. Eastern Coast.—Somaliland.—Coll. Brit. Mus.
  *b*1. Eastern Interior.—Abyssinia.—Coll. Brit. Mus. "Khartoum (Coll. Wallace)."—Butler. "Nubia."—Hopffer.
  bb. Eastern Islands.—"Socotra."—Coll. Brit. Mus.

IV. Asia.
Arabia: Aden, Lahej, Shaik Othman.—Coll. Brit. Mus. "Syria."—Staudinger.

Sub-Family 2.—PAPILIONINÆ.

Papilioninæ and *Parnassinæ*, Swainson, Cab. Cyc., Hist. and Nat. An. Ins., pp. 87 and 90 (1840).
Papilionides, Boisduval, Sp. Gen. Lep., i. p. 171 (1836).
Papilionidæ, Doubleday, Gen. D. Lep., i. p. 1 (1846).
Papilionidæ, Trimen, Rhop. Afr. Aust., i. p. 10 (1862).
Papilioninæ, Bates, Journ. Ent., 1864, p. 177.

IMAGO.—*Head* of moderate size, rather or decidedly broad, usually hairy; *eyes* round or ovate, prominent, smooth; *palpi* usually very short (but of moderate length in *Thais* and *Sericinus*, and long in *Teinopalpus*), scaly and hairy; *antennæ* variable in length, thickness, and elevation, but in most genera short, with a rather abruptly-formed, curved, sub-cylindrical club.

Thorax mostly rather stout (robust in *Parnassius*, *Doritis*, and *Teinopalpus*, but slender in *Thais* and *Sericinus*), more or less hairy (densely so in *Parnassius* and *Doritis*). *Wings* very variable in outline, but mostly elongate; discoidal cells always closed. *Fore-wings* (except in *Parnassius*, *Thais*, and *Doritis*) usually more or less produced apically; subcostal nervure five-branched (except in *Parnassius* and *Hypermnestra*, where it is four-branched),—the first and second nervules given off before extremity of discoidal cell, the fourth and fifth branching off about midway between end of cell and apex; three discocellular nervules almost always well-developed (but the first very short in *Sericinus*, *Thais*, and *Doritis*, and wanting in *Parnassius*),—the third so inclined as to look like a continuation of median nervure, and making the lower radial nervule appear to be a fourth median nervule; near base, between median and submedian nervures, a transverse internomedian nervule in *Ornithoptera*, *Papilio*, *Euryades*, and *Eurycus*; internal nervure, running free to inner margin, present in all genera except *Doritis*. *Hind-wings* usually more or less prolonged in analangular region, and very often conspicuously tailed on third median nervule (extraordinarily so in *Leptocircus*); inner margin hollowed so as to leave abdomen perfectly free, and often folded back on itself: precostal nervure forked, its lower branch joining costal nervure so as to form a small prediscoidal cell (except in *Thais*, *Doritis*, and *Parnassius*, where the precostal nervure is simple); submedian nervure more or less curved; internal nervure wanting. *Legs* rather thick (in *Papilio* and *Eurycus* long also); fore-tibiæ with an elongate process or thickened spur on the middle of the inner edge; terminal spurs of middle and hind tibiæ strong; tarsi long, their terminal claws large, simple, without appendages.

Abdomen usually of moderate size and length (very large in *Ornithoptera*, and short, thick, and very hirsute in *Parnassius* and *Doritis*); anal plates in ♂ well developed, usually more or less conspicuous (in *Ornithoptera* and *Parnassius* spined or hooked at tip); in impregnated ♀ of four genera (*Parnassius*, *Eurycus*, *Euryades*, and *Luehdorfia*) an inferior corneous appendage, variable in form, usually constituting a pouch, open posteriorly.[1]

LARVA.—Stout, usually smooth, but sometimes with numerous

---

[1] It is mainly on account of these singular appendages to the female abdomen that Mr. H. J. Elwes, in his very interesting paper "On the Genus *Parnassius*" (*Proc. Zool. Soc. Lond.*, 1886, pp. 17, 18), has proposed to form a distinct Family, "*Parnassiidæ*," of these four genera. But when it is considered that this appendage or pouch has been shown (as Mr. Elwes fully details in his paper) by Von Siebold and Mr. Arthur Thomson, in the case of *Parnassius*, and by Burmeister in the case of *Euryades*, to be no structural part of the insect, but simply the adhering coagulated and dried condition of a secretion poured out by the ♂ during *coitus*, it seems quite impossible to recognise the "pouch" as a distinctive family or even genus character. As regards *Euryades*, the two species it contains are so near *Papilio*, that probably the Felders would not have separated it from the latter but for the female's possessing the appendage in question.

That *Parnassius* and *Doritis* are very aberrant forms of the *Papilioninæ* is, however, evidenced in all their stages, and more especially in their rounded blunt pupa enclosed in a slight web (and in *Parnassius* covered with a bluish efflorescence), which is even more like those of many moths than those of most of the *Hesperidæ*. But some approach to them is afforded by the pupa of *Thais*, which is also enclosed by a few silken threads; and although their larvæ are of unusual form, the presence of the Y-shaped neck tentacle is too close and remarkable a link to admit of their severance from the *Papilioninæ*.

spine-like tubercular processes; second (prothoracic) segment containing dorsally an exsertible forked glandular fleshy organ, emitting a penetrating odour.

Pupa.—Very variable in shape, but usually more or less curved or bent backward superiorly, and angulated; head usually more or less bifid, but sometimes sub-truncate, blunt, or rounded; back of abdomen often tuberculated.

This Sub-Family is very distinct from the *Pierinæ*, the approach made to the latter by the aberrant *Parnassius*[1] being very slight, and such genera of *Pierinæ* as *Mesapia*, Gray, *Davidina*, Oberth., and *Styx*, Staud., not exhibiting any structural affinity, but only a superficial resemblance to *Parnassius*.[2] The salient features in the *Papilioninæ* are, as regards the perfect insects, the arrangement of the disco-cellular and radial nervules of the fore-wings, making the lower radial apparently a fourth median nervule; the development of the interno-median nervule and of the free internal nervure in the same wings; and in the hind wings the hollowed inner-margin (with absence of the internal nervure), and formation of a small prediscoidal cell by the branched precostal nervure; while the first pair of tibiæ bear on their inner edge a conspicuous process or spur. The caterpillars stand alone among those of the entire Sub-Order[3] in possessing the strongly scented exsertible fleshy organ or "tentacle" in the neck, which is instantaneously protruded and directed vibratingly towards any part where the insect may be touched. This organ is usually of some red or crimson tint, but sometimes blue or yellow; and its sudden appearance, menacing motion, and penetrating disagreeable odour, combine to make its possessors alarming and repulsive to their foes.

Papilio, with some 400 species, is beyond comparison the dominant genus of the Sub-Family, extending throughout the globe, but very poorly represented in the Palæarctic, Nearctic, and Australian Regions. The other twelve genera, including *Ornithoptera* (about twenty species), together muster only sixty-one species, and of these *Parnassius* alone counts twenty-three, five of the genera being monotypic. In curious contrast to its poverty in species of *Papilio* is the richness of the Palæarctic Region in generic forms, *Sericinus*, *Thais*, *Hypermnestra*, *Doritis*, and *Luehdorfia* being all peculiar to it, and *Parnassius* peculiar except for two or three North-American forms;

[1] See Doubleday, *Gen. Diurn. Lep.*, i. p. 1 (1846). The singular Turkestan genus *Hypermnestra*, however, has in outline and hind-marginal markings of the wings, and especially the pattern and colouring of the under side of the hind-wings, a striking resemblance to *Anthocharis* and the *Daplidice* group of *Pieris*.

[2] It is worth mention, however, that the chrysalis of *Zegris* (a Southern Palæarctic genus close to *Anthocharis*) is enclosed in a delicate silken web after the manner of that of *Parnassius*.

[3] The whole Order *Lepidoptera* is known to yield only one other genus, viz., *Dicranura* (*Cerura*) among the Bombyces, whose larvæ have an equally developed organ of this remarkable description; and in these the organ is double instead of merely forked, enclosed in special sheaths, and occupies the posterior instead of the anterior end of the body.

while *Teinopalpus* and *Armandia* are just on the adjacent northern boundary of the Oriental Region.

Thais and *Doritis* are limited in range to the Mediterranean Sub-Region, and *Luehdorfia* to the Siberian eastern shore.

The Australian Region has one peculiar genus (*Eurycus*), and the Neo-Tropical another (*Euryades*); the African Region has hitherto yielded species of *Papilio* only.

The *Papilioninæ*, though not a very numerous Sub-Family, are pre-eminent for their combination of large size, variety of form, and richness of colouring. The largest and (with the exception perhaps of the Nymphaline genus *Morpho*) most splendid of butterflies are the *Ornithopterœ* of the Malayan Archipelago—a few species extend the genus to India on the one hand and Australia on the other,—magnificent creatures, whose males are velvety-black with crimson-marked thorax, long golden-yellow abdomen, and wings banded and patched with vivid green (changing into golden or blue in some species) or rich yellow, and whose duller whitish-barred and spotted females are even larger than their mates, measuring from 7 to $8\frac{1}{2}$ inches across the expanded fore-wings. The very isolated *Papilio Antimachus* of Western Africa, though smaller in body, owing to the extraordinary elongation of the fore-wings has a still wider expanse, varying from $7\frac{1}{2}$ to $9\frac{1}{2}$ inches. Nearly all the *Papilioninæ* are above the middle size, except the genera *Thais*, *Doritis*, and part of *Parnassius*; and the only really small forms are the species of *Leptocircus* (exp. $1\frac{1}{2}$—$1\frac{3}{4}$ inches), which make the most of their diminutive stature by the extraordinary length of their caudate hind-wings.

Familiar representatives of the group are the "Swallow-Tails" of Europe, *P. Machaon* (found in England) and *P. Podalirius*, and the beautiful "Apollo" butterflies of the Alps (*Parnassius Apollo* and *P. Delius*.) Nearly all the *Papilioninæ* (except certain protected groups of *Papilio* in South America and Indo-Malaya) are strong on the wing, and many are remarkably swift and lofty flyers. Many species of the genus *Papilio*—especially those of the *Podalirius* group or "Swallow-Tails" *par excellence*—are attracted by the moisture at the margins of streams and pools, and observers in tropical regions record with admiration the often immense assemblages of these lovely insects at such drinking stations.

Genus PAPILIO.

Papilio, Linn. (part), Syst. Nat., i. 2, p. 744 (1767); Fab. Syst. Ent., p. 442 (1775).
Papilio, Latreille, "Hist. Nat. Crust. Ins., xiv. p. 108 (1805);" and Enc. Meth., ix. p. 9 (1819).
Papilio, Boisduval, Sp. Gen. Lep., i. p. 183 (1836).
Papilio, Doubleday, Gen. Diurn. Lep., i. p. 5 (1846).

Imago.—*Head* large, clothed with short hair, often with a frontal tuft of longer hair; *eyes* ovate, very prominent, smooth; *palpi* very short, not rising above half height of head, pressed close against face, clothed with longish hair and scales inferiorly,—basal and middle joints about equal in length, terminal one minute; *antennae* of moderate length or rather long, variable in thickness,—the club long, gradually formed, usually curved upward and outward; *haustellum* long.

Thorax rather robust; prothorax forming a distinct neck. *Wings* exceedingly varied in outline, large, with strong nervures. *Fore-wings* usually sub-triangular, often more or less produced apically, sometimes very narrow and elongate, rarely falcate; costa slightly or moderately convex; costal nervure very thick, and extending about three-fifths length of wing; subcostal nervure five-branched, very close to costal nervure as far as its second nervule,—its first and second nervules closely approximate, long, given off not very near to each other, but far before extremity of discoidal cell (first in *Leonidas* and *Pylades*, and allies, greatly abbreviated and running into costal nervure),—third rather long, curved, given off at end of cell, running close to second, and ending at apex,—fourth and fifth branching off far beyond end of cell, but nearer to it than to apex, rather short (fourth much arched and remote from third); disco-cellular nervules all fully developed,—the first and second almost vertical, or slightly inclined inwardly or outwardly, but the third very oblique inwardly, and so directly continuous of median nervure that lower radial nervule has all the appearance of an additional fourth median nervule; an interno-median nervule uniting median and submedian nervures near their origin, so as to form a small sub-discoidal cell; internal nervure well developed, free, terminating on inner margin at some distance from base; discoidal cell long and broad; ♂ in some species with the lower radial and three median nervules, and the submedian nervure on disc, bordered above and below by a long space of closely-appressed fine silky short hairs, narrowing to a point both inwardly and outwardly. *Hind-wings* always more or less elongate inferiorly (sometimes extremely so), commonly tailed at extremity of third median nervule; costa moderately arched; apex much rounded off; hind-margin more or less dentate, sometimes so strongly as to bear several short tails; inner margin hollowed (in ♂ sometimes folded back and bearing a coating or fascicle of hairs); precostal nervure forked, its lower branch united to costal nervure so as to form a small prediscoidal cell; costal nervure ending at apex; first subcostal nervule given off far before extremity (at from one-third to two-thirds length) of discoidal cell; disco-cellular nervules straight, upper one longer (sometimes much longer) than lower, oblique; discoidal cell short and much narrower than in fore-wing; submedian nervure incurved, ending at anal angle; internal nervure wanting. *Legs* long, usually thick; fore-coxae free, clothed with short hair; femora without hair, or with a very little inferiorly,—first pair longer, second as long as, third pair shorter than, tibiae; all tibiae strongly spinulose,—terminal spurs of second and third pairs rather long, slender,—first pair, at about middle of their inner edge, with a stout (often acuminate) process; tarsi very long (especially first joint), considerably longer than tibiæ, densely spinulose beneath,—the claws large, rather straight, simple.

Abdomen rather stout, of moderate length or rather long, thickened posteriorly; anal plates in ♂ usually large and conspicuous.

LARVA.—Rather thick, often more or less swollen on third thoracic segment, and thence abruptly attenuated to head; usually smooth, but in some groups armed with numerous curved fleshy tubercular processes; head small, smooth; penultimate segment often more or less bifid dorsally into small acute prominences; exsertile Y-shaped tentacle on back of first thoracic segment usually rather long; thoracic segments sometimes each with a pair of short filamentous processes superiorly and laterally.

PUPA.—Very variable in form; rarely almost straight, usually with abdomen more or less curved, and thorax and head more or less bent backward. Head usually bifid (often very deeply), sometimes truncate, round, and blunted. Dorso-thoracic prominence sometimes produced into a forward-pointing process. Abdomen often dorsally armed with tubercles, which are sometimes developed into conspicuous processes.

For number, diversity, and beauty of its species, the great genus *Papilio*, even if we withdraw from it the magnificent and barely separable group *Ornithoptera*, stands unrivalled. So conspicuous and prominent a feature are its members in the butterfly life of the tropics and adjacent latitudes, that they have been for considerably more than a century more extensively collected and better known generally than any others of their tribe. Though modern investigations have shown that structurally their affinity to the Heterocera prevents their any longer being regarded as the highest or most specialised butterflies, they stood for so long at the head of the Order, that they have received more study and examination than any other Lepidoptera.

After comparing the catalogues published by Boisduval (1836), Doubleday (1847), G. R. Gray (1852 and 1856), C. and R. Felder (1864), Kirby (1871 and 1877), and C. Oberthür (1879), I think that the species of *Papilio* may fairly be held to exceed four hundred in number. Notwithstanding their very great diversity in outline, colouring, and pattern, the generic characters throughout this large assemblage offer but little modification; and I agree with the great majority of lepidopterists that it is impossible to break up *Papilio* into satisfactory genera. At the same time, as remarked by Mr. Distant (*Rhop. Malay.*, p. 323), the genus is readily divisible into well-marked groups; and Boisduval, the Felders,[1] and Wal-

---

[1] *Act. C. R. Soc. Zool.-Bot. Vindob.*, xiv. (1864). This is the most elaborate of the published investigations of the structural characters; but it is, in my opinion, carried into a minute analysis more refined than natural. There are no fewer than seventy-five sections

lace,[1] have treated it with more or less detail from this point of view.

Tropical South America is richest in these fine insects, and next to it stands the Malayan Archipelago, the two regions together yielding nearly half the known species. Tropical Asia is considerably less productive, except along the Himalaya, having produced about seventy species. The Ethiopian Region (including twelve from Madagascar and the Mascarene Islands) has hitherto yielded sixty-two. In comparison with the foregoing, the Palæarctic and Nearctic Regions and the Australian continent are exceedingly poor, each producing under twenty species.

Nearly all the species of *Papilio* are above the middle size, and many of them very large butterflies. The smallest known kind appears to be *P. Triopas*, Godt., from Cayenne and the Lower Amazons, which expands rather under $2\frac{1}{2}$ inches; while the largest, *P. Antimachus*, Drury, from the Gaboon and Cameroons, measures $7\frac{1}{2}$ to $9\frac{1}{2}$ inches across the wing.

The African species are conveniently arranged in nine groups, represented by the following well-known species, viz., *P. Podalirius*, Linn. (not Ethiopian;—*P. Policenes*, Cram., typical in Africa), *P. Leonidas*, Fab., *P. Demoleus*, Linn., *P. Nireus*, Linn., *P. Merope*, Cram., *P. Hesperus*, Westw., *P. Zenobia*, Fab., *P. Antenor*, Drury, *P. Antimachus*, Drury. All but the last two groups have representatives in South Africa, and may be distinguished as follows:—

GROUP 1.—*Policenes*, Cram., representative. Sexes alike. Head broad, tufted frontally; antennæ short, with a thick abruptly-formed club. Fore-wings very produced apically, so as to have a long hind-margin and a short inner margin; hind-wings very produced inferiorly, and with a long, almost straight, sword-shaped tail on third median nervule; inner-marginal fold of hind-wings in ♂ often supporting a long brush of radiating hair. Black, with common sub-basal stripes, a discal macular band, and a submarginal row of spots, pale green.

(Four species: *P. Policenes*, Cram., *P. Antheus*, Cram., *P. Porthaon*, Hewits, *P. Colonna*, Ward.)

GROUP 2.—*Leonidas*, Fab., representative. Near Group 1. Sexes alike. Head broader; antennæ stouter, with broader more abruptly-formed club. Fore-wings with apical portion much elongated, but rounded off; hind-margin much more concave in middle; inner margin considerably longer; first branch of subcostal nervure very short and slender, running into costal nervure. Hind-wings rounded, not (or but

---

and exceedingly numerous sub-sections; and the small value of some of these may be estimated from the fact that the ♂ *Papilio Merope* is placed in Sub-section C. of Section lv., while its ♀, *P. Hippocoon*, figures in Sub-section B. of Section lvii.

[1] "On the Phenomena of Variation and Geographical Distribution, as illustrated by the *Papilionidæ* of the Malayan region" (*Trans. Linn. Soc. Lond.*, xxv., 1865).

very slightly) produced inferiorly; never tailed. Black, with greenish or white discal and submarginal series of spots, and basal patch in hind-wings, and often in fore-wings also.

(Four species: *P. Leonidas*, Fab., *P. Brasidas*, Feld., *P. Corinneus*, Bertol., *P. Morania*, Angas.)

GROUP 3.—*P. Demoleus*, Linn., representative. Sexes alike. Head smaller, not tufted frontally; antennæ longer, with a gradually-formed, usually slender club. Fore-wings intermediate in form between that of Group 1 and that of Group 2, rather rounded apically, but not (or slightly) concave hind-marginally. Hind-wings a good deal prolonged inferiorly, bearing on third median nervule either a more prominent dentation or a broad (in *P. ophidicephalus* also long) yellow-marked spatulate tail. Black with a common discal stripe (macular in fore-wings), and submarginal series of spots pale-yellow; hind-wings usually with a blue, black, and red ocellus at apex and anal angle; fore-wings beneath with several longitudinal very pale yellow streaks diverging from base.

(Three species: *P. Demoleus*, Linn., *P. ophidicephalus*, Oberthür, *P. Constantinus*, Ward.)

GROUP 4.—*P. Lyæus*, Doubl., representative. Sexes alike. Structurally close to Group 3, but fore-wings more pointed, and hind-wings with a broad blunt production inferiorly, bearing on third median nervule either a very prominent dentation or a short spatulate tail. Black with a common discal band and submarginal spots of submetallic greenish-blue or bluish-green.

(One species: *P. Lyæus*, Doubl.)

GROUP 5.—*P. Cenea*, Stoll, representative. Sexes extremely different. Structurally very near Group 3; hind-wings without inferior blunt prolongation inferiorly of Group 4; in ♂ bearing on third median nervule a long, somewhat spatulate broad tail; in ♀ tailless. ♂ very pale sulphur-yellow, with a black border in fore-wings and a black discal band (usually more or less broken) in hind-wings; ♀ black, with ochre-yellow, ochre-red, or white discal and basal spots or patches, and small white submarginal spots, mimicking three species of *Danainæ*.

(One species: *P. Cenea*, Stoll.)

GROUP 6.—*P. Euphranor*, Trim., representative. Sexes alike, or not much unlike. Structurally near Group 4, but in colouring and in outline of hind-wings more like Group 3. Fore-wings bluntly sub-falcate; hind-wings not produced as in Group 4, but bearing at end of third median nervule a rather short, more or less broadly-spatulate tail. Black, with a common discal stripe (macular in fore-wings) and a series of submarginal spots, sulphur-yellow (in *Euphranor* ♂ this series only in hind-wings, in ♀ also in fore-wings and a second parallel series in hind-wings).

(One species: *P. Euphranor*, Trim.)

GROUP 7.—*P. Echerioides*, Trim., representative. Sexes very dissimilar. Structure generally much as in Group 5. Fore-wings with costa much arched, apex bluntly protuberant, rounded; hind-wings rounded, only slightly dentate, not produced inferiorly, without tail. Black; ♂ with a common sulphur-yellow discal band, broad in hind-wings, macular and narrowing inferiorly in hind-wings; ♀ with white discal spots in fore-wings, an ochre-yellow sub-basal patch in hind-wings; and a common submarginal series of rounded white spots:— mimicking *Amauris Echeria*.

(One species: *P. Echerioides*, Trim.)

Of these fifteen South-African species, only the two last named (*Euphranor* and *Echerioides*) and *Morania* appear to be peculiar to the sub-region; indeed, *Echerioides* is reported by Plötz to have been found in the Cameroons in West Africa; but it is possible that the nearly allied *P. Zoroastres*, Druce, may have been mistaken for it. Of the rest, seven species extend over the Southern Tropical belt (*Brasidas* apparently only on the Western, and *Colonna*, *Porthaon*, *Cenea*, and *Ophidicephalus* only on the Eastern side); while the six others range into both tropics, *Demoleus* alone appearing to occupy the entire Ethiopian Region.[1]

The LARVÆ, from their size and colouring, and disregard of concealment, and also from the attachment of many of them to orange-trees and other cultivated plants, are better known than those of most other important genera, especially in the case of the Indian and Malayan species. They are mostly of some shade of green, here and there varied with transverse or longitudinal or oblique paler or darker markings of subdued tints; but the Indian *P. dissimilis* has a dark-grey tuberculated caterpillar, varied with bright sulphur-yellow bands and crimson spots; and the strongly tuberculated larvæ of the Oriental groups, represented by *P. Nox*, *P. Coon*, and *P. Polydorus*, are of a purplish or purplish-red colour.[2] They are exceedingly inactive, only moving from one leaf to another as food is required. The food-plants of the several groups are characteristic; the true "Swallow-Tails" (Group 1, *supra*, *Policenes*, &c.), and the section next to them (Group 2, *supra*, *Leonidas*, &c.), being found on *Anonaceæ*; the *Demoleus* group on *Umbelliferæ* and *Rutaceæ*, and the *Nireus* and *Cenea* groups on *Rutaceæ*.

[1] It is noteworthy that this dominant *Papilio* is the only African species of the genus that has a close ally in the Oriental Region, *P. Erithonius* representing it in India, China, the Malayan Archipelago, and (under a slight variation) Australia.

[2] These latter species feed on *Aristolochia*, and it is curious to find that the larvæ of the genera *Ornithoptera* and *Thais*, which live on the same group of plants, are similarly coloured and tuberculated.

The PUPÆ exhibit various shades of green or brown, and many diversities of form. Some of the green ones (in South Africa notably those of *P. Brasidas* and *P. Cenea*), which are suspended on or among leaves, are both in form and colour unmistakably modified to resemble the surrounding foliage. The chrysalis of *P. Lyæus* was discovered by Mrs. Barber (*vide infra*, p. 237) to have a singular faculty of assimilation to the colour of immediately surrounding objects; and I have found that of *P. Demoleus* to present the same phenomenon to a less extent.[1] The chrysalis of the Indian *P. dissimilis* closely resembles a withered twig broken off short.

Mimicry is well illustrated in this genus. Mr. Bates in 1862 called attention to several striking cases in South America,—two in which species of *Papilio* mimic respectively a *Lycorea* and a *Heliconius*, and three in which species of *Papilio* are themselves mimicked by a *Euterpe* (*Pierinæ*), a *Castnia*, and a *Pericopis* (both the latter moths). Mr. Wallace in 1865 tabulated fifteen cases known to him in the Indian and Malayan Region, viz., three in which species of *Papilio* imitate species of *Danais*, three in which they imitate species of *Euplœa*, two in which they imitate species of *Hestia*, and one in which a ♀ form of *Papilio* copies a *Drusilla* (Sub-Family *Morphinæ*). He also cited six cases in which slow-flying *Papiliones* of the *Polydorus* and *Coon* groups are themselves simulated by females of species of *Papilio* belonging to other groups of the genus. Africa has not hitherto afforded any instances of the last named very curious mimicries, but it presents some surprisingly exact imitations, of which the seven following are known to me, viz.: *P. Brutus* ♀ closely copies *Amauris Niavius*; *P. Cenea* ♀ in its three pronounced forms copies *A. Echeria*, *A. dominicanus*, and *Danais Chrysippus*; *P. Echerioides* ♀ copies *A. Echeria*; *P. Cynorta* ♀ copies *Acræa Gea* ♀; *P. Ridleyanus* ♂ and ♀ copy *Acræa Egina* ♂ and ♀; *P. Leonidas* copies *Danais Limniace*, var.;[2] and *P. Rex* copies *D. formosa*. These deceptive simulations are too detailed and exact to admit of their protective purpose being misunderstood, more especially in those cases where the ♀ only is concerned, that sex departing in the most startling manner, alike in colouring, pattern, and outline, from the facies of the ♂, as well as from that of its ♀ congeners in the same group.

As regards their local distribution, the fifteen known South-African species are all found on the South-Eastern Coast. *P. Porthaon* and *Colonna* do not seem to occur south of Delagoa Bay; *Leonidas* extends to Zululand; *Corinneus*, *Morania*, *Antheus*, *Policenes*, and *Constantinus* range to Natal; *Euphranor* has been found in Natal, Kaffraria Proper, and Transvaal; *Brasidas*, *Ophidicephalus*, and *Echerioides* extend from

---

[1] Mr. G. F. Mathew (*Trans. Ent. Soc. Lond.*, 1888, p. 176) notes a similar power of assuming the colour of the objects to which they may be respectively attached in the chrysalides of the Australian *P. Erectheus* (*Ægeus*).

[2] The very closely allied *P. Brasidas* imperfectly but obviously mimics *Amauris Echeria*.

Natal through Kaffraria into the eastern districts of Cape Colony; *Lyæus* and *Cenea* descend as far southward and westward as the Knysna district; and *Demoleus* alone penetrates to the Cape peninsula itself. The Transvaal has not hitherto been much worked by collectors, but the following seven species have been received by me from that great tract of country, viz.: *Corinneus, Constantinus, Ophidicephalus, Demoleus, Euphranor, Lyæus,* and *Echerioides.*

The borders of woods are the chief resort of this genus, and it is there that all the South-African species are to be found,—the widely-spread *P. Demoleus,* however, occurring freely also in open country. The ♂s are strong fliers; *P. Lyæus* often soars over lofty trees; and the rare *Euphranor* appears to have habitually rather a high flight. *Policenes,* like the rest of the true "Swallow-Tails," is remarkably swift, but is fond of settling to drink at muddy places. *P. Cenea* and *Ophidicephalus*,[1] and probably others, have the habit of following for hours in the forenoon a set course through the woods, apparently in rivalry and in search of the females, the latter being comparatively slow and inactive, and more commonly taking wing in the afternoon. On their frequent visits to flowers, the species of *Papilio* keep their wings in rapid vibration, not closing them while feeding, but appearing to be constantly on the alert to be off again. When basking in the sun on leaves, they have, however, the habit of holding the wings horizontally or slightly deflected, with the hind-wings half hidden by the overlapping fore-wings. This posture is a specially favourite one with the African *P. Demoleus.*

300. (1.) Papilio Policenes, Cramer.

Papilio Policenes, Cram., Pap. Exot., i. pl. xxxvii., ff. A, B (1779).
Papilio Pompilius, Herbst. and Jabl., Ins. Natursyst., Schmett., iii. t. xlix. ff. 5, 6 (1788).
Papilio Agapenor, Fab., Ent. Syst., iii. 1, p. 26, n. 76 (1793).
Papilio Policenus, Godt., Enc. Meth., ix. p. 52, n. 77 (1819).
Papilio Policenes, Boisd., Sp. Gen. Lep., i. p. 261, n. 84 (1836).
Papilio Agapenor, Westw., Arc. Ent., i. p. 149 (1845).
Papilio Policenes, Trim., Rhop. Afr. Aust., i. p. 14, n. 4 (1862).
♂ *Papilio Policenes,* Staud., Exot. Schmett., i. pl. 7 (1884).

Exp. al., (♂) 2 in. 9 lin.—3 in. 3 lin.; (♀) 3 in.—3 in. 6 lin.

♂ *Brownish-black, transversely striped and spotted with pale-green; common to both wings—a rather obscure basal stripe; a moderately wide sub-basal stripe; a discal band of variously-shaped, partly-connected spots; and a submarginal series of smaller, narrow (in hind-wing thin and lunulate) spots.* Fore-wing: basal stripe and costal origin of sub-basal stripe tinged with yellow; latter stripe from close to costal edge to inner-

---

[1] This is the largest of the South-African butterflies, the ♂ attaining an expanse of 5¼ inches, and the ♀ one of rather over 5½ inches.

marginal edge; discoidal cell crossed by three almost equidistant moderately wide bars (of which the middle and outer ones are more or less slightly bisinuated), commencing close to costal edge, but not extending beyond median nervure; in upper part of cell, close to its extremity, a small spot, immediately surmounted by a much smaller one above costal nervure; a little beyond cell, two similarly placed but smaller spots, the lower of which immediately surmounts second spot of discal band; spots of this incurved band, eight, all separate except the two lowest, which form a large irregular marking (between first median nervule and inner margin), emitting inwardly a narrow upward projection, almost meeting lower extremity of first disco-cellular base; first three spots of discal band moderate-sized, quadrate,—the first beyond the rest,—next three increasing in size downward, inwardly rounded, outwardly truncate; spots of submarginal series pretty even in size, but in form the first three are roughly rounded, while the rest are lunulate. *Hind-wing:* basal band directed obliquely, short, attenuated to a point just below median nervure; below it, along inner-marginal fold, a streak of white scales, sometimes prolonged to anal angle; subbasal band wider than in fore-wing, continuous, obliquely running from costa to between first and second median nervules, near anal angle, where it abruptly and bluntly terminates,—its commencement white as far as subcostal nervure; discal band commencing broadly on costa, narrowing and the spots becoming more distinctly separated and rounded downward, strongly angulated between second and third median nervules; the first and second spots forming one large costal marking, which is white as far as subcostal nervure; the last spot smallest, lunulate, a little beyond extremity of sub-basal band; at anal angle, below and a little before last-named spot, a small bright crimson-red spot; six lunules of submarginal series irregular, sublinear—the first rather indistinct,—the fourth, fifth, and sixth immediately succeeded externally by ill-defined whitish-grey scaling, broadly lunulate in form; cilia white between third median nervule and anal angle, except about extremities of nervules and along superior edge of tail; tail rather broad, long, nearly straight, black, rather widely tipped with white. UNDER SIDE.—*Much paler, glistening, dull-brown; main pattern similar, but markings generally rather wider; discal band and submarginal macular series greenish-white; space between these mostly hoary-grey; in hind-wing, narrow brown space between sub-basal band (also greenish-white) and discal band outwardly bordered throughout, from costa to anal angle, by a sublinear series of eight dull crimson-red marks internally white-edged and externally fuscous-edged. Fore-wing:* space between discal band and submarginal macular series brown from costa to fifth subcostal nervure, and blackish-brown from second median nervule to inner margin; submarginal series inwardly thinly brown-edged and enclosed by hoary-grey, which narrowly extends to apex. *Hind-wing:* basal band wider, white, and narrowly extending almost to the last crimson mark (at anal angle); between costal nervure and second subcostal nervule, discal band (which is enlarged, almost continuous, and greenish-white) succeeded by two conspicuous brownish-black spots; submarginal lunular series very indistinct, but with brown edging internally; the space beyond the fourth, fifth, and sixth lunules black, with hoary-grey as on upper side. *Abdomen*: black above, white beneath; a narrow lateral longitudinal black stripe separated from dorsal black by a yellowish one. ♀ *Like* ♂, *but duller and paler*. *Hind-wing*: *basal stripe much enlarged, forming a moderately wide border along inner margin to not far from crimson-red mark, which is larger and lunulate*.

The South-African specimens above described are throughout paler than the West-African type-form, the ground being less dark and the markings less green; and they present larger spots in the submarginal row of the fore-wings. On the under side, also, the crimson markings of the hind-wings are narrower and less continuous, and the dark markings of both fore and hind wings are less developed. A ♂ from the Zambesi (taken by the Rev. H. Waller), though nearer to the West-African form in nearly all respects, has the submarginal spots of the fore-wings rather larger; but all the other green markings of the upper side are remarkably reduced, especially the cellular bars and discal spots 1–6 of the fore-wings.

? LARVA.—Back ferruginous-red, transversely striped with black. Head sandy-yellow. On each segment, except second, third, and fourth (thoracic), a median transverse streak of greyish-blue, black-edged both anteriorly and posteriorly; on thoracic segments no blue streak, but the transverse black edging streaks strongly marked; second segment bright yellow; on each thoracic segment, situate dorso-laterally, a pair of forward- and outward-pointing short acute black spines. Lower part of sides (separated from dorsal ferruginous by a lateral black stripe) pale-bluish, fading into whitish on the under surface. On anal segment a pair of short, acute, backward- and outward-pointing spines, superiorly yellow at base and mesially bluish, but at tip and inferiorly black.

I received in 1878 four living larvæ, collected near D'Urban by Colonel Bowker, and was led to think that they probably were those of *Policenes* because one of them became an imperfect pupa, agreeing in the main with some from which I reared this species, and also because they closely resembled a pencil-drawing by Mr. W. D. Gooch of a larva taken in the same locality, and referred by him doubtfully to *Policenes*. One of the four examples sent by Colonel Bowker was much smaller than the others, and evidently in a much earlier moult, but it differed in no particular except in the proportionally larger thoracic spines.

Food-plant: *Artabotrys*, *n. sp.*, one of the *Anonaceæ*, native of D'Urban, Natal, and of Delagoa Bay. (Determined for Colonel Bowker, on reference to Kew, by Mr. J. Medley Wood, Curator of the Botanic Gardens, D'Urban.)

PUPA.—Flattened dorso-abdominally; thoracic dorsal projection much prolonged, pointing forward and upward, and extending as far forward as, or even a little farther than, the head; cephalic prominences short, acute, widely apart; lateral margin with two small acute projections, one behind the other; a little beyond these, abdominal margin considerably expanded laterally (almost foliated).

Bright pale-green; roughened lateral edge of dorsal projection down to lateral abdominal expansion, median frontal line of that projection, and cephalic and thoracic lateral margin, all creamy-ferruginous, which tint forms an irregular lateral patch where edges of base of dorsal projection and those of thorax meet; two spots of the same colour on back of abdomen (third segment); spiracles superior, fuscous.

PLATE II. fig. 4.

I obtained twelve *Policenes* (six of each sex) from pupæ sent by Colonel Bowker from D'Urban, Natal. They reached Cape Town on 30th November 1878, and the butterflies emerged during the succeeding fortnight, with the exception of one ♂, which did not make its appearance until 12th February 1879. Seven of these had the date of their becoming pupæ attached, and I was thus able to record that (with the exception just noted) the chrysalis state lasted from seven to fifteen days. All these pupæ were suspended to leaves, in a position near the mid-rib.[1]

I have not seen *P. Nyassæ*, Butler (*Ann. and Mag. Nat. Hist.*, 1877, p. 459), from Lake Nyassa, but it is described as intermediate between *Policenes* and *Antheus*, and, from the particulars given by the describer, appears to be nearer to the latter than to the former.

I was unfortunate as regards meeting with this *Papilio* during my summer stay in Natal. I had previously observed a specimen flying rapidly near D'Urban on the 4th August 1865, and looked forward to obtaining many other examples in 1867, but only once (on 20th February) came across a solitary individual, which evaded my attempts to capture it. This example looked very green on the wing; it flew swiftly, but settled twice on damp sand. Mr. W. D. Gooch (*Entomologist*, 1880, p. 230) notes that the butterfly is common, especially in November and December, and "can be taken freely, sipping the moisture from damp mud on the margins of rain-pools in the sandy paths near bush, all along the coast." Most of the specimens received from Colonel Bowker were taken in the months mentioned, but he sent one pair *in copula* on 5th April 1879. The latter are unusually small, the ♂ expanding only 2 in. 7 lin. and the ♀ 3 in. 2 lin.

---

[1] The figures purporting to be those of the larva and pupa of *Policenes* which are given by W. W. Saunders (from R. W. Plant's drawings of Natal specimens) in *Trans. Ent. Soc. Lond.*, 2nd Series, iv. pl. 13, ff. a, b, c., 1857, unquestionably illustrate the earlier stages of another *Papilio*—either *P. Demoleus* or a near ally.

Localities of *Papilio Policenes*.

I. South Africa.
  E. Natal.
    a. Coast Districts.—D'Urban. Pinetown (*J. H. Bowker*). "Lower Umkomazi."—J. H. Bowker.

II. Other African Regions.
  A. South Tropical.
    a. Western Coast.—"Angola (*J. G. Monteiro*)."—Druce. "Chinchoxo (*Falkenstein*)."—Dewitz.
    b. Eastern Coast.—Zambesi (*Rev. H. Waller*). "Mt. Schimba, Coast opposite Zanzibar (*Raffray*)."—Oberthür.
  B. North Tropical.
    a. Western Coast.—"Gaboon (*Theorin*)."—Aurivillius. "Fernando Po (Coll. Hewitson)."—Kirby. "Cameroons; Mungo and Victoria (*Buchholz*)."—Plötz. "Calabar and Oware."—Felder. "Benin (*Ménager*)."—Oberthür. Gold Coast (*F. M. Pask*). "Between Mansu and River Prah (*A. S. Bell*)."—Hewitson. Sierra Leone.—Colls. Brit. Mus. and Hewitson.

301. (2.) Papilio Antheus, Cramer.

Papilio Antheus, Cram., Pap. Exot., iii. pl. ccxxxiv. ff. n, c (1782).
Papilio Antharis, Godt., Enc. Meth., ix. p. 52, n. 78 (1819).
Papilio Agapenor, Boisd., Sp. Gen. Lep., i. p. 255, n. 79 (1836).
Papilio Antheus, Westw., Arc. Ent., i. p. 150 (1845).
    „   „   Trim., Rhop. Afr. Aust., i. p. 13, n. 3 (1862).

Exp. al., (♂) 3 in. 6–10½ lin.; (♀) 4 in. 2 lin.

Closely allied to *P. Policenes*, Cram., but considerably larger.

♂ *Brownish-black, transversely striped and spotted with pale-green; common basal and sub-basal stripes as in Policenes; other common markings very like those of Policenes in pattern and disposition. Fore-wing:* no yellow tinge at beginning of sub-basal stripe; *disco-cellular transverse bars strongly (the outermost one very strongly) bisinuated;* no spot in upper part of discoidal cell close to extremity; two small costal spots a little beyond cell better marked; spots of discal band larger, more rounded, the lower four especially extended on the inner side so as to form a continuous band, and *the wide marking formed by the last two subquadrate and internally approaching much closer to sub-basal stripe;* spots of submarginal series rounder, the lower ones hardly lunulate, the lowest more distinctly geminate. *Hind-wing: inner edge of discal band of spots much closer to sub-basal stripe;* in this band the component spots differ widely in form and position from the corresponding ones in *Policenes,* viz., the fourth is considerably enlarged and almost wholly *in* discoidal cell (the very small quite separate portion beyond cell being sometimes minute or even wanting),—the fifth is considerably smaller,—the *sixth (between first and second median nervules) is greatly elongated inwardly, so as to be only separated from the fourth by median nervure*, and the seventh is much reduced and attenuated (rarely obsolete), *being immediately preceded by a conspicuous crimson-red lunule* (additional to the inferior one also present in *Policenes*); spots of submarginal series much larger and broader, especially the upper non-lunulate ones; hoary-grey scaling beyond three lower spots of this series more conspicuous and of a more pronounced lunulate form. UNDER SIDE.—*Much paler, glistening dull-brown or yellowish-brown, with a slight violaceous lustre; nearly all the greenish-white markings somewhat enlarged; no perceptible hoary-grey suffusion between discal and submarginal series of spots.* Fore-wing: on costa, at base, a narrow edging of crimson-red; *first and third disco-cellular bars outwardly, and second inwardly, edged with black*; a series of black lunules inwardly edging submarginal series of spots; inner-marginal area, on both sides of broad lower portion of discal band, blackish; along hind-margin a series of very indistinct small blackish nervular marks. *Hind-wing:* a very narrow crimson-red mark at base; a black spot on costa between sub-basal stripe and broad commencement of discal macular band, and a black diffused outer border to the same band in its upper part; *between costal and subcostal nervures, on inner side of discal macular band, a conspicuous elongated black spot, inwardly edged with crimson-red;* below this, in discoidal cell, a very much smaller similar spot, usually ill-defined; spots of submarginal series distinct, all lunulate, inwardly edged with black (densely bluish-scaled between third and first median nervules); on hind-margin a series of black sub-lunulate marks, the three lower ones large, conspicuous, inwardly bounded by bluish; crimson-red anal-angular mark inwardly white-edged and bounded by black on both sides, immediately surmounted by a narrower similarly-coloured mark.

Abdomen black above, white beneath; on sides *transversely* striped with cream-colour and black,—the two creamy stripes next base tinged with crimson-red.

♀ Like ♂; but (as in ♀ *Policenes*) with the basal stripe in hind-wing enlarged and forming a rather wide inner-marginal border.

The large size of *Antheus*, its want on the under side of the hind-wings of all but the two uppermost and lowermost of the long series of crimson-red marks, and the transverse striping of the abdomen, are conspicuous features which at once distinguish the species from *Policenes*.[1] The very nearly allied *P. Erombar*, Boisd., from Madagascar, may be readily recognised by the confluence in both wings (in the fore-wings only on inner margin) of the enlarged central band and the sub-basal stripe, and the absence of any crimson spot on the upper side of the hind-wings. It should, however, be noted that

[1] I have not seen *P. Lurlinus*, Butl. (*Ann. and Mag. Nat. Hist.*, 5th Ser., xii. p. 106, 1883), from Victoria Nyanza, but from Mr. Butler's description (of the ♂ only), its differences from *Antheus* appear to be very slight, consisting chiefly in its larger size (exp. about 4 in.) and greater width of all the green markings.

the crimson spot is sometimes wanting in West-African examples, and the South-African Museum possesses a Gold Coast ♂ in which not only is this the case, but the confluence of the band and stripe of the hind-wing takes place about the middle of the discoidal cell.

As a rule, South-African specimens have the greenish markings paler and the dark markings of the under side less developed, especially the hind-marginal nervular marks in the fore-wings and the disco-cellular spot in the hind-wings.

This *Papilio* is essentially a Tropical form, and only enters the South-African Sub-Region on the north-east boundary. It is, however, as Mrs. Monteiro informs me, numerous at Delagoa Bay. On the Natal Coast it seems to be very rare, Colonel Bowker having met with but two or three specimens. One of these, a very fine ♂, presented by him to the South-African Museum, is noted as "obtained, just out of pupa, from Mrs. Watts, of D'Urban, end of October 1878."

Localities of *Papilio Antheus*.

I. South Africa.
  E. Natal.
    a. Coast Districts.—D'Urban (*J. H. Bowker*).
  II. Delagoa Bay.—Lourenço Marques (*Mrs. Monteiro*).
    1. "Inhambane."—Felder.

II. Other African Regions.
  A. South Tropical.
    a. Western Coast.—"Angola (*J. J. Monteiro*)."—Druce.
    b. Eastern Coast.—"Zanguebar and Zanzibar (*Raffray*)."—Oberthür.
    b1. Eastern Interior.—"Nyassa (Hewitson Coll.)."—Kirby.
  B. North Tropical.
    a. Western Coast.—"Gaboon (*Theorin*)."—Aurivillius. Fernando Po (*J. Bourke*). Gold Coast (*J. M. Pask*). Sierra Leone.—Coll. Brit. Mus.

302. (3.) Papilio Porthaon, Hewitson.

♂ *Papilio Porthaon*, Hewits., Exot. Butt., iii. pl. 2, ff. 21, 22 (1865).

Exp. al., (♂) 3 in. 4 lin.; (♀) 3 in. 4–5 lin.

Closely allied to *P. Policenes* and *P. Antheus*, Cram.

♂ *Brownish-black, with dull-whitish, slightly greenish-tinged stripes and spots*; common basal stripe as usual; *sub-basal stripe very narrow in fore-wing, but rather wide and with a marked disco-cellular expansion in hind-wing*; submarginal series of spots well developed (as in *Antheus*); all those in hind-wing outwardly bounded with well-marked hoary-grey lunular marks. *Fore-wing: three disco-cellular transverse stripes, extremely narrow, strongly ward, the outermost one rather sharply angulated*; spot in cell near extremity, as in *Policenes*; discal band of spots more as in *Antheus*, the upper four being larger than in either *Antheus* or *Policenes*, but the marking formed by the combination of the two lowest much smaller and narrower, and without any prolongation to median nervure; usually a very small spot just below and beyond first spot of discal band. *Hind-wing:* discal band represented mainly by a broad costal bar, the lower curved end of which (between subcostal nervules) approaches but does not meet the disco-cellular expansion of sub-basal stripe; *remaining spots of discal series* disposed as in *Policenes*, but *very much smaller and more widely apart* (and occasionally obsolete for the most part). UNDER SIDE.—Glistening pale dull-brown, with a slight bronzy lustre; pale markings of upper side reproduced; no hoary-grey between discal band and sub-marginal series of spots, as in *Policenes*, but *a series of dark marks, inconspicuous and ill defined in fore-wing, but black and well defined in hind-wing;* a tinge of crimson-red over ground-colour between basal and sub-basal stripes. *Fore-wing:* ground blackish near inner margin. *Hind-wing: space between sub-basal stripe and broad costal bar of discal series occupied by a crimson-red, inwardly finely white-edged stripe, bordered on both sides with black;* a series of four similarly coloured but very much smaller markings, forming a disjointed streak just beyond lower part of sub-basal stripe, between radial nervule and anal angle (where it is wider and better marked); *two of the black marks above mentioned, much longer than the rest, extend from the upper part of the crimson-red disjointed streak to two of the discal series of greenish-white spots;* an ill-defined blackish spot on costa between basal and sub-basal stripes.

Abdomen longitudinally striped on the sides as in *Policenes*, but with a basal lateral crimson stain as in *Antheus*.

♀ Like ♂, except that (as in the allied species) the basal band is broadly developed in hind-wing, forming an inner-marginal border. UNDER SIDE.—*Hind-wing:* crimson broken discal streak commences very thinly just below second subcostal nervule, instead of below radial nervule.

Among the distinguishing marks of this species, the most prominent are the tenuity of the sub-basal stripe and disco-cellular bars in the fore-wings, and the strong angulation of the outermost of the latter; the disco-cellular expansion of the sub-basal stripe, and the close approximation to it of the upper expanded portion of the discal series of markings in the hind-wings; the absence, or very faint indication, in the latter wings of the anal-angular crimson spot; and on the under side, the widely-interrupted and broken character of the central series of crimson marks in the hind-wings. Altogether, except for the sinuosity and angulation of the disco-cellular bars of the fore-wings, *Porthaon* is nearer to *Policenes* than to *Antheus*.

PUPA.—Form almost as in *Policenes*, but dorso-thoracic anterior projection shorter, and marginal abdominal expansion less prominent. Pale-brownish, finely speckled with fuscous; on upper side, two blackish spots on head, one at apex of thoracic projection, and a row on each side of abdomen, about half-way between middle and lateral margin.

Described from two specimens (which yielded a ♂ and a ♀ of *Porthaon*), brought from Delagoa Bay by Mrs. Monteiro in July 1886. The butterflies did not emerge till the following December, and were both dwarfs. Other pupæ of this species and of *P. Colonna*, Ward, which were taken on to England by Mrs. Monteiro, appear to have come out very satisfactorily in the Insect House at the Zoological Society's Gardens.

This butterfly was originally described by Hewitson from Zambesian specimens of the ♂. Mr. Butler (*Ann. and Mag. Nat. Hist.*, 4th Ser., xix. pp. 458 and 460, 1877) mentions that many specimens of it and the allied *P. Nyassæ* were contained in a collection made at Lake Nyassa. It appears to occur extra-tropically only at Delagoa Bay, where Mrs. Monteiro informs me it is common.

Localities of *Papilio Porthaon*.

I. South Africa.
    II. Delagoa Bay.—Lourenço Marques (*Mrs. Monteiro*).
II. Other African Regions.
    A. South Tropical.
        *b*1. Eastern Interior.—" Zambesi River (— *Dickinson*)."—Hewitson. " Lake Nyassa (*F. A. Simons*)."—Butler.

303. (4.) Papilio Colonna, Ward.

Papilio Colonna, Ward, Ent. M. Mag., x. p. 151 (1873).
Papilio tragicus, Butl., Ent. M. Mag., xiii. p. 56 (1876).
Papilio Colonna, Oberthür, Etudes d'Ent., iii. p. 15 (1878).

PLATE XI. fig. 5 (♂).

Exp. al., (♂) 3 in.—3 in. 5 lin.; (♀) 3 in. 5 lin.

♂ *Black, with very narrow pale bluish-green transverse stripes and spots; discal band of spots bent outwardly to posterior angle in fore-wing, and wanting in hind-wing;* common basal stripe yellowish-tinged; common sub-basal stripe much as in *Policenes*, but attenuated to a point at each extremity. *Fore-wing:* first very narrow, almost straight, disco-cellular bar prolonged to inner margin on to submedian nervure (with a small separate spot immediately below the nervure); second disco-cellular bar still narrower, very straight; third sinuated inferiorly, still narrower; spot in upper part of cell, almost at extremity, very small or wanting; in discal band upper four spots almost as in *Policenes*, but excavated externally, but the remaining three (especially the sixth and seventh) greatly reduced and narrowed, quite separate, and situated so as to deflect the lower part of the band to posterior angle (instead of inwardly, as in the three allied species above described); last spot of this band very narrow, and appearing like an enlarged eighth spot in submarginal series; seven spots of this series as in *Porthaon*, but narrower. *Hind-wing:* only trace of discal band two exceedingly small (sometimes scarcely perceptible) narrow spots near costa, *so that between long sub-basal stripe and submarginal lunules there is a broad unbroken area of black*; submarginal lunules rather large,—the first (on costa) white, the fourth obsolete inferiorly, the fifth obsolete, the sixth (last) sometimes obsolete, but occasionally linear and distinct; *a little before this last lunule a crimson-red one, variable in size*; also a crimson-red rather small spot at anal angle; basal half of tail black, outer half white, with a black median streak and thin black edges. UNDER SIDE.—*Glistening dull-brown, rather darker than in Porthaon, with a bronzy gloss; greenish markings as on upper side, but paler and rather broader, and with rather blurred edges, especially common basal and sub-basal stripes.* Hind-wing: crimson-red marks of central series arranged generally as in *Policenes*, but much more irregularly placed and completely separated, their inner white edgings fainter, but their outer black borders greatly enlarged into spots much broader than in *Porthaon*; of these red marks, the fourth or the fifth (at extremity of discoidal cell and immediately beyond it respectively) is occasionally obsolete, and the seventh (lunulate and white-edged both inwardly and outwardly) is far beyond anal-angular red spot, instead of contiguous to the latter, as in *Policenes* and *Porthaon*; on costa, about middle, immediately beyond outer black border of second crimson mark, an ill-defined whitish transverse streak,—sometimes a similar smaller streak beyond third mark,—rarely similar fainter spots beyond fourth and fifth marks; sub-basal stripe broader, shorter, blunter at extremity; submarginal lunules larger; each inwardly bounded by a large ill-defined black spot; succeeded by a hind-marginal series of thin sublinear lunules.

Abdomen black dorsally and laterally, hoary-grey beneath; on each side a series of five or six oblique creamy-white streaks, conspicuous on the black ground.

♀ Like ♂, but ground-colour not so deep a black; basal stripe of hind-wing enlarged as in ♀ *Porthaon*.

Described from specimens (only one ♀) collected at Delagoa Bay by Mrs. Monteiro. In my figure of a ♂ (Plate XI.) the green of the markings is given of too yellowish a tint, the actual colour in nature being decidedly glaucous. Though in many respects closely allied both to *P. Policenes* and *P. Porthaon*, the singularly different direction and reduced development of the discal macular band in the fore-wings, and its total or nearly total suppression in the hind-wings, very markedly separate *Colonna* from both those species, and still more from *P. Antheus* and *P. Ecombar*. As Mr. Oberthür has remarked (*Études d'Ent.*, liv. iii. p. 15, 1878), the black field of the hind-wings approximates this butterfly in aspect to the otherwise not remotely allied *P. Pailolaus*, Boisd., of Northern and Central America; but in the latter the common discal band is continuous and normal in position as regards the fore-wings, and in the hind-wings superiorly developed, much as in *Porthaon*.

LARVA.—Dull yellowish-green, with an interrupted brownish longitudinal streak along each side of the back. Sides of posterior half dark-brown; in one place (apparently on ninth segment) the strongly sinuated dorsal margins of this dark-brown closely approximate. Head brownish-sandy. Each of the three thoracic segments bearing dorso-laterally a pair of short forward- and laterally-pointing tubercular projections,—the second and third pairs clustered with small spines.

PUPA.—Pale bluish-green. Margin expanded laterally about wing-covers, but constricted in two places towards head, so that the outline viewed dorsally or ventrally is sinuated. Dorsal thoracic projection and upper part of edge of wing-covers inferiorly roughened with brown granules, which in the latter part are inferiorly edged conspicuously with white.

These descriptions of the earlier stages were made from drawings by Mrs. Monteiro, which she kindly lent to me in the year 1883.

This strikingly handsome "Swallow-Tail," originally described from examples taken at Ribè near Mombas, has been since recorded from other more southern stations, but seems limited in range to the Eastern Coast of Africa, where it occurs extra-tropically at Delagoa Bay. Oberthür notes that it was collected in some abundance by M. Raffray on the coast opposite Zanzibar. Mrs. Monteiro informed me that it was common at Delagoa Bay; she noticed that it seldom flew high, but kept about two feet from the ground, pursuing a circular course in and about dense bush.

Localities of *Papilio Colonna*.

I. South Africa.
   II. Delagoa Bay.—Lourenço Marques (*Mrs. Monteiro*).
II. Other African Regions.
   A. South Tropical.
      b. Eastern Coast. — "Zambesi." — Butler. "Bagamoyo; Mts. Schimba and Mombayo (*Raffray*)."—Oberthür. Mombas: "Ribè."—Ward.

304. (5.) Papilio Leonidas, Fabricius.

Papilio similis, Cram., Pap. Exot., i. pl. ix. ff. B, C (1779).
Papilio Leonidas, Fab., Ent. Syst., iii. 1, p. 35, n. 103 (1793).
    ,,   ,,   Godt., Enc. Meth., ix. p. 44, n. 56 (1819).
    ,,   ,,   Boisd., Sp. Gen. Lep., i. p. 242, n. 66 (1836).
    ,,   ,,   Westw., Arc. Ent., i. p. 149 (1845).
    ,,   ,,   Staud., Exot. Schmett., i. pl. 7 (1884).
VAR. A. *Papilio Pelopidas*, Oberth., Etudes d'Ent., iv. p. 55, pl. v. f. 1 (1879).

Exp. al., (♂) 3 in. 0-3 lin.; (♀) 3 in. 5-6 lin.

♂ *Brownish-black, with numerous very pale glaucous-greenish spots; a large basal patch of the same colour in hind-wing.* Fore-wing: three good-sized oblique marks in discoidal cell, of which the smallest (near base) is almost longitudinal,—the largest (very oblique and wider inferiorly) not far beyond the first and completely crossing cell,—and the third (rounded superiorly and exteriorly) more remote and close to upper part of extremity of cell; an irregular incurved discal series of spots of very different sizes and shapes, viz., the first and second small, rather elongate, between third and fifth subcostal nervules, the lower one placed obliquely slightly beyond the upper one; the third and fourth extremely small (fourth often wanting), nearer to end of cell; the fifth elongate, or little larger than the first; the sixth (the largest in the series), between second and third median nervules, continuous in an oblique line of middle mark in cell; the seventh very small, sublinear (very rarely wanting); and the eighth (between first median nervule and submedian nervure) in size next to the sixth, but narrower; a submarginal row of eight small rounded spots, of which the third is much more elongate than the rest and out of line *before* them, so as to appear the third in an oblique subapical row with the first and second spots of the discal series. *Hind-wing:* base narrowly black; basal patch white above subcostal nervure, lying between costal and submedian nervures, divided into five unequal portions by blackish subcostal nervure and nervules and median nervure and its first nervule, roughly subquadrate (its farthest projection between subcostal nervules), not occupying cell to its extremity; a short discal series of four small or very small rounded spots not far beyond outer edge of patch, between second subcostal and first median nervules; a submarginal row of seven rounded spots, six of them of about the size of the corresponding row in fore-wing, but the seventh (close to inner margin) minute; a hind-marginal row of minute inter-nervular white spots, with the cilia touching them also white; inner-marginal fringe of long hair white where it borders basal patch, but ochre-yellow beyond. UNDER SIDE.—*Hind-wing and apical area of fore-wing pale-brown, somewhat glistening; markings much as on upper side, but whiter and with less defined edges; basal patch of hind-wing with a faint yellowish tinge; bases stained with deep dull-red. Hind-wing:* at base an inferior black spot enclosing two small white ones.

Head black, with two white spots on front, two smaller ones on vertex, and one on each palpus; *thorax* black, with two dorsal and two lateral white spots on collar, a white streak on each pterygode, and numerous large white spots on breast; *abdomen* black dorsally, cream-coloured laterally with a black longitudinal streak and black segmented incisions, whitish inferiorly, with an ochre-yellow basal tuft and median stripe, and two rows of small black spots on each side.

♀ *Like ♂, but with all the markings paler (sometimes almost white); in hind-wing, hind-marginal white dots a little larger, especially on under side.*

VARIETY A. (♂) *Pelopidas*, Oberthür.—*Several of the principal markings much enlarged. Fore-wing:* median oblique disco-cellular mark twice as broad as in *Leonidas*; sixth spot of discal series rather, and fifth greatly enlarged, so that these three markings form a broad oblique band; the seventh spot is considerably and the eighth moderately enlarged and lengthened; and the first and second are much longer; first and third disco-cellular marks and eight spots of submarginal row, on the contrary, considerably smaller. *Hind-wing*: basal patch considerably wider, especially along inner margin, and in discoidal cell reaching nearly to extremity; submarginal spots much reduced in size. The basal red appears to be wanting on the under side in both wings.

(*Hab.*—"Zanguebar: Tchouacka (*Raffray*)."—Oberthür.)

This marked variety was named and figured by Mr. C. Oberthür (*op. cit.*) from a single example; it is possibly, as he suggests, entitled to species rank, but not having examined the specimen, I am unable to give any decided opinion on this point.

P. Leonidas has long been well known from the Western Coast of Tropical Africa, and later as a native also of the Eastern Coast. It was not until 1878 that I learned the existence of the typical form to the south of the tropic, where its prevalent representative is the doubtfully distinct "Variety A." of Mr. G. R. Gray (= *Brasidas*, Feld.). Specimens received in that year and subsequently from Delagoa Bay (Mrs. Monteiro) quite agree with West-Coast examples; and just recently—in November 1887—I have received a ♂ and a ♀ taken as far south as Etshowe in Zululand by Captain A. M. Goodrich. The ♀ shows a tendency in the direction of *Brasidas* by the almost obsolete condition of the basal cellular mark in the fore-wings.

A pupa received from Mrs. Monteiro resulted in a ♂ on the 9th December 1886. I could not discover that this Delagoa Bay specimen differed at all from the chrysalides of *Brasidas* from Natal, which are described below.

As Boisduval (*op. cit.*) and Mrs. Butler (*Cat. Fab. D. Lep. in Brit. Mus.*, p. 243) have pointed out, *P. Leonidas* in pattern and colouring presents unmistakable resemblance to the variety (*Petiverana*, Doubl. = *Leonora*, Butl.) of *Danais Limniace*, Cram., inhabiting Western Tropical Africa.[1] This mimicry is not close as regards the basal patch of the hind-wings, but is accentuated by the similarity in the outline of the wings, the white spotting of the head and thorax, and the median ochreous-yellow colouring on the under side of the abdomen.

Localities of *Papilio Leonidas*.

I. South Africa.
    F. Zululand.—Etshowe (*A. M. Goodrich*).
    II. Delagoa Bay.—Lourenço Marques (*Mrs. Monteiro*).

---

[1] Mr. J. Morton Pask, R.N., who presented specimens of this *Danais* to the South-African Museum, informed me that it was a very abundant species at Cape Coast Castle.

II. Other African Regions.
  A. South Tropical.
    a. Western Coast.—Angola: Bembe Mines (*J. J. Monteiro*).—Coll. Brit. Mus. "Chinchoxo (*Falkenstein*)."—Dewitz.
    b. Eastern Coast.—Zambesi River (Rev. H. Rowley).—Coll. Oxford Mus. Zanguebar: "Tchouacka (*Raffray*)."—Oberthür.
    b1. Eastern Interior.—Bamangwato Country (*H. Barber*).
  B. North Tropical.
    a. Western Coast.—"Cameroons."—Oberthür. Cape Coast Castle (*J. M. Pask* and *J. Bourke*). Sierra Leone (*A. R. Innes*). "Rio Nunez."—Oberthür. "Gambia (*A. Moloney*)."—G. E. Shelley. "Senegal."—Felder.

305. (6.) Papilio Brasidas, Felder.

♂ *Papilio Leonidas*, Var. a., G. R. Gray, Cat. Lep. Brit. Mus., p. 7, n. 21, pl. v. f. 3 (1852).
Papilio Leonidas, Trim., Rhop. Afr. Aust., i. p. 21, n. 9 (1862).
Papilio Brasidas, Feld., Verh. Zool.-Bot. Gesellsch. Wien, 1864; Sp. Lep., p. 307 [19], n. 249, and p. 353 [65], note 144.

Exp. al., (♂) 3 in. $2\frac{1}{2}$–7 lin.; (♀) 3 in. $3\frac{1}{2}$–$7\frac{1}{2}$ lin.

Very closely allied to *P. Leonidas*, Fab., *but the spots paler (not of so green a tint), smaller, and in fore-wing (especially in ♀) fewer.*

♂ *Brownish-black, with dull white (sometimes slightly greenish-tinged) markings, in pattern and arrangement as in P. Leonidas.* Fore-wing: first or basal oblique mark in discoidal cell, and small elongate seventh spot of discal series (between first and second median nervules), wholly wanting; second or median disco-cellular mark highly variable, sometimes as large as, or larger than, in *Leonidas*, but usually narrower and divided transversely, occasionally into three, but more often into two parts,—rarely reduced to one very small (upper) spot; third disco-cellular mark, spots of discal series, and spots of submarginal row nearly always smaller,—the last-named sometimes minute. Hind-wing: basal patch almost always narrower, especially on inner margin, the neuration crossing it only black near its outer edge; spots of discal series reduced to three, two, or one, or not seldom wanting altogether; spots of submarginal row much smaller, the first often obsolescent or wanting; inner-marginal fringe of long hair yellow-ochreous throughout. UNDER SIDE.—Ground-colouring darker; markings smaller than on upper side, especially basal patch of hind-wing, which has a decided tinge of ochre-yellow; basal red usually much duller than in *Leonidas*, sometimes obsolescent.

♀ *Like ♂, but markings whiter and (with exception of basal patch of hind-wing) very frequently much smaller and fewer.* Fore-wing: small third spot of discal row almost always, and fifth very often, wanting; disco-cellular marks usually much reduced,—rarely represented by a median and terminal indistinct very small spot,—and in one example absent except for two scarcely perceptible median dots. UNDER SIDE.—As in ♂, except for usually smaller and fewer spots.

LARVA (second moult).—Yellowish-green above, paler on the sides, pale-green beneath; back of segments 3, 4, and 5 occupied by a lozenge-shaped space of rich dark-green. A median dorsal narrow stripe, whitish on thoracic segments, but beyond that dull pinkish-yellow, edged on both sides with dark-green. Second segment reddish-brown superiorly, with a short but acute tubercular process on each side; third and fourth segments each bearing dorso-laterally a pair of short acute yellow tubercular processes, edged inwardly and basally by an ovate black spot, and connected by a transverse reddish-brown band; a blackish line runs longitudinally from each ovate black spot on third segment to the reddish-brown second segment. On each side, from sixth to last segment, a series of seven oblique slender dark-green lines; also a longitudinal (spiracular) dull-whitish stripe, commencing superiorly on third segment, curving downward on fifth, and thence running straight to extremity. Anal segment bearing superiorly a pair of rather long acute divergent processes. Retractile forked scent organ of second segment deep indigo-blue.

(Adult).—Paler and brighter green above, and of a bluer tint beneath. Dorsal markings, except reddish-brown of second segment and black spots of third and fourth segments, apparently very much fainter and partly obsolete; and the dorsal and lateral stripes thinner dark-green. Segmental incisions marked, shining.

Food-plant *Popowia Caffra*, a climbing shrub of the family *Anonaceæ* (H. C. Harford, *in epist.*).

PUPA.—Semi-transparent. Bright yellowish-green. On each side of back, from apex of dorso-thoracic projection to anal extremity, a narrow ochrey-yellow marginal ridged or elevated narrow stripe; on summit of back a pair of similar ridged stripes, commencing abruptly and considerably apart near base of dorso-thoracic projection, and converging till they meet at anal extremity; a slender dark-green dorsal median line. On each side, a series of oblique thin streaks, rather darker than ground-colour, extending from near dorsal median line to marginal yellow ridge (and beyond it on cephalic area).

Very thick and robust, especially towards head, which is remarkably blunt and sub-rotund. Dorsally much concaved and widened about middle; dorso-thoracic peak prolonged upward and forward a good deal beyond front of head, and slightly curved; anal extremity elongate and attenuated. Attached to mid-rib or to foot-stalk and mid-rib of under side of leaves; usually head downward.

The above description of the larva is made from drawings and notes of Natalian examples furnished by Captain H. C. Harford in 1869 and 1872; that of the pupa is from living Natalian specimens received from Colonel Bowker in 1878.[1] With the latter arrived two

---

[1] In 1879 I obtained Mr. W. D. Gooch's excellent figures and notes of Natalian lepidopterous larvæ, among which are pencil outlines (with remarks) of the early stages of this *Papilio* quite confirmatory of Captain Harford's.

larvæ, one already suspended for pupation, the other in the stage described as the "second moult" by Captain Harford. Only the former of these larvæ produced the perfect insect, but I obtained ten butterflies from the pupæ received. Of these, the dates of pupation of nine being on record, I am able to note that the duration of the chrysalis state in November and December varied from twelve to eighteen days. These eleven pupæ were (with the exception of two on the twigs of the food-plant, and one—from the larva that pupated after arrival—attached to the side of a box) suspended on the under side of the leaves, the tail being fixed at the base of the mid-rib, or, in a few cases, just free of the beginning of the leaf, to the short foot-stalk. In this position, the difficulty of distinguishing the pupa from the adjacent leaves is considerable, as the concavity and depression of the dorsal aspect, from the attenuated tail to the prominent and curved thoracic peak (which intercepts the view of the blunted head), and the outline of this aspect, the resemblance of the four raised dorsal stripes converging at the tail to a mid-rib or short foot-stalk, and the slightly darker oblique streaks and minute flecks representing the venation, all combine to assimilate the pupa to the surrounding foliage in the most deceptive manner.

In my former work on South-African Butterflies, I followed Mr. G. R. Gray in treating this *Papilio* as an austral form of *P. Leonidas*, Fab.; but the accession of more material, and the fact of the unmistakable modification of *P. Brasidas* (more especially the ♀) in mimicry of a different Danaine (viz., *Amauris Echeria*) from that imitated by *P. Leonidas*, have led me to keep the two forms apart, notwithstanding their very intimate alliance. The occurrence of both *Leonidas* and *Brasidas* in Zululand marks up till now (January 1888) the southern limit of the former and the northern one of the latter on the East Coast; but according to Mr. Druce (*Proc. Zool. Soc. Lond.*, 1875, p. 416), the two forms both appear in the late Mr. Monteiro's collection, formed in Angola, very much farther to the northward on the West Coast.

Brasidas appears to have rather a limited range in South Africa, not being known to me to exist to the westward of King William's Town and East London, where Mr. D'Urban reported it to be rare. Colonel Bowker did not meet with many examples in the Trans-Kei country; but on and near the coast of Natal and Zululand the butterfly is by no means uncommon, and often numerous. I took a good many specimens about D'Urban at the end of January and in February 1867; they frequent woods and their vicinity, flying briskly but not very swiftly, and often visiting *Lantana* and other flowers. In relation to this *Papilio*'s marked resemblance to *Amauris Echeria*, Stoll,[1] both in pattern and outline of wings. I was much interested to see that it was in the habit of settling precisely in the way affected by the *Amauris*, viz., on a projecting leaf or twig, with the wings closed and hanging downward, and in this exposed position remaining motionless for a considerable time. On more than one occasion I have mistaken *Brasidas* when so posed for *Echeria*. Captain Goodrich's Zululand specimens were captured in October and November 1886, and January and April 1887.

---

[1] See vol. i. pp. 37 and 59; and *Trans. Linn. Soc. Lond.*, xxvi. p. 507, and note (1869).

Localities of *Papilio Brasidas*.

I. South Africa.
    B. Cape Colony.
        b. Eastern Districts.—King William's Town and East London (*W. S. M. D'Urban*).
    D. Kaffraria Proper.—Bashee River (*J. H. Bowker*).
    E. Natal.
        a. Coast Districts.—D'Urban. "Lower Umkomazi."—*J. H. Bowker.*
        b. Upper Districts.—Maritzburg (*S. Windham*).
    F. Zululand.—Etshowe (*A. M. Goodrich*).

II. Other African Regions.
    A. South Tropical.
        a. Western Coast.—"Angola (*Monteiro*)."—Druce.

306. (7.) Papilio Corinneus, Bertoloni.

Papilio Corinneus, Bert., "Mem. Acad. Sci. Bologn., 1849, p. 9, t. 1, ff. 1-3."
Papilio Pylades, Var. a., G. R. Gray, Cat. Lep. Brit. Mus., i. p. 8 (1852).
Papilio Anthemenes, Wallengr., K. Sv. Vet.-Akad. Handl., 1857; Lep. Rhop. Caffr., p. 6, n. 5.

Exp. al., (♂) 3 in. 0–3 lin.; (♀) 3 in. 1–4½ lin.
Very nearly allied to *P. Pylades*, Fab.

♂ *Black, with a large white patch and white spots in both wings.* Fore-wing: in discoidal cell four spots of very different shape and size, viz., one very small, more or less rounded, before middle of cell,—the second, a little beyond, of moderate size, oblique, wedge-shaped, reaching about half-way across cell; the third very much larger, irregularly bi-sinuated, slightly oblique, narrow superiorly, but much widened inferiorly, extending quite across cell; and the fourth, about the size of the second, but rounded, in upper part of cell close to extremity; just beyond extremity of cell a short transverse row of three very small spots, of which the uppermost is the largest and inclining to be quadrate; a discal macular band composed of nine spots very different in form and size, viz., the first and second subapical, of moderate size, the second, the larger and longer, and projecting outwardly considerably beyond the first; the third and fourth very small, not far beyond, and rather larger than, the second and third spots of the short transverse row just beyond cell; the fifth, large, ovate, between lower radial and third median nervules next to largest cellular spot; the sixth, small, or very small, elongate, ovate; and the seventh, eighth, and ninth, greatly enlarged and confluent into a large patch, which occupies inner margin for a little beyond base to a little before posterior angle, is bounded superiorly by median nervure and its second nervule, and presents a very prominent, rounded, outward projection of outline between second and first median nervules; close to hind-margin, a series of eight rather small rounded spots, of which the lowest is geminate; base with a white suffusion over the black extending into discoidal cell nearly as far as, or sometimes a little beyond, first spot. *Hind-wing:* white patch occupying all basal half of wing, except a strongly white suffused, narrow, blackish base, and streak along submedian nervure, its outer edge rather irregular crossing discoidal cell very near extremity, slightly impinging narrowly and diffusedly on the black between nervules, and dentated by the black on nervules; close to hind-margin, a series of seven rather small spots, of which the first is very small, often indistinct, and sometimes wanting,—the second and third rather broad sub-lunulate,—the fourth thin and lunulate,—the fifth and sixth thin and sagittiform,—and the seventh minute and linear, close to anal angle; midway between outer edge of white patch and hind-marginal series of spots, a row of three rather small rounded spots (inter-nervular) from radial to first median nervules,—rarely a minute fourth spot above radial nervule; at anal angle, usually a very indistinct ochreous-yellow spot. *Cilia* white, interrupted with black at the end of the nervules. UNDER SIDE.—*Pattern similar, but apical area of fore-wing and outer area of hind-wing ochreous-yellow, and bases of both wings lakered; many of the white markings rather enlarged. Fore-wing:* basal red fills discoidal cell as far as origin of first median nervule, and extends superiorly as far as third cellular spot; apical ochreous-yellow enters cell as far as the third spot, and extends narrowly along costa as far as first spot; second cellular spot produced by an incurved greyish prolongation quite across cell; sixth spot of discal series enlarged by a greyish ring; black of ground limited to a small basal space below median nervure, the lower part of outer half of discoidal cell, and a space bounded by third median nervule, hind-margin, and outer edge of white patch. *Hind-wing:* neuration and costal edge of white patch ochreous-yellow; basal red narrow, outwardly black-edged, extending very narrowly along inner margin; patch externally with a black edging; strongest from second subcostal nervule to anal angle, and along that inferior half immediately bounded by a series of four small lunulate white marks; of the seven spots near hind-margin, the upper three are inwardly, the lower four both inwardly and outwardly, black-edged; anal-angular spot large, ovate, of a brighter tint than the adjacent ochreous-yellow.

Head black, with two large white spots on front, two smaller ones on vertex, a white outer fringe bordering the eyes, and white palpi. *Thorax* black, clothed above with white hair, except on collar, which bears two white spots; beneath with two large median and seven good-sized lateral white spots on each side. *Legs* white, with the femora above black. *Abdomen* above mesially black, relieved on each side with white; a conspicuous lateral band of ochreous-yellow from close to base almost to extremity; beneath white, with a gradually-narrowing median black streak, the segmental incisions also marked with well-defined transverse black streaks in basal half.

♀ Like ♂, but the hind-wing usually with a rather broader black outer field, and the ochre-yellow spot at anal angle well marked or even conspicuous. UNDER SIDE.—*Hind-wing*: red streak along submedian nervure obscured by blackish from near base; a moderately wide white border between this streak and inner margin, which has itself a well-marked black edging.

I am indebted to Mr. W. L. Distant for a tracing of Bertoloni's insect and some notes on its special characters, which enabled me to decide on keeping *Corinneus* distinct from both *Pylades*, Fab., and *Morania*, Angas, and identifying it with the *Authemenes* of Wallengren. In outline the wings of *Corinneus* differ from those of *Pylades*, the fore-wings being less produced apically, and the hind-wings less so inferiorly, besides being much less prominently dentated at extremity of nervules. In pattern *Corinneus* differs very markedly in the fore-wings by the limitation of the large white patch, which does not rise above the median nervure and its second branch, whereas in *Pylades* it encroaches considerably on the discoidal cell, is confluent with the second and third cellular spots, and also includes the much-enlarged sixth spot in the discal series; the outward projection of the patch between first and second median nervules is also much stronger in *Corinneus*. The head in front is white with a median black stripe, instead of black with two large white spots, as in *Pylades*; and on the under side of the abdomen the median black streak is a character wholly wanting in *Pylades*, and the transverse black streaks on its segmented incisions are scarcely indicated in the latter species. In the black-filled discoidal cell and space between third and second median nervules, and in the black abdominal markings, *Corinneus* approaches *Morania*, but in all other respects is closer to *Pylades*.[1]

The range of *Corinneus* is found to differ considerably from that of *Pylades*, for while the latter prevails on the Western Coast of North Tropical Africa, the former seems limited to regions (both east and west) to the south of the Equator, and extends to the coast of Natal. At Delagoa Bay it appears to be common, in company with the strictly Southern allied species *Morania*, but farther southward seems to be scarce. In February 1867 I captured a ♀ at Verulam, and Colonel Bowker sent another from Pinetown in 1884, while two specimens occurred in the collection of Natalian insects formed by my colleague for the Colonial and Indian Exhibition of 1886. On the confines of the tropic it is evidently a frequent form, many specimens from the tracts between the Transvaal and Zambesi having reached me from Mr. F. W. and Mr. H. Barber, Mr. F. C. Selous, and Mr. Eriksson.

Localities of *Papilio Corinneus*.

I. South Africa.
    E. Natal.
        a. Coast Districts.—Verulam. Pinetown (*J. H. Bowker*).
II. Delagoa Bay.—Lourenço Marques (*Mrs. Monteiro*).

---

[1] From the particulars given by Dewitz (*Nov. Act. Leop.-Carol.-Deutsch. Akad. Naturf.*, xli. pp. 187 and 209, 1879), it is quite clear that the specimens from Angola (*Pogge*) and Chinchoxo (*Falkenstein*), recorded as *Pylades*, are referable to *Corinneus*.

K. Transvaal.—Lydenburg District (*T. Ayres*). Limpopo River (*F. C. Selous*).

II. Other African Regions.
   A. South Tropical.
      a. Western Coast.—"Angola (*J. J. Monteiro*)."—Druce. Congo.—Coll. Brit. Mus. "Chinchoxo (*Falkenstein*)."—Dewitz [as *Pylades*, Fab.].
      b. Eastern Coast.—"Mozambique."—Bertoloni. "Zanguebar."—Oberthür [as *Anthemenes*, Wallengr.].
      b1. Eastern Interior.—Matabeleland (*H. Barber*). Gubulewayo (*F. C. Selous*). Daka River (*F. W. Barber*).

307. (8.) Papilio Morania, Angas.

Papilio Morania, Ang., Kafirs Illustr., pl. xxx. f. 1 (1849).
Papilio Pylades, Trim., Rhop. Afr. Aust., i. p. 22, n. 10 (1862).

Exp. al., (♂) 2 in. 9–11 lin.; (♀) 2 in. 11 lin.—3 in. 1 lin.
Very nearly allied to *P. Corinneus*, Bertol.

♂ *Black, with a large white patch and white spots in both wings.* Forewing: white patch bounded as in *Corinneus*, but decidedly broader in its *inferior part*, beginning nearer to base and extending nearer to posterior angle, so that the outward projection of the portion between first and second median nervules is slight; *only three spots in discoidal cell*,—the first not so small and rounder than in *Corinneus*,—the second much enlarged, curved, and prolonged inferiorly quite across cell (as on under side in *Corinneus*),—*the third (representing both third and fourth of Corinneus) large, very broad*, excavate inwardly and rounded outwardly; remaining (subapical and hind-marginal) spots in number, position, form, and relative size as in *Corinneus*, except that the hind-marginal spots are closer to the edge. Hind-wing: *white patch considerably wider, so that hind-marginal black border is correspondingly narrower*,—the white completely filling discoidal cell, and projecting beyond it both superiorly and inferiorly; white spots as in *Corinneus*, but the inner row of three or four is of course nearer to the edge of the white patch in consequence of the greater extension of the latter; three upper spots of outer row closer to hind-marginal edge; anal-angular ochre-yellow spot more distinct. UNDER SIDE.—Very like that of *Corinneus*. Forewing: apical ochre-yellow darker, not invading discoidal cell, but itself slightly obscured by a fuscous suffusion immediately beyond extremity of cell, and *bounded externally by a narrow black border*. Hind-wing: basal red narrower, but *its inner-marginal extension (inferiorly much obscured with ochreous-brown) much wider*, filling, or occasionally almost filling, the space between submedian nervure and median nervure and its first nervule; outer black edging of white patch broader, inferiorly diffused, where the two or three lunulate whitish marks bounding it are less conspicuous, being mixed with ochre-yellow; these marks and the anal-angular spot (more orange than in *Corinneus*) are only separated from the inner row of white spots by the diffused black edging of white patch; ochre-yellow of border darker and rather broadly and regularly edged with black throughout.

♀ Like ♂; hind-wing with ochre-yellow anal-angular spot better developed. UNDER SIDE.—*Hind-wing:* white inner-marginal border not black-edged, but with a blackish longitudinal streak running near its inner edge about as far as end of abdomen.

Head with frontal white spots smaller. *Abdomen* above more decidedly black down the middle, with the white on each side more sharply separated, and so widened in apical half that *the ochre-yellow lateral band does not extend beyond fourth segment;* beneath with the black middle and lateral streaks and segmental incision streaks more strongly marked.

In addition to the various differences above pointed out, *Morania* presents shorter and more rounded wings, and much slighter dentation of the hind-wings.

LARVA.—Dull-green with a yellowish tinge; two parallel darker transverse lines across back of each segment; incisions of segments pale bluish-grey. First thoracic segment and lateral stripe bright-yellow, the latter deepening to orange on last segment, and edged inferiorly throughout by dark-green. Lower portion of sides pale bluish-green, with a bluish-white stripe immediately above legs. Head pale-green. Each thoracic segment with a pair of short, pointed, laterally projecting spines, of which the first pair is shortest and mainly black, but sandy-yellow at base,—the second ferruginous basally, but thence black,—and the third whitish tinged with blue; between the bases of the second and third pairs a transverse ferruginous-brown streak (incomplete on second thoracic segment). Anal segment bearing dorsally a pair of terminal, short, acute, yellow spines.

A small example in an earlier moult is very different, having three dorsal longitudinal blackish lines, of which the middle one is thinnest, the transverse darker dorsal lines more developed, and a blackish transverse bar on the penultimate segment. The lower sides and the under side are fuscous, the head is sandy-yellow, the thoracic spines are proportionally larger, and the transverse bars between the second and third pairs broader and much darker.

Food-plant, *Artabotrys*, n. sp., one of the *Anonaceæ*, native of D'Urban, Natal, and Delagoa Bay. (Determined for Colonel Bowker, on reference to Kew, by Mr. J. Medley Wood, Curator of the Botanic Gardens, D'Urban.)

PUPA.—Stout, rounded, but tapering posteriorly to rather a long point; not depressed or widened dorsally, as in *Brasidas*. Not nearly so blunt and rounded anteriorly as in *Brasidas*, the cephalic prominences, though short, being acute and widely divergent. Dorso-thoracic peak not so long as in *Brasidas*, and more ascendant, so as not to project so far in front of the head. About one inch in length. Bright-green; a well-defined, raised, pale-yellow line curving along each side from apex of dorso-thoracic peak, and meeting at anal extremity. On back a similar thinner line, starting from same peak (but near its base), almost immediately branches into two gently divergent lines, which from third abdominal segment gradually converge, to meet again at anal extremity. Between these lateral and dorsal lines, on each side a series of very thin V-like yellow marks (with the angles directed backward), and all truncated by the segmental incisions. On third abdominal (seventh) segment, a small ferruginous spot in a dull-creamy ring adjoins each dorsal line on its inner side about the middle. About midway between the lateral line and the cephalic prominence on each side, a somewhat sinuous, less distinct, pale-yellow line, which on wing-cover marks position of median nervure, and emits thin indications of positions of nervules. In addition to these principal pale-yellow markings, there are numerous minor reticulations and dottings of the same colour distributed over the surface. Frontal line of dorso-thoracic peak, from its apex to head, minutely roughened, and coloured dull-ferruginous, with a few creamy specks.

The larva is described from four Natalian specimens (one of an early moult) forwarded by Colonel Bowker in November 1878. I did not rear the perfect insect from them, but two of them became pupæ quite like those of *Morania*; and coloured drawings (accompanied by the wings of that species) lent me by Mrs. Monteiro in 1883 (made from Delagoa Bay larvæ), agreed closely with my description above given.[1]

Of the pupa, I received from Colonel Bowker thirteen living specimens, and obtained the imago from each one, ten butterflies appearing between 24th November and 16th December 1878. Of the remaining three, two did not make their appearance before a year afterwards (14th November and 9th December 1879), and the last not until 14th March 1880. Colonel Bowker noted the date of pupation in five instances, so that I can record the normal duration of the chrysalis state in those few cases as varying from twelve to sixteen days. It is noticeable that in outline this pupa does not simulate a complete leaf so well as that of *P. Brasidas*, but the anterior extremity, with the concave, roughened, and ferruginous-tinted line from the tip of the thoracic peak to the head, gives precisely the effect of the irregular edge of a leaf gnawed and partly eaten away by insects. The under side of the mid-rib of a leaf is usually the situation of the suspended

---

[1] In 1886 Mrs. Monteiro succeeded in taking alive to England various pupæ of Delagoan *Papiliones*, and wrote to me in November of that year that *P. Morania* was appearing at the Insect House of the Zoological Gardens "from two distinct caterpillars." There can be little doubt that the two different larvæ were those of *Corinneus* and *Morania* respectively, as both butterflies resulted from pupæ left at the South-African Museum by Mrs. Monteiro while I was away from the Colony. As far as the *cauvie* of the *Corinneus* pupa go, I can see no difference between it and *Morania* pupæ, except the greater prominence of both the dorso-thoracic peak and the cephalic projections.

pupa; in two instances only, among the specimens sent by Colonel Bowker, was the insect attached to the twigs of the food-plant. As regards the surface of the pupa, the yellow lines, V-streaks, and minuter markings admirably represent the aspect of the leaves, and effectually conceal the insect among the foliage.

The characters above described as distinguishing this butterfly from *P. Corinneus* are remarkably constant; the only sign of instability that I have found among a large number of specimens being the rare occurrence, in the large terminal disco-cellular spot in the fore-wings, of a small, superior, blackish spot (in one instance extended into a crossing streak), being the merest indication of the broad black bar that in *Corinneus* completely divides the marking into two unequal spots.

A dwarf ♀ that I captured at D'Urban, Natal, in January 1867 expands only 2 in. 4 lin., and has the under side unusually dull.

I did not find *Morania* at all numerous about D'Urban in 1867; most of my specimens were taken in February, but others occurred at the end of January, end of March, and beginning of April. They flew low, and by no means rapidly, keeping about low trees and shrubs on the edge of the woods, sometimes settling on leaves, but very rarely on flowers. Colonel Bowker took a good many examples in November; he sent me the sexes netted *in copulâ* in March 1879.

The known range of this species is much more restricted than that of *Corinneus*, being apparently limited to the South-East Coast from Natal to Delagoa Bay.

Localities of *Papilio Morania*.

I. South Africa.
  E. Natal.
    a. Coast Districts.—D'Urban. Verulam.
  F. Zululand.—Etshowe (*A. M. Goodrich*).
II. Delagoa Bay.—Lourenço Marques (*Mrs. Monteiro*).

308. (9.) **Papilio Demoleus**, Linnæus.

Papilio Demoleus, Linn., Mus. Lud. Ulr. Reg., p. 214, n. 33 (1764); and Syst. Nat., i. 2, p. 753, n. 46 (1767).
  „ „ Cram., Pap. Exot., iii. pl. ccxxxi. ff. A, B (1782).
  „ „ Wulfen, Capens. Ins., p. 29, n. 28 (1786).
Princeps dominans Demoleus, Hübn., Samml. Exot. Schmett., i. pl. 117 (?1806).
Papilio Demoleus, Godt., Enc. Meth., ix. p. 43, n. 52 (1819).
  „ „ Boisd., Sp. Gen. Lep., i. p. 237, n. 60 (1836).
  „ „ Trim., Rhop. Afr. Aust., i. p. 17, n. 6 (1862).
♂ *Papilio Demoleus*, Staud., Exot. Schmett., i. pl. 13 (1884).
LARVA and PUPA (South African), Trim., *op. cit.*, i. p. 18, and ii. pl. 1, ff. 1, 1a (1866).

Exp. al., (♂) 3 in. 9 lin.—4 in. 3 lin.; (♀) 2 in. 11 lin.—4 in. 5 lin.

♂ *Black, with pale sulphur-yellow stripes and spots. Fore-wing:* a yellow streak from base just above subcostal nervure, about half the length of discoidal cell; basal portion, to some distance, thickly streaked with short, thin, yellow transverse lines, and irrorated with yellowish scales; in outer portion of discoidal cell two good-sized spots, the lower one farther from base than upper, and a transverse yellow streak touching the closing nervules; above the latter, outside cell, a short longitudinal streak; beyond this a subquadrate spot on costa commences a transverse row of ten irregular-shaped spots, the first three curving outwards towards hind-margin, the remainder inclining inwardly, increasing in size to about middle of inner margin,—the eighth and ninth much larger than the rest; two rows of spots along hind-margin, the inner row of nine spots commencing with an elongate outwardly-curving mark on costa, the outer row on hind-marginal edge consisting of eight lunular spots. *Hind-wing:* a little before middle a transverse stripe continuous of the median row of spots in fore-wing, narrowing to beyond middle of inner margin, and divided very unequally into six by the crossing nervures; on costa touching, and half encircled by, outer edge of stripe, an ovate, bronzy, blue-ringed spot in a black ring; on inner margin, just before anal angle and almost touching stripe, a black inwardly densely blue-scaled spot, bounded outwardly by a very broad dark-red crescent; two rows of hind-marginal spots as in fore-wing, but larger and more lunular, the inner row of seven (of which the first and seventh touch the costal and inner-marginal ocelli respectively), the outer of six spots; all the black portion of wing, as far as inner row of marginal spots, thickly irrorated with yellow dots; dentation on first median nervule on hind-margin more produced than the rest (indicating the generic inclination to a *tail*). UNDER SIDE.—*Considerably paler than upper side.* *Fore-wing:* spots as on upper side; yellow streak above subcostal nervure extends to beyond middle; four yellow streaks radiating from base in cell (besides two very fine bounding lines along subcostal and median nervules respectively), and two below cell; no irrorations. *Hind-wing:* base widely yellowish, divided into three curved transverse stripes by the black-clouded costal and precostal nervures; stripe before middle as on upper side; ocelli similar; within discoidal cell, at extremity, a dull-golden inwardly bluish-scaled crescent, edging a semicircular black marking; inner row of hind-marginal spots (which are larger than on upper side) closely paralleled on their inner side by a row of five sublunular marks, coloured like the crescent in cell, and, like it, bounding each a blackish mark; the dark space between cell and these latter markings is thickly irrorated with dull-golden scales.

Head black, with palpi, a streak on each side from palpus to back of vertex, and edging round eyes, pale sulphur-yellow. *Thorax* above black, irrorated with sulphur-yellow, except pterygodes (which have, however, an internal yellow streak continuous of that on each side of the head); breast sulphur-yellow with two oblique black stripes on each side; legs black, more or less densely scaled with yellow except on tarsi. *Abdomen* black above irrorated with sulphur-yellow; laterally and beneath yellow, with a lateral and an inferior longitudinal black streak on each side.

♀ Like ♂, but slightly duller and paler. *Hind-wing*: first spot of submarginal row, bounding costal ocellus, more or less stained with dull-reddish on its inner side.

LARVA.—*Pale yellowish-green, marbled with purple or purplish-grey*, running in irregular transverse, and in places irregularly confluent, markings on the sides. Numerous pale-ferruginous, small, ocellate spots sprinkled about purple markings. A broad longitudinal white stripe above spiracles. Head and legs pale sandy-brown, as well as two small pointed tubercles on segment next head, from between which is protruded, when the animal is irritated, a crimson Y-shaped tentacle-like organ, emitting a very peculiar pungent odour. Two similar smaller tubercles on anal segment. A very sluggish larva, and very variable in the distribution of its colours. The young caterpillar differs strikingly from the full-grown one, being very dark, without green colouring, and clothed with short spines. Feeds on *Umbelliferæ*, *Bubon galbanum* and *gummiferum*, and in gardens on the fennel. Among trees it is common on the orange and lemon;[1] and Mrs. Barber noted that near Grahamstown it also fed on *Tepris lanceolata* (the "white ironwood") and *Hippobromus alata*; while in Natal Colonel Bowker found it on *Calodendron capense* (the "wild chestnut").

PUPA.—Elongate, rather slender anteriorly; head bluntly but deeply bifid, the projections irregularly dentate on the inner edge, and with a denticulated superior ridge inclining outward; thorax moderately angulated laterally, its dorsal projection rather acute, considerably elevated as to its anterior edge; abdomen widening from base to hind part of third segment, where it is very slightly angulated, and thence narrowing gradually to tail. On back of abdomen four rows of tubercles, of which those of the two middle rows (especially on fourth, fifth, and sixth segments) are larger and more prominent; also a solitary mesial tubercle on back of second segment, and a tuberculated ridge margining from base to widest parts. Surface generally rough, with here and there minute acute tubercule, anteriorly on under side.

Variable in colouring; usually ashy-grey or brownish-grey (when, as usual, attached to the old stems of its food-plants, which it closely resembles), but often much tinged with pale dull-sandy or ochreous-yellow, and more rarely with greenish. Prominences of head and peak

---

[1] I noticed at Highlands, near Grahamstown, that the larvæ feeding on the orange were all of a darker green than the umbellifer-feeders above described, and with the purple bands (though strongly and broadly marked) limited to the three thoracic and six succeeding segments. The head was almost ferruginous and the pro-legs pale greenish-grey. Mrs. Barber wrote to me in 1869 that she once, in a season of great drought, when most insects were unusually scarce, found a cuckoo in a very weakly state, with its crop full of *Demoleus* larvæ, and she considered that necessity had compelled the bird to swallow these distasteful creatures, which are in the habit of feeding fully exposed on flowers and leaves, but appear, as a rule, to be entirely unmolested by insectivorous animals.

of thorax, a dorsal median stripe from meta-thorax to fourth abdominal segment, an irregular streak at each side of back of same abdominal segments, some streaks roughly indicating main neuration of wings, and some longitudinal streaking on under side of head and leg-cases, all darker and in parts approaching to blackish.[1]

ABERRATION (♂).—In *fore-wing*, the disco-cellular spots and all the spots of discal series (except the small third one) enlarged and confluent into a broad band, the outer half of cell being completely filled with sulphur-yellow. In both wings the spots of the submarginal series are almost entirely effaced, and those of the hind-marginal edge exceedingly small; and in hind-wing the red of the anal-angular ocellus is replaced by black. On the under side the same peculiarities are reproduced, but the submarginal and hind-marginal spots are of nearer the normal size, though very obscured in tint.—Hab. Plettenberg Bay, Cape Colony. (Sent to me in 1880 by the late W. H. Newdigate.)

A ♀ presenting some approach towards the "sport" just described was bred near Cape Town from a fennel-eating larva in January 1869,

---

[1] Soon after receiving Mrs. Barber's account (*Trans. Ent. Soc. Lond.*, 1874) of the remarkable adaptation of colour in the pupa of *P. Lyaeus* (*Nireus*, auct., part) to its immediate surroundings, I endeavoured to ascertain if the chrysalis of *P. Demoleus* (which in nature I knew to be variable) was similarly susceptible. I found that this was the case, although to a less extent; but unfortunately I kept no notes of the experiment. Lately (in February 1888) I have repeated my endeavour, causing the larvæ to pupate in boxes respectively lined with paper of different colours, but covered with glass only, and fully exposed to daylight. The results were as follow, viz.:—

1. Pupa formed on *deep-green* paper: above pale sandy-ochreous with a reddish tinge; beneath paler, but with the wing-covers tinged with greenish-grey. All the customary darker streaks and markings almost obsolete.

2. Pupa formed on *brown* paper: not unlike the above, but all the back with a more reddish tinge; in one case with the front and sides of thoracic peak, the lateral thoracic angles, wing-covers, and the sides and under surface of abdomen, all varied with bright pale-green. [*N.B.*—Some of the food-plant was accidentally left in the box with this latter pupa.]

3. Pupa formed on *white* paper: paler than others, especially on back.

4. Pupa formed on *bright chrome-yellow* paper: above creamy yellow-ochreous; beneath dull greenish-grey, except abdomen, which was pale flesh-colour.

5. Pupa formed on *vermilion-red* paper: above bright greenish-yellow, with ordinary darker markings pale creamy-ferruginous; beneath wholly very pale greyish-green. Another example was very different in tint, not unlike pupa on bright chrome-yellow paper, but of a deeper and more rufous yellow above.

6. Pupa formed on *rich deep-blue* paper: above pale sandy-ochreous, beneath of almost the same colour; usual markings dull fuscous-grey.

A pupa formed in complete darkness nearly resembled those formed on the deep-green paper, but was paler and more cream-coloured above.

A larva left to choose its place of suspension in a large new breeding-cage selected a wooden bar across the arched top, and the resulting pupa much resembled the pale ochrey-yellow tint of the freshly-planed deal to which it was attached.

In these cases it will be noticed that although there was considerable variation in the colouring of these pupæ (most noticeable in those on the red and yellow papers), the tints for the most part by no means reproduced or even approached those of the surfaces immediately about the insects. The adaptability of the pupa to the hues of its ordinary natural surroundings is, however, very observable, though it appears to be but slightly and irregularly susceptible to colours beyond those narrow limits.

by Captain Sandford, R.E. In the left wings of this example the two disco-cellular spots of the fore-wing are enlarged and confluent, but the terminal streak in cell is almost obsolete, while the spots of the discal series, though with diffused edges and somewhat widened inwardly, are not enlarged, but smaller than usual. The submarginal spots are all but obsolete in the fore-wing and wholly wanting in the hind-wing,— and in the latter the transverse bar is widened and diffused inwardly. The right wings are normal, except for the absence of the small third spot in discal row of fore-wing, and of the first and second spots in submarginal row of hind-wing.

Apart from the two individuals here noticed, I have found *Demoleus* to be remarkably constant throughout its wide range as regards both pattern and colouring.[1] It exhibits occasionally, however, a variation in the transverse bar of the hind-wings, the second division of which is often more or less produced along the lower edge of the costal ocellus so as partly to enclose the latter, and in one ♂ (which I captured at Kimberley in 1872) does completely so enclose the ocellus by uniting with the first spot of the submarginal row immediately beyond that marking.

In size the butterfly seems more impressible by surrounding conditions, being noticeably smaller in dry upland interior districts than near the coast. Some specimens that I took on the wing near Kimberley are under $3\frac{1}{2}$ inches in expanse, and Colonel Bowker frequently met with dwarfed specimens in Basutoland (one ♀ that he sent expanding only 3 in. 1 lin., and a ♂ not more than 2 in. 7 lin.), though he informed me that others of various sizes up to the ordinary one were to be seen in that territory. The largest South-African examples I have noticed inhabit the Natal Coast (where the species is remarkably abundant); but the finest individual I ever measured was taken at Fernando Po by Lieut. Bourke, R.N., of H.M.S. *Druid*, in 1873; it expanded 4 in. 9 lin.[2]

The Indian *P. Erithonius*, Cram., is a very near ally of *Demoleus*; in the fore-wings it scarcely differs, except in the want of the fifth spot of the discal series; in the hind-wings, the transverse bar is much broader, and constantly very nearly completely encloses costal ocellus, and includes a small additional portion (sometimes, however, isolated) between the bases of second and third median nervules, and the dark-

---

[1] As Oberthür notes (*Etudes d'Ent.*, liv. iii. p. 14, 1878), specimens of *Demoleus* occur in which all the pale-yellow markings are stained with a dull ochrey-reddish tint. The same aberrant stamp of colour has been noticed in *Papilio Machaon*, and I believe in some other congeners with pale-yellow markings. The two or three *Demoleus* of this unusual tint that have come under my notice were all very worn and battered; and I have conjectured that possibly, under certain circumstances, the pale-yellow in aged specimens is changed into the dingy-reddish tinge in question.

[2] It is worth noting that in Mr. Bourke's collection (made at many points along the West African Coast) there were Fernando Po examples of three other species of *Papilio*, which were much larger than the specimens captured on the mainland, viz., *P. Nireus* and *P. Charopus*, both 4 in. 9 lin. across the fore-wings, and *P. Merope*, slightly over 5 in.

red of the anal-angular ocellus extends baseward, so as to occupy nearly the whole marking. On the under side the fore-wings present a subapical short series of dull golden-yellow spots between the discal and submarginal series; and in the hind-wings all the yellow markings are so much enlarged as to occupy nearly the whole ground of the wing except the discal series of dull golden-yellow blue- and black-edged markings, commencing with the costal ocellus.[1]

P. Demoleus is an active and powerful flyer, but does not attain the elevation exhibited on the wing by both *P. Nireus* and *P. Cenea*. It is fond of sporting about rocky knolls on the slopes or summits of hills. It occurs commonly over open ground generally, and is a frequent visitor to garden-flowers. On Table Mountain I have noticed that it seemed to favour *Agapanthus*, when that flower was in bloom; and at Plettenberg Bay it sometimes entered the woods to feast on the honey of *Plumbago*. The butterfly usually first appears in the Western districts about the end of September (but I have seen it as early as the 6th), and remains out until the beginning or middle of April. There appears to be a succession of broods during the hot months, as I have taken full-grown larvæ at different dates in November, December, and February, and obtained the perfect insects after a pupation varying from fourteen to twenty-seven days; but from larvæ full-grown in April the butterflies do not emerge until the succeeding September or October.

This butterfly has an exceedingly wide Ethiopian distribution, and is the only *Papilio* that has penetrated to the Cape peninsula. It is common in Madagascar, and has been recorded by Mr. Butler from the Comoro Islands; but it was not known in Mauritius when I visited that island in 1865. But in March and April 1870 it appeared in Port Louis in some numbers, and was in July of the same year successfully introduced (by means of five larvæ and three pupæ sent by M. Bouton to Dr. Vinson) into Réunion, becoming numerous there in 1871. (See *Annual Report of Royal Society of Arts and Sciences, Mauritius*, 1871, p. v.; and *Transactions* of the same body, N.S., vol. vi. p. 30, 1872). Mr. Bourke informed me that the butterfly abounded all along the coast of Tropical Western Africa.

Localities of *Papilio Demoleus*.

I. South Africa.
  B. Cape Colony.
    a. Western Districts.—Cape Town. Hout Bay and Simon's Town, Cape District. Stellenbosch. Paarl. Bain's Kloof. Michell's Pass, Tulbagh District. Robertson. Montagu. Swellendam (*L. Taats*). Oudtshoorn (— *Adams*). Knysna and Plettenberg Bay.

---

[1] *Erithonius* has a very wide range through India, the Malayan Islands, China, &c., and (as the very slight variation *Sthenelus*, Mach., cannot be regarded as a good species) Australia. Judging from the figures given in Moore's *Lepidoptera of Ceylon* (i. pl. 61), the larva of *Erithonius* is very like the darker larvæ of *Demoleus*, but has two distinct collars of small sub-ocellate pale-ferruginous spots on third thoracic and first abdominal segments, and the markings on the sides blacker and (except for some scattered small spots) restricted to abdominal segments 7-9. The pupa is represented as dull-green, with a greenish-yellow back.

  b. Eastern Districts.—Port Elizabeth. Uitenhage. Grahamstown and Mitford Park, Albany District. King William's Town. " Windvogelberg, Queenstown District."—W. D'Urban. Murraysburg (*J. J. Muskett*). Colesberg (*D. Arnot* and *A. F. Ortlepp*).
  c. Griqualand West.—Kimberley.
 D. Kaffraria Proper.—Butterworth and Bashee River (*J. H. Bowker*).
 E. Natal.
  a. Coast Districts.—D'Urban. Verulam. Mapumulo. " Lower Umkomazi."—J. H. Bowker.
  b. Upper Districts.—Maritzburg. Estcourt (*J. M. Hutchinson*).
 F. Zululand.—Etshowe (*A. M. Goodrich*). St. Lucia Bay (the late Colonel *H. Tower*).
 G. "Swaziland."—E. C. Buxton.
 K. Transvaal.—Potchefstroom and District (*T. Ayres*). Lydenburg (*A. F. Ortlepp*).
 L. Bechuanaland.—Motito (the late Rev. *J. Frédoux*).

II. Other African Regions.
 A. South Tropical
  a. Western Coast.—Damaraland (*J. A. Bell*). "Angola (*Monteiro*)."—Druce. Congo: "Kinsembo (*Ansell*)."—Butler. "Chinchoxo (*Falkenstein*)."—Dewitz.
  b. Eastern Coast.—Zambesi River (*Rev. H. Rowley*) and Shire River (*Rev. H. Waller*). "Zanguebar and Zanzibar (*Raffray*)."—Oberthür. "Endara (*Kersten*)."—Gerstäcker.
  b1. Eastern Interior.—"Kilima-njaro (*H. H. Johnston*)." Godman and Salvin.
  bb. Eastern Islands.—Madagascar: Tamatave (*Caldwell*) and Murundava (*Grevé*). "Mauritius."—Bouton. "Comoro Islands: Johanna (*Bewsher*)."—Butler.
 B. North Tropical.
  a. Western Coast.—"Gaboon (*Theorin*)."—Aurivillius. Fernando Po (*Bourke*). "Old Calabar."—Oberthür. "Lower Niger (*W. A. Forbes*)."—Godman. Sierra Leone (*A. R. Innes*). "Rio Nunez."—Oberthür. "Gambia (*Moloney*)."—G. E. Shelley. "Senegal."—Boisduval and Oberthür.
  b1. Eastern Interior.—Abyssinia: "Shoa (*Antinori*)."—Oberthür. "Nubia."—Felder.
 C. Extra-Tropical North-Africa.—"Egypt."—Felder and Gerstäcker.

309. (10.) Papilio ophidicephalus, Oberthür.

 Papilio Menestheus, Trim., Rhop. Afr. Aust., ii. p. 320, n. 214, pl. 2, f. 1 [♂], (1866).
Papilio ophidicephalus, Oberthür, Études d'Ent., iii. p. 13 (1878).
  "   " Trim., Trans. Ent. Soc. Lond., 1879, p. 345.

Exp. al., (♂) 4 in. 7 lin.—5 in. 3 lin.; (♀) 5 in. $3\frac{1}{2}$—7 lin.

 ♂ Tailed. *Black, with very pale sulphur-yellow bands and spots;* the ground thinly irrorated with pale-yellowish scales; *common to both wings,*—a transverse macular band, commencing broadly on costa of *fore-wing* near apex, narrowing and with its spots more separate as it approaches inner margin beyond middle, and thence crossing *hind-wing* in a continuous stripe, straight on its inner, dentate on its outer edge, rather before middle, from costa to inner margin,—a submarginal row of large sub-lunulate spots (eight in fore-wing, seven in hind-wing), and a hind-marginal row of seven lunules (larger in hind-wing) marking excavations between nervules. *Fore-wing:* in cell, near extremity, an oblique elongate marking, formed of two confluent spots,— and also four thin longitudinal streaks of pale sulphur-yellow scales from base (the lower three branching from a common origin) nearly to extremity; *on disc*, especially about median nervules and submedian nervure, *a clothing of short, cotton-like hair.* *Hind-wing:* two superiorly blue-edged and blue-scaled ocellate spots, placed quite as in *Demoleus* but more elongate,—that on costa *not* reddish itself, but the first lunule of submarginal row bounding it externally, like that on inner margin, dull-red; between the two spots more or less indistinct indications of similar spots, of which, however, only one (that next to inner-marginal spot) is distinctly blue-scaled; some reddish-yellow scaling along median nervules; *tail* long, spatulate, marked on each side near its extremity with an elongate, pale-yellow inwardly-convex spot. UNDER SIDE.—*Similar, paler, all the markings with more or less diffused edges;* all nervures more or less completely edged with pale-yellowish on both sides, *especially lower edge of costal nervure of hind-wing.* *Fore-wing:* a longitudinal streak between median and submedian nervures; streaks in cell better defined. *Hind-wing:* three streaks in cell like those of fore-wing; traces of ocelliform spots between the two ocelli much better marked, all more or less blue-scaled.

Head, thorax, legs, and abdomen coloured and marked quite as in *Demoleus*, Linn., but the superior pale sulphur-yellow stripes on head and pterygodes less conspicuous. The antennæ (as Felder has pointed out) are much more slender, especially as regards the club, which is scarcely recurved; but the agreement in other structural points (as well as in pattern and colouring) is close in the two species.

♀ *Like ♂, but decidedly duller, the ground more fuscous than black, and the yellow markings considerably deeper in tint.* *Fore-wing:* band broader, especially at costal commencement, the lower component spots larger, almost touching (in one example confluent into a continuous band); space between band and submarginal row of spots irrorated with pale-yellow; lowest spot of submarginal row tinged with dull-red. *Hind-wing:* sixth as well as seventh lunule of submarginal row dull-red, except at its extremities. Under side as in ♂.

This is the Southern representative of *Papilio Menestheus*, Drury, but is a much larger form, no example of either sex of the West-African butterfly that I have measured expanding more than $5\frac{1}{3}$ inches across the wings.[1] Apart from size, *Ophidicephalus* is best

---

[1] The form found in Madagascar (*P. Lormieri*, Distant) is nearer to the West-African type, having the spots of the fore-wing band all small and separate, the markings generally of small size, and the disc of hind-wings beyond middle scarcely irrorated. The band of the fore-wings is, however, straighter, as in *Ophidicephalus*, not incurved costally, as in *Menestheus*.

recognised by (1st) the more decided yellow, inclining to sulphureous, of the markings; (2d) the greater size of all the markings, *but especially the width of the transverse band of fore-wings near costa and the contiguity and outwardly-truncate form of its component spots;* (3d) the more conspicuous ocelli of the hind-wings and irroration of the disc between those markings; (4th) *the much longer and basally much broader tails.* In the ♂, the discal silky clothing is barely seen on the third median nervule of the fore-wings; and the disco-cellular oblique marking of the same wings in both sexes is not separated into two distinct spots. The dentation of the stripe of the hind-wings which borders the costal ocellus inferiorly is much more prolonged and acuminate.

Three male examples, which were taken by Mr. T. Ayres in the Lydenburg District of the Transvaal, are in some respects intermediate between the Southern and Tropical Western forms, though nearer to the former. In size, colouring, and development of hind-wing ocelli, and tails, they are quite like *Ophidicephalus;* but in the fore-wings the transverse band is as narrow as in *Menestheus* (except at its costal commencement, where it is somewhat broader), and its component spots are all separated from each other except the first three, though they preserve the outwardly truncate form characteristic of *Ophidicephalus*. In one specimen, moreover, the oblique marking of the discoidal cell in the fore-wings is divided into two parts, but the upper part remains much larger than the corresponding mark in *Menestheus*.

This very fine *Papilio*, the largest of the South-African butterflies, was discovered in Kaffraria Proper by Colonel Bowker, who forwarded specimens from the Trans-Kei territory as long ago as 1862, and has subsequently met with the species in the King William's Town and East London districts and in Natal. He notes it as common in its favourite haunts, which are deep wooded kloofs, where it follows a regular line of flight along the course of a stream, keeping usually about five or six feet from the ground. One of the localities where the species occurs in abundance is the Perie Bush near King William's Town, where Mr. Mansel Weale took it in March 1873. Colonel Bowker notes that it first makes its appearance at the end of September or beginning of October. Both he and Mr. A. D. Millar inform me that near D'Urban, the ♀ has been observed ovipositing on a species of *Zanthoxylon*, and the latter writes that the young larvæ are very similar to those of *P. Demoleus*. In Natal Colonel Bowker has personally observed the butterfly all along the coast from the Tugela to the Umkomazi, and inland at Karkloof and in woods above Maritzburg. Mr. W. D. Gooch has published a graphic account (*Entomologist*, 1880, pp. 228-229) of the difficulty of capturing this apparently easy prey on its course through the bush, the butterfly having a knack of evading the sweep of the net just at the critical moment. During my visit to Natal in 1867, I saw but one example, which was flying rapidly across open land on the road between D'Urban and Verulam.

It is not recorded how far to the eastward typical *Menestheus* extends in Africa, but it may possibly occur side by side with *Ophidicephalus*, which Oberthür records from Zanguebar, and which, Mr. Butler informs me, has lately (1887) been brought from Kilima-njaro.

Localities of *Papilio ophidicephalus*.

I. South Africa.
  B. Cape Colony.
    b. Eastern Districts.—King William's Town and Perie Bush (*J. H. Bowker* and *J. P. Mansel Weale*). "Gonubie River, Izeli, Buffalo River, and East London."—J. H. Bowker.
   D. Kaffraria Proper.—Bashee River (*J. H. Bowker*). "Baziya (*Baur* and *Hartmann*)."—Möschler. "Boolo Forest and Tsomo River."—J. H. Bowker.
  E. Natal.
    a. Coast Districts.—D'Urban (*M. J. M'Ken*). Pinetown (*J. H. Bowker*). Between D'Urban and Verulam. "Verulam."—W. D. Gooch. "From Tugela River to Umkomazi."—J. H. Bowker.
    b. Upper Districts.—Maritzburg (*Colonel Scott, R.A.*). "Ulundi, sources of Bushman's River."—J. M. Hutchinson.
  K. Transvaal.—Lydenburg District (*T. Ayres*)—Var.

II. Other African Regions.
  A. South Tropical.
    b. Eastern Coast.—"Zanguebar: Schumba Mountains (*Raffray*)."—Oberthür.
    b1. Eastern Interior.—"Kilima-njaro (*F. J. Jackson*)."—Butler.

310. (11.) Papilio Constantinus, Ward.

Papilio Constantinus, Ward, Ent. M. Mag., viii. p. 34 (1871); and African Lepid., pt. i. pl. i. ff. 1, 2 [♂], (1873).
    "   "   Oberth., Etudes d'Ent., iii. p. 12, pl. i. f. 1 [♀], (1878).

Exp. al., (♂) 3 in. 8–11 lin.

♂ *Brownish-black, with a very pale sulphur-yellow transverse stripe (macular in fore-wing, unbroken in hind-wing), and submarginal spots of the same colour; hind-wing with a short, broad, spatulate tail.* Fore-wing: basal area very finely irrorated with yellowish, extending through two-thirds of discoidal cell; transverse stripe consisting of eight moderate-sized separated spots, of different shapes, but more or less narrowed and pointed externally,—the uppermost spot (between fourth and fifth subcostal nervules) remarkably narrow, elongate, and acuminate, and shaped like an axe with its handle pointing outward,—the seventh rather deeply bifid externally, and only separated from the eighth by black submedian nervure; in discoidal cell, immediately adjacent to fourth spot of transverse stripe, an irregularly rounded or sub-ovate large pale sulphur-yellow spot; just above and beyond upper part of extremity of cell, an elongate mark of the same colour, externally deeply bifid, and crossed near its inner end by base of third subcostal nervule; submarginal spots small, but increasing in size downward, arranged in inter-nervular pairs from below fifth subcostal nervule to above submedian nervure,—one of the minute uppermost pair often wanting, and one or more of the lower pairs usually coalescent; sexual discal badges of silky appressed hairs on median nervules and submedian nervure well developed. *Hind-wing*: transverse stripe well before middle at its origin on costa, narrow, or rather narrow, running obliquely as far as submedian nervure, and terminating acuminately beyond middle; along inner margin, and also on external border of lower part of stripe, a little pale sulphur-yellow irroration; submarginal spots larger than in fore-wing (except uppermost pair), the upper spot in each of the five pairs considerably larger than the lower one; sometimes traces of a sixth pair of spots between first median nervule and submedian nervure; sometimes a small indistinct anal-angular pale sulphur-yellow spot; tail black, edged on each side of its wider part towards extremity by a narrow elongate pale sulphur-yellow spot tinged with ochre-yellow. *Common to both wings*, an inter-nervular series of pale sulphur-yellow lunular marks, more or less tinged with ochre-yellow, along all hind-marginal edge. UNDER SIDE.—*Hind-wing and apical area of fore-wing pale-brown, with blackish inter-nervular rays; yellowish markings all paler; submarginal spots enlarged, with diffused edges.* Fore-wing: basi-disco-cellular area yellowish-white, with four blackish longitudinal rays (of which the three lower branch from one stem); outer part of costal margin clouded with whitish as far as apex. *Hind-wing*: neuration black; *inter-nervular rays beyond transverse stripe broadly but irregularly clouded with brownish-black;* rays before stripe (viz., one between costal and subcostal nervures, and three with common stem in discoidal cell) blackish but unclouded; above costal nervure, from base to middle, a yellowish-white margin, edged near base first with black and then with chrome-yellow; small anal-angular spot, and marginal lunulate mark just beyond it, chrome-yellow; immediately beyond outer boundary of discoidal cell, between origins of first subcostal and second median nervules, a more or less indistinct whitish streak, interrupted on nervules.

Head black, with palpi, two frontal stripes, four spots on vertex, a spot at the back of each eye, and a lateral inferior half-ring round eye, pale sulphur-yellow; *antennæ conspicuously tipped with ochre-yellow.* *Thorax* black above, with a clothing of yellowish hairs, and two spots on prothorax; beneath yellowish-white, with a black stripe down the coxa of each leg. *Abdomen* black above, irrorated with yellowish; beneath and laterally, yellowish-white, with two longitudinal black stripes on each side.

I have not seen the ♀, but, from Oberthür's note and figure above cited, that sex does not appear to differ from the other except as regards the ground-colour, which is duller and browner; the somewhat broader transverse stripe; and the rather larger and much duller-coloured submarginal spots. The under side is depicted as generally darker.

This *Papilio* is not distantly related to *P. Thersander*, Fab.,[1] from West Africa, but differs markedly in the form and superiorly outward curve of the discal macular band in the fore-wings, in possessing a large disco-cellular spot in the same wings, and in the broader, shorter tails of the hind-wings; while on the under side the basal yellowish-white of the fore-wings and discal blackish clouding of the hind-wings are quite wanting in *Thersander*. The very singular elongate hatchet-shaped first spot of the fore-wing band in *Constantinus* is evidently formed by the extension and coalescence of two widely separate spots which lie between the corresponding nervules in *Thersander*.

This plainly-coloured but strikingly marked butterfly was first described from East-African examples, and has since been found to occur at several points on that side of the continent; but Herr Möschler has now recorded it as inhabiting also the remote Gold Coast. Mrs. Monteiro informed me that the species was not uncommon at Delagoa Bay, and had much the same habits as *P. Demoleus*. Stragglers have been met with in the Eastern Transvaal and upper districts of Natal; and in the latter country Mr. J. M. Hutchinson, who met with five specimens on the Bushman River, a few miles below Estcourt, in the year 1881, informs me that the butterfly was confined to tracts known as the "Thorns." He notes that its flight was comparatively weak, and that it frequently settled on low flowering plants, and was much more easily captured than *Demoleus*.

Localities of *Papilio Constantinus*.

I. South Africa.
  E. Natal.—Estcourt (*J. M. Hutchinson*).
  II. Delagoa Bay.—Lourenço Marques (*Mrs. Monteiro*).
  K. Transvaal.—Lydenburg District (*T. Ayres*).

II. Other African Regions.
  A. South Tropical.
    b. Eastern Coast.—"Bagamoyo."—Oberthür. Mombasa: "Ribé."—Ward.
    b1. Eastern Interior.—Lotsani River (*F. C. Selous*). "Kilimanjaro (*F. G. Jackson*)."—Butler.
  B. North Tropical.
    a. Western Coast.—Accra: "Aburi (*Weigle*)."—Möschler.

---

[1] I had for long supposed, with Mr. Kirby, that this butterfly would prove to be the ♀ of *P. Phorcas*, Cram. (= *Doreus*, Fab.); but Mr. Distant (*Proc. Zool. Soc. Lond.*, 1879, p. 648) states that two ♂ ♂ of *Thersander* had been received by Mr. Horniman from Aburie, near Accra, and that they differed from the ♀ in having all the macular markings pale-yellow instead of creamy-white. This discovery would leave the ♀ of the green-banded *Phorcas* still unknown; but I find that Herr Möschler (*Abhandl. Senckenberg. Naturf. Gesellsch.*, 1887, p. 51) mentions having received both sexes of *Phorcas* from Accra and Aburie, and although he quotes *Thersander*, Fab., as the ♀, his note on two (apparently aberrant) examples of that sex, in which the "*grüne Grundfarbe*" is more or less tinged or replaced by ochre-yellow, appears to indicate that the normal colouring of the central band is green in the ♀ as well as in the ♂.

311. (12.) **Papilio Euphranor**, Trimen.

♂ ♀ *Papilio Euphranor*, Trim., Trans. Ent. Soc. Lond., 1868, p. 70, pl. v. ff. 1, 2.

Exp. al., (♂) 3 in. 6–11 lin.; (♀) 3 in. 9 lin.—4 in. 2 lin.

♂ *Brownish-black, with a pale sulphur-yellow transverse stripe (macular in fore-wing); hind-wing only with a submarginal series of spots of the same colour, and with a rather broad, slightly spatulate, unspotted black tail.* Fore-wing: costa narrowly and finely irrorated with sulphur-yellow as far as extremity of discoidal cell, where the irrorations become so dense as to form a more or less distinct longitudinal streak; this streak, with a more conspicuous one immediately below it, forming the beginning of the macular transverse stripe of eight spots, gradually increasing in width to its termination a little beyond inner margin; this band is rather abruptly angulated on upper radial nervule, where its second spot closely approximates, or (usually) is confluent with the last of a short series of three smaller spots of the same colour (in line with the six other spots of discal stripe) from costa not far before apex; in cell, at lower edge of extremity, sometimes a small, ill-defined sulphur-yellow spot; on hind-marginal edge, below apical projection, a series of six inter-nervular lunular marks, gradually increasing in size downward. *Hind-wing*: transverse stripe before middle continuous (nearly straight along its inner edge, but outwardly irregular, and sharply denticulated with black on the crossing nervures), and extending to submedian nervure, along which it runs a narrow termination to beyond middle; adjoining middle part of stripe, but separated from it by a black bar at extremity of discoidal cell, a curved row of three pale sulphur-yellow spots, of which the uppermost is largest; a small sub-lunular orange spot a little before anal angle, sometimes touching extremity of transverse stripe; submarginal spots arranged in six inter-nervular pairs, the upper spot in each pair the larger, and the spots of the fourth (and usually also of the fifth) pair confluent; a series of thin, sulphur-yellow, inter-nervular, lunulate marks along hind-marginal edge. UNDER SIDE.—*General pattern like that of upper side; hind-wing and apical area of fore-wing chocolate-brown, clouded with pale ochreous-brown (which latter has in parts a faint violaceous lustre).* Fore-wing: upper spots of discal stripe and those of row from near apex represented by ill-defined yellowish-white marks; cellular spot larger and more distinct, and succeeded by a very small spot immediately beyond cell; other spots of discal stripe as on upper side, but much more acuminate externally. *Hind-wing*: basal area unclouded by paler brown, but discal hind-marginal area with two wide clouds, one apical, the other on median nervules; *transverse stripe almost white, much narrower than above, and about its middle greatly attenuated;* the adjacent spots, just beyond cell, smaller, almost white; submarginal series of spots obsolete, or but faintly indicated; orange anal-angular spot wanting, or represented by a few scales; in discoidal cell an indistinct bifurcate black streak.

Head black, with two well-marked frontal pale sulphur-yellow spots, two (less distinct) at bases of antennæ, and two posterior to these; inferior part of ring round eyes and palpi pale sulphur-yellow, the latter with a black ring. *Thorax* brownish-black above, with three pairs of very indistinct yellowish spots on prothorax and an almost imperceptible yellow streak on pterygodes; beneath chocolate-brown; legs black, the femora yellowish-white beneath. *Abdomen* brownish black above, pale-brown laterally, chocolate-brown beneath.

♀ *Duller; ground-colour not so dark, markings yellower, broader; in fore-wing a submarginal series of five additional yellow spots, and in hind-wing an inner row of yellow spots running parallel to the submarginal series.* Fore-wing: uppermost of the five additional rounded spots (which prolong the short apical series of three to just above submedian nervure) almost always confluent with third spot of transverse stripe. *Hind-wing*: inner row of spots not far before submarginal row, similarly arranged, but the spots in every pair usually confluent, —the last of the series forming a rather conspicuous orange anal-angular lunulate spot; hind-marginal lunule beyond this sometimes tinged with orange. UNDER SIDE.—Like that of ♂, but in fore-wing five additional spots well defined, and in hind-wing some of the inner row of spots tolerably distinct, including the orange lunulate spot.

In size and general appearance *P. Euphranor* is not unlike *P. Constantinus*, Ward, but may at once be distinguished by its differently-formed, wider, and yellower discal stripe; unspotted tails; chocolate-brown under side without black inter-nervular rays; brown under surface of body; and antennæ not tipped with yellow. Its true alliance, however, is not with *Constantinus* or the *Thersander* group, but with the large and singular West-African *P. Hesperus*, Westw., and *P. horribilis*, Butl., as will readily be recognised on comparing the colouring and pattern of the under side, the neuration and outline of the wings, and the form and colouring of the body, notwithstanding the much smaller size of *Euphranor* and its very different upper side pattern.

This fine *Papilio*, which appears to be peculiar to Southern Africa, was discovered by Colonel Bowker at the end of the year 1865, near the River Tsomo, a tributary of the Kei, in Kaffraria Proper. It was found numerously at the Boolo Forest about the end of November and again in February; its flight being described as "like that of *P. ophidicephalus*, but higher, and often extended from the forest to the open."[1] A living specimen of the pupa was forwarded to me, but the perfect insect emerged on the journey, so that only the pupal

---

[1] I am almost sure that it was this butterfly—if not, it was some very near ally—of which I saw several examples in a forest at Tunjumbili, on the Tugela frontier of Natal, early in March 1867. Their flight was limited to a small open space, across which they constantly passed at a considerable height, often settling on the lofty trees on either side.

skin reached me. Colonel Bowker describes its colouring as "bright green beneath and pale green with bright spots above." In shape it somewhat resembles the pupa of *P. Nireus* (L.), but is more attenuated anteriorly (with the cephalic prominences much shorter, and scarcely bent outwardly at all), and very much broader abdominally; the superior part of abdominal margin not angulated, but prominently convex. As far as I have been able to trace its distribution, *Euphranor* seems to be very local, and limited to high-lying forest tracts.

Localities of *Papilio Euphranor*.

I. South Africa.
    D. Kaffraria Proper.—Tsomo River (*J. H. Bowker*). "Baziya (*Baur and Hartmann*)."—Möschler.
    E. Natal.
        b. Upper Districts.—Karkloof (*J. H. Bowker* and *A. D. Millar*).
    K. Transvaal.—Lydenburg District (*T. Ayres*).

312. (13.) **Papilio Lyæus**, Doubleday.

♂ *Papilio Nireus*, Cram., Pap. Exot., iv. pl. ccclxxviii. ff. F, G (1782).
♂ *Papilio Nireus*, Swains., Zool. Illustr., 1st Ser., iii. pl. 125 (1822).
♂ ♀ *Papilio Nireus*, Boisd., Sp. Gen. Lep., i. p. 224, n. 42 (1836).
Papilio Lyæus, Doubl., "Ann. Nat. Hist., xvi. p. 178 (1845);" and Gen. D. Lep., p. 13, n. 98 (1846).
Papilio Nireus, G. R. Gray [part], Cat. Lep. Brit. Mus., i. p. 25, n. 119 (1852); and List Lep. Brit. Mus., i. p. 35. n. 126 (1856).
Papilio Lyæus, Wallengr., K. Sv. Vet.-Akad. Handl., 1857; Lep. Rhop. Caffr., p. 1.
♂ ♀ *Papilio Nireus*, Trim., Rhop. Afr. Aust., i. p. 15, n. 5 (1862).
Papilio Lyæus, Oberth., Etudes d'Ent., iii. p. 13 (1878), and iv. p. 54, n. 109 (1879).
Larva and Pupa (South-African), Mrs. Barber, Trans. Ent. Soc. Lond., 1874, pl. ix.

Exp. al., (♂) 3 in. 5 lin.—4 in. 2 lin.; (♀) 3 in. 9 lin.—4 in. 5 lin.

♂ *Deep-black, with a rather narrow metallic bright pale-greenish-blue median stripe common to both wings, and a submarginal series of spots of the same colour in hind-wing; cilia black, interrupted with white between nervules.* Fore-wing: median stripe variable in width, but never broad, and always narrowing inferiorly to inner margin,—its commencement close to costa, at extremity of discoidal cell, consisting of a small portion completely separated from the main stripe and divided into three unequal spots by subcostal nervure and upper disco-cellular nervule;—main stripe divided into six inwardly more or less hollowed, outwardly more or less projecting portions by crossing neuration,—the uppermost portion being in discoidal cell; close to apex two rather small spots, obliquely placed, of which the inner and inferior one is the larger; in some examples, close to hind-margin, lying between upper radial nervule and submedian nervure, a more or less incomplete series of very small inter-nervular spots (the lower ones divided by inter-nervular folds). *Hind-wing*: median stripe variable in width, running obliquely from costa about middle across outer third of cell to beyond first median nervule, a little before and above anal angle,—its inner edge almost straight (only slightly indented on nervures), its outer edge emitting more or less prominent nervular dentations; commencement of stripe, above first subcostal nervure, very much narrower than rest of it,—middle part of stripe extending beyond extremity of discoidal cell,—termination below first median nervule elongate, narrow, more or less acuminate inferiorly, not approaching near inner margin; submarginal series of inter-nervular spots lying between apex and submedian nervure, the spots below radial nervule in pairs (of which that between second and third median nervules is usually more or less enlarged), while those above it are almost always single; a small anal-angular white marginal lunule very near termination of median stripe. UNDER SIDE.—*Hind-wing and apical area of fore-wing dark-brown (sometimes with more or less of a reddish-ochreous tinge), nearly always clouded mesially with shining pale-greyish; neuration black; inter-nervular rays indistinctly blackish; in hind-wing a conspicuous sub-marginal, sub-metallic, very pale creamy stripe. Fore-wing*: dull brownish-black; brown of apical area clouded with pale-greyish costally, both a little beyond extremity of discoidal cell and a little before apex,—the two clouds, in the best-marked examples, indistinctly converging on radial nervules; hind-marginal series of very small spots usually better developed than on upper side, seldom altogether absent, but sub-metallic pale-creamy instead of greenish-blue. *Hind-wing*: submarginal stripe rather variable in width, and more perceptibly submacular when narrower; it occupies a corresponding position to that of the series of spots on upper side, and, like it, begins marginally at apex, and ends marginally between first and second median nervules; basi-costal space strongly tinged with reddish-ochreous above costal nervure; three longitudinal black streaks in discoidal cell from a common stem,—the upper one much shorter than the others; anal-angular marginal lunule rather larger than on upper side, and metallic-creamy instead of white; a larger spot of the same tint, variable in size and form, but usually sub-lunulate, at some distance before anal angle, above submedian nervure, close to inner margin; median shining-greyish fascia sometimes faint or partly evanescent, rarely almost wholly obsolete except for a pale scaling about bases of median nervules,—when best developed, extending slightly before middle from costa to inner margin, and externally somewhat diffused and radiating on median nervules (and rarely on subcostal nervule). *Head* black, with two white spots in front, two at bases of antennæ, and two (wider apart) behind these on vertex; palpi black with a white spot at base and a white tip; an outward and inferior half-ring round eyes white. *Thorax* black, with three successive pairs of well-separated small white spots dorsally on the fore part; breast with a frontal whitish tuft, and six or eight indistinct whitish spots on each side; legs black, the femora and tibiae outwardly whitish-scaled. *Abdomen* black above, dull greyish-brown on the sides and beneath, with two lower lateral series of minute white spots, sometimes scarcely visible.

♀ *Black duller, more sooty than in ♂; stripe and spots much duller and paler, more greenish, scarcely metallic.* Fore-wing: median stripe always widened (and sometimes very much wider) on inner margin, instead of being narrowed there, as in ♂,—its general width variable; often a third very small spot under the second sub-apical one, and sometimes a more or less incomplete submarginal series of similar small spots divided by inter-nervular folds; hind-marginal series of minute spots often mixed with whitish and touching white interruptions of cilia. *Hind-wing:* beginning of median stripe little if at all narrower than the rest, and near costa dull-greyish; its edges somewhat diffused, and its lower extremity narrowly prolonged to a point just above anal-angular white lunule; an inner-marginal spot as on under side of ♂, but smaller and greenish, and sometimes touching median stripe; submarginal spots larger, and so shaped that each pair below second sub-costal nervule forms a strongly-curved crescent with the points more or less prolonged. UNDER SIDE.—*Hind-wing and apical hind-marginal area of fore-wing of a paler, more rufous brown than in ♂, with the shining pale-greyish fascia and clouding always strongly developed, and in parts quite whitish;* inter-nervular hind-marginal small paired whitish spots better developed in fore-wing, and also present (and larger) in hind-wing, except (usually) between radial and second median nervules. *Fore-wing:* costal pale clouding broadly developed, almost silvery, completed, confluent on radial nervules (so as to isolate a small brown costal space), and thence extending downward in a much-narrowed submarginal ray, becoming macular and whitish below third median nervule. *Hind-wing:* basal reddish-ochreous much more developed, extending over all the area as far as median shining-greyish fascia; submarginal stripe very different from that of ♂, being expanded into a wide shining-grey band, bright and silvery superiorly, but becoming more or less indistinct and diffused between third and first median nervules; a little before this band, and at points touching it, a more or less connected series of shining-greyish lunular marks, often indistinct.

White spots of head and body more apparent than in ♂, especially the abdominal lateral spots.

M. C. Oberthür notes two aberrant ♂s from Shoa, Abyssinia, and another from Tchouacka in Zanguebar, in which the upper-side stripe was almost entirely obliterated in the fore-wings, and sensibly reduced in the hind-wings. His doubtful reference of this aberration to *P. pseudonireus*, Feld. (founded on a single ♂ from Bogos), seems borne out by Felder's description [1] of his type, in which, however, the band

[1] *Reise der Novara, Zool., Lep.,* i. p. 94 (1864).

is described as less reduced than it appears to be in M. Oberthür's specimens.

An aberrant ♀, taken by Mrs. F. W. Barber near Grahamstown, shows just the opposite tendency, the stripe in the fore-wing being fully twice as broad as usual.

LARVA.—Deep yellowish-green. Widest portion of back, on third thoracic and first abdominal segments, occupied by a subovate patch or shield of pale bluish-green, crossed mesially by a thin whitish line, closely irrorated generally with white dots, and bearing near its posterior edge a transverse row of four small pinkish-lilac spots. Anteriorly a greenish-yellow, posteriorly a narrow ochreous-yellow, edging borders this patch, the anterior edging being marked with four double (concentric) very small, thin, blackish rings, and at each extremity by a small black yellow-pupilled ocellus, surrounded by a separate, thin, imperfect blackish ring. All the green of the back posterior to the patch is mottled irregularly with greenish-yellow. From second to last abdominal segment a lower-lateral white stripe; on each side of back more or less distinct traces of a thin yellowish stripe, which in some examples is crossed obliquely by a short whitish inferiorly dark-bordered streak. Projections on back of first thoracic segment yellow, shorter and blunter than in *P. Demoleus*, but those on back of anal segment pale-yellow, longer, farther apart, and with a straight creamy-white connecting streak between their bases. Head and thoracic legs pale bluish-green; pro-legs very pale-greyish with a greenish tinge.

Described from larvæ found on the Orange at Highlands, near Grahamstown. Mrs. Barber informed me that the native food-trees in that locality were *Vepris lanceolata* and *Calodendron capense*. The dorsal green (excepting the ovate patch of a blue tint) assimilates very closely to that of the upper surface of the leaves of the orange, upon which the caterpillar is always found resting; but Mrs. Barber (*Trans. Ent. Soc. Lond.*, 1874, p. 519) has pointed out that when the larva feeds on *Vepris* it is of a lighter green, so as to resemble in tint the leaves of that tree.

PLATE II. fig. 5, 5.

PUPA.—Length, 1 in. 3 lin. Somewhat attenuated anteriorly, cephalic processes short and directed laterally outward (not obliquely forward, as in *Demoleus*), so that frontal line of head is widened and but slightly concave. Thoracic lateral angles moderately acute; dorsal prominence also elevated acutely, but not inclined forward. Sides of abdomen widely flattened, and so extended as to form a very marked angle on each side of third abdominal segment; whence the abdomen narrows very rapidly and greatly to the extremity. Infra-pectoral region, where wing-covers meet, very strongly convex. A marked constriction dorsally at junction of thorax and abdomen. In its natural position, attached vertically or nearly so, head uppermost, the anterior portion of this pupa is seen to be very much more bent backward than it is in *Demoleus*.

Usually pale bluish-green, inclining to yellowish on under surface of abdomen, but very variable in tint. Point of dorso-thoracic prominence, two spots below it (at abdominal base), and edge of abdominal lateral angles, creamy-reddish. A row of minute indistinct blackish spots on each side of back of abdomen.

This chrysalis, though usually pale-green, is variable in colouring, specimens that I reared near Grahamstown pupating in the same wooden box (on the sides) within a day or two of each other, varying from that tint to a more or less ochreous-tinged, much duller hue. Mrs. Barber was subsequently so fortunate as to observe the extreme susceptibility of this pupa to colour influences, as pointed out in her paper above quoted. From these most interesting observations it appears that the colour of the object on which a larva pupated was very closely reproduced in the pupa. Pupæ among orange twigs were of the ordinary green colour; others, among half-dried leaves of the "bottle-brush," were pale yellowish-green; one attached to the wooden frame of the case was of the yellowish tint of the wood; and another, attached to a part of the frame where wood and purplish-brown brick joined each other, was coloured on the under surface like the wood and on the upper surface like the brick. The experiment of causing a larva to pupate on scarlet cloth had no effect except that the ordinary small red spots were brighter than usual; but this is not to be wondered at, considering that the environment of these insects could never, through endless generations in the past, have rendered the assumption of a scarlet colour of any advantage in concealment.[1]

Colonel Bowker, in 1874, sent me from King William's Town four pupæ, three of the ordinary green colour, and the fourth (which had been purposely placed when changing on the mud-mortar of a wall) of a dull greenish-yellow, much clouded dorsally with dull creamy ferruginous-grey. These were winter pupæ; two became perfect insects in July (1st and 24th), one on October 1st, and the last on December 21st. I set three of the butterflies loose in the Museum

[1] A most remarkable instance of pseudo-mimicry recently came to my notice in connection with the pupa of this *Papilio*. In September 1887 I received from the Rev. N. Meeser, of George, Cape Colony, a small box containing what I took at the first glance for three ordinary green chrysalides of *P. Lyæus*. Only one of these objects, however, was a veritable chrysalis, the two others being the seed-capsules of a plant, stated by Mr. Meeser to be a species of *Hakea*. The tint of green, the general lateral outline (especially the bulging ventral convexity of the wing-covers), the projections of the bifid head, the attenuated form of the posterior abdomen and anal extremity, and even the slight ferruginous tips of the projections on the head, are all reproduced in the seed-capsules to a very deceptive extent. The chrysalis was found "in the neighbourhood of a hedge of the *Hakea*;" and if this plant had been a native of South Africa, it can scarcely be questioned that a strong case of mimicry would readily have been admitted by observers. As a recent introduction from Australia, however, it is clear that the *Hakea* cannot have been the model for the pupa of a *Papilio* of a specially African group.

enclosure, but they were promptly seized by the ever-watchful butcher-bird, *Lanius collaris*.

It has been with considerable hesitation that I have separated this *Papilio*, under Doubleday's name of *Lyæus*, from the well-known *Nireus*, Linn., notwithstanding that Doubleday has been followed in this by several authors. There can be no doubt that Linné's original description of *Nireus* applies to the well-known butterfly from the coast of Western Africa figured by Clerck and Drury, and by Cramer on his Plate 187, ff. A, B.[1] The form figured by the latter as the ♂ of the *Nireus* depicted in the plate named—which is itself a ♂—will be found on Plate 378, ff. F, G, and is a rather small ♂ from South Africa, exhibiting very clearly the differences more or less distinctly presented in specimens inhabiting that part of the continent. These differences are : 1°, shorter wings, the fore-wings not so pointed apically, the hind-wings not so produced inferiorly (and with shorter and rounder dentations); 2°, the common stripe decidedly bluer, usually rather broader in fore-wing, and considerably narrower and shorter at its termination in hind-wing (where in *Nireus* the *portion between second and first median nervules is much enlarged and elongated, and extends along upper side of first median nervule considerably beyond the portion below that nervule*[2]—just the reverse of what is seen in *Lyæus*); 3°, the submarginal spots in hind-wing usually larger, more elongate, and more numerous; and, on the under side of the ♂, 4°, the paler hind-wing and apical area of fore-wing, and the presence in both of some broad shining-greyish clouding (wholly wanting in *Nireus*); and 5°, the wider and continuous shining-creamy submarginal stripe (which in *Nireus* is throughout broken into spots both by nervules and inter-nervular folds).

The upper-side differences are very constant, and so also is the under-side one of the continuous hind-wing stripe; but specimens of the ♂ occur in Natal, Zululand, Transvaal, and especially Delagoa Bay, in which the shining-greyish clouding of the under side is more or less reduced, and sometimes almost obsolete.

M. C. Oberthür (*Ann. Mus. Civ. di Genova*, xv. p. 147, 1880), while distinguishing between *Nireus* and *Lyæus*, and rightly pointing out the application to them respectively of Cramer's different figures, still supposes (as he had previously intimated in his *Études d'Entomologie*, liv. iii. p. 13, 1878) that *Nireus* (Cramer's Pl. 187, A, B) is the form spread through all the southern parts of Africa, and that *Lyæus* (Cramer's Pl. 378 F, G) is peculiar to countries north of the Equator; but, as I have abundantly shown, it is the latter that inhabits all South Africa proper, and as M. Oberthür records that all the Abyssinian

---

[1] The late Mr. G. R. Gray evidently took the opposite view; for in both his British Museum Catalogues of Papilionidæ (1852 and 1856) he gave the Linnean *Nireus* as the Southern form, and proposed the new name of *Erinus* for the West-Coast one.

[2] This character is also prominent in the closely-allied *P. Bromius*, Doubl., from West Africa, distinguished from *Nireus* by its much wider bluish-green stripes.

specimens taken by both Raffray and Antinori were of the *Lyæus* form, we may conclude that this is emphatically the form of the East and South, as *Nireus* is of the West.

This very beautiful species is common and widely spread over Eastern South Africa, wherever native woods extend; its southern and western limit appearing to be Knysna in Cape Colony, where I found it abundantly. It has a rapid and powerful flight, often soaring over the highest forest trees, but its frequent visits to flowers render it tolerably easy of capture. In the woods *Plumbago capensis* seemed its principal attraction, and in gardens the flowers of the orange and the common and Indian periwinkles. At D'Urban, Natal, where it attains its largest size and is very numerous, I often took it at rest on leaves of trees at the edges of woods. There appear to be a succession of broods from September to May, and I have records of various stragglers taken during the three winter months, but the butterfly is at this season seldom met with. The paired sexes were captured by Colonel Bowker at King William's Town in November 1876, and at D'Urban in January 1888.

Localities of *Papilio Lyæus*.

I. South Africa.
  B. Cape Colony.
    a. Western Districts.—Knysna and Plettenberg Bay.
    b. Eastern Districts.—Uitenhage. Grahamstown, Highlands, and Mitford Park, Albany District. King William's Town. East London (*P. Borcherds*). Windvogelberg, Queenstown District (*G. E. Bulger*).
  D. Kaffraria Proper.—Bashee River (*J. H. Bowker*).
  E. Natal.
    a. Coast Districts.—D'Urban, Mapumulo. "Lower Umkomazi." —J. H. Bowker.
    b. Upper Districts.—Umqueqa Falls. Greytown. Maritzburg (*Miss Colenso* and *A. Windham*).
  F. Zululand.—Etshowe (*A. M. Goodrich* and *T. Vachell*).
  G. "Swaziland."—The late *E. C. Buxton*.
  H. Delagoa Bay.—Lourenço Marques (*Mrs. Monteiro*).
  K. Transvaal.—Lydenburg District (*T. Ayres*).

II. Other African Regions.
  A. South Tropical.
    b. Eastern Coast.—River Shiré (*Rev. H. Waller*). ? "Tchouacka, Zanguebar (*Raffray*)."—Oberthür—["Var. *Nireus*."]
  B. North Tropical.
    b1. Eastern Interior.—Abyssinia, "Lake Tsana (*Raffray*)," and "Shoa (*Antinori*)."—Oberthür.

313. (14.) **Papilio Cenea**, Stoll.[1]

♂ *Papilio Brutus*, Godt. (pars), Encyc. Méth., ix. p. 69, n. 122 (1819).
  „ „ Donov., Nat. Repos., iii. pl. 77 (1825).
  „ „ Boisd. (pars), Faune Entom. de Madag., &c., p. 12 (1833).
  „ „ Boisd., Var. A., Spec. Gen. Lep., p. 221, n. 39 (1836).
  „ „ Chenu, Encyc. d'Hist. Nat., Pap., pl. 2, f. 1 (?1852).
Papilio Merope, Doubl. and Westw. (pars), Gen. Diurn. Lep., i. p. 13, n. 92 (1846).

---

[1] See some remarks on this species and its allies in vol. i. pp. 38, 39.

Papilio Merope, G. R. Gray (pars), Cat. Lep. Brit. Mus., Pap., p. 25 (1852).
,, ,, Trimen (pars), Rhop. Afr. Aust., i. p. 11 (1862).
,, ,, Trimen, in Trans. Linn. Soc., xxvi, tab. 43, f. 1 (1869).
,, ,, J. P. Mansel Weale, Trans. Ent. Soc. Lond., pl. i. f. 3 (1874).

♀ (Form 1). *Papilio Cenea*, Stoll, Suppl. Cramer, Pap. Exot., p. 134, pl. xxix. f. 1, 1A (1791).
Danais Rechila, Godt., Encyc. Meth., ix. p. 183, n. 24 (1819).
? *Papilio Trophonius*, Westw. (pars), Arc. Entom., i. p. 153 (1845).
Papilio Cenea, Doubleday and Westw. (pars), Gen. Diurn. Lep., i. p. 20, n. 255 (1846).
,, ,, G. R. Gray (pars), *op. cit.*, p. 70, n. 322 (1852).
,, ,, Trimen (♂), Rhop. Afr. Aust., i. p. 20, n. 8 (1862).
Papilio Merope, Trimen (first form of ♀), in Trans. Linn. Soc., *loc. cit.*, f. 3 (1869).
,, ,, Butler (♀, Form a), in Trans. Ent. Soc., 1869, p. 275.
,, ,, Kirby (♀), Synon. Cat. Diurn. Lep., p. 563, n. 305 (1871).
Variety.—*Papilio Merope*, Trimen (1st Form of ♀ var.), in Trans. Linn. Soc., *loc. cit.*, f. 4, and p. 521.
Papilio Merope, Butler (♀, Form aa), in Trans. Ent. Soc., *loc. cit.*, p. 276.

♀ (Form 2). *Papilio Merope*, Trimen (2nd Form of ♀, *Hippocoon*, Fab., var.), in Trans. Linn. Soc., *loc. cit.*, f. 6 (1869).

♀ (Form 3). *Papilio Trophonius*, Westw., "Ann. Nat. Hist., ix. p. 38 (1842)," and Arcan. Entom., i. pl. 39, ff. 1, 2 (1845).
Papilio Cenea, Doubl. and Westw. (♀), Gen. Diurn. Lep., i. p. 20, n. 255 (1846).
,, ,, G. R. Gray (♀), Cat. Lep. Brit. Mus., Pap., p. 70, n. 322 (1852); and List Lep. Brit. Mus., Pap., p. 82, n. 339 (1856).
,, ,, Trimen (♀), Rhop. Afr. Aust., i. p. 20, n. 8 (1862).
Papilio Merope, Trimen (4th Form of ♀), in Trans. Linn. Soc., *loc. cit.*, f. 5.
,, ,, Butler (♀, Form b = "Merope, true"), in Trans. Ent. Soc., *loc. cit.*, p. 276.
,, ,, Kirby (♀, Var. b), *op. cit.*, p. 563, n. 305.

♂ and ♀ (Form 2), *Papilio Tibullus*, or *Merope*, Cram., var. ?, Kirby, Proc. R. Dublin Soc., 1880, p. 48.
♂ and ♀ (Form 1), *Papilio Cenea*, Trimen, Trans. Ent. Soc. Lond., 1881, p. 169, pl. ix. ff. 1, 2.
Larva and Pupa (*South African*), J. P. Mansel Weale, Trans. Ent. Soc. Lond., 1874, pp. 133–134, pl. i.

Exp. al., (♂) 3 in. 7 lin.—4 in. 3 lin.; (♀) 3 in. 7 lin.—4 in. 3 lin.[1]

♂ Very pale creamy sulphur-yellow; fore-wing with a broad hind-marginal black border, hind-wing with an irregular (often broken) discal black band, and black hind-marginal lunulate marks; a rather broad,

---

[1] Two dwarfed examples, a ♂ and ordinary form of ♀, taken at D'Urban, Natal, by Colonel Bowker in August 1878, expand only 3 in. 3 lin.; and another ♀ from the same district is only 3 in. 4 lin. across the wings.

spatulate tail of moderate length on third median nervule. Fore-wing: costa with a black border, rather wide from base to a little before middle, thence much narrower till it meets hind-marginal border; the latter broad apically, and enclosing an elongate sub-ovate spot of the ground-colour between fourth and fifth subcostal nervules, but narrowing very considerably to posterior angle,—its inner edge usually more or less irregular, and only slightly and unequally dentating the ground-colour on nervules, but in some of the more strongly-marked specimens emitting deep and almost regular nervular dentations; along its outer edge on hind-margin itself, a series of very small thin inter-nervular spots of the ground-colour—rarely obsolescent; middle disco-cellular nervule occasionally marked by a very small black dot or a black line. *Hind-wing: discal black band exceedingly variable as regards width and continuity,*—when completely continuous, it is always narrower between second and third median nervules (where the wider and more complete interval occurs in the interrupted state; the other interval, less frequent than the first, being between second subcostal and radial nervules); inner edge of this band variably uneven, slightly diffused, sometimes denticulating ground-colour on nervules; but not unfrequently more or less pierced itself by small nervular dentations of the ground-colour; outer edge always presenting large and prominent nervular dentations; along hind-margin a series of narrow more or less lunulate black markings, which in the most strongly-expressed specimens form a thoroughly continuous serpentine streak (touching on most nervules the outer dentations of discal band, and leaving inter-nervular hind-marginal lunules of ground-colour), but in others are partly or altogether and widely separated; tail with a median black line of variable definition, and almost always with a thin black edging on each side,—in strongly marked specimens these streaks are more or less diffused and confluent, making more than basal half of the tail black or blackish; along hind-marginal edge (including tail) a more or less decided tinge of ochre-yellow; just before anal angle, in discal black band, a small lunate spot of the ground-colour, sometimes very indistinct. UNDER SIDE.— *Hind-wing and borders of fore-wing dull ochreous-yellow,—the former paler or much paler in basal half, and with discal band represented by ferruginous-brown.* Fore-wing: costal border more or less blackish inwardly as far as middle, especially at and near base; from lower radial nervule, the inner side of hind-marginal border is blackish and this dark edging increases in width downward till at posterior angle it occupies all the border· inter-nervular folds sometimes defined by brown rays crossing yellow-ochreous part of border. *Hind-wing:* throughout marked with more or less distinct inter-nervular brown rays (in some of the more strongly marked specimens becoming blackish and often diffused where crossing discal band); three similar but usually thinner and darker longitudinal rays in discoidal cell, divergent from base,—the upper and middle ones having a common stem; discal band irregularly varied with very pale grey, most prominent between second subcostal and radial nervules (where on upper side occurs usually the wider of the two gaps often met with in discal band); hind-marginal lunulate markings brown; neuration in the more strongly marked examples dark-brown or blackish except near base.

Head and body above black, thorax and abdomen beneath and laterally ochreous-yellow. *Head* with two large frontal spots, six superior spots, half-ring round eyes, and palpi pale sulphur-yellow. *Thorax* above with four pale sulphur-yellow spots,—two on prothorax and two (larger) on bases of pterygodes; legs black, femora pale sulphur-yellow in front. *Abdomen* with two rows of black spots on each side (the upper row marking the line of junction of the dorsal blackish and the ventral yellow), and with two black longitudinal streaks beneath.

♀ (*Cenea*, Stoll).[1] *Sooty-black, with ochreous-yellow and white (or all white) spots in both wings, and a broad ochreous-yellow pre-median band in hind-wings. Fore-wing*: in discoidal cell, towards its extremity, an oblique, narrow marking, usually with a more or less marked hook at each end, crossing about half the width of cell, white, often tinged inferiorly with ochreous-yellow; beyond cell, a discal curved row of four spots,—three rather small (third longer than first and second) between third subcostal and third median nervules,—the fourth, much larger, elongate-ovate, between first and second median nervules (sometimes diffusedly extending below first median nervule); of these spots, the first is always white, the second and third usually ochreous-yellow, but not rarely whitish or white, and the fourth nearly always ochreous-yellow or pale ochreous-yellow, but sometimes white; a subapical spot as in ♂, but white (very rarely wanting); a submarginal row of three (rarely two) smaller, rounded spots between third median nervule and submedian nervure, usually more or less tinged with ochreous-yellow, but in some specimens quite white; in some of the white-spotted examples, an indistinct narrow ochreous-yellow mark on inner margin before middle. *Hind-wing*: ochreous-yellow band somewhat paler and broader in specimens having the spots of the fore-wing white, a little diffused on both inner and outer edges,—the former being not very far from base, and the latter a little beyond middle; a submarginal series of twelve small spots arranged in inter-nervular pairs, of which ordinarily only the uppermost two are white, while the rest are ochreous-yellow, but in some examples (and always in those which have white-spotted fore-wing) all are white. UNDER SIDE: *Ground-colour of fore-wing black, and the pale markings in both wings answering to those of upper side; but ground-colour of hind-wing and apical hind-marginal area of fore-wing lighter or deeper ochreous-brown, with inter-nervular rays agreeing with those*

---

[1] Mimics *Amauris Echeria*, (Stoll).

of ♂. *Fore-wing*: disco-cellular streak and all the spots (except the largest discal one) pure white, larger than on upper side, their edges being diffused; costa narrowly edged with ochreous-brown. *Hind-wing*: inner edge of pale ochreous-yellow-band ill-defined, the basal dark-brown or ochreous-brown being diffused,—outer edge rather nearer middle, so that the band is narrower than on upper side; three streaks in cell as in ♂; inter-nervular streaks usually rather faintly marked; greyish mark between second subcostal and radial nervules (which in ♂ answers to ordinary wider break in band of upper side) almost always present (and in the white-spotted examples often conspicuous), but very faint in two specimens, and wanting in two others; submarginal spots rather larger than on upper side, all white, and besides the full number of six pairs (which seems not to vary) there is a smaller seventh pair close to anal angle.

Coloration and markings of head and body like those of ♂, except that the thorax beneath, instead of being uniformly clothed with ochreous-yellow, is black, varied with eight large whitish spots on each side, interspersed with a few orange-ochreous markings; and that the abdomen is on the sides of a deeper and duller ochreous-yellow.

♀ Second Form (near *Hippocoon*, Fab.).[1] *Sooty-black, with all the markings pure-white, and the chief ones in both wings greatly enlarged.* *Fore-wing*: disco-cellular oblique streak and subapical spot as in ordinary ♀,—the spot sometimes wanting; instead of the small widely-separated second and third spots of discal row, a broad oblique continuous subapical bar, with which superiorly the first spot of discal row is sometimes incorporated; instead of the isolated fourth spot a broad patch occupying nearly all inner margin, not entering discoidal cell, but rounded superiorly and rising as far as second median nervule. *Hind-wing*: band so much widened as to constitute a very large central patch occupying all the area except some very narrow basal black and a broad hind-marginal black border; externally the white patch is deeply pierced by black nervules and also (less conspicuously) by inter-nervular blackish rays; the remainder of the inter-nervular rays so marked on the *under side* indistinctly visible. UNDER SIDE.—*Borders of wings sometimes of the same ochreous-brown as in ordinary ♀, but usually paler, and always narrower owing to the broad white markings.* *Hind-wing*: basal brown always very much narrower, and often almost obsolete,—the white extending almost to base itself; externally the white does not extend quite so far as on upper side, except between second subcostal and radial nervules, where it is prolonged into the scarcely distinguishable very pale-greyish mark; neuration beyond middle black or blackish; inter-nervular rays and streaks varying from ochreous-brown to black.

Coloration and markings of head and body as in ordinary ♀, except that the abdomen has the sides much paler and is almost white beneath.

---

[1] Mimics *Amauris dominicanus*, Trim.

♀ Third Form (*Trophonius*, Westw.).[1] *Pattern of Second Form, but large patch in both wings wider, and pale brick-red instead of white. Fore-wing*: disco-cellular oblique white streak usually longer and wider; small subapical white spot rarely well expressed, sometimes very minute, often altogether wanting; subapical oblique white bar as in Second Form, but sometimes with a creamy-reddish tinge; inner-marginal brick-red patch usually extending close to base and over median nervure into discoidal cell. *Hind-wing*: patch usually extending to base itself (but rarely with a very narrow blackish suffusion there), and externally nearer to hind-margin,—its outer edge not so conspicuously marked with dark neuration and inter-nervular rays; a few of the submarginal spots sometimes tinged with creamy-reddish. UNDER SIDE.—*Borders of wings as in Second Form; in other respects the pattern and coloration of upper side is reproduced, with the brick-red rather paler and duller; dark neuration and inter-nervular streaks not strongly marked.*

Coloration and markings of head and body as in ordinary ♀.

In addition to the three prominent and very distinct forms above described, the ♀ *P. Cenea* presents numerous intermediate and other variations, of which the following are the principal instances that have come under my notice, viz.:—

  A. Between Forms 1 (*Cenea*) and 2 (analogue of *Hippocoon* Fab.).
    a. P. Merope, Butler (♀, *P. Cenea*, var.), in *Trans. Ent. Soc., loc. cit.*, p. 275.
      This individual is very close to the typical *Cenea*, but in the shape and position of the very restricted patch in the hind-wings resembles the individual (*c*) following hereunder.—*Hab.* Grahamstown, Cape Colony.
    b. P. Cenea, Trim. (♀, variation), *Trans. Ent. Soc. Lond.*, 1874, p. 150, note.
      All the upper-side markings in this specimen are white, and though answering to those of the *Hippocoon*-like form, are so reduced and attenuated as (with the single exception of the very much narrowed and dentated subapical bar of the fore-wings) more to resemble those of the white-spotted variety of *Cenea*.—*Hab.* Knysna, Cape Colony.
    c. P. Merope, Trim. (♀, variation), in *Trans. Linn. Soc., loc. cit.*, f. 2.
      All the markings in this individual are dull white. The fore-wings have the subapical bar of the *Hippocoon*-like form, and an inner-marginal patch strictly intermediate in size and shape between those of the latter form and of the *Cenea* form respectively. The patch of the hind-wings is much narrowed by a fuscous basal suffusion.—*Hab.* Tsomo River, Kaffraria (*J. H. Bowker*).
    d. P. Cenea, Trim. (♀, variation) [= *P. Merope*, peculiar *Hippocoon* form, J. P. Mansel Weale], *Trans. Ent. Soc. Lond.*, 1874, p. 144.
      In the fore-wings both the subapical white bar and the inner-marginal white patch are considerably smaller and narrower than in the ordinary Southern *Hippocoon*, the latter marking

---

[1] Mimics *Danais Chrysippus*, (Linn.).

being interiorly clouded with blackish. It most nearly resembles the variation figured in the second plate accompanying my paper in the *Linnean Society's Transactions* (vol. xxvi. tab. 43, f. 2), and, like that example, wants the apical spot of the fore-wings; but (as far as I can make out in its very damaged state) it has more resemblance to *Hippocoon* in the wider white space of the hind-wings.—*Hab.* King William's Town, Cape Colony.

    e. Two specimens near the simple variation of ordinary ♀, but having all the white spots of the fore-wing considerably enlarged, as well as the ochreous-yellow mark on inner margin before middle; while the ochreous-yellow band of the hind-wing is increased to a patch as large as in the *Hippocoon*-like Second Form.—*Hab.* Delagoa Bay (*Mrs. Monteiro*).

    f. An example like ordinary ♀ in all respects, except that in fore-wing it has a continuous white subapical bar as in Second Form, and that in hind-wing the ochre-yellow band (though scarcely extending beyond middle) reaches almost to base itself.—*Hab.* King William's Town (*J. H. Bowker*).

    g. An individual like the last (*f*) in the subapical bar, but differing in that there is a good-sized very pale-yellow inner-marginal marking on fore-wing, and that the large discal spot is white, and diffusedly extended downward so as almost to meet it.—*Hab.* Delagoa Bay (*Mrs. Monteiro*).

B. Between Forms 2 (analogue of *Hippocoon*, Fab.) and 3 (*Trophonius*, Westw.).

    h. P. Merope, Trim. (♀, variation), in *Trans. Linn. Soc.*, loc. cit., p. 510, note.

    This specimen and another received in 1884 have the ordinary markings of the forms which they link, excepting that the patch of the hind-wings, though not obscured at the base, is decidedly narrower. All the markings are tinged with faint, dull, ochreous-yellow.—*Hab.* St. Lucia Bay (the late *Colonel H. Tower*), and Delagoa Bay (*Mrs. Monteiro*).

C. Form 3 (*Trophonius*, Westw.).

    i. P. Cenea, Trim. (♀, variation), *Trans. Ent. Soc. Lond.*, 1874, p. 153.

    This example has the subapical bar of the fore-wings considerably broader than usual, and *yellowish brick-red* instead of white. The field of red common to both wings differs from that ordinarily presented in being darker (inclining to ferruginous) and smaller, in the fore-wings not reaching to the median nervure, and clouded with fuscous between that nervure and the submedian nervure.—*Hab.* Bathurst, Cape Colony (*Miss M. Barber*).

LARVA.—"1st *Stage.* Black, with white filamentous tubercles on second segment and anal segment.

"2d *Stage.* Two pairs of filamentous tubercules on same segments, the first and last pair longest; a white transverse lunular band, connected with the head laterally, across sixth and seventh segments. Laterally a broad white band above spiracles. Last two segments whitish.

"From this growth to the last change but one, the filamentous tubercles grow longer, and the ground colour changes from greenish-brown to greenish, and the white markings grow less distinct.

"*Full-grown Larva.*—Bluish-green, like larva of *Philognoma Varanes*. Tubercles very short, those next head yellowish, on anal segment whitish; very much like a slug in shape. Y-like organ crimson-lake at base, tapering to greenish-white. On fourth segment, two small black spots, bordered by a narrow white line; sixth and seventh segments festooned with delicate whitish zigzag lines. A double row of bluish-white dots along back. Lateral borders above spiracles white. Head and true legs green, false legs pale-ochreous. $1\frac{1}{4}$ to $1\frac{3}{8}$ inch long."

"Feeds on *Tepris lanceolata*."—J. P. Mansel Weale.

PUPA.—Anteriorly much attenuated, but about middle greatly expanded laterally; posteriorly tapering rather abruptly to a point. Inferior side strongly convex, especially about broadest part of wing-covers; superior side moderately concave from tail to disco-thoracic prominence, which is acute but not much elevated; lateral line along expansion forming a thin sharp ridge. Head very prominent; the two ordinary points not being divergent, but directed straight forward with their inner edges closely contiguous, so as to form a single forward projection tapering together to one point.

Bright yellowish-green on back and rather dull-green on under side. Along back, a median whitish-ferruginous streak, commencing with a small spot on thoracic elevation, and gradually becoming slightly wider and better defined until it reaches anal extremity. On each side of this streak on fifth abdominal segment, a subquadrate reddish-white spot, external to which is a dot of the same colour (the second and most apparent in a longitudinal row of four). Expansion of lateral margin bounded by yellow (here and there varied with whitish-ferruginous) along the ridge separating the dorsal and ventral aspects, from head to tail. Spiracles situated just above this thin lateral ridge. On ventral aspect a narrow longitudinal yellow streak defines the line of the median convex ridge. On wing-covers, besides some very fine pale lines indicating the neuration, some similar transverse lines on each side of median streak. On sixth segment of abdomen, on each side of median streak, a sub-ovate whitish ferruginous spot. Length, 1 in. $3\frac{1}{2}$ and 1 in. $4\frac{1}{2}$ lin.; greatest width (across third abdominal segment) 6 lin.; greatest depth (at junction of second and third abdominal segments) from breast to back 4 lin. and 5 lin.

(Described from two pupæ forwarded by Mr. J. P. Mansel Weale from the neighbourhood of King William's Town.)

Mr. Weale (*loc. cit.*, p. 134) observes "that the larva about to pupate generally fixes its anal legs below the axil of a leaf-stalk, and fastens itself below sixth" [actually metathoracic] "segment with a double thread to the petiole." In this position the ventral or under side of the pupa, which is darker than the dorsal or upper side, is uppermost; and the general resemblance of the insect to the leaflets of its food-plant, which in outline is very remarkable, is thus completed, *V. lanccolata* having the leaflets darker above and paler below. Mr. Weale further points out that such minutiæ as the more glossy upper surface of the leaflets, the slight inflexion of their margins, the slightly ferruginous tint of the mid-rib, and the reticulated venation, are all to some extent imitated in this chrysalis. The modifications of shape and outline which combine with the colouring to complete this deceptive resemblance are unusually great, when the pupa is compared with those of other species of *Papilio*. Not only is the whole pupa much flattened, and the convexity of the ventral and pectoral region balanced by an unusual concavity of the dorsal region (with almost a suppression of the dorso-thoracic prominence), but the development and expansion of the lateral longitudinal ridges is very pronounced. The cephalic projections, however, exhibit the most unique form. If these had retained the customary conspicuous divergence into two prominent processes, as in *P. Demoleus*, *P. Lycæus*, &c., it is obvious that the general resemblance to a leaf would have been greatly lessened, and the object of concealment to some extent frustrated. These projections are, however, brought closely together, so that their inner edges touch throughout their length to the very extremity, and their outer edges converge to a common point; and in this manner the top of the leaf is accurately represented.

P. Cenea is a very near ally of *P. Merope*, Cram. (*Pap. Exot.*, tab. 151, A, B, and 378 D, E), from West Africa,[1] and it was not until 1873 that I became convinced that the two forms should be treated as distinct species. As regards the ♂, *Cenea* is to be distinguished by its shorter wings, darker and more rufescent under-side colouring, shorter tail on hind-wings, much fainter inter-nervular dark rays on the under side in both fore and hind wings, and broader and more continuous (ferruginous-ochreous, not fuscous) discal band on the under side of the hind-wings. With respect to the ♀ s, the *first* or ordinary Southern form (*Cenea*, Stoll)—which is imitative of a Southern species of *Danainæ*, *Amauris Echeria*—appears to have no analogue in Tropical West Africa; but the much rarer *second* form is very like the ordinary West-Coast ♀ *Merope* named by Fabricius *P. Hippocoon*, being separable very readily, however, by its smaller size, shorter wings, narrower subapical white bar in the fore-wings, and much larger white patch in the hind-wings. The equally scarce *third* form (*Trophonius*, Westw.) was evidently originally figured from a Southern example; and its apparently rare Western analogue, figured by Hewitson (*Exot. Butt.*, iv. Papilio xii. f. 40), presents as regards the fore-wings a broader, more oblique, almost wholly brick-red subapical bar, but (unlike the form *Hippocoon* of the same region) has quite as broad a patch in the hind-

---

[1] Fabricius' name of *Brutus* was given to the same butterfly in 1781, two years later than Cramer's publication of his earlier figures (vol. ii.).

wings as *Trophonius* exhibits, and is little if at all larger than the latter.[1]

As I pointed out in 1873 (*Trans. Ent. Soc. Lond.*, 1874, p. 149), the circumstances under which the several forms and variations of this most interesting *Papilio* occur in South Africa do not warrant the assignment of certain variations of the ♂ to separate ♀ forms proposed by Mr. Butler (*op. cit.*, 1869, p. 275). The ♂s, not only from the same district, or from the same locality, but even from the same wood, vary indefinitely as to their black markings within certain limits. An instance of this is given by seven examples reared by Mr. Weale from larvæ of one season found in the same spot.[2] I possess five examples, taken by Mrs. Barber, Mr. F. Barber, jun., and myself, in the same little copse at Highlands, near Grahamstown, which present great variation in the discal upper-side band of the hind-wings,[3] and a noticeable difference in the width of the hind-marginal band of the fore-wings, as well as in the dentation of its interior edge. A very remarkable specimen, taken by Mrs. Barber at the mouth of the Kleinemond River, recalls, in the character of the spots which represent the hind-wing bands, the ordinary West-African ♂, but is also signalised by *a very narrow black border to the fore-wings*, only slightly denticulated on its inner edge. The other extreme form in the Southern ♂ is that described by Mr. Butler under the head of "(aa.) *Cenea*, var.," from

---

[1] The explanation of this discrepancy seems obvious. The Western *Hippocoon* closely mimics the largest of Western *Danaides* (*Amauris Niavius*), which has a small white patch in hind-wings; while *Trophonius* is modified in imitation of the considerably smaller *Danais Chrysippus*, in which nearly the whole field of hind-wings is brick-red. In both the Western and Southern *Trophonius* form of ♀ the subapical bar of fore-wings is sometimes almost as red as the other markings. This variation appears to be in imitation of the *Dorippus* variety of *Danais Chrysippus*.

Another remarkable form of the ♀ is that named by Doubleday (*Gen. Diurn. Lep.*, pl. 3, f. 4), *P. Dionysos*. It is peculiar to Western Africa, and, in company with the curious allied variation figured by Mr. Hewitson (*loc. cit.*, f. 39), is of high interest, not only as combining the features of *Hippocoon* and *Trophonius*, but as indicating, in its possession of *merely a trace of black* between the white subapical bar and inner-marginal space of the fore-wings, the mode in which (as suggested by me in the *Transactions of the Linnean Society, loc. cit.*, with reference to the ♀ *Meriones* of Madagascar), the extraordinary modification of the fore-wing markings of the ♀s was most probably initiated. *Dionysos* is, in fact, of all the West-African ♀s, the least profoundly modified form as compared with the ♂. All the Western ♀s, like the ♂s (but more so in the outer portion of the hind-wings), are distinguished from Southern examples by the strongly marked fuscous rays between the nervures.

[2] The seven males present the customary amount of variation in the transverse black markings of the upper side of the hind-wings,—from three sub-quadrate discal blotches to a continuous irregular bar,—and in these particular markings no two of them nearly agree. It is the same with the amount of black marking on the tails of the hind-wings, which varies from a simple median streak, with an accompanying short suffused stripe bounding the basal half of the tail interiorly, to a black space absorbing almost the whole basal two-thirds of the tail. Four of the seven specimens possess, more or less faintly, the blackish line defining the second disco-cellular nervule of the fore-wings.

[3] The most imperfect condition of this band that I am aware of is exhibited by a specimen which I captured at Knysna, Cape Colony, in which the three patches representing the band are reduced to widely-separated, irregular, attenuated spots, smaller (especially that at anal angle) than in the Western race.

Port Natal, in which all the black markings are strongly developed, especially the discal band of the hind-wings, which in some examples is quite unbroken. This form is most prevalent in Natal, the adjacent Coast country, and Delagoa Bay, but also occurs near Grahamstown. It is (except, perhaps, in size) the farthest removed from the ordinary *Merope* ♂. I know of no locality in South Africa in which the ♂s are constant to any particular pattern; but, amid all their variation, I have noticed no example that approaches the West-African ♂ in the strongly-marked inter-nervular rays of the under side, except where (in some of those in which the black markings are most developed) the rays cross the discal band in the hind-wings.

Looking to the Southern ♀s, it is equally observable that the several well-defined forms are not restricted to particular localities. *Cenea* (typical) and *Trophonius* were taken by me in the same spots at Knysna and Plettenberg Bay respectively, and I have since received the *Hippocoon*-like form from the former locality. Mr. Weale has bred *Cenea* (variety), *Trophonius*, and a variation closer to *Hippocoon* than to *Cenea*, from larvæ taken in one spot near King William's Town; and Mrs. Barber has sent me the three forms, as well as a variation (very near that delineated on fig. 2 of the second plate accompanying my paper in the *Linnean Society's Transactions* already referred to), all of which were taken at Highlands, near Grahamstown.

In Kaffraria Proper, as well as near D'Urban, Natal, Colonel Bowker has met with *Cenea* (var.), *Trophonius*, and the *Hippocoon*-like form; and I have recorded above the singular linking variations found by Mrs. Monteiro at Delagoa Bay.

In a most interesting series lately (1887) collected by Miss Newdigate about Forest Hall, in the Zitzikamma Forest, Plettenberg Bay, there are ten ♂s presenting similar variations to those of the Grahamstown examples above mentioned, the hind-wing band especially showing every gradation from three separate spots to complete continuity;[1] twelve ordinary ♀s of the typical *Cenea* form, but varying a good deal in the size of the spots and of the hind-wing patch; one ♀ of the Second Form, but with an exceedingly narrow subapical white bar in the fore-wings; and two ♀s of the Third Form (= typical *Trophonius*).

[1] One ♂, with the hind-wing band continuous throughout, has in the dark border of the fore-wings not only a larger than ordinary subapical spot, but a series of six diffused pale sulphur-yellow markings, of which the first (subcostal) is a conspicuous longitudinal ray, the second (interrupted by black-ringed subapical spot) and third shorter rays, and the rest rather small spots.

An indication of aberration in the same direction is afforded by two other ♂s (one from the George District of the Cape, and the other from Pinetown in Natal), which alike exhibit two very small pale sulphur-yellow spots in the fore-wing border between first radial and third median nervules.

In *Proc. Ent. Soc. Lond.*, 1873, p. xxii., I have recorded, in connection with other instances of aberrant neuration, the case of a Kaffrarian ♂ *Cenea*, in which the subcostal nervules of both hind-wings are united by a transverse additional nervule, so that a completely closed second cell is formed immediately adjoining the ordinary discoidal cell, and extending beyond it.

The species-identity of the conspicuous pale-yellow tailed male *Papilio* known as *P. Merope* or *P. Brutus* with the sombre-tinted tail-less females *P. Cenea*, *P. Trophonius*, *P. Hippocoon*, and other unnamed varieties, was pointed out by myself as in the highest degree probable in 1868 (*Trans. Linn. Soc. Lond.*, 1869), and was proved in 1873 by Mr. J. P. Mansel Weale, who reared the ♂ and the three prominent forms of ♀ — eight of each sex — from exactly similar larvæ (*Trans. Ent. Soc.*, 1874). In 1881 I was enabled to place on record Colonel Bowker's capture of the paired sexes at D'Urban on the 22d February in that year (*Trans. Ent. Soc.*, 1881, p. 169, pl. ix.); and I received from him in the following year a second pair (still united), taken on the 3d July. I one case the ♂ was of the prevalent continuous-banded Natalian variation, but in the other had the hind-wing band reduced to three rather small and widely separated blotches; the ♀ being in both cases the ordinary one of the white-spotted variation. Colonel Bowker noted that in flight the ♂ carried the ♀.[1]

This fine species is quite confined to wooded districts, but is numerous in such tracts during the summer, the ♀ being, however, much less frequently seen on the wing than the ♂. The latter is exceeding conspicuous in flight, taking a rapid irregular course over and about the underwood, and following usually a set circuitous route, only interrupted by occasional halts at flowers, especially those of *Plumbago*. At Knysna I noted a specimen with peculiarly torn wings return repeatedly over the same ground, and twice captured it. Mr. Mansel Weale and Colonel Bowker have both noted this behaviour of the ♂, and agree in their observation that it is related to the presence of the ♀, who remains settled in some shady spot among the weeds and bushes. Colonel Bowker has seen a ♂ with a broken wing return many times to the same ♀, passing on after making "two or three dips with half-closed wings." Mr. Weale further records that "as the afternoon draws on the females leave their retiring spot and flutter slowly about, sometimes coming out into the open, but more apparently to show themselves than for the sake of food. On one occasion I saw four males busily courting a female, but unfortunately I disturbed them." The ♀ is always, as far as I have observed, much slower on the wing than the ♂, and stays much longer at the flowers she visits. I noticed at Knysna that she specially affected the small white flowers of a low-growing labiate.

The brown and ochre-yellow colouring of the under side of *P. Cenea* serves well to protect the butterfly from observation when at rest among withered foliage.

Mrs. Barber, at the beginning of the year 1871, was fortunate enough to observe this protective resemblance in nature, and sent me the following note on the subject, viz. :—

"I caught a fine *Merope* with my finger and thumb the other day. It was just beginning to rain, and, though it was not late, *Merope* thought proper to seek a resting-place, which he wisely chose upon a shrub which resembled his own under-side colouring. It was a splendid match. When he closed his wings among the yellow and brown seeds and flowers of the shrub, no bird would ever have distinguished him. I had no net with me, and my first attempt was a failure. However, the butterfly took a turn round the neighbourhood, examined several other shrubs (which he found were not so good, I suppose), and eventually returned to the same perch."

The mimicry of *Amauris Echeria*, *A. dominicanus*, and *Danais Chrysippus* respectively by the three forms of the ♀ is very apparent. It is closest in the first of these, the simulation of *Echeria* by the smaller specimens of the ordinary ♀ (*Cenea*, Stoll) being exact to deception, but is quite near enough to be effectual in the two others, as I have noticed at Knysna and D'Urban respectively.

---

[1] The late Mr. Hewitson has put on record (*Ent. M. Mag.*, 1874, p. 113) his receipt from Fernando Po of *P. Merope* and *P. Hippocoon*, captured *in copulâ* by Mr. Rogers.

The range of the species beyond South Africa is not clearly known, but as it occurs (♂ of continuous-banded marking, and ♀ of *Hippocoon*-like form) at Zanguebar and Zanzibar, it is not improbable that it is native to the intervening country north of Delagoa Bay. Many years ago I noted a ♂ *Merope*, brought by Mr. H. Waller from Mount Morambala, on the River Shiré, an important lower tributary of the Zambesi, but did not record whether it agreed with the more Southern examples.

Localities of *Papilio Cenea*.

I. South Africa.
 B. Cape Colony.
  a. Western Districts.—Oakhurst, George District (*Sir A. Scott*). Knysna and Plettenberg Bay.
  b. Eastern Districts.—Kleinemond River, Bathurst District (*H. J. Atherstone* and *Miss M. Barber*). King William's Town (*W. S. M. D'Urban* and *J. P. Mansel Weale*). Grahamstown. Seymour, Stockenstrom District (*W. C. Scully*). Fort Warden, Kei River (*J. H. Bowker*).
 D. Kaffraria Proper.—Bashee River and Manubie Forest (*J. H. Bowker*). "Baziya (*Baur* and *Hartmann*)."—Möschler.
 E. Natal.
  a. Coast Districts.—D'Urban. Tongaati River. Pinetown (*J. H. Bowker*). "Lower Umkomazi."—J. H. Bowker.
  b. Upper Districts.—Maritzburg (*Colonel Scott, R.A.*).
 F. Zululand.—Etshowe (*T. Vachell*). St. Lucia Bay (the late *Colonel Tower*).
 H. Delagoa Bay.—Lourenço Marques (Mrs. Monteiro).

II. Other African Regions.
 A. South Tropical.
  b. Eastern Coast.—Zanzibar.—Coll. W. Distant. Zanguebar (*R. P. Le Roy*).

314. (15.) **Papilio Echerioides**, Trimen.

♂ *Papilio Messalina*, Trim., Rhop. Afr. Aust., ii. p. 329 (1866).
♂ ♀ *Papilio Echerioides*, Trim., Trans. Ent. Soc. Lond., 1868, p. 72, n. 2, pl. vi. ff. 1, 2.

Exp. al., (♂) 3 in. 3–6 lin.; (♀) 3 in. 3–8 lin.[1]

♀ *Black, with a common transverse yellowish-white band, regularly interrupted in fore-wing but continuous (and much wider) in hind-wing; a row of yellowish-white spots on hind-margin of hind-wing.* Forewing: band discal, commencing near apex, and composed of eight very distinct well-separated spots, gradually increasing in size (both vertically and longitudinally) from the first irregularly-ovate small spot between fourth and fifth subcostal nervules to the seventh between first median nervule and submedian nervure; the eighth (inner marginal) is nearly as long as the seventh, but not half as broad, and only

---

[1] A dwarf ♂ expands only 2 in. 9 lin., another 2 in. 10 lin., and a dwarf ♀ only 2 in. 8 lin. All three examples are from the Tsomo River.

separated from it by black nervure; the sexual discal coating of hairs somewhat woolly, specially apparent on median and radial nervules separating spots of band; cilia with very minute white inter-nervular dots, except just below apex, where a very small white spot on hind-marginal edge (between third and fourth subcostal nervules) with adjoining white cilia is rather conspicuous. *Hind-wing:* band median, but nearer to base than to hind-margin, its inner edge sharply defined, nearly straight,—its outer edge somewhat beyond middle, rather diffused, pierced slightly by black inter-nervular rays; a series of six conspicuous, moderate-sized, more or less rounded spots (each crossed by inter-nervular fold), usually touching, but sometimes just separate from, white marks of cilia, which are very much larger than in fore-wing; the first and second of the hind-marginal spots smaller than the rest. UNDER SIDE.—*Outer half of hind-wing and apical area of fore-wing dark ochreous-brown; in hind-wing, median band greatly reduced, and space between it and base dull yellow-ochreous with black markings.* *Fore-wing:* ground brownish-fuscous; most spots of band larger, on each side more rounded, less widely separated than on upper side,—the second spot almost obsolete and (together with the first, which is indistinctly defined) confused with some oblique greyish-white ill-defined apical clouding; subcostal nervure at base edged beneath by white scaling. *Hind-wing:* band greatly narrowed, its outer half (except on costa) being obscured by pale-brownish, and further pierced by inter-nervular brown rays and marked by two brown striæ in discoidal cell; basal yellow-ochreous with the crossing nervures, a rounded spot, and thick end of ray between costal and subcostal nervures, a three-branched disco-cellular streak, and thick ends of rays on each side of submedian nervure, all black; a small white spot at origin of median and submedian nervures; just above radial nervule, about midway between end of cell and hind-margin, an ovate whitish spot, variable in size and definition; hind-marginal spots much smaller than on upper side,—the third distinctly elongated in direction of discal spot, and the lower ones usually distinctly divided by blackish inter-nervular rays.

Head and body above black. *Head* with basal and terminal joints (hirsute) of palpi, a pair of frontal and three pairs of vertical spots, and a spot outwardly edging each eye, yellowish-white. *Thorax* above with two pairs of similar spots on prothorax, and another pair (larger and hirsute) on pterygodes; breast black, with nine or ten large ochre-yellow, yellowish, and yellowish-white spots on each side; legs black, with the femora whitish or yellowish in front. *Abdomen* laterally dusky-brownish; inferiorly pale-yellowish or yellowish-white, with two parallel longitudinal streaks and segmental incisions black.

♀ Quite unlike ♂, closely resembling ordinary ♀ of *Papilio Cenea,* Stoll, and, like it, mimicking *Amauris Echeria,* Stoll. *Brownish-black with white spots; hind-wing with a pale ochreous-yellow broad band (or rather patch before middle.*) *Fore-wing:* the following white spots, viz., in discoidal cell, not far from extremity, and adjoining subcostal nervure, a short oblique white mark of variable size and shape (but always much smaller than in *P. Cenea*, and sometimes almost obsolete); beyond cell a short oblique subapical bar crossed by upper radial nervule; a large ovate spot below cell, between second and first median nervules (its upper edge sometimes extending beyond second median nervule); a subapical hind-marginal spot as in ♂, but very much larger; and a submarginal series of four similar spots between first radial nervure and submedian nervure. *Hind-wing:* central band or patch (rarely dull yellowish-white) considerably wider than in ♂, except costally (where it is much narrower) and inner-marginally (where it is rather narrower),—its inner edge being nearer base, and its outer edge much more rounded; hind-marginal spots white, larger than in ♂, the lower ones usually not quite so close to marginal edge, and sometimes a small additional spot close to anal angle. UNDER SIDE.—*As in ♂, except for the totally different white marks of the fore-wing, which agree with those of the upper side, but are rather larger.* Fore-wing: uppermost spot of submarginal row of four obsolete. *Hind-wing:* basal ochreous-yellow narrower than in ♂; some indistinct greyish-white scaling between discal white spot and third spot of hind-marginal row; of the latter the second and fourth spots are often very small and obsolescent, and rarely wanting altogether.

In the dwarf ♀ above noted, all the usually white spots are of the same pale ochreous-yellow as the hind-wing patch, except that the marginal ones are rather paler.

Head and body marked and coloured as in ♂, except that the sides and back of abdomen are not so dark.

As regards the ♂ *P. Echerioides* differs from its near ally the West-African *P. Cynorta*, Fab. (*Zeryntius*, Boisd.), as follows, viz.: (1) The macular band of the *fore-wing* is much narrower, its inner edge being at some distance from discoidal cell, and its outer edge much farther from hind-margin; (2) this band is uninterrupted, the separate spots gradually widening from its origin, while in *Cynorta* it is abruptly interrupted, the second spot being absent or quite obsolescent, and the third very large; (3) the outer contour of the band is slightly concave instead of convex, and the component spots are narrower and less blunt internally, the separating black rays being broader; (4) in the *hind-wing*, the continuous band is usually narrower, its inner edge being further from the base, while its outer edge is almost even, instead of being sharply and regularly dentated by the inter-nervular rays; and (5) the hind-marginal spots are a conspicuous difference, being either very small or (oftener) altogether wanting in *Cynorta*. On the under side, in the *fore-wing*, (6) the inner edge of the band is rather farther from the discoidal cell than it is on the upper side, whereas in *Cynorta* it is so close as partly to touch the extremity of the cell; (7) the small whitish spot in the upper part of the cell, near its extremity, sometimes found in *Cynorta*, is wanting;[1] in the *hind-wing*, (8) the basal ochreous is darker and more rufous in tint, extending farther costally; (9) the band is considerably narrower, its inner edge not so even, the brownish clouding of its outer border better defined; (10) the inter-nervular rays are not so strongly marked; (11) the discal whitish spot is not found in *Cynorta*. It is also to be noted that *Echerioides* is considerably the larger of the two species.

The ♀ s of these two very similar ♂ s are surprisingly different, for while that sex in *P. Echerioides* (as mentioned above) so closely mimics *Amauris Echeria* as to be in life indistinguishable from the ♀ *P. Cenea*, the ♀ of *Cynorta* (described by Westwood under the name of *Boisduvallianus*[2]) closely copies the very differently marked *Planema Gea*, (Fab.) ♀,—a well-known native of Western Africa. The feature common to these two ♀ s—by which the collector can at once distinguish *Echerioides* from *Cenea*—is the ochre-yellow black-spotted basal patch on the under side of the hind-wings. This character is in the ♂ s of *Cynorta* and *Echerioides* even more developed than in the ♀ s, and is in direct mimicry of the *Planema*; and its continued existence in the ♀ *Echerioides*—which mimics a Danaine butterfly not possessing this peculiar marking—points to the inference that mimicry of the *Planema* group was in both these *Papiliones* the earlier tendency, and has only more recently been diverted in the direction of *Amauris* in the case of the Southern species.

Archdeacon Kitton, of King William's Town, first brought this interesting *Papilio* to my notice, having in April 1863 forwarded a ♂ example captured by him in the Perie Bush.[3] Colonel Bowker in 1865 found the butterfly numerously in the Sogana and Boolo Forests, near the Tsomo River in Kaffaria, and sent me a number of specimens of both sexes, including two pairs taken *in copula*. He noted that there were two broods in the year, one in November, not lasting for more than about four weeks, and the other in January, continuing to appear until far into March. "The ♂ s," he wrote, "take a constant course through the forest, returning regularly by the same route; while the ♀ s keep about the place, but fly at a lower elevation, and do not appear to take the rounds of the ♂ s. The sexes disappeared together at the end of November, and did not appear again until early in January, when they both came out on the same day. When the ♂ and ♀ meet, they whirl about together among the tops of plants in the forest, and as soon as united, disappear down under the leaves."

I only once chanced upon living *Echerioides*, on the 8th March 1867, in high-lying woods at Tunjumbili, on the Tugela border of Natal. The ♂ s were tolerably numerous, but I saw only two ♀ s, and at first mistook one of them for *P. Cenea*. The flight of the insects quite confirmed Colonel Bowker's

---

[1] A single example from Natal has a very faint trace of this spot in each fore-wing.

[2] Strong evidence of the species-identity of *Cynorta* and *Boisduvallianus* was adduced in 1878 by the late Mr. D. G. Rutherford, who exhibited at the June meeting of the Entomological Society of London a specimen showing on the left side the wing-markings of *Cynorta*, and on the right side those of *Boisduvallianus*.

[3] This specimen reached me in a very damaged state; and I erroneously referred it to *P. Messalina*, Stoll, and so recorded it in *Rhopalocera Africae Australis*, ii. p. 329.

description, the ♂ s coursing rather rapidly and irregularly over the underwood, while the ♀ s hovered flutteringly near the ground, and often settled on leaves.

The species seems by no means generally distributed in forest tracts, evidently preferring, if not being limited to, woods lying at some elevation. Several examples have been sent to me by Mr. T. Ayres from the Eastern Transvaal.

Localities of *Papilio Echerioides*.

I. South Africa.
   B. Cape Colony.
      b. Eastern Districts.—King William's Town (*Archd. Kitton*).
   D. Kaffraria Proper.—Tsomo River (*J. H. Bowker*).
   E. Natal.
      b. Upper Districts.—Tunjumbili, Tugela River.
   K. Transvaal.—Lydenberg District (*T. Ayres*).

II. Other African Regions.
   B. North Tropical.
      a. Western Coast.—"Cameroon Mountains; Bonjongo (*Buchholz*)."—Plötz.[1]

Family V.—HESPERIDÆ.

Hesperides, Latreille, "Cons. Gen. Crust. Arachn. et Ins. (1810);" Encyc. Meth., ix. p. 706 (1819), [excl. *Urania*].
Hesperidæ, Leach, "Samouelle's Comp., p. 242 (1819)."
Hesperides, Boisduval, Sp. Gen. Lep., i. p. 167 (1836).
Hesperidæ, Swainson, Hist. and Nat. Arr. Ins., p. 97 (1840).
Hesperiidæ, Westwood, Intr. Mod. Class. Ins., ii. p. 360 (1840); and Gen. Diurn. Lep., ii. p. 505 (1852).
Hesperidæ, Trimen, Rhop. Afr. Aust., ii. p. 285 (1866).
Hesperiina, Herrich-Schäffer, Corresp.-Blatt. Zool.-Min. Ver. Regensb., 1869, pp. 28 and 50.
Urbicolæ (Fab.), Scudder, Bull. Buff. Soc. Nat. Sci., 1874, p. 195.
Hesperiina, Plötz, Stett. Ent. Zeit., 1879, p. 175; and Mitth. Nat. Ver. Neu-Vorpomm. u. Rügen, 1884, p. 1.
Hesperiidæ, Distant, Rhop. Malay., p. 366 (1886).

IMAGO.—*Head very broad; eyes* large, very prominent, smooth; *palpi* set apart from each other, broad, thick, short, appressed to face, densely clothed with scales and hair on basal and middle joints,—terminal joint very slender, usually very short (often minute and almost hidden by hair of middle joint), clothed with extremely fine and short appressed hairs; *antennæ wide apart at origin, with a more or less elongate club, usually curved, often reflexed or attenuated into a slender hooked tip; haustellum (maxillæ) much elongated; usually a compact tuft or pencil of stiff hairs between base of each antenna and margin of eye*.

Thorax usually very robust, sometimes broader than head. *Wings*

---

[1] This is probably *P. Zoroastres*, Druce (*Ent. M. Mag.*, 1878, p. 226), from Fernando Po, described as nearly allied to *P. Echerioides*, but larger and with the bands pure white (♂).

small, thick; *fore-wings* usually more or less elongated apically; subcostal nervure five-branched,—*nervules all near together, originating at about equal distances apart, the fifth (and sometimes the fourth also) given off at extremity of discoidal cell* (in *Cyclopides* the first nervule in some species is very short, and runs into costal nervure); upper radial nervule originating close to fifth subcostal at extremity of cell,—*lower one more or less attenuated*, springing from junction of disco-cellular nervules; discoidal cell long and narrow, the closing nervules very slender; *first median nervule often given off not far from base. Hind-wings* short and blunt apically, longer, and usually more or less prominent or lobate (rarely broadly and lengthily caudate) about anal angle; *a longitudinal fold between submedian and internal nervures; subcostal nervure usually angulated near base, so as to touch costal nervure and to shut off from discoidal cell a small prediscoidal cell;* disco-cellular nervules exceedingly attenuated, sometimes imperceptible; *radial nervule much atrophied, occasionally wanting,*—the space between second subcostal nervule and third median nervule being always narrow; *internal nervure very long, extending to anal angle. Legs perfect in both sexes;* fore-legs only differing from the others in their much shorter tibiae, which bear inwardly (except in *Cyclopides*) beyond middle a flattened acute appendage or spur; *hind-legs almost always bearing, besides terminal spurs, a pair of similar (often shorter) spurs beyond middle;* all the tarsi long and stout, more or less spinulose inferiorly, their terminal claws and appendages small.

Abdomen attenuated posteriorly, usually short (especially in ♀), seldom as long as inner margin of hind-wings, and very rarely longer.

LARVA.—Elongate, cylindrical, markedly attenuated anteriorly (thoracic segments); head large, much widened superiorly. Usually naked, but sometimes thinly hairy.

PUPA.—Elongate, rounded (rarely sub-angulated bluntly anteriorly); head with a median frontal more or less elongate acute projection; haustellum-case sometimes freely extending beyond wing-cases to a point far past extremity of abdomen. Attached by tail and by a free silken girth round middle, and also usually enclosed by other silken threads or in an imperfect open thin web.

As may be seen from the characters italicised in the foregoing diagnosis, the *Hesperidæ* constitute a family very distinct from all other butterflies; their structure is indeed so peculiar, and in the main so constant, that they seem almost entitled to rank as a Sub-Order of the Lepidoptera. It is very noteworthy that, although their most striking characters—such as the large head, wide apart antennæ, and four-spurred tibiæ of the hind-legs—certainly approximate them to moths (*Heterocera*) generally,[1] it is not found, upon close examination, that the

---

[1] The ♂ of the Australian *Euschemon Rafflesia* (Macleay)—an unquestionable Hesperide—alone possesses the striking Heterocerous feature of a fully-developed bristle and loop linking the fore and hind wings near the base. (See vol. i. p. 15.)

Hesperidæ are nearly related to any particular Heterocerous group or family. The connection so long supposed—from the resemblance of the antennæ and general *facies*—to exist between them and the singular moths of the *Castnia* group has been proved illusory, as no wider disparity in the whole Order can be found than that shown by the simple almost unbranched condition of the neuration of the wings in the *Hesperidæ* and its highly complicated arrangement in *Castnia* and allied forms.[1] Throughout the *Hesperidæ* the peculiar neuration is so remarkably constant, that it affords little or no means for discriminating genera; and even Speyer, who has published[2] by far the most thorough investigation on this portion of the structure of the family that has yet appeared, is constrained to regard as the most momentous point for observation so slight a matter as the position of the lower radial nervule in the fore-wings, viz., whether it originates exactly midway between the upper radial and third median nervules, or nearer to one or the other of them.

In strong contrast to the constancy of the wing-neuration is the irregularity with which the numerous secondary sexual characters presented by the ♂ *Hesperidæ* are distributed. These conspicuous badges, consisting of a long costal or discal groove or fold, lined with peculiarly formed scales in the fore-wings; a felt-like patch of scales in the same wings; a tuft or pencil of very long stiff hairs on the tibiæ of the hind pair of legs, or on the coxæ of the front pair; a very dense stiff fringe of hairs and scales on the tibiæ and part of the tarsi of the first pair of legs; a great enlargement of the first joint of the tarsi of the hind pair, or a pair of long curved appendages attached to the thorax posteriorly on the under side;—occur or are totally wanting in species which are so intimately related as to be unquestionably congeneric.

The characters of most value in this extremely difficult group (apart from the rarely-occurring *absence* of such salient ones as the additional pair of spurs on the hind-tibiæ, the solitary spur on the fore-tibiæ, or the extra-antennal tufts on the head), are those presented by the clavation of the antennæ, and by the terminal joint of the palpi. Beyond these, one has to depend on such general features as the relative robustness and length of the body, the size and shape of the wings, and the particular pattern of markings and system of colouring.

It is most probable that when this entire Family has been as thoroughly studied as the few Palæarctic members of it have been by Speyer,[3] a large proportion of the very numerous proposed and tacitly accepted genera will be no longer recognised; for in no group of the Lepidoptera has there been more random and careless creation of generic names. The most moderate of recent writers on the *Hesperidæ*, Herrich-Schäffer, tabulated 34 genera in 1869; Mr. W. F. Kirby's

---

[1] See Westwood, *Trans. Linn. Soc. Lond.*, Second Series, *Zool.*, i. p. 155, &c., pl. 28-33 (1877).

[2] *Stett. Ent. Zeit.*, 1879, p. 477, &c.   [3] *Stett. Ent. Zeit.*, 1878 and 1879.

Catalogue (1871–77) gave 60; and the late C. Plötz recognised (1886) no fewer than 67.

The large broad head and narrow neck of the Hesperide larvæ give them a very different aspect from the caterpillars of other butterflies; but they are in most cases not easily discoverable, owing to their habit of living and feeding in leaves which they roll or curl up by means of silken threads, much after the manner practised by the larvæ of moths of the *Tortrix* tribe. The chrysalis state is usually assumed within leaves so curled; and besides the ordinary attachments, there are often various enclosing threads and a slight cocoon.

In comparison with other groups of butterflies, the *Hesperidæ* exhibit a very limited area of wing-surface, and this makes them appear to be much smaller insects than on observing their large bodies they prove to be. In this respect, and, to a certain extent, in the slender hooked termination of their antennæ, they bear some resemblance to many of the *Sphinges* (hawk-moths), which, too, they approach in the great length of their trunk. The largest of them, the West-African *Hesperia Iphis*, the Indian *Erionota Thyrsis* ♀, the Central-American *Pyrrhopyga spatiosa*, and the South-American *Goniloba Astylus*, do not expand across the fore-wings much more than three inches; but the bodies of these species equal in bulk those of the largest *Charaxes*. Many extensive genera, however, such as the South-American *Eudamus, Goniloba, Tamyris*, and *Proteides*, and the Old-World *Hesperia* (*Ismene*, auct.), consist almost entirely of species expanding from two to two and one-third inches; but the majority of the Family is of smaller stature, with an expanse ranging from three-fourths of an inch to an inch and a half. The minimum size is exhibited by the South-American *Ancyloxypha melanoneura*, Felder, and the South African *Pyrgus nanus*, Trim., scarcely exceeds this in the smaller male examples.

There is not much variety in the shape of the wings, except in the case of many species of *Eudamus*, where the hind-wings have a great (in some instances very great) prolongation, broad and ending bluntly, at the anal angle. In *Caprona*, some species of *Pterygospidea*, and especially in *Helias* and *Achylodes*, the hind-wings are more or less angulated, and in parts excised, and in the genera last named the apex of the fore-wings has an acute projection.

For the greater part, the *Hesperidæ* are not brightly coloured, and glittering or metallic hues are of extreme rarity among them. The prevalent ground-colour is brown of a darker or lighter shade, and spaces or spots of ochre-yellow are very frequent. Very many are marked with transparent or semi-transparent spots or macular bands. There exists, however, a considerable proportion of more gaily-tinted species; the South-American genus *Tamyris* (*Pyrrhopyga*, auct.) standing first in the varied colouring of its members,—of which *T. versicolor*, Latr., with its red, blue, and yellow markings, is perhaps the most striking. Metallic blue and green occur in several of the *Gonilobæ* (such as *G. Mercatus*, (Fab.), *Alector*, Feld., &c.), and are still more developed in the West-African *Hesperia Bixæ*, (L.), and *Chalybe*, Westw.; and submetallic similar hues adorn *Abantis Zambesina* and *paradisea*, of Tropical and Extra-Tropical Southern Africa.

The New World is the metropolis of this family, about two-thirds of the recorded species and a majority of the genera being peculiar to it. Even North America (Extra-Tropical), which is poor in butterflies generally, has yielded over 100 species—mostly belonging, however, to the two genera *Pamphila* and *Nisoniades*. Tropical America has produced nearly three-fifths of the known species. The Palæarctic region is exceedingly poor in *Hesperidæ*, the researches of very many collectors and of lepidopterists having resulted in a series of 46 species only, of which 28 belong to Europe proper. Next to but far behind the overwhelmingly rich Neo-Tropical region comes the Oriental, which appears to have produced considerably over 200 species; while under half this number are noted as inhabiting the Australian region. Africa and its islands do not hitherto seem to have yielded more than about 160 species, the majority of which are only known to me by Plötz's brief descriptions. The South-African list embraces 64, thus distributed among the following genera, viz.:— *Cyclopides* 9, *Pyrgus* 14, *Thymelicus* 7, *Pamphila* 12, *Ancyloxypha* 3, *Abantis* (including *Leucochitonea*, Wallengren) 5, *Caprona* 2, *Pterygospidea* 7, and *Hesperia* (= *Ismene*, Swainson) 5. None of these genera are peculiar to the Ethiopian region except *Abantis* and *Caprona*, but these two do not appear to be represented north of the Equator. Thirty of the sixty-four South-African species at present seem to be peculiar to the Extra-Tropical area, but this number will probably be reduced when the South-Tropical belt is better explored by collectors.

The flight of the *Hesperidæ* is very peculiar; though always swift, and usually of great velocity, it is exceedingly short, being continually arrested by the abrupt settling of the insect. This curious action on the wing has gained for the few species inhabiting England the common name of "Skipper." When resting, these butterflies exhibit various modes of holding the wings, some—such as *Cyclopides*, *Hesperia*, and *Pamphila*—having their wings either quite erect in the ordinary manner, or with the hind-wings partly open; others (*Caprona*, *Pterygospidea*, &c.), keeping all the wings fully expanded; and *Tamyris*, according to Swainson,[1] sitting with wings lowered, but (as in some species of *Papilio*) so that the fore-wings overlap and almost conceal the hind-wings. As mentioned above (vol. i. p. 30), the European *Nisoniades Tages*, when completely in repose, holds its wings deflected in the attitude of a Bombycide or Nocturide moth; and Speyer mentions[2] that Zeller had noticed this as long ago as 1847, not only in *Tages*, but also in newly-emerged *Pyrgus Malvarum* (= *Alceæ*, Esp.). Those *Hesperidæ* that keep the wings fully expanded have the habit of

[1] *Zool. Illustr.*, i. pl. 33 (1820-21).   [2] *Stett. Ent. Zeit.*, 1878, p. 168.

settling on the under side of leaves, and they do this so quickly, that to the inexperienced collector they seem to have unaccountably vanished. A good many species are both early and late on the wing, but several do not make their appearance till about sunset. They are rather eager and greedy honey-suckers, and the length of their trunk enables them to rifle the stores of many tubular flowers unvisited by shorter-tongued butterflies.

GENUS CYCLOPIDES.[1]

Steropes, Boisd., "Voy. Astrol., Lep., p. 167 (1832)."
Cyclopides, (Hübner, 1816), Westw., Gen. Diurn. Lep., ii. p. 520 (1852).
Carterocephalus, Lederer, "Verh. Zool.-Bot. Ges., ii. p. 26 (1853)."
Cyclopides, Trim., Rhop. Afr. Aust., ii. p. 292 (1866).
Cyclopides and *Carterocephalus*, Speyer, Stett. Ent. Zeit., xxxix. pp. 181–82 (1878), and xl. p. 487 (1879).
Carterocephalus and *Cyclopides* (part), Plötz, Stett. Ent. Zeit., 1884, pp. 386–389.

IMAGO.—*Head* as wide as thorax, hairy; *eyes* round, smooth; a thin compact tuft or pencil of longish stiff hairs springing from between base of each antenna and border of eye; *palpi* long, porrected horizontally with dense bristly hair,—terminal joint rather long, tapering, slightly depressed at tip, with very short appressed hairs, usually almost hidden by bristly dense hair of middle joint; *antennæ* short, with a pronounced, rather thick, somewhat compressed, slightly-curved club, blunt (very rarely acuminate) at tip.

Thorax short and narrow, moderately or sparsely hairy. *Wings* rather long apically. *Fore-wings* somewhat truncate; costa beyond basal curve almost straight, or slightly concave beyond middle; costal nervure terminating at a little beyond middle; subcostal nervure five-branched,—three nervules originating before, fourth and fifth at, extremity of discoidal cell,—first nervule in three species (*Willemi*, *Ægipan*, and *Meninx*) very short, running into costal nervure; upper radial nervule originating near or very near to origin of fifth subcostal nervule; disco-cellular nervules slender, slightly oblique; first median nervule given off at a point considerably before middle and far apart from the other median nervules. *Hind-wings* very slightly prominent at anal angle; costa beyond basal lobe almost straight or very slightly arched; costal nervure terminating at a little before apex; subcostal nervure branching at a considerable or a little distance before extremity of discoidal cell,—latter being variable in length; disco-cellular ner-

[1] Mr. Kirby (*Cat. D. Lep.*, 1871, p. 623) adopted Dumeril's name *Heteropterus* (1806) for this genus. It is true that Dumeril happened to give the species *Aracinthus*, Fab. (= *Morpheus*, Pall.) as representing *Heteropterus*; but as his extremely vague and brief definition of the genus applies, and was certainly meant to apply, to the whole group of *Hesperidæ*, and not to what modern zoologists mean by a genus, it appears to me that *Heteropterus* cannot be retained as a generic name.

vules and radial nervule extremely slender; internal nervure very long, running not far from inner margin, terminating at anal angle. Legs of moderate size, first pair smaller than rest; tibiæ of first pair very short, without appendage; tibiæ of hind pair usually with two pairs of spurs, but in some species (several Palæarctic and three South-African, viz., *Willemi*, *Ægipan*, and *Meninx*) with the terminal pair only,—no ♂ tuft in any species; tarsi stout and rather long, especially first joint, and usually more or less finely spinulose beneath.

Abdomen long (in ♂ very slender), and extending as far as or a little beyond anal angle of hind-wings), tufted at tip; basal half more or less hairy.

LARVA.—Moderately slender (*Morpheus*) or stout (*Palæmon*); head smaller than usual in this Family.

PUPA.—Slender, elongate, very narrowly acuminate posteriorly; frontal spine long, straight, projecting horizontally (*Morpheus*).

(These characters of the larvæ and pupa of the European species named are from figures by Duponchel and Guenée, and by Boisduval, Rambur, and Graslin.)

I do not think that *Carterocephalus* can be held distinct from *Cyclopides*, the only peculiar characters being the slightly lower origin of the lower radial nervule of the fore-wings, and the longer discoidal cell of the hind-wings; for, as Speyer (*Stett. Ent. Zeit.*, 1878, p. 181) points out, the absence of the second pair of spurs on the hind-tibiæ— upon which Lederer chiefly relied—is an unstable distinction, occurring in *Cycl. ornatus*, Brem. This want of the second pair of spurs also characterises (as mentioned in my diagnosis above) the South-African *Willemi*, *Ægipan*, and *Meninx*, and these three species are further linked by the short first subcostal nervule of the fore-wings running into the costal nervure; but these forms are very unlike in facies and pattern of markings. The most aberrant of the known species seems to be *Lepeletierii*, Godt., in which the club of the antennæ ends in an acute point, the palpi are rather longer, the fore-wings more pointed, and the abdomen is thicker; but in all these features it is approached by *Tsita* and *Inornatus*, Trim., which also resemble it in their unspotted plain brown colouring. The South-African conspicuously yellow-spotted species are *Metis*, Cr., and *Malgacha*, Boisd.; those more faintly and sparsely spotted are *Willemi*, Wallgrn., *Syrinx*, Trim., and the ♀ of *Ægipan*, Trim.; the little *Meninx*, Trim., is gaily yellow-spotted along the hind-margins on the under-side, and it, *Syrinx*, and *Lepeletierii* alike have two longitudinal white stripes on the under side of the hind-wings. *Willemi* presents on the under side of the hind-wings a very peculiar pattern of black neuration and cross-streaks on a pale yellow-ground.

The prominent distinguishing characters of this genus are the absence of any appendage to the fore-tibiæ, the not infrequent absence of the second or middle pair of spurs on the hind-tibiæ, the slenderness and length of the body, the bristly hairiness of the palpi, and the usually more or less bluntly-ending club of the antennæ.

A good many species seem to have been incorrectly recorded under this genus, especially various Tropical South-American forms described and figured by the Felders in the third volume of the Lepidoptera of the *Reise der Novara* (1867), pp. 521–523, pl. lxxiv. Besides the five Palæarctic species, there are two from North America and two from Chili, and as many as eleven from the Ethiopian Region (two peculiar to Madagascar). Of the African species, nine are found in South Africa, and of these five—viz., *Syrinx*, *Ægipan*, *Meninx*, *Tsita*, and *Inornatus*—appear to be peculiar to the Sub-Region. One, *Malgacha*, is also a native of Madagascar; *Lepeletierii* ranges far along the western coast of the Southern Tropical belt; *Willemi* occupies great part of that belt, and is also recorded from Somaliland, in North-East Africa; while *Metis*, the most richly-coloured of the genus, is not known to occur beyond the Sub-Region except in Angola.

In South Africa *Metis* is the most numerous and widely distributed species, and *Malgacha* is only second to it. *Lepeletierii* is apparently confined to a few localities in Cape Colony, particularly towards the western side. *Tsita* and *Inornatus* are, on the contrary, widely spread on the eastern side of the country, and come but little westward of the Kei River. *Willemi* and *Meninx* are Transvaal natives; and the exceedingly local *Syrinx* and *Ægipan* have been found only in a few very elevated stations in the eastern districts of Cape Colony, Basutoland, and Natal.

315. (1.) Cyclopides Metis, (Cramer).

♂ *Papilio Metis*, Linn., Mus. Lud. Ulr. Reg., p. 325, n. 143 (1764); and Syst. Nat., i. 2, p. 792, n. 245 (1767).
♂ ,, ,, Dru., Ill. Nat. Hist., ii. pl. xvi. ff. 3, 4 (1773).
♂ ,, ,, Cram., Pap. Exot., ii. pl. clxii. f. G (1779).
♂ ,, ,, Wulf., Capens. Ins., p. xxxiii. n. 32 (1786).
♂ ♀ *Hesperia Metis*, Latr., Enc. Meth., ix. p. 776, n. 129 (1823).
♂ ♀ *Cyclopides Metis*, Trim., Rhop. Afr. Aust., ii. p. 293, n. 182 (1866).
♂ *Heteropterus Metis*, Staud., Exot. Schmett., i. pl. 100 (1888).

Exp. al., (♂) 1 in. 1½–2½ lin.; (♀) 1 in. 1–2½ lin.

♂ *Dark purplish-brown, with orange-yellow spots*. *Fore-wing*: basal area irrorated with yellow scales in three longitudinal streaks, viz., on costa, below median nervure, and on inner margin; two spots on costa about middle, the upper just above and partly beyond the lower (which is in cell); beyond middle, a transverse row of five spots, of which the second is beyond the line of the others, the third and fourth only separated by second median nervule, and the fifth (just above submedian) smallest and sometimes obsolete. *Hind-wing*: basal irroration confined to neighbourhood of median and submedian nervures;

a large subquadrate spot at end of cell, with a smaller spot just below it; a submarginal row of seven or eight spots, of which the first, fourth, and fifth are the largest, and the last three ill-defined. *Cilia* of the ground-colour, but in hind-wing yellow near and at anal angle. UNDER SIDE.—*Paler, more glossy. Fore-wing:* spots larger, paler; basal irroration forming a distinct streak on costa, and another in cell (the latter confluent with cellular spot). *Hind-wing:* unicolorous, excepting only a small yellow spot on costa near base. *Cilia* as above.

♀ *Brown, without purplish tinge; spots paler, larger. Fore-wing:* spots in and above cell forming one marking; more irroration in cell. *Hind-wing:* the small spots of submarginal row very small, some or all of them occasionally wanting (in one example, the first spot is very small, and the only two others, the fourth and fifth, are scarcely visible as dots). *Cilia* yellow throughout in hind-wing, and at anal angle of fore-wing. UNDER SIDE.—*Hind-wing and apex of fore-wing marked with ferruginous-ochreous, which in the former broadly indicates the position of the spots of upper side Fore-wing:* no cellular streak from base; cellular spot often confluent with third spot of transverse row.

Head and body of ground-colour intermingled with golden-yellow. Antennæ half-ringed (inwardly) with alternate dark-brown and golden-yellow, and tipped with the latter; palpi with mixed hairs of the same two colours. Abdomen superiorly rather inconspicuously half-ringed with the two colours and with a terminal tuft of golden yellow. In ♀ the yellow is throughout paler and duller.

A ♂ taken at Etshowe in Zululand by Mr. T. Vachell, of the 27th (Inniskillen) Regiment, has on the upper side the spots of the fore-wings unusually small, and the lowest spot of the discal row wanting; the under side is normal.

The very deep rich colouring of this well-known species—the golden-yellow spots being very conspicuous on the dark purple-brown ground—at once separates *Metis* from its congeners; and in the ♂ the very dark unmarked under side of the hind-wings is a characteristic feature. It seems to present no variation over its extensive South-African range.

At Cape Town this species is common, especially about hedgerows and in gardens, where it is fond of settling on leaves in sheltered situations. Though it often when settled keeps the wings fully expanded, I have noticed that it sometimes holds them all vertically erect, the wings of the right and left sides not touching, but standing parallel quite apart. On the wing it is active, but with a somewhat fluttering motion. It occurs throughout the year, but is rare during the winter months. In the Botanic Gardens at Cape Town this butterfly is one of the most frequent victims of the climbing South-American Asclepiad, *Physianthus albens*, which nips with a vice-like tenacity the proboscis of any insect attempting to rifle its nectaries, *Metis* with its long trunk thus sharing the fate of several Noctuæ and Sphinges, and being held prisoner till it dies.

Localities of *Cyclopides Metis*.

I. South Africa.
  B. Cape Colony.
    a. Western Districts.—Cape Town. Genadendal. Caledon District (*G. Hettarsch*). Swellendam (*A. C. Harrison*). Knysna and Plettenberg Bay.
    b. Eastern Districts.—Uitenhage. Tharfield, Bathurst District (*Mrs. Barber* and *Miss M. L. Bowker*). Grahamstown. King William's Town (*W. S. M. D'Urban*). Windvogelberg, Queenstown District (*Dr. Batho*).
  D. Kaffraria Proper.—Bashee River (*J. H. Bowker*).
  E. Natal.
    a. Coast Districts.—D'Urban (*J. H. Bowker*).
    b. Upper Districts.—Tunjumbili.
  F. Zululand.—Etshowe (*A. M. Goodrich* and *T. Vachell*).
  K. Transvaal.—Lydenburg District (*T. Ayres*).

II. Other African Regions.
  A. South Tropical.
    a. Western Coast.—"Angola (*J. J. Monteiro*)."—Druce.

316. (2.) Cyclopides Malgacha, (Boisduval).

Steropes Malgacha, Boisd., Faune Ent. Madag., &c., p. 67, n. 1 (1833).
Hesperia Limpopona, Wallengr., K. Sv. Vet.-Akad. Handl., 1857; Lep. Rhop. Caffr., p. 50, n. 5.
♂ ♀ *Cyclopides Malgacha*, Trim., Rhop. Afr. Aust., ii. p. 294, n. 183, pl. v. f. 10 [♂], (1866).
♂ *Cyclopides Malgacha*, Grandidier, Hist. Madag., xix., Lepid., ii. 1, pl. lii. ff. 6, 6a (1885).

Exp. al., (♂) 1 1 lin.—1 in. 2 lin.; (♀) 1 in. 1–2 lin.
Allied to *C. Metis*, (Cram.), but smaller and paler.

♂ *Greyish-brown, with small yellow spots arranged as in Metis.* Fore-wing: basal irroration more evenly spread, not distributed in rows; two costal spots more separate; *an indistinct submarginal row of small yellowish spots*. Hind-wing: spots of submarginal row *all* small, those near costa commonly wanting. Cilia of *fore-wing* of the ground-colour slightly mixed with yellowish, of *hind-wing* yellow. UNDER SIDE.—*Hind-wing and apex of fore-wing creamy-greyish or yellowish-ochreous, the former with spots of upper side faintly reproduced in paler yellow.* Fore-wing: spots much larger than above; costa and cell closely irrorated with yellow.

♀ *Duller, paler, spots larger:* cilia pale-yellowish throughout. UNDER SIDE.—*Hind-wing:* spots more distinct.

A ♂ in my collection, captured near Cape Town, *wants all the spots in fore-wing* except the minute ones of submarginal row; on the *under side*, however, the spots near costa are observable, but small and ill-defined.

Six Basutoland examples, of which two are ♀ s, have the under side colouring of the hind-wings and apices of fore-wings considerably paler than in Cape specimens. One of the females is remarkable for the well-defined rows of submarginal spots on the upper side (especially in the hind-wings), and for the vivid orange of the spots on the under side.

A further and more decided variation occurs in a ♂ from the Potchefstroom District, Transvaal, in which the upper-side spots are in both wings much reduced in size, but the under side of the hind-wing is very much darker and browner than usual, with all its yellow spots remarkably distinct and sharply defined.

About Cape Town and in the Cape Peninsula this is a rarer insect than *Metis*, and does not seem to occur during the winter months, first appearing about the middle of September. In habits it nearly resembles the species named, but is much less conspicuous, and flies more slowly, always keeping near the ground.

Apart from its being a native of Madagascar, *Malgacha* appears to have much the same distribution as *Metis*, but I have not seen any specimens from Zululand.

Localities of *Cyclopides Malgacha*.

I. South Africa.
   B. Cape Colony.
      a. Western Districts.—Cape Town; and Hout Bay, Noord Hoek, and Simon's Town, Cape District. Eerste River and Stellenbosch. Paarl. Robertson. Swellendam (*A. C. Harrison*). Plettenberg Bay.
      b. Eastern Districts.—Between Somerset East and Murraysburg (*J. H. Bowker*). Stormbergen (*Mrs. Barber*). Burghersdorp (*D. R. Kannemeyer*).
      c. Basutoland (*J. H. Bowker*).
   C. Orange Free State.—Vet River (*C. Hart*).
   D. Kaffraria Proper.—Tsomo River and Heads of St. John's River (*J. H. Bowker*).
   E. Natal.
      b. Upper Districts.—Karkloof (*J. H. Bowker*).
   K. Transvaal.—Potchefstroom District (*T. Ayres*).

II. Other African Regions.
   A. South Tropical.
      bb. Eastern Islands.—" Madagascar: Tamatave."—Boisduval.

317. (3.) Cyclopides Syrinx, Trimen.

♂ *Cyclopides Syrinx*, Trim., Trans. Ent. Soc. Lond., 1868, p. 93, pl. v. f. 8; and ♀, ibid., 1870, p. 387.

Exp. al., (♂) 1 in. 2–3 lin.; (♀) 1 in. 2½–4 lin.

♂ *Dull-greyish brown, with small ill-defined dull pale-yellowish spots; basal areas sparsely irrorated with pale-yellowish. Fore-wing*: a spot in discoidal cell near extremity, sometimes surmounted by two yellowish costal dashes; a discal series of six or seven spots, viz., three or four forming an oblique subcostal outward-inclining streak between end of cell and apex, and three others in an inward-inclining line between third median nervule and submedian nervure; inner margin edged rather faintly with the same dull pale-yellowish from near base to beyond middle; in some specimens a submarginal series of six very indistinct elongate yellowish marks. *Hind-wing:* a clothing of yellowish-grey hairs from base over cell and along inner margin; the trace of a spot in cell; two or three discal spots (the uppermost one rather large) between second subcostal nervule and submedian nervure; a submarginal series of marks as in fore-wing, but longer, usually obsolete near apex. *Cilia* of the ground-colour slightly mixed with yellowish. UNDER SIDE.—*Hind-wing and costal and apical border of fore-wing brownish yellow-ochreous; the hind-wing with two longitudinal whitish stripes. Fore-wing:* lower discal spots usually, cellular spot rarely, obsolete; a rather brighter tinge of yellow along costa; no trace of submarginal marks. *Hind-wing:* upper longitudinal stripe, from base through discoidal cell to hind-margin, broad, pretty even in width, creamy or yellowish-white; lower stripe narrow, sublinear near base, but somewhat widening to hind-margin, running between median and submedian nervures, white without yellowish tinge; no other markings.

♀ *Spots more decidedly yellow. Fore-wing:* only two spots in costal part of discal series, the first and fourth being obsolete. In both wings submarginal marks almost obsolete. UNDER SIDE.—*Hind-wing (except inner-marginal fold) and costal and apical border of fore-wing much paler, inclining to greyish;* both longitudinal stripes of hind-wing broader than in ♂, the superior one yellower, and the inferior one whiter.

In this *Cyclopides* the ordinary spots are pallid and poorly defined, and in some specimens obscured and sub-obsolete. One ♂ is entirely devoid of spots on both surfaces except for a trace of the superior discal spot on the upper side of the hind-wings.

C. Syrinx to a great extent combines the characters of *C. Malgacha,* (Boisd.), and *C. Lepeletierii,* (Latr.), differing from the former in having conspicuous pale stripes on the under side of the hind-wings, and from the latter in being spotted with yellowish.

This appears to be not only a strictly mountain butterfly, but one confined to the highest elevations and to particular spots. Colonel Bowker is the only collector that to my knowledge has taken this insect; he discovered it on the 19th January 1867 on the summit of Gaika's Kop, the highest point of the Amatola Mountains (Queenstown District of Cape Colony), being estimated as about 6500 feet above the sea. He noted it as being very numerous among a plant named "Mountain Bamboo," flitting about in hundreds, but not occurring anywhere beyond the immediate vicinity of that plant. The only other recorded locality for the species is the Maluti Mountains in Basutoland, where, in January 1869, Colonel Bowker again met with it, frequenting the same sort of Mountain Bamboo as on the Amatolas. Among eleven specimens received, two only are females.

Localities of *Cyclopides Syrinx*.

I. South Africa.
   B. Cape Colony.
      b. Eastern Districts.—Gaika's Kop, Amatola Mountains, Queenstown District (*J. H. Bowker*).
      c. Basutoland.—Maluti Mountains (*J. H. Bowker*).

318. (4.) Cyclopides Ægipan, Trimen.

♂ ♀ *Cyclopides Ægipan*, Trim., Trans. Ent. Soc. Lond., 1868, p. 94, pl. vi. f. 9 [♂].

Exp. al., (♂) 1 in. $1\frac{1}{2}$-$2\frac{1}{2}$ lin.; (♀) 1 in. $2\frac{1}{2}$ lin.

♂ Glossy *dark-brown* (in *five examples with a submetallic bronzy-greenish gloss*); *cilia paler brown*, very glossy, mixed with yellow in hind-wing. Fore-wing: some inconspicuous ochre-yellow irroration about base, chiefly on costa; beyond middle, a short costal mark of two or three very small indistinct yellowish spots (sometimes obsolete). UNDER SIDE.—*Hind-wing and costal border of fore-wing very glossy yellowish-grey.* Fore-wing: ground-colour much paler than on upper side, not much darker than hind-wing in some examples; costal mark usually better developed, chrome-yellow; *a hind-marginal chrome-yellow border* from apex (where it is of moderate width) narrowing to a point about second median nervule. *Hind-wing:* inner-marginal fold darker, about the same tint as ground-colour of fore-wing.

Palpi with a dense admixture of yellow hairs, with fewer fuscous ones; *abdomen* with lateral stripe, under side of anal segment, and conspicuous terminal tuft chrome-yellow.

♀ *Paler, the brown ground-colour with a yellower tinge; fore-wing with three chrome-yellow marks.* Fore-wing: in discoidal cell a transversely elongate spot near extremity; costal mark beyond middle much larger than in ♂, well defined, tripartite; beneath the latter, between second and third median nervules, a small spot. *Cilia* of fore-wing mixed with yellowish-grey; of hind-wing, wholly pale chrome-yellow. UNDER SIDE.—Hind-wing yellower; fore-wing with costal mark conspicuous, but other spots indistinct.

This species is a near ally of *Malgacha*, Boisd., but readily distinguishable by its larger size, total want of spots in the hind-wings, and yellow apical hind-marginal edging on the under side of the fore-wings. The spotless, or all but spotless, upper side resembles that of *Lepeletierii*, Latr., but from that species and from *Syrinx*, Trim., this butterfly is easily known by the absence of white stripes on the under side of the hind-wings.

Like *C. Syrinx*, Trim., this Hesperid frequents high elevations, Colonel Bowker having discovered it on the Hog's Back in the Amatola range on 20th January 1867, where he reported it to be not uncommon among long grass near water. With the exception of a single example from the Transvaal from Mr. T. Ayres in 1879, the insect had not again been received by me until February 1888, when four fine ♂ s arrived from Mr. J. M. Hutchinson of Estcourt, Natal. These specimens were taken on Table Mountain, Ulundi, in Weenen County, at a height of 6000 feet; and Mr. Hutchinson wrote that he took thirty examples, and could have easily taken many more without moving from one place, as they flew low and were easily captured. Only a single ♀ has reached me, accompanying five ♂ s from the Amatola station.

Localities of *Cyclopides Ægipan.*

I. South Africa.
  B. Cape Colony.
    b. Eastern Districts.—Hog's Back, Amatola Mountains (*J. H. Bowker*).
  E. Natal.
    a. Upper Districts.—Ulundi, Weenen County (*J. M. Hutchinson*).
  K. Transvaal.—Lydenburg District (*T. Ayres*).

319. (5.) Cyclopides Meninx, Trimen.

♂ *Cyclopides Meninx*, Trim., Trans. Ent. Soc. Lond., 1873, p. 121, pl. i. f. 12.

♂ ♀ *Thymelicus Meninx*, Wallengr., Ofv. K. Vet.-Akad. Förh., 1875, p. 92.

♂ *Cyclopides argenteostriatus*, Plötz, Stett. Ent. Zeit., 1886, p. 110, n. 19b.

Exp. al., (♂) $10\frac{1}{2}$—$11\frac{1}{2}$ lin.; (♀) $11\frac{1}{2}$ lin.

♂ *Dark-brown, rather glossy, without markings; cilia paler, more glossy.* UNDER SIDE.—*Paler; a hind-marginal series of chrome-yellow spots; hind-wing with three longitudinal silky-white stripes. Fore-wing:* four or five hind-marginal spots, small, triangular, inter-nervular, between apex and third or second median nervule, diminishing in size downward; on costa, beyond middle, three or four inter-nervular yellow dashes, and immediately beneath the outermost but one of these, usually two small similar marks, one above the other. *Hind-wing:* upper silky-white longitudinal stripe broad, commencing abruptly in discoidal cell near extremity, and reaching to hind-margin; lower one much narrower, running from base between median and submedian nervures, and fining off into a thin line before reaching hind-margin, and the lowest one forming a narrow inner-marginal edging from base to beyond middle; six hind-marginal yellow spots, larger than in fore-wing,—the first sublinear, elongate,—the third at extremity of upper white stripe; some indistinct thin inter-nervular yellowish striæ on disc, and a short disco-cellular one at base.

Head and body all very dark brown; among hairs of palpi a few dull-yellowish ones; antennæ very dark brown above, but white beneath.

♀ *Paler, and with a slight tinge of ochreous-yellow. Fore-wing:* yellow subcostal marks beyond middle larger, better defined, three together forming a short transverse streak; below and beyond these a fourth very small yellow spot. UNDER SIDE.—As in ♂, but all hind-marginal spots larger, and the subcostal marks of fore-wing as on upper side, but larger.

This little *Cyclopides* is on the upper side in both sexes like a miniature *C. Ægipan*, Trim., but is strikingly different on the under side, the marginal yellow spots and the shining white stripes of the hind-wings giving it a very varied and unique aspect. Except for the under-side stripes, *Meninx* bears a near resemblance to a South-American species, figured under the name of *Menes* by Cramer (*Pap. Exot.*, iv., pl. ccccxciii. ff. H, 1) and Stoll (*Suppl.*, pl. vii. f. 6 G).

Mr. W. Morant, in December 1868, first made this butterfly known to me, forwarding a ♂ which he had taken near Potchefstroom, Transvaal, and which I described as the type of *C. Meninx* in 1872. Later in the latter year, I received from him seven ♂s and a ♀ from the same locality, as well as a ♂ from Mr. T. Ayres, also taken at Potchefstroom. In the collection acquired from Mr. Ayres in 1879, the South-African Museum received four fine ♂s, all noted as taken in the Potchefstroom District. Mr. Morant described the insect as occurring early in December and flying feebly, and both he and Mr. Ayres stated that it was confined to marshy ground and the banks of streams. Wallengren (*Öfv. K. Vet.-Akad. Förh.*, 1875, p. 92) mentions that Mr. N. Person found *Meninx* pretty commonly at Potchefstroom in damp spots during the months December to February.

Localities of *Cyclopides Meninx*.

I. South Africa.
    E. "Natal."—Plötz.
    K. Transvaal.—Potchefstroom and District (*W. Morant* and *T. Ayres*).

320. (6.) Cyclopides Willemi, (Wallengren).

♂ *Heteropterus Willemi*, Wallengr., K. Sv. Vet.-Akad. Handl., 1857; Lep. Rhop. Caffr., p. 47, n. 2.
♂ *Cyclopides? Willemi*, Trim., Rhop. Afr. Aust., ii. p. 296, n. 186 (1866).
Cyclopides Chiles, Hewits., Descr. 100 N. Sp. Hesp., ii. p. 42, n. 5 (1868); and Exot. Butt., v. pl. 59, ff. 12, 13 (1874).

Exp. al., (♂) 1 in. 3–3½ lin.

♂ *Dark-brown, very slightly glossy; fore-wing with an outer-discal and also a submarginal series of dull pale-yellowish spots.* Fore-wing: spots of discal series elongate, the first three forming (with a minute superior costal dot of the same colour) a short transverse subcostal streak, the fourth and fifth nearer hind-margin, and almost touching third and fourth spots of submarginal series, and the ninth and seventh (of which the latter is small and sometimes obsolete) almost in a line below first three; eight spots in submarginal row, of which the four upper are linear and indistinct, but the four lower much broader and distinct. *Hind-wing* without markings. UNDER SIDE.—*Hind-wing and apical hind-marginal border of fore-wing yellowish-white*, with *strongly blackish-clouded neuration. Fore-wing:* ground a little paler than on upper side; costal edge and upper side of costal nervure with a line of yellowish-white; only first three spots of discal series represented, and those smaller and more separate than on upper side; apical hind-marginal yellowish-white border wide at apex, but narrowing to a point above submedian nervure, divided into seven or eight separate, elongate, inwardly acuminate spots by the blackish-clouded nervules. *Hind-wing:* neuration, a streak from base through discoidal cell to hind-margin, two very irregular median and two less irregular submarginal transverse blackish streaks from costal to submedian nervure, and inner-marginal fold except along inner edge, all strongly blackish.

Head and palpi clothed with a mixture of yellow and blackish hairs. Thorax blackish above, with mixed yellow and blackish hairs frontally and laterally; yellowish-white beneath. Legs yellowish-white, with a black middle streak longitudinally. Abdomen blackish above, yellowish-white beneath, and with a small terminal yellow tuft.

The blackish neuration and irregular transverse stripes, conspicuous on the almost white ground of the under side of the hind-wings, present a very peculiar reticulated appearance, and give *Willemi* some resemblance to the larger and beneath much more conspicuously ornamented *C. Steropes* (W. V.) of Europe; but the butterfly has no very near congener known to me.

Very few examples of this species have come under my notice, and I have not seen the female. As far as known, it would appear to inhabit chiefly the country skirting the Southern Tropic, Mr. Hewitson recording it from Damaraland, while a few have been taken in Bechuanaland and the Eastern Transvaal. In 1886 Mr. A. G. Butler recorded its having occurred in the remote region of Somaliland.

Localities of *Cyclopides Willemi.*

I. South Africa.
   K. Transvaal.—Lydenburg District (*T. Ayres*) and Barberton (*C. F. Palmer*). Crocodile River (*F. H. Barber*). Marico River (*F. C. Selous*).

II. Other African Regions.
   A. South Tropical.
     a. Western Coast.—" Damaraland."—Hewitson.
     b1. Eastern Interior.—Between Transvaal and Gubulewayo (*A. W. Eriksson*).
   B. North Tropical.
     b. Eastern Coast.—" Somaliland (*Thrupp*)."—Butler.

321. (7.) Cyclopides Lepeletierii, (Latreille).

Hesperia Lepeletier, Latr., Enc. Meth., ix. p. 777, n. 134 (1823).
Cyclopides Lepeletierii, Trim. [part], Rhop. Afr. Aust., ii. p. 295, n. 184 (1866).

Exp. al., (♂) 1 in. 1½ lin.; (♀) 1 in. 2 lin.
  ♂ *Rather glossy dark-brown, unicolorous, without marking of any*

kind; cilia greyer, more glossy. UNDER SIDE.—*Hind-wing and costal and broad apical hind-marginal border of fore-wing reddish-brown with a tinge of yellow-ochreous. Fore-wing:* ground rather paler and greyer than on upper side. *Hind-wing: two white longitudinal streaks,*—the upper one very conspicuous, sharply defined, running from base through discoidal cell to hind-margin,—the lower one duller and thinner, running from base to hind-margin between first median nervule and submedian nervure, not well-marked before middle; inner-marginal fold dull greyish-brown.

Head and body fuscous-brown above, the thorax beneath clothed with greyish hair, and the abdomen with a whitish median stripe. Antennæ blackish above, barred with brown and white beneath; palpi with fuscous hairs above, but beneath with whitish basal scales and mixed sandy and grey hairs.

♀ Like ♂, but fore-wings rather longer and acuter, and cilia greyer. UNDER SIDE.—Paler, and with a stronger tinge of ochreous-yellow than in ♂. *Hind-wing:* upper stripe thinner and not so purely white or so sharply defined; lower stripe almost obsolete.

VARIETY A.—Much larger; *exp. al.* (♂) 1 in. $3-3\frac{1}{2}$ lin.; (♀) 1 in. 4–6 lin.

Paler; cilia greyer. Under side paler; in hind-wing some more or less developed white irroration immediately above lower white stripe. In both sexes, but especially in the ♀, the fore-wings are more pointed at apex. (*Hab.*—Piketberg, Tulbagh, and Swellendam Districts, Cape Colony.)

I think that the smaller examples above described are really referable to Latreille's species, the expanse of which is given as about an inch; but I am doubtful as to the much larger Variety A. The only known locality of the smaller (typical) form is Uitenhage in the Eastern Cape Colony, while the variety belongs to the Western Districts. The Kaffrarian specimens included under this species in my former catalogue must, I think, be removed to the next species, *C. Tsita*, mihi, although the presence of the reduced white streaks on the under side of the hind-wings approximates them to *Lepeletierii*.

Colonel Bowker took a few of the typical form (including one ♀ example) at Uitenhage in October 1879 and in 1881. I received from Mr. S. D. Bairstow a ♂ captured at the same place. I first met with Var. A. at Vogel Vley, Tulbagh District, in October 1863; and subsequently (September 1869) took a good many specimens of both sexes in the Piketberg District. The insect frequents broken shrubby slopes and banks of streams, preferring stony places; it settles constantly on the ground and on stones, and is easily captured.

Localities of *Cyclopides Lepeletierii*.

I. South Africa.
    B. Cape Colony.
        a. Western Districts.—Vogel Vley, Tulbagh District. Berg River Bridge and Piketberg. Swellendam (*L. Taats*)—Var. A.

  b. Eastern Districts.—Uitenhage (*J. H. Bowker* and *S. D. Bairstow*).

II. Other African Regions.
 A. South Tropical.
  a. Western Coast. — Congo: "Kinsembo (*H. Ansell*)."—Butler. "Angola (*Pogge*) and Chinchoxo (*Falkenstein*)."—Dewitz.

322. (8.) Cyclopides Tsita, Trimen.

Cyclopides Tsita, Trim., Trans. Ent. Soc. Lond., 1870, p. 386, pl. vi. f. 13.

Exp. al., (♂) 1 in. 1–4 lin.; (♀) 1 in. 3–3½ lin.
Allied to *C. Lepeletierii,* Latr., and *C. inornatus,* Trimen.

 ♂ *Glossy greyish-brown, unicolorous, without markings of any kind; cilia paler, with a silky lustre.* UNDER SIDE.—*Hind-wing (except fuscous-brown inner-marginal fold) and costal and apical-hind-marginal border of fore-wing paler brown with a reddish tinge, and with the neuration usually finely whitish in parts; hind-margin of both wings with a fine whitish edging line.* Hind-wing: two longitudinal white streaks (as in *Lepeletierii*) but more attenuated,—in some specimens quite faint or reduced to mere whitish lines; neuration generally whitish, but seldom so above disco-cellular fold; between the two white streaks (even when the streaks are scarcely represented) some more or less developed diffuse white irroration.

 ♀ Like ♂, but rather paler on both surfaces, and with longer and more pointed fore-wings. UNDER SIDE.—Fine white neuration general and better defined, but white stripes reduced to mere lines like the white nervures.

 It is in specimens from King William's Town and Kaffraria Proper that the white stripes and irroration of the under side of the hind-wings are best expressed. A constant feature distinguishing *Tsita* from *Lepeletierii* is the fine white neuration more or less prevalent on the under side, to which should be added the fine white hind-marginal edging line. The Natal examples that I have met with are smaller than usual, and, like the Basutoland ones on which I founded the species, have the white stripes on the under side of the hind-wings no wider and no more conspicuous than the adjacent white nervures.

 Mr. W. S. M. D'Urban noted this obscurely tinted species as abundant in the King William's Town District; and Colonel Bowker found it commonly about grassy spots in Kaffraria Proper, and in similar places by river-banks in Basutoland. The examples that I captured on the Natal coast haunted similar localities, flitting about long grass in February and March. A specimen taken in Zululand by Captain Goodrich is ticketed "November 1886," and three from Weenen County, Natal, were captured by Mr. J. M. Hutchinson in January 1888.

Localities of *Cyclopides Tsita*.

I. South Africa.
  B. Cape Colony.
    b. Eastern Districts.—King William's Town (*W. S. M. D'Urban*).
    d. Basutoland.—Koro-Koro (*J. H. Bowker*).
  C. Orange Free State.—Parijs, Upper Vaal River (*E. G. Alston*).
  D. Kaffraria Proper.—Tsomo and Bashee Rivers (*J. H. Bowker*).
  E. Natal.
    a. Coast Districts.—D'Urban (*J. H. Bowker*). Verulam and Itongati River.
    b. Upper Districts.—Ulundi, Weenen County (*J. M. Hutchinson*).
  F. Zululand.—Etshowe (*A. M Goodrich*).
  K. Transvaal.—Potchefstroom and District (*T. Ayres*). "Schoman's Farm, Vaal River (*N. Person*)."—Wallengren.

323. (9.) Cyclopides inornatus, Trimen.

Cyclopides inornatus, Trim., Trans. Ent. Soc. Lond, 3rd Ser., ii. p. 179 (1864); and Rhop. Afr. Aust., ii. p. 295, n. 185, pl. 5, f. 11 (1866).

Exp. al., (♂) 1 in.—1 in. 1 lin.; (♀) 1 in. 1–1½ lin.

♂ *Dull-brown, rather glossy; cilia paler, greyish externally.* Forewing: beyond middle, about midway between apex and extremity of discoidal cell, an oblique subcostal row of three exceedingly indistinct pale-greyish dots. UNDER SIDE.—*Hind-wing, and costal border, and broad apical hind-marginal area of fore-wing pale reddish-brown, with a tinge of ochreous-yellow;* in both wings an indistinct terminal discocellular dull greyish-white dot, and a discal series (elbowed in forewing and very strongly curved in hind-wing) of six and eight similar dots. *Hind-wing*: inner marginal fold pale-grey; a greyish white dot near base, between costal and subcostal nervures.

♀ *Like ♂, but fore-wing longer and more pointed apically, and discocellular and lower discal dots of under side (which are better expressed than in ♂) sometimes reproduced in fore-wing.*

This species is well distinguished from *Tsita*, Trim., by its smaller size, minute subcostal spots in the fore-wings, and redder under side (without whitish neuration or streaks, but with minute pale discal spots).

Colonel Bowker discovered this remarkably dull-coloured and inconspicuous little butterfly in Kaffraria Proper. In February and March 1867 I took several examples on the coast of Natal, flitting about long grass; but it was certainly very local in its distribution. The only example known to me from within the boundary of the Cape Colony is a ♂ captured by Colonel Bowker in 1873 at Fort Warden, Kei River. At the end of October 1878 the same collector met with several fine specimens at Northdene, near D'Urban, Natal.

Localities of *Cyclopides inornatus.*

I. South Africa.
  B. Cape Colony.
    b. Eastern Districts.—Fort Warden, Kei River (*J. H. Bowker*)
  D. Kaffraria Proper.—Tsomo and Bashee Rivers (*J. H. Bowker*).
  E. Natal.
    a. Coast Districts.—D'Urban and Pinetown (*J. H. Bowker*). Verulam. Umvoti. Mapumulo.
  F. Zululand.—St. Lucia Bay (the late *Colonel H. Tower*).

Genus PYRGUS.

Pyrgus (Hübner, 1816), Westw., Gen. D. Lep., ii. p. 516 (1852).
Syrichthus and *Spilothyrus*, Boisd., "Gen. Ind. Méth., p. 35 (1840)."
Spilothyrus and *Scelothrix*, Rambur, "Cat. Lep. Andal., i. p. 63 (1858)."
Pyrgus, Trim., Rhop. Afr. Aust., ii. p. 286 (1866).
Hesperia, Kirby, Syn. Cat. D. Lep., p. 611 (1871).
Pyrgus and *Scelothrix*, Speyer, Stett. Ent. Zeit., 1878, pp. 187, 189.
Pyrgus, Speyer, *loc. cit.*, 1879, p. 492.
Hesperia, Moore, Lep. Ceylon, i. p. 182 (1881).
Pyrgus and *Carcharodus* (Hübner, 1816), Plötz, Mitth. Nat. Ver. Neu-Vorpomm., &c., 1884, pp. 2 and 23.

IMAGO.—*Head* not so broad as thorax, densely hairy; *palpi* rather long, with basal and middle joints densely hairy and sometimes bristly, and with terminal joint of variable length, obliquely or horizontally porrect, blunt, clothed with short appressed hairs; *antennæ* short, with a stout sub-cylindrical, rather gradually-formed, slightly compressed, usually straight (sometimes slightly bent) club, blunt at tip; tufts at bases of antennæ long, conspicuous.

Thorax robust, more or less hairy (especially posteriorly and beneath [1]); pterygodes with long hair. *Wings* blunt, not (or very slightly) prominent apically. *Fore-wings* with hind-margin usually rather convex (especially in ♀); neuration as in *Cyclopides*; in ♂ of some species (only *P. Elma* in South Africa), costa from near base to about middle recurved so as to form a groove or deep fold. *Hind-wings* prominent at anal angle; hind-margin usually entire, but moderately dentate in a few species (only *P. Elma* in South Africa); neuration as in *Cyclopides*.

Legs rather long; fore-tibiæ with well-developed acutely-pointed appendage; middle and hind tibiæ with very long and acute terminal spurs,—the latter also with a second pair of spurs well developed, hairy superiorly (and in ♂ of some species—none South-African—with a conspicuous tuft of very long hair springing from its base on

[1] The ♂ in a considerable section of European species (of which *Malvæ*, Linn., and *Alveus*, Hübn., are well-known members) presents a singular pair of scabbard-shaped scaly and hairy appendages, springing posteriorly from the breast at the base of the hind-legs, and about one-third the length of the abdomen. They partly cover a very deep longitudinal groove which occupies the basal portion of the under side of the abdomen.

inner side); femora finely hairy beneath; tarsi long and stout, spinulose.

Abdomen stout, of moderate length; in ♂ terminally tufted and laterally compressed, and beneath with a more or less deeply excavated median groove, widest at base and extending to beyond basal half.

LARVA.—Stout, somewhat tapering anteriorly, very sparsely and shortly pubescent; head of moderate size, granulated.

PUPA.—Rather stout; head prominent; thorax rounded and elevated dorsally; covered generally with a bluish or whitish powder.

(These characters of the earlier stages are from the descriptions and figures in Duponchel's *Iconographie, &c., des Chenilles*, i. (1849), of those of *P. Malvæ*, L.)

The butterflies of this genus are further distinguished by the length of the cilia of the wings, which is emphasised by their being white with conspicuous black interruptions at the extremities of the nervules.

Although, as indicated in the foregoing diagnosis, there exist considerable diversities in *Pyrgus* as far as the secondary sexual characters of the ♂ are concerned, I agree with Dr. Speyer's later view (*l. c.*, 1879) that these are insufficient to warrant the separation of the respective sets of species presenting them in the proposed distinct genera *Carcharodus* and *Scelothrix*. It is remarkable that, with the exception of *P. Elma*, Trim. (a member of the *Alceæ* group, which possesses the costal fold in the fore-wings), none of the South-African species presents any of the secondary ♂ characters in question.[1]

These small but robust butterflies have been well studied in Europe, where no fewer than seventeen species are found (besides several marked varieties), and representatives of the genus occur throughout, but are more numerous in the southern countries. The recorded *Pyrgi* amount in all to about seventy-six; the Palæarctic Region yielding twenty-two; the Nearctic fourteen; Central and South America fifteen; India and China six; Africa nineteen; and Australia, according to Plötz, one (*Argina*, H. Schäff.). There can be little doubt that many more species remain to be discovered; the small stature, plain colouring, and swift flight (combined with the very restricted localities of some of the forms) rendering these insects very likely to escape the collector's observation.

The black or blackish-brown (rarely greyish) ground-colour of *Pyrgus* is marked with white (rarely also with transparent) spots; and on the under side a similar pattern prevails, but the ground-colour (except the blackish disk of the fore-wings) is of some tint of yellowish-, reddish-, greyish-, or greenish-ochreous. The fourteen known natives of South Africa are for the most part (nine species) of this prevalent

---

[1] Plötz (*loc. cit.*, p. 6) notes of *P. Sataspes*, Trim., "der Umschlag der Vfl. ist orange;" but in this he appears to have been misled by the colourist of the figure in *Rhop. Afr. Aust.* (ii. pl. 5), who has given the costa too warm a tint, and probably conveyed the impression that this stripe of colour indicated the presence of a costal fold.

conspicuously white-spotted black pattern, and belong to the European group of *P. Sao* and *P. orbifer*, (Hübn.); but *P. Elma*, Trim. (a member of the *Alveæ* group), has the fore-wings with vitreous spots only; *P. Sandaster*, Trim., has the white spots exceedingly small, and the under side of the hind-wings dark-brown with two very sharply-defined dentate white stripes; and the more aberrant group (apparently peculiar to Africa) which contains *Mohozutza*, Wallengr., *Chaca*, Trim., and *Tucusa*, Trim., presents on the upper side, besides white spots on the fore-wings, a common submarginal row of pale fulvous spots, and on the under side a greyish-creamy ground-colour varied with fulvous and spotted with black. In South Africa the most numerous and widely distributed species are *Vindex*, Cram., *Mafa*, Trim., *Diomus*, Hopff., and *Elma*, Trim.,—*Vindex* and *Diomus* being common about Cape Town. *Asterodia* and *Saluspes*, Trim., seem next in width of range, the latter occasionally occurring near Cape Town. *Mohozutza*, Wallengr., has an extensive eastern distribution through Kaffraria Proper, Natal, Zululand, and Transvaal; *Dromus*, Plötz, has been received from Natal and Delagoa Bay, and *Tucusa*, Trim., from Natal and Transvaal. *Nanus*, Trim., is not uncommon about Cape Town, and extends over the Western and Central Districts of the Cape Colony and into Griqualand West. *Chaca*, Trim., is apparently very rare,—I only know of its occurrence near Grahamstown and in Kaffraria Proper; *Agylla*, Trim., is known from a few scattered stations in the Eastern Cape Colony and Griqualand West; and *Transvaalia*, Trim., is represented by two examples only from the Potchefstroom district.

Three species, *Diomus*, *Elma*, and *Chaca*, are recorded from localities within the South-Tropical belt of Africa, and two, *Vindex* and *Dromus*, from the western side of the North-Tropical belt in addition.

The members of this genus are of active habits; they have a short quick flight near the ground, stopping very abruptly, and settling with wings fully expanded, usually on the bare earth or on stones. They mostly frequent open ground, delighting in dry hill slopes and waste spots. *P. Elma* is the only South-African species known to me which is almost confined to wooded localities.

324. (1.) Pyrgus Vindex, (Cramer).

Papilio Vindex, Cram., Pap. Exot., iv. pl. ccliii. ff. G, H (1782).
Hesperia Vindex, Latr., Enc. Meth., ix. p. 785. n. 148 (1823).
? *Pyrgus Vindex*, Westw., Gen. D. Lep., pl. lxxix. f. 6 (1852).
    „ „ Trim., Rhop. Afr. Aust., ii. p. 287. n. 177 (1866).
? *Aberr.*—*Papilio Spio*, Linn., Mus. Lud. Ulr. Reg., p. 338, n. 156 (1764); and Syst. Nat., i. 2, p. 796, n. 271 (1767).

Exp. al., 10 lin.—1 in. 1 lin.

Brownish-black, with rather large white spots. *Fore-wing*: on costa an elongate whitish spot near base, and a thin white edging as far as apex, interrupted in four places; three spots in discoidal cell, the outer one lunular and closing cell, that nearest base subquadrate, the central one (nearer to basal than outer spot) quadrate, elongate; below first cellular spot, touching submedian nervure, a whitish spot divided by longitudinal fold; a discal row of four spots, of which the first is rather large and crossed by two subcostal nervules, the second small, between third and second median nervules, the third large and just below outer cellular spot and crossed by first median nervule, the fourth small and semicircular (above submedian nervure), and the nearest to base of the row; between this row and cilia two rows of white dots, the inner one of eight dots usually complete and well marked, the outer imperfect, with its dots between nervules and touching cilia. *Hind-wing*: a small round spot in discoidal cell near base; a rather, broad, short, white, transverse, discal band, abruptly commencing on second subcostal nervule, crosses end of cell and extends irregularly and more narrowly to submedian nervure; rows of dots as in fore-wing, but outer row more regular and perfect than inner, which is obsolete towards costa, but of which the first dot (on fold beyond cell) is enlarged into a good-sized spot. *Cilia* broad, white, broadly interrupted with black on nervules. UNDER SIDE.—*Costa and apex of fore-wing and ground-colour of hind-wing paler or darker yellowish-ochreous; black of fore-wing less dark. Fore-wing*: spots as on upper side, but all rather larger; costa from base broadly whitish. *Hind-wing*: costa edged with whitish, especially on basal prominence; discal band longer than on upper side, commencing on costa, but widely interrupted on subcostal nervules, its extremity joining a longitudinal white space on inner margin from base to anal angle, bounded by submedian; before middle a similar transverse white stripe, commencing with a large costal patch, more or less interrupted on subcostal nervure, and narrowing to join inner-marginal white on submedian. *Cilia* tinged with yellowish, the interruptions with ochreous.

Head above blackish, with a white spot on each side next the eye, but frontally clothed with a mixture of blackish, whitish, and yellowish hairs; pencil of hairs at base of each antenna mostly black, but partly yellowish; antennæ black above but white beneath, with the extremity of the club beneath ferruginous-red; palpi above with mixed hairs like those of front, but beneath with creamy-white hairs and scales only, except as regards terminal joint, which is black. *Thorax* above black, rather densely clothed frontally and laterally with yellowish-grey, posteriorly with hoary-grey hair; beneath very densely clothed with white or creamy-white hair; legs creamy, tinged with brownish superiorly. *Abdomen* above black, with very distinct white segmental incisions; laterally tinged with ochreous-yellow; inferiorly white.

Specimens from Kaffraria Proper, Natal, Zululand, and Griqualand West seem always to be larger, and with paler, larger-spotted under side than the typical form, figured roughly by Cramer, which prevails in the south of the Cape Colony.

I quote Westwood's figure (*op. cit.*) with some doubt, because only the upper side is depicted, and the narrower hind-wing band there shown makes it possible that the illustration may apply to the nearly allied *Mafa*, mihi.

As regards the *Spio* of Linnæus, Mr. C. Aurivillus, in his excellent *Recensio Critica Lepidopterorum Musei Ludovicæ Ulricæ*, published in 1882 in the *Transactions of the Royal Swedish Academy*, gives (Tab. i. ff. 3, 3*a*) copies of figures by Clerck (previously unpublished), which there appears to be no reason to doubt were made from Linné's type specimen. As stated in my letter quoted (pp. 124–125) by Mr. Aurivillius, I consider these figures to warrant the opinion that *Spio* is an aberration of *Vindex*, in which the largest discal spot in the fore-wing was confluent with the small spot which in ordinary examples is just below the former, and was also so much extended superiorly as to unite with the middle disco-cellular spot. The figure of the under side is rough and obscure—a fault very common with the older iconographers when dealing with small species—but is more like that of *Vindex* than any other species.

Vindex and its near ally *Dromus*, Plötz, are distinguishable among their South-African congeners by the broader and more sharply-defined white markings of the upper side, especially the median band of the hind-wings. Among European species, the southern *P. Sao*, Hübn., seems nearest to *Vindex*, but it has much smaller spots, and the underside ground-colour is ochreous-red.

This is a widely-distributed and common South African *Pyrgus*, frequenting open ground and waste places. About Cape Town it is numerous at the base and on the lower slopes of the mountains, settling with expanded wings on the ground, on stones, or on low plants. It appears almost throughout the year, June and July being the only months in which I have not noticed it, but is more abundant from August to October than at other seasons.

Localities of *Pyrgus Vindex*.

I. South Africa.
   B. Cape Colony.
      a. Western Districts.—Cape Town.   Muizenberg, Cape District. Hex River, Worcester District.   Robertson.   Oudtshoorn (—*Adams*).   Knysna and Plettenberg Bay.   Van Wyk's Vley, Carnarvon District (*E. G. Alston*).
      b. Eastern Districts.—Near Grahamstown (*W. S. M. D'Urban*). King William's Town (*W. S. M. D'Urban*).
      c. Griqualand West.—Vaal River (*J. H. Bowker*).   Kimberley (*H. L. L. Feltham*).
   D. Kaffraria Proper.—Bashee River (*J. H. Bowker*).
   E. Natal.
      a. Coast Districts.—D'Urban.   Itongati River.
      b. Upper Districts.—Udland's Mission Station.   Greytown.   Estcourt (*J. M. Hutchinson*).   Rorke's Drift (*J. H. Bowker*).

F. Zululand.—Napoleon Valley (*J. H. Bowker*). Etshowe (*A. M. Goodrich* and *T. Vachell*).
H. Delagoa Bay.—Lourenço Marques (*Mrs. Monteiro*).

II. Other African Regions.
  A. South Tropical.
    a. Western Coast.—" Angola (*Pogge*)."—Dewitz.
    b1. Eastern Interior.—Between Transvaal and Gubulewayo (*A. W. Eriksson*).
  B. North Tropical.
    a. Western Coast.—Gaboon River (*G. Geynet*). " Lower Niger (*W. A. Forbes*)."—Godman and Salvin.

325. (2.) Pyrgus Dromus, Plötz.

Pyrus Dromus, Plötz, Mitth. Naturw. Vereine Neu-Vorpomm. und Rügen, 1884, p. 6, n. 13.

Exp. al., 1 in.—1 in. 1½ lin.

Closely allied to *P. Vindex*, (Cram.). *White spots similarly arranged, but mostly somewhat larger, notably so (in fore-wing) last spot of discal series and third of submarginal series, and (in hind-wing) median band.* Fore-wing: third spot of discal series not (or but very slightly) extending below first median nervule, but fourth (last) spot larger (very much larger than in *Vindex*) and subquadrate; submarginal series of small spots more sinuate. *Hind-wing:* median band much wider inferiorly, being of almost even width throughout, much more curved, not marked by crossing nervures. UNDER SIDE.—*Colouring as in Vindex, but hind-wing with one rather wide continuous white median band from costa to submedian nervure.* Fore-wing: spots as on upper side. Hind-wing: close to base, four white spots, forming a more or less disconnected short transverse band; both inner and outer edges of median band irregularly dentated, the band itself being widest on costal edge and bisinuated, with a rather acute inward projection between median nervure and its first nervule.

The characters italicised appear to be quite constant, and serve well to distinguish this species from *Vindex*. It is common about D'Urban in Natal, where I took it in June 1865, and again in February and March 1867. I have received a dwarf ♂ from the Gaboon which expands only ten lines across the fore-wings.

Localities of *Pyrgus Dromus*.

I. South Africa.
  E. Natal.
    a. Coast Districts.—D'Urban.
  H. Delagoa Bay.—Lourenço Marques (*Mrs. Monteiro*).

II. Other African Regions.
  A. South Tropical.—" Congo."—Plötz.
  B. North Tropical.—Gaboon River (*G. Geynet*).

326. (3.) Pyrgus Mafa, Trimen.

Pyrgus Mafa, Trim., Trans. Ent. Soc. Lond., 1870, p. 386, pl. vi. f. 12.

Closely allied to *P. Vindex*, Cram.

Exp. al., 11½ lin.—1 in.

Black spotted with white; the spots in number and arrangement quite as in P. Vindex, but mostly smaller, and very sharply defined; the median band of hind-wing narrower, interrupted inferiorly and denticulate externally. UNDER SIDE.—*Hind-wing: the sub-basal and central white stripes rather narrow, not oblique, interrupted more or less markedly in two places*, viz., the former on costal and median nervures, and the latter on second subcostal and first median nervules; of the separate spots or portions of the stripes, the largest is the middle one of the central stripe, *which is denticulate both inwardly and outwardly, but much more strongly outwardly.*

It is doubtful whether this form should be regarded as a species distinct from *Vindex*, for the characters above indicated, though well-marked through a good series of specimens from different localities, are less distinctly expressed in three examples that I captured at Eerste River,—one of which has the upper side discal spots of the fore-wing as large as in ordinary *Vindex*.

Colonel Bowker found *Mafa* pretty numerously in Basutoland, but elsewhere it appears to be rather scarce, though with a wide distribution through the interior, and especially the North-Eastern districts. The few specimens that I have noticed in life were on the wing during September, with the exception of one which I took at Greytown in Natal on the 15th March.

Localities of *Pyrgus Mafa*.

I. South Africa.
  B. Cape Colony.
    a. Western Districts.—Eerste River, Stellenbosch District.
    b. Eastern Districts.—Burghersdorp (*D. R. Kannemeyer*). Kraai River, Barkly District (*J. H. Bowker*).
    c. Griqualand West.—Kimberley, Vaal River (*J. H. Bowker*).
    d. Basutoland.—Maseru and Koro-Koro (*J. H. Bowker*).
  C. Orange Free State.—Parijs, Upper Vaal River (*E. G. Alston*).
  E. Natal.
    b. Upper Districts.—Greytown. Estcourt (*J. M. Hutchinson*).
  K. Transvaal.—Potchefstroom District (*T. Ayres*).

327. (4.) Pyrgus Asterodia, Trimen.

Pyrgus Asterodia, Trim., Trans. Ent. Soc. Lond., 3rd Ser., ii. p. 178 (1864); and Rhop. Afr. Aust., ii. p. 289, n. 178, pl. 5, f. 6 (1866).

Exp. al., 10 lin.—1 in.

Allied to P. Vindex, Cram., *but ground-colour paler, more glossy; and spots smaller, not so purely white. Fore-wing:* three spots in cell, the *central largest and nearer to outer than inner spot* (the reverse being the case in *Vindex*); *above central spot is one on costa, composed of two short lineolæ*, no trace of which exists in any example of *Vindex*; row of spots beyond middle, including an additional dot (just above third median nervule), and more curved inwardly, so that the two lowest spots come in line with central cellular spot and that above it,—the lowest spot largest, while in *Vindex* the last but one is *invariably* the largest in the row; an interrupted submarginal row of dots, *but no restige of the row immediately before cilia*; spot beneath first cellular one wholly wanting. *Hind-wing:* transverse stripe commencing on costa, interrupted on first subcostal nervule, very much attenuated inferiorly; dot in cell near base small, indistinct; row of dots as in fore-wing. UNDER SIDE.—*Fore-wing:* apical colour warmer, less inclining to greyish. *Hind-wing: ground-colour more inclining to fulvous* (markedly so in some specimens); first stripe near base narrow, but joining basal white edging on costa; second irregular and denticulated, but continuous from costa to inner-marginal stripe, which is greyish rather than white; some fuscous variegation of ground-colour, especially on edges of stripes and spots of submarginal row, which latter are somewhat enlarged, though ill defined.

In the disposition and relative sizes of the discal series of spots in the fore-wing, and in the form and direction of the two white bands on the under side of the hind-wing, *Asterodia* is not unlike *Dromus*, Plötz; but, on the upper side, the much smaller and duller white markings (and especially the very much narrower and longer median stripe of the hind-wing) give it a totally different aspect, and the differences above emphasised seem constantly to separate it from both *Vindex* and *Dromus*. Its smaller size is also a distinction.

I discovered this *Pyrgus* at Plettenberg Bay as long ago as 1859. It frequented marshy plains about a stream called the Bitouw River, settling constantly on the flowers of low plants. It has not since occurred to me in life, and but very few examples have reached me from correspondents, although the species seems to be of tolerably wide distribution.

Localities of *Pyrgus Asterodia*.

I. South Africa.
    B. Cape Colony.
        a Western Districts.—Breede River, Swellendam District (*L. Taats*). Plettenberg Bay.
        b. Eastern Districts.—Between Zwartkops and Coega Rivers, Uitenhage District (*J. H. Bowker*). Grahamstown (*M. E. Barber*). Burghersdorp (*D. R. Kannemeyer*).
        d. Basutoland.—Koro-Koro (*J. H. Bowker*).
    K. Transvaal.—Potchefstroom (*W. Morant* and *T. Ayres*).

328. (5.) Pyrgus Agylla, *sp. nov.*

Exp. al., 10—11½ lin.

Closely allied to *P. Asterodia*, Trim., but smaller, *with the markings of a purer white, and the discal ones larger;* hind-wing with basi-cellular spot distinct and well developed, and submarginal spots larger. UNDER SIDE.—*Hind-wing and apex of fore-wing darker in ground-colour*, less strongly tinged with ochreous-yellow; *nervular interruptions of cilia conspicuously black*, without admixture of ochreous yellow. *Hind-wing: sub-basal and median white bands much more irregular and denticulate on both edges, and more sharply defined by darker fuscous edging in parts;*—the opposite projections of these bands all but touch each other in two places, viz., at origin of subcostal and of median nervules respectively; submarginal white spots enlarged, and near anal angle sometimes partly confluent.

Though so very near *Asterodia* in most respects, the under side, with its strongly black-marked cilia, and very sharply-defined, irregularly-shaped white bands on a ground darker than usual, gives this form a peculiar aspect, rendering it easily recognisable.

This little *Pyrgus* seems confined to the dry upland interior districts of the Cape Colony. It was first brought to my notice in 1871 by Colonel Bowker, who took a few examples (three at Hope Town on the Orange River and one between Somerset East and Murraysburg) in April of that year; and a few months later Mrs. Barber and he sent three examples from Griqualand West. In September 1872 I took a single specimen at Kimberley.

In two of Mrs. Barber's specimens the middle spots of the fore-wing are so enlarged as to give the effect of a straight white fascia from costa to submedian nervure.

Localities of *Pyrgus Agylla*.

I. South Africa.
  B. Cape Colony.
    b. Eastern Districts.—Between Somerset East and Murraysburg (*J. H. Bowker*). Murraysburg (*Mrs. Muskett*). Hope Town (*J. H. Bowker*).
    c. Griqualand West.—Kimberley. Between Riet and Modder Rivers (*Mrs. Barber*). Vaal River (*J. H. Bowker*).

329. (6.) Pyrgus Transvaaliæ, *sp. nov.*

Exp. al., 11 lin.

Allied to *P. Vindex*, Cram., and *Dromus*, Plötz.

♂ *Brownish-black, with very distinct, mostly rounded, white spots. Fore-wing:* spots in disposition and relative size most like those of *Dromus*, but *basi-cellular spots very narrow and longitudinally elongate,* terminal cellular spot more quadrate, and costal spot of discal row narrower and more distinctly tripartite; spots of strongly-sinuated submarginal series very distinct (in one specimen larger than in the other) *and of equal size throughout ;* no spot beneath basi-cellular one. *Hind-wing :* a distinct basi-cellular spot; median band resembling that of *Diomus,* Hopff., but not curved, more oblique, more macular (the crossing nervules forming strongly-marked interruptions); *submarginal series of spots more complete than in any of the allied forms,* there being a series of three (rather smaller than the rest) somewhat obliquely placed between first subcostal and third median nervules. UNDER SIDE.— *Hind-wing and costal and apical border of fore-wing glossy yellowish-brown ; all spots very distinct; hind-wing with median band not so macular as on upper side, but continuous from costa to submedian nervure, and with a sub-basal curved series of four conspicuous spots. Fore-wing :* an elongate small whitish spot immediately below discoidal cell near base (followed in one specimen by a second similar spot); *four upper spots of submarginal series much larger and more elongated longitudinally* than on upper side. *Hind-wing :* a white spot at base; first and second spots of sub-basal row rather large and pyriform, third spot very small, fourth spot narrow and elongate, adjoining greyish-white of inner-marginal fold; seven spots in median band (which is irregular in outline on both edges), the two lowest spots almost touching the fourth spot in sub-basal series; spots of irregular submarginal row considerably larger and more elongated than on upper side.

This is a very distinct little species, at once to be recognised by the thickly maculated under side of the hind-wings, owing chiefly to the unusual separation and definition of the sub-basal spots, and (to a less degree) of those forming the median band, and to the number and large size of the spots of the submarginal series.

Only two examples have come under my notice; one sent to me from the Transvaal by Mr. T. Ayres in 1878, and the other in a collection made by the same naturalist in that country, and acquired by the South-African Museum in the following year.

Locality of *Pyrgus Transvaaliæ.*

I. South Africa.
  K. Transvaal.—Potchefstroom District (*T. Ayres*).

330. (7.) Pyrgus Diomus, Hopffer.

Pyrgus Diomus, Hopff., "Monatsb. K. Akad. Wissensch. Berlin, 1855, p. 643;" and Peters' Reise Mossamb., Ins., p. 420, t. xxvii. ff. 9, 10 (1862).
Syrichtus ferax, Wallengr., Wien. Ent. Monatschr., 1863, p. 137.
Pyrgus Vindex, Cram., ? Var. Trim., Rhop. Afr. Aust., ii. pp. 287–288 (1866).
Hesperia (Syrichtus) Diomus, Wallengr., Sv. Vet.-Akad. Förh., 1872, p. 50.

Pyrgus Diomus, Möschl., Verh. Zool. Bot. Gesellsch. Wien, 1883, p. 286.
Hesperia Sandaster, Staud., Exot. Schmett., ii. pl. 100 (1888).[1]

Exp. al., 1 in.—1 in. 2 lin.

Aspect of *P. Vindex,* Cram., *but not of so deep a black, with the spots of a duller white and mostly rather narrower; cilia yellowish-white, with the black interruptions duller and not so sharply defined. Fore-wing:* basi-costal whitish more diffused, not forming a distinct white mark inwardly; cellular spots narrower, the inner one occasionally wanting; *in discal row there is no separate fourth spot, that marking being immediately and vertically beneath, and completely united with third spot. Hind-wing:* basi-cellular spot usually small and indistinct, and sometimes wanting; median band narrower, duller, and more oblique than in *Vindex,* crossed rather conspicuously by end of black median nervure and origins of its second and third nervules, and often narrowly and indistinctly prolonged towards costa. UNDER SIDE.—*Hind-wing and apical area of fore-wing pale greyish or yellowish-ochreous, usually with a slight greenish tinge in ♂; hind-wing with sub-basal and median white stripes very oblique,* well separated—the former rather vaguely defined internally and inferiorly,—the latter with well-defined edges (not much denticulated), continuous, and varying little in width from costa to where it joins anal-angular white; *white spots of submarginal row confluent into an irregularly-denticulate rather ill-defined streak, also obliquely placed and running almost parallel with median band;* beyond this streak, the hind-marginal border is much paler than the rest of the ground, and often nearly white in ♂. Cilia beneath rather indistinctly interrupted with greyish-ochreous, especially in hind-wing.

A fine ♀ example taken in Basutoland by Colonel Bowker varies in possessing much enlarged discal spots in the fore-wings.

This species is readily recognised by the peculiarly oblique white bands on the under side of the hind-wings; and Herr Möschler (*loc. cit.*) was undoubtedly right in dissenting from my suggestion in 1866 that it was probably a variety of *Vindex;* but I had as long ago as 1870 (*Trans. Ent. Soc. Lond.,* p. 385) corrected that suggestion, and stated that I had seen reason to hold *Diomus* as distinct.

About Cape Town this butterfly is numerous; its habits are precisely those of *P. Vindex,* and the two forms fly in company. Though having an extensive range in the Southern Tropical Region, *Diomus* does not appear to have been recorded from any locality north of the Equator. Its distribution in South Africa proper seems to be general, but it has appeared more rarely than *Vindex* in collections received.

---

[1] Dr. Staudinger's figure has no resemblance to *P. Sandaster,* Trim. (see below, p. 291), but represents an apparent ♀ of *P. Diomus.*

Localities of *Pyrgus Diomus*.

I. South Africa.
  B. Cape Colony.
    a. Western Districts.—Cape Town. Ookiep, Namaqualand District (*L. Péringuey*).
    c. Griqualand West.—Kimberley. Between Riet and Modder Rivers (*Mrs. Barber*).
    d. Basutoland.—Maseru (*J. H. Bowker*).
  D. Kaffraria Proper.—Tsomo River (*J. H. Bowker*).
  E. Natal.
    a. Coast Districts.—D'Urban (*J. H. Bowker*).
  F. Zululand.—Napoleon Valley (*J. H. Bowker*).
  H. Delagoa Bay.—Lourenço Marques (*Mrs. Monteiro*).
  K. Transvaal.—Potchefstroom (*W. Morant* and *T. Ayres*). Eureka City, near Barberton (*C. F. Palmer*).

II. Other African Regions.
  A. South Tropical.
    a. Western Coast. — Damaraland: "R. Kuisip (*Wahlberg*)." — Wallengren.
    b. Eastern Coast.—"Querimba (*Peters*)."—Hopffer.
    b1. Eastern Interior.—Between Transvaal and Gubulewayo (*A. W. Eriksson*).

331. (8.) Pyrgus Sataspes, Trimen.

♂ ♀ *Pyrgus Sataspes*, Trim., Trans. Ent. Soc. Lond., 3rd Ser., ii. p. 178 (1864); and Rhop. Afr. Aust., ii. p. 290, n. 179, pl. 5, f. 7 (1866).

Exp. al., (♂) 10 lin.—1 in.; (♀) 11½ lin.—1 in. 1 lin.

♂ *Fuscous-brown, with small, dull-whitish spots*. Fore-wing: only two, rather widely separated, spots in discoidal cell, one of which closes it; spots in transverse row beyond middle less separated than in *Vindex* group, the row of three on costa more in line with the rest, not lying so obliquely towards hind-margin; a row of dots immediately before cilia tinged with ochreous, often indistinct, but always present; between it and row of spots a sinuate line of similar dots; base more or less clouded with dull-ochreous scales. Hind-wing: no spot near base; median band interrupted at origin of second and third median nervules, very much narrower than in *Vindex* group; two rows of dots as in fore-wing, but inner one not sinuate and often almost obliterated. Cilia dull greyish-yellow, inconspicuously varied with fuscous. UNDER SIDE.—*Fore-wing*: markings as above, but outermost row of dots wanting; costa and apex widely coloured with pale greyish-yellow, the latter clouded with pale reddish-brown. *Hind-wing: pale fuscous, tinged with reddish, crossed by two yellowish-white stripes*, commencing on costal nervure,—the first short, often indistinct, narrow before middle,—*the second median, conspicuous, slightly oblique, rather narrow (being widest costally)*, uniting with an inner-marginal longitudinal stripe to form a white space at anal angle; hind-margin widely tinted with dull-reddish, and edged with a pale line; rows of dots obsolete.

♀ *Paler, more conspicuously spotted; under side of hind-wing reddish-ochreous, only tinged with fuscous next to stripes.*

In some of the more distinctly-marked examples of both sexes, there are in the fore-wing two short sub-basal whitish longitudinal streaks or dashes—one in discoidal cell just above, and the other immediately below, median nervure.

This little *Pyrgus* is readily known by its brownish upper side with small dull spots; and by the dull reddish-brown of the under side of the hind-wings, crossed by a very distinct continuous median white stripe, only a little denticulate on its edges, and not nearly so oblique as in *Diomus*.

P. Zebra, Butl. (*Ann. and Mag. Nat. Hist.*, 6th Ser., i. p. 207, March 1888). is described as "nearest to *P. Sataspes*;" it is a native of India ("Campbellpore and Chittur Pahar, *Major Yerbury*"), and, judging from the description given, differs in presenting a greyish-brown under side of the hind-wings, marked with an additional outer narrow white stripe from apex to lower part of hind-margin. This additional stripe is a point of agreement with *P. Diomus*, Hopff.

I first met with the species at Knysna, Cape Colony, in 1858; it was there not uncommon, and remained out from September to February 1859. About Cape Town I have since constantly taken it, but only in the summer and later spring months. In the neighbourhood of Grahamstown I took a few specimens during February. There is nothing remarkable in its habits; like its congeners, it is fond of resting with expanded wings on the ground or on stones.

I have received this butterfly from but few localities, and usually singly; it is inconspicuous, and probably passed over among its immediate allies.

Localities of *Pyrgus Sataspes*.

1. South Africa.
    B. Cape Colony.
        a. Western Districts.—Cape Town. Knysna and Plettenberg Bay.
        b. Eastern Districts.—Between Zwartkop and Coega Rivers (*J. H. Bowker*). Kleinemond River, Bathurst (*H. J. Atherstone*). Grahamstown (*Miss H. L. Bowker*). Mitford Park, Albany District.
    E. Natal.
        a. Coast Districts.—D'Urban (*J. H. Bowker*).
        b. Upper Districts.—Estcourt (*J. M. Hutchinson*).

332. (9.) Pyrgus nanus.

Pyrgus Sataspes, VAR. A., Trim., Rhop. Afr. Aust., ii. p. 290 (1866).

Exp. al., (♂) $9\frac{1}{2}$—$11\frac{1}{2}$ lin.; (♀) 11 lin.—1 in.

Nearly allied to *P. Sataspes*, Trim.

♂ *Brownish-black, with small white spots; spots in number and position almost as in Sataspes, but of a purer white and more sharply defined; hind-wing median band divided as in Sataspes, but not at all oblique.* UNDER SIDE.—*Hind-wing:* ground-colour much paler; base with a whitish tinge; both transverse bands whiter, and beginning on costal edge,—the sub-basal one not extending below subcostal nervure, —*the median one outwardly dark-edged, biangulated (just above third and just upon first median nervules), and not oblique, its attenuated lower extremity joining white of inner-marginal fold at some distance before anal angle; an indistinct bisinuated submarginal series of whitish dots.*

♀ White spots and white of cilia purer, more conspicuous. UNDER SIDE.—All the markings more sharply defined.

A dwarf ♂ that I captured near Cape Town in April 1872 expands only 8 lin.

The characters emphasised in the above description appear to be thoroughly constant, and warrant the separation of this form from *Sataspes*.

I am not aware of the occurrence of this little species beyond the limits of the Cape Colony, but within those limits it appears to be of wide distribution. Near Cape Town I have not found it at all numerous; there seem to be two broods, one in September and October, and the other in March and April. Mr. L. Péringuey took some examples in the Namaqualand District in October and November 1885; and I took a specimen at Robertson in January 1876.

Localities of *Pyrgus nanus.*

I. South Africa.
   B. Cape Colony.
      a. Western Districts.—Cape Town. Malmesbury. Vogel Vley, Tulbagh District. Robertson. Ookiep, Namaqualand District (*L. Péringuey*).
      b. Eastern Districts.—Uitenhage (*J. H. Bowker*). Between Somerset East and Murraysburg (*J. H. Bowker*).
      c. Griqualand West.—Kimberley (*H. L. L. Feltham*).

333. (10.) Pyrgus Sandaster, Trimen.

♂ *Pyrgus Sandaster*, Trim., Trans. Ent. Soc. Lond., 1868, p. 92, pl. v. f. 9.

Exp. al., (♂) 10—11 lin.; ♀ 11 lin.—1 in.

♂ *Glossy brownish-black, with minute white spots; cilia black, with narrow white inter-nervular interruptions. Fore-wing:* five minute inter-nervular white dashes on costa; a linear transverse white mark in discoidal cell towards extremity; discal row of eight or nine spots irregular, strongly bisinuate,—spots one to three forming a straight transverse costal line, spots four and five a little beyond them, spots six and seven lying very obliquely (almost longitudinally) before those above, and spots eight and nine (one or both sometimes obsolete) vertically below seventh spot; in some specimens, an indistinct transverse linear white mark before middle, between first median nervule and submedian nervure; immediately before cilia a partly obsolescent series of internervular white dots. *Hind-wing*: usually a whitish spot in discoidal cell towards extremity; a discal row of two or three (usually rather indistinct) spots between second subcostal and first median nervules; two or three indistinct pre-ciliary whitish dots near apex. UNDER SIDE.—*Dark-brown, with all the borders of hind-wing and costa and apex of fore-wing densely but unequally irrorated with whitish; hind-wing with two very narrow transverse stripes sharply defined by blackish edges. Fore-wing*: base of cell irrorated with whitish; costal dashes represented by more distinct small spots; a good-sized quadrate white spot on costa before middle; a faint transverse whitish streak at extremity of discoidal cell; other spots larger and better defined than on upper side; ground-colour paler near inner margin. *Hind-wing*: base whitish; sub-basal stripe from costa to submedian nervure, angulated and more slender inferiorly; discal stripe widely interrupted on first subcostal nervule, irregularly dentate on its inner edge, more regularly and acutely on its outer edge, attenuated inferiorly and ending on submedian nervure.

♀ Like ♂, but all the white markings (except discal spots of hind-wing) better developed and more sharply defined, especially transverse stripes on under side of hind-wing, the edgings of which are quite black.

This is a very distinct species, differing from all its congeners in its very small white spots, and (on the under side) in its dark ground-colour and very narrow sharply-defined black-edged transverse stripes of the hind-wings.

Only ten specimens (three ♀ s) have come under my notice. The first was taken near Murraysburg, about the centre of the Cape Colony, in 1864, by Dr. J. J. Muskett; three were captured by Mrs. Barber on the Fish River in November 1871. On the 18th and 20th August 1873 I met with three examples in Little Namaqualand; and Colonel Bowker took three at Uitenhage on the 6th October 1879.

I took all the few examples that I saw in Namaqualand; they settled on the ground in dry stony places, and were not easily taken. Mrs. Barber wrote that it was very numerous at Fort Brown on a particular hillock covered with *Barleria* shrubs, but that she did not notice it elsewhere in the vicinity. Colonel Bowker found his Uitenhage specimens "at the foot of stony hills."

Localities of *Pyrgus Sandaster*.

I. South Africa.
    B. Cape Colony.
        a. Western Districts.—Oograbies, and between Komaggas and Spectakel, Namaqualand District.
        b. Eastern Districts.—Uitenhage (*J. H. Bowker*). Murraysburg (*J. J. Muskett*). Fort Brown, Fish River, Albany District (*Mrs. Barber*).

334. (11.) **Pyrgus Elma**, Trimen.

Pyrgus Elma, Trim., Trans. Ent. Soc. Lond., 3rd Ser., i. p. 288 (1862); and Rhop. Afr. Aust., ii. p. 291, n. 180, pl. 5, f. 8 [♂]. (1866).

Exp. al., (♂) 1 1½ lin.—1 in. 1 lin. ; (♀) 1 in. 1-3 lin.

♂ *Dark-brown, varied with lighter, and with vitreous and white spots. Fore-wing*: dark to before middle, as far as a transverse curved blackish streak ; beyond streak the wing is light-brown, with some darker shades and dashes; a disco-cellular transverse vitreous streak contiguous to outer edge of blackish streak ; near it, between first and third median nervules, two small vitreous spots, the lower one larger, quadrate ; a transverse line of three contiguous, minute, vitreous spots on costa, not far from apex ; cilia fuscous, unequally varied with brownish-whitish between nervules. *Hind-wing: almost black ;* a small round white spot in discoidal cell, near base ; about middle a conspicuous rather wide white band, attenuated at inferior extremity, from first subcostal nervule to submedian nervure ; traces of a submarginal row of whitish dots ; anal-angular region hoary ; cilia whitish, with broad fuscous-brown nervular marks on lower half. UNDER SIDE.—*Rather glossy, much paler ; dull brownish-ochreous,* with a tinge of yellow ; base and margins of fore-wing clouded with whitish. *Fore-wing:* vitreous marks as above ; a whitish spot at extremity of cell ; an indistinct submarginal row of whitish dots. *Hind-wing:* an additional white dot above that in cell, the two forming a short sub-basal transverse streak ; white band commences very narrowly on costa ; whitish at anal angle, extending along inner-marginal fold, sometimes conspicuous ; submarginal whitish spots not apparent except between second median nervule and submedian nervure, where they form a short continuous streak.

♀ Like ♂, but generally rather duller and paler ; cilia more distinctly varied, especially in fore-wing ; the vitreous spots larger.

This butterfly belongs to the well-known European group containing *Alcea*, Esp., and *Lavatera*, Esp. ; it is nearer to the former in size, colouring, and pattern of the fore-wings, but as regards the white spot and band on the upper side of the hind-wings, exhibits some resemblance to the latter. A much closer ally, however, would appear (judging from the description and a figure of the upper side in the *Lepidoptera of Ceylon*, p. 183, pl. 71, f. 7) to be the Cingalese *Albofasciata*, Moore, in which there is little difference observable except in the more uniform ground-colour of the wings, and the narrower, longer, more regular, and more curved white band of the hind-wings.[1]

---

[1] Perhaps a still more closely allied form is one collected on Mount Sinai by Mr. J. K. Lord, of which I possess a pair, kindly presented to me by Prof. Meldola. These specimens are worn, but evidently approach *Elma* very nearly, size and pattern being almost identical, but the upper side and under side being alike much paler ; the cellular transparent mark in the fore-wings, as well as those near costa, wider ; the spot and band in the hind-wings less distinct, the latter being narrower and prolonged to submedian nervure ; the submarginal row of spots in the hind-wings more distinct and sinuate, and the discal stripe on the under side

Pupa.—"Clothed with a minute pubescence; reddish-brown, with greenish-brown abdomen. Back of thorax slightly protuberant; two minute dark projections on base of wing-covers. Enveloped in leaf of *Abutilon Sonneratianum*, and attached by anal segment."—J. P. Mansell Weale, *in epist.*, 1877.

This is a species widely distributed over Eastern South Africa; to the westward it is common on the coast of the Knysna District of the Cape Colony, and I once met with it as far to the west as Robertson. It is prevalent in wooded spots, preferring pathways and open bits of ground, where it settles with expanded wings, the white bar of the hind-wings making it rather conspicuous. The summer months are its chief season, but I have noticed it on the wing as early as the middle of September. Mrs. Barber informed me that the eggs are deposited on a species of *Sida*.

Localities of *Pyrgus Elma*.

I. South Africa.
  B. Cape Colony.
    a. Western Districts.—Robertson. Mossel Bay. Knysna and Plettenberg Bay.
    b. Eastern Districts.—Port Elizabeth. Uitenhage. Grahamstown (*Mrs. Barber*), and Zwaartwater Port, Albany District. Mouth of Kowie River, Bathurst District (*J. L. Fry*). King William's Town (*W. S. M. D'Urban*). Burghersdorp (*D. R. Kannemeyer*).
  D. Kaffraria Proper.—Bashee River (*J. H. Bowker*).
  E. Natal.
    a. Coast Districts.—D'Urban and Pinetown (*J. H. Bowker*). "Lower Umkomazi."—J. H. Bowker. Verulam.
    b. Upper Districts.—Greytown.
  F. Zululand.—Etshowe (*A. M. Goodrich*).
  K. Transvaal.—Potchefstroom District (*T. Ayres*). Upper Limpopo River (*F. C. Selous*).

II. Other African Regions.
  A. South Tropical.
    a. Western Coast.—"Angola."—Kirby (Cat. Hewits. Coll.).
    b1. Eastern Interior.—"Lake Nyassa."—Kirby (Cat. Hewits. Coll.).

335. (12.) Pyrgus Mohozutza, Wallengren.

♂ *Hesperia Mohozutza*, Wallengr., K. Sv. Vet.-Akad. Handl., 1857; Lep. Rhop. Caffr., p. 50, n. 7.
Pyrgus Mohozutza, Trim., Rhop. Afr. Aust., ii. p. 291, n. 181, pl. 5, f. 9 [♀], (1866).

Exp. al., (♂) $11\frac{1}{2}$ lin.—1 in. $1\frac{1}{2}$ lin.; (♀) 1 in. 1–3 lin.

♂ *Blackish-brown, with creamy-white and fulvous spots Fore-wing:* of the hind-wings not straight across, but angulated and more irregular on edges. The ♀ has, moreover, a reddish tinge, especially on the under side of the hind-wings, where the markings are almost obliterated.

This is doubtless the insect which the late Mr. F. Walker (*Entomologist*, 1870, p. 56) referred to the Spanish *Marrubii*, Rambur (regarded by Kirby and Plötz as a variety of *Althææ*, Hübn.); but the specimens I have do not agree at all closely with Herrish-Schäffer's figures (*Schmett. Europ.-Hesp.*, ff. 14, 15) of *Marrubii*, in which the under side is represented as decidedly greenish, with pale-yellowish neuration interrupting sharply the transverse white stripes.

costa and main nervures irrorated with ochreous to about middle; a quadrate creamy spot at extremity of discoidal cell, succeeded by a sinuate transverse row of six smaller spots of the same colour (of which the three upper are separate from the others, forming one elongate streak on subcostal nervules); bordering hind-margin a row of five or six elongate fulvous spots, becoming obsolete below second median nervule. *Hind-wing*: some whitish-ochreous hairs on median and submedian nervures; usually a very indistinct whitish spot at extremity of discoidal cell. Cilia broad, creamy-white, with thin nervular black interruptions, irregular in hind-wing. UNDER SIDE.—*Costa and hind-marginal row of spots (which latter is in both wings a continuous broad band) bright orange-fulvous. Fore-wing*: ground-colour not so dark; creamy spots less distinct. *Hind-wing: ground-colour pale greyish-creamy*, excepting for a wide fuscous space at anal angle, gradually narrowing along submedian nervure to base: two black spots at base, one on costa, the other between costal and subcostal nervures; discoidal cell pale orange-fulvous, divided about its middle by a black V, and exteriorly tipped with black; a stripe of the same colour, also black-tipped, extends from base to beyond middle, between median and submedian nervures; hind-marginal fulvous band internally edged with black, interrupted by creamy nervures. *In both wings*, a thin submacular black streak along hind-marginal edge.

♀ *Not so dark; hind-wing with a hind-marginal series of seven fulvous spots; cilia more creamy. Fore-wing*: fulvous spots of hind-marginal row larger, better developed inferiorly, with occasionally an indistinct seventh spot below first median nervule. *Hind-wing*: disco-cellular spot enlarged, pale-fulvous, always distinct, and sometimes conspicuous; seven spots in hind-marginal series, usually larger (more elongate) than those in fore-wing. UNDER SIDE.—*Fulvous colouring everywhere deeper and brighter; creamy area of hind-wing duller*, and fuscous of inner-marginal fold and anal angle not so deep, shading off into more or less developed cream-colour along inner-marginal edge.

This handsome Hesperid was discovered by Wahlberg; and the first specimens brought to my notice were taken in Kaffraria Proper in 1863, by Colonel Bowker, who wrote to me that it occurred in open country near the Bashee River from January to April, frequenting low flowers among the grass. In March 1867 I took a good many specimens in the Umvoti district of Natal; it frequented flowers on the open hills, and was on one occasion rather numerous about a violet-flowered papilionaceous shrub near Greytown. It is not very alert, and easily captured while feeding.

Localities of *Pyrgus Mohozutza*.

I. South Africa.
    D. Kaffraria Proper.—Tsomo and Bashee Rivers (*J. H. Bowker*).
    E. Natal.
        a. Coast Districts.—D'Urban (*A. D. Millar*). Pinetown (*J. H. Bowker*). "Lower Umkomazi."—*J. H. Bowker*.

b. Upper Districts.—Udland's Mission Station, Great Noodsberg, Tunjumbili, Hermansburg, and Greytown. Maritzburg (*J. H. Bowker*).

F. Zululand.—Etshowe (*A. M. Goodrich* and *T. Vachell*).

K. Transvaal.—Lydenburg District (*T. Ayres*).

336. (13.) Pyrgus Chaca, Trimen.

♂ ♀ *Pyrgus Chaca*, Trim., Trans. Ent. Soc. Lond., 1873, p. 118, pl. i. ff. 9 [♂], 10 [♀].

Exp. al., (♂), 1 in. 3½ lin. ; (♀) 1 in. 5½ lin.

♂ *Fuscous, with white and ochreous spots.* Fore-wing: irrorated thinly in parts with some yellowish scales; a broad, sub-reniform, white spot at extremity of discoidal cell; a very irregular row of eight white spots, of which the first three are contiguous, and form a short, rather wide, costal and subapical stripe, the fourth and fifth are nearest hind-margin and minute, the seventh largest and just under disco-cellular spot, and the eighth indistinct and just above submedian nervure; a hind-marginal row of seven indistinct small ochreous spots, of which the lowest is suffused and almost obsolete. *Hind-wing:* some hoary hairs along line of median and submedian nervures; an indistinct, good-sized, ochreous spot at extremity of discoidal cell; vague traces of three or four small ochreous spots near upper half of hind-margin. Cilia white, varied with black at ends of nervures. UNDER SIDE.—*Hind-wing and costal and apical region of fore-wing mingled creamy and fulvous-ochreous*, the fulvous predominating in hind-wing. Fore-wing: white spots as above, but rather larger, and relieved by black edgings from the ochreous colouring round them; a small black spot on costa close to base; a row of seven small black spots on hind-margin from apex to anal angle, the lowest two suffused and enlarged, so as to join inner-marginal blackish ground-colour. *Hind-wing:* between costal and submedian nervures, two transverse white bands (one before, the other beyond middle) edged on both sides with macular black streaks; a hind-marginal row of six small black spots; space between submedian nervure and inner-marginal edge blackish, irrorated anteriorly with creamy-ochreous; costa at base with a black curved mark, about and beyond middle bordered with grey.

♀ *Not so dark as* ♂; spotting similar, but *all the yellow-ochreous spots larger and more distinct*; cilia creamy instead of white, more broadly varied with black. UNDER SIDE.—As in ♂, but the fulvous brighter, especially in hind-wing.

Apart from the very considerable difference in size (*P. Mohozutza* not exceeding about 13 lines), both sexes of *P. Chaca* may, as regards the markings of the upper side, be recognised by the possession of two small additional white spots (the fourth and fifth) in the transverse row beyond the middle of the fore-wings, and the duller and proportionally much smaller ochreous spots of the hind-marginal row in the fore-wings of both sexes and in the hind-wings of the ♀. On the upper side of the hind-wings of the ♂ *Chaca* there are, however, an indistinct central spot, and traces of three hind-marginal spots not found in the ♂ *Mohozutza*. The under side in both sexes differs from that of *Mohozutza* by being paler; by having a row of distinct black spots, instead of a black streak, along the hind-marginal edge of the wings; and by presenting a whitish transverse band before the middle in the hind-wings.

Mr. Bowker took this butterfly in several localities near the Tsomo River, Kaffraria, in December, and notes it as occurring in "open grassy glades between forests upon high mountains, sitting on flowers with closed wings." Mrs. Barber, however, who met with the species near Bathurst in March 1870, writes that it frequented "long grass, in rather a low and damp locality, among trees." Mr. Schiffman's specimens were captured on one of the hills round Grahamstown, but I am not aware in what kind of station.

The species appears to be rare and very local; I have seen only two examples of each sex.

Localities of *Pyrgus Chaca*.

I. South Africa.
   B. Cape Colony.
      b. Eastern Districts.—Grahamstown (— *Schiffman*). Bathurst (*Mrs. Barber*).

II. Other African Regions.
   A. South Tropical.
      a. Western Coast.—"Angola (*Pogge*)."—Dewitz.
      b. Kaffraria Proper.—Tsomo River (*J. H. Bowker*).

337. (14.) Pyrgus Tucusa, Trimen.

Pyrgus Tucusa, Trim., Trans. Ent. Soc. Lond., 1883, p. 359.

Exp. al., 1 in. 1–3 lin.

♂ *Fuscous, with semi-transparent white spots. Fore-wing*: a dull yellowish-white scaling from base to before middle along costa, median nervure, and inner margin; at extremity of discoidal cell two elongate spots, one above the other; a discal continuous row of nine spots, strongly curved outwardly in its upper portion, but thence sharply deflected inwardly to submedian nervure a little beyond middle; of this row the largest spot is the seventh (immediately below disco-cellular spots), and the smallest the eighth; close to upper part of hind-margin a row of five almost obsolete dull fulvous-ochreous spots. *Hind-wing*: median and submedian nervures clothed with dull yellowish-white scales and hairs; at extremity of cell an ill-defined whitish spot. *Cilia* white, in *fore-wing* with narrow fuscous nervular interruptions, in *hind-wing* with scarcely visible similar interruptions. UNDER SIDE.—*Dull creamy white, varied with very pale fulvous and with small black spots. Fore-wing:* white spots as on upper side, and edged with fuscous; inner-marginal area pale fuscous; hind-marginal border creamy, faintly tinged with fulvous, becoming macular towards posterior angle; along hind-marginal edge a row of small inter-nervular black spots. *Hind-wing:* a thin costal edging near base, discoidal cell, and a small space beyond, and a widish hind-marginal border, very pale fulvous; near base two small black spots above discoidal cell, two in cell (with a small space of the ground-colour between them), and one below cell; two parallel discal rows of small black spots (seven in each row), parallel at a little distance apart; inner-marginal fold tinged with fuscous at its broad anal-angular extremity; hind-margin thinly edged with fuscous, interrupted (like the fulvous border) by nervules of the ground-colour.

♀ *Fore-wing: white spots inclining to creamy, all smaller than in ♂ (especially the seventh, while the eighth is wanting or exceedingly minute); hind-marginal spots distinct, pale fulvous-ochreous, seven in number. Hind-wing: disco-cellular spot large, pale fulvous-ochreous; close to hind-margin a row of seven spots of the same colour,* not extending below submedian nervure. *Cilia* creamy white, with wider fuscous interruptions, of which in *hind-wing* there are only three, at the extremities of median nervules. UNDER SIDE.—*As in ♂, but the fulvous markings much brighter. Hind-wing:* two additional fulvous marks immediately before inner discal row of black spots, viz.. one between costal and subcostal, the other between median and submedian nervures.

This *Pyrgus* closely resembles both *P. Mohozutza* (Wallengr.) and *P. Chaca*, Trim.; in size it is nearer to the former, but in pattern to the latter. The continuous character of the discal row of white spots, due to the constant presence of two spots between the costal three and those below median nervure, distinguishes it in both sexes from *Mohozutza*. The ♂ is further separable by the purer white of the discal spots and the cilia, and by the obsolete condition of the hind-marginal ochreous spots, which in *Mohozutza* are well marked. On the under side both sexes present in the hind-wing a regular inner discal row of seven black spots instead of the three or four widely-separated and irregularly-placed spots in *Mohozutza;* paler (and on costa and submedian nervure much reduced) fulvous markings, and a whitish instead of fuscous inner-marginal fold. Apart from its very much smaller size, *Tucusa* in both sexes is very like *Chaca* on the upper side, but on the under side it wholly wants in the hind-wings the broad transverse median fulvous band and fulvous base, separated by a creamy-white black-edged band, so conspicuous in *Chaca.*

An aberrant ♀ sent from the Transvaal in 1873 by Mr. H. Barber has the fore-wings longer than usual, with the discal spots all smaller and tinged with yellow-ochreous, but three ♂ s accompanying that specimen are normal.

Localities of *Pyrgus Tucusa*.

I. South Africa.
 E. Natal.
  b. Upper Districts.—Estcourt (J. M. Hutchinson).
 K. Transvaal.—(H. Barber).

Genus THYMELICUS.

Pamphila, Westwood (part), Gen. Diurn. Lep., ii. p. 521 (1852).
Pamphila, Trimen (part), Rhop. Afr. Aust., ii. p. 297 (1866).
Thymelicus (Hübner, 1816), Herrich-Schäff., Corr.-Blatt. Zool.-Min. Ver. Regensb., 1868, p. 44.
Thymelicus, Speyer, Stett. Ent. Zeit., 1878, p. 182, and 1879, p. 488.
Thymelicus, Plötz, Stett. Ent. Zeit., 1884, p. 284.

IMAGO.—*Head* with usual small tuft of hairs close to origin of each antenna : *palpi* long,—the second joint densely hairy,—the third long, ascendant or erect, or horizontally porrected, clothed with short appressed hairs ; *antennæ* short, with a gradually-formed sub-cylindrical club, thinner and slightly curved towards tip (which is blunt).

Thorax robust and hairy. *Wings* with neuration as in *Pamphila*, Fab. *Fore-wings* more or less pointed apically ; discoidal cell about three-fifths the length of the wing, its very slender closing nervules forming an inwardly oblique line ; ♂ in some species with a lower discal linear groove running obliquely from second median nervule near its origin to submedian nervure before middle. *Hind-wings* small, bluntly rounded, usually rather prominent (sometimes much so) anal-angularly ; discoidal cell very short, from one-third to less than half the length of the wing,—its closing nervule angulated inwardly in the middle. *Legs* rather short and thick ; femora (especially hind ones superiorly) more or less hairy ; appendage to tibiæ of first pair inconspicuous, often with difficulty perceptible among the scaly hairs ; tarsi spinulose beneath.

Abdomen rather long,—in ♂ laterally compressed and slightly tufted at extremity.

LARVA.—Slender, gradually attenuated anteriorly, rather abruptly pointed posteriorly ; head small ; no hairs on tubercles.

PUPA.—Slender and elongate ; frontal spike of head short ; extremity of abdomen very attenuated and acute ; case of haustellum extending freely to about half the length of abdomen.

(These characters are those of the early stages of *Thaumas* (*Linea*) and *Lineola* (European species), as figured and described by Hübner, Duponchel, and Boisduval.)

There is little to distinguish this genus from *Pamphila*, Fab., except the blunt unhooked club of the antennæ, and the much longer and usually more erect terminal joint of the palpi. In the former feature it resembles *Pyrgus*, but, unlike that group, the ♂ in *Thymelicus* never exhibits a costal fold in the fore-wings, a tuft on the hind-tibiæ, or posterior thoracic appendages in any species.

About thirty species are recorded, pretty evenly distributed among the great zoological regions, with the exception of Australia. The Ethiopian Region has yielded eight species, the Palæarctic, Nearctic, and Oriental Regions six each, and the Neotropical Region five.

The African species have a very different facies from that of the European ones; their hind-wings are apically shorter, and the terminal joint of their palpi is not erected. Five are natives of South Africa; of these two, *Lepenula*, Wallengr., and *Macomo*, Trim., are brown, with ochre-yellow markings; two, *Niveostriga* and *Wallengrenii*, are brown, with a few whitish spots in the fore-wings; and the fifth, *Barberæ*, Trim., has conspicuous pure-white spots on a brown ground, and black-and-white cilia as in *Pyrgus*. All five inhabit the eastern side of the country, and the only one known to me to extend into the tropical belt is *Lepenula*. *Macomo*[1] appears to be the least rare; of the scarcest species, *Barberæ*, I have seen only four examples.

338. (1.) Thymelicus Lepenula, (Wallengren).

♂ *Hesperia Lepenula*, Wallengr., K. S. Vet.-Akad. Handl., 1857; Lep. Rhop. Caffr., p. 50, n. 6.

♂ *Pamphila? Lepenula*, Trim., Rhop. Afr. Aust., ii. p. 298, n. 189 (1866). *Cyclopides Chersias*, Hewits., Ann. and Mag. Nat. Hist., 4th Ser., xx. p. 327 (1877).[2]

PLATE XI. fig. 6 (♂).

Exp. al., (♂) 1 in. 1–2 lin.; (♀) 1 in. 2½ lin.

♂ Brown, with wide pale ochreous-yellow markings. *Fore-wing*: a large basal marking filling all discoidal cell and a shorter area between median and submedian nervures, from base itself; costa to about middle and inner margin very densely to beyond middle irrorated with ochreous-yellow; a short transverse subapical and subcostal mark, externally tridentate on nervules; beneath this, and between third median nervule and submedian nervure, an oblique wide discal band, externally quinquedentate and internally strongly indented by ground-colour below first median nervule. *Hind-wing*: basi-cellular and inner-marginal area clothed with some pale-yellow hairs; in discoidal cell, near extremity, a good-sized spot; beyond it a wide irregular discal band on patch, between subcostal and submedian nervures, inwardly rather vaguely defined, outwardly better defined, more even, forming short

---

[1] A near but quite distinct ally is *T. Capenas* (Hewits.), from the Zambesi, at once recognised by having the nervules yellow near the superior half of the hind-margin of the fore-wings on upper side.

[2] I examined the six specimens of *Chersias* in the Hewitson Collection, and found them unquestionably identical with *Lepenula*, Wallengr.

nervular dentations. *Cilia* of fore-wing brownish-grey, except about posterior angle, where pale-yellowish prevails; of hind-wing wholly pale-yellow. UNDER SIDE.—*Hind-wing, and all fore-wing except part of central and inner-marginal area, pale chrome-yellow or sulphur-yellow, without marking. Fore-wing:* in cell and on lower disc a faint indication of the upper-side pattern and colouring; from base, below median nervure and its first nervule, a conspicuous blackish suffusion, extending to rather beyond middle only along first median nervule, but below submedian nervure to posterior angle.

Head and body above blackish with yellowish hairs, beneath pale chrome-yellow or sulphur-yellow. Antennæ black; palpi with mixed yellow and brown hairs above, and pale chrome-yellow hairs beneath,— the terminal joint black; vertex with a median yellow line and a yellow spot above each eye. Abdomen above with yellow segmental half-rings.

♀ Like ♂, *but in fore-wing basal yellow marking is much narrower inferiorly, and further reduced by one dark-brown ray from base in discoidal cell (widening outwardly), and by another between median and submedian nervures.*

This species is well characterised by the wide development of the yellow markings of the upper side, and by the almost (in the hind-wings wholly) spotless pale-yellow under side. Its nearest ally in these respects is *Harona*, Westw.,[1] a species not hitherto recorded from south of the Tropic, in which the yellow of the upper side is still more extended, occupying the costa of the fore-wings and nearly the whole of the hind-wings, but in which the under side of the hind-wings is of a paler yellow.[2]

Lepenula appears to belong to the interior tracts of South Africa, especially the Transvaal country; but in 1881 I was surprised to receive from Mr. S. D. Bairstow a ♂ example captured by him at Uitenhage. At the end of 1871 Colonel Bowker found this butterfly very sparingly on a hillside at Klipdrift (now Barkly), on the Vaal River, and sent me a specimen for determination. About Potchefstroom it seems to be not uncommon.

Localities of *Thymelicus Lepenula*.

I. South Africa.
   B. Cape Colony.
      b. Eastern Districts.—Uitenhage (*S. D. Bairstow*).
      c. Griqualand West.—Vaal River: Barkly (*J. H. Bowker*).
   K. Transvaal.—Potchefstroom and District (*T. Ayres* and *W. Morant*).

II. Other African Regions.
   A. South Tropical.
      b1. Eastern Interior (special locality not stated : *A. W. Ericksson*).

---

[1] App. Oates' *Matabele Land*, p. 353 (1881).

[2] On the upper side *Lepenula* bears a close resemblance to *Maro* (Fab.), a Cingalese species, but wants the fuscous spotting and brownish clouding of the latter on the under side.

339. (2.) Thymelicus Macomo, (Trimen).

Cyclopides Macomo, Trim., Trans. Ent. Soc. Lond., 3rd Ser., i. p. 405 (1862).

♂ ♀ *Pamphila Macomo*, Trim., Rhop. Afr. Aust., ii. p. 297, n. 188, pl. 6, f. 6 [♂], (1866).

♀ *Thymelicus Macomo*, Staud., Exot. Schmett., i. p. 100 (1888).

Exp. al., (♂) 1 in. $0\frac{1}{2}$–2 lin.; (♀) 1 in. 2–$3\frac{1}{2}$ lin.

♂ *Brown, with mostly sharply-defined, sub-quadrate, pale ochreous-yellow spots; cilia dull yellowish-white, mixed with greyish in fore-wing. Fore-wing:* basal third scaled with pale ochreous-yellow; a good-sized terminal disco-cellular spot; between it and submedian nervure, obliquely placed towards base, two smaller spots, of which the upper one is crossed by beginning of first median nervule; a discal row of six sharply-defined spots, of which the upper three are considerably smaller and united to form a short and narrow subapical bar,—the fourth and fifth united between third and first median nervules,—and the sixth separated just above submedian nervure. *Hind-wing:* over discoidal cell and upper part of inner-marginal fold some sparse yellowish hairs; a small and rather indistinct terminal disco-cellular spot; an irregular discal band composed of five spots, the second of which (just beyond end of cell) is much the largest, and the fifth rather large and indistinct. UNDER SIDE.—*Hind-wing, and all fore-wing except fuscous inner-marginal area, creamy-yellow, with small black spots. Fore-wing:* a short black longitudinal streak in basal two-thirds of discoidal cell, and a small black spot at upper corner of extremity of cell; only lower three spots of discal row apparent, much paler than on upper side, but inner edge of upper three indicated by three minute black dots, and outer edge of all except the lowest by a series of five inter-nervular short black rays (thicker inwardly) extending to hind-margin; two similar but longer and more linear rays from just beyond extremity of cell; costa and hind-margin with a linear black edging. *Hind-wing:* an indistinct fuscous streak along costa from base, a little within the edge; a spot close to base just above subcostal nervure, and another (before middle) below it; an irregular discal row of five spots, of which the third and fourth are considerably beyond the others; and a sub-marginal row of five, of which the fifth is considerably before the others; inner-marginal fold and anal-angular area of a deeper yellow, and marked superiorly by a strong longitudinal black ray from base; hind-margin with a linear black edge.

Head black, with some creamy-white hairs in front, two white spots on middle line of vertex, a larger spot behind base of each antennæ, and a yellowish half-ring behind eyes; *antennæ* black, the basal half of club beneath and some of the articulations marked with pale-yellowish; *palpi* above black with a conspicuous white mark near extremity of second joint, beneath (except terminal joint) creamy-yellow. *Thorax* and *abdomen* above blackish, beneath creamy-yellow; the former above with some pale-yellowish hairs frontally, laterally, and posteriorly; the latter above with last four segmental incisions and terminal tuft yellow, beneath whitish edged on each side by a series of small black spots. *Legs* with femora and tibiæ yellow tipped with black, and tarsi brownish scaled with yellowish-white.

♀ Like ♂, but rather paler; the wings blunter—the fore-wing not so pointed apically, and the hind-wing less prominent about anal angle.

A very distinct species, well characterised by its pronounced and sharply-defined spotting on the upper side, and by the conspicuous black spots and streaks of the under side.

Mr. W. S. M. D'Urban first made known to me this pretty species; he took a single example at King William's Town on the 8th December 1860. Colonel Bowker has since forwarded several specimens from Kaffraria Proper and the Coast Districts of Natal; it does not appear to be at all numerous in the latter region, and during my visit in 1867 I fell in with but one specimen, on 21st March, between Itongati and Verulam. At Pinetown Colonel Bowker took the species in December 1884. Only four ♀s have reached me, the largest being from Delagoa Bay.

Localities of *Thymelicus Macomo*.

I. South Africa.
   B. Cape Colony.
      b. Eastern Districts.—King William's Town (*W. S. M. D'Urban*). West bank of Kei River (*J. H. Bowker*).
   D. Kaffraria Proper.—Butterworth, Tsomo, and Bashee Rivers (*J. H. Bowker*).
   E. Natal.
      a. Coast Districts.—Lower Umkomazi, D'Urban, and Pinetown (*J. H. Bowker*.) Between Verulam and Itongati.
II. Delagoa Bay.—Lourenço Marques (*Mrs. Monteiro*).

340. (3.) Thymelicus niveostriga, (Trimen).

♂ *Pamphila? niveostriga*, Trim., Trans. Ent. Soc. Lond., 3rd Ser., vol. ii. p. 179 (1864); Rhop. Afr. Aust., ii. p. 298, n. 190, pl. 6, f. 7 (1866); and (♀) Trans. Ent. Soc. Lond., 1870, p. 389.

Exp. al., (♂) 1 in. 1½–3 lin.; (♀) 1 in. 3–4 lin.

♂ *Fuscous-brown; fore-wing with semi-transparent white spots.* *Fore-wing:* costa tinged with dull-ochreous; an elongate, sub-reniform spot at extremity of discoidal cell; beyond it, a strongly inward-curved row of six more or less quadrate spots, commencing below costa, of which the two first spots are small and contiguous, the third all but obsolete, the fifth largest, and just below spot in cell, and the sixth (above submedian nervure) somewhat suffused and invaded by ground-colour. *Hind-wing:* darker, spotless. UNDER SIDE.—*Wholly, excepting inner-marginal area of fore-wing, pale reddish-ochreous, marked with paler creamy nervures; on inner-margin of hind-wing a glistening, snow-*

white border, bounded interiorly (along deep fold of wing) with blackish. *Fore-wing*: only three spots, viz., that in cell, and the fourth and fifth of transverse row; fuscous greyer, paler. *Hind-wing*: from base to hind-margin, traversing discoidal cell, a straight very indistinct narrow creamy streak. *Cilia* broad, above and below dull greyish-white.

♀ *Similar to* ♂, *but hind-wing not, or very slightly, darker than fore-wing. Fore-wing*: last spot of discal row very small and indistinct. UNDER SIDE.—Rather more reddish; fuscous of fore-wing paler.

Colonel Bowker discovered this curious species in the year 1863, on the Bashee River in Kaffraria, and subsequently sent several specimens from that locality and from the Tsomo River. In 1869 he sent a ♂ and two ♀s from Basutoland. These examples are a little smaller than the Kaffrarian ones, and the fore-wings are less acuminate apically. In the ♂ the hind-wings are scarcely darker than the fore-wings, and in this example as well as in one of the ♀s the first subapical and lowest discal spots of the fore-wings are wanting, and the other spots on both upper side and under side much reduced. The snow-white inner-marginal stripe on under side of the hind-wings is narrower and less shining than in the Kaffrarian specimens and those taken by myself in Natal.

Colonel Bowker noted this very local "Skipper" as occurring among long grass and rushes in the neighbourhood of water. Near the Hermansburg Mission Station, in Natal, on 10th March 1867, I captured six examples on the summit of a lofty hill-ridge; they were flitting about the purple flowers of a leguminous shrub of moderate height (which was numerous in one spot), in company with *Pyrgus Mohozutza* and other butterflies.

Localities of *Thymelicus niveostriga*.

I. South Africa.
    B. Cape Colony.
        a. Basutoland.—Koro-Koro, Maluti Mountains, and Makaleng River (*J. H. Bowker*).
    D. Kaffraria Proper.—Tsomo and Bashee Rivers (*J. H. Bowker*).
    E. Natal.
        b. Upper Districts.—Hermansburg.
    F. Zululand.—Etshowe (*A. M. Goodrich*).

341. (4.) Thymelicus Wallengrenii, Trimen.

Thymelicus Wallengrenii, Trim., Trans. Ent. Soc. Lond., 1883, p. 361.

PLATE XI. fig. 7 (♀).[1]

Exp. al., (♂) 1 in. 2–4 lin.
Closely allied to *T. niveostriga*, Trim.

♂ *Dull brown; the hind-wing much darker, almost fuscous; fore-wing with almost transparent disco-cellular terminal spot and six discal*

---

[1] In this figure the under side is depicted as much darker than it should be, and not reddish enough, and the white stripe of hind-wing is not defined with sufficient sharpness, and not carried near enough to the base of the discoidal cell.

spots. Fore-wing: spot at extremity of cell divided by disco-cellular fold; first three spots of discal row minute but distinct, forming a thin outward-curving costal ray about midway between disco-cellular spot and apex; other spots of discal row larger, forming a strongly inward-curving row between third median nervule and submedian nervure,—the middle spot of these three being the largest, subquadrate, and immediately below disco-cellular spot. *Hind-wing*: spotless. *Cilia* broad, dusky-whitish. UNDER SIDE.—*Hind-wing and costa and apical area of fore-wing pale ochreous-brown, with a reddish tinge. Fore-wing*: spots as on upper side, except that those in costal part of discal row are usually less distinct and sometimes obsolete; *a hind-marginal whitish cloud over lower part of ochreous-brown of apical area*. *Hind-wing: a very conspicuous, rather broad, white longitudinal stripe*, commencing at a little distance from base and running along disco-cellular fold to hind-margin; a duller, much less conspicuous, wider whitish stripe along inner-marginal edge; space of inner-marginal fold brownish grey.

♀ *Paler, duller; hind-wing not darker than fore-wing;* spots of fore-wing rather larger, the three lower ones of discal row more widely separated from each other. UNDER SIDE.—*Hind-wing*: grey of inner-marginal fold paler; inner-marginal white stripe very narrow and ill-defined; on disc the minute traces of a row of whitish spots.

From *T. niveostriga*, Trim., this species may at once be recognised by (1.) the conspicuous longitudinal central white stripe on the under side of the hind-wing. Other distinctions are (2.) three spots instead of two in costal part of discal row; and on under side (3.) the nervures not paler than the ground-colour; (4.) the whitish cloud on hind-margin of fore-wing; and (5.) the dull and ill-developed state of the inner-marginal white stripe of the hind-wing, which in *niveostriga* is shining pure white.

It gave me much pleasure to name this new butterfly in honour of Pastor H. D. J. Wallengren, the distinguished Swedish lepidopterist, whose good fortune it was to be the first to make known to science a large number of the South-African Rhopalocera.

Like so many other novelties among the *Hesperidæ*, this curious *Thymelicus* rewarded the unremitting researches of Colonel Bowker, who discovered it in Zululand in 1880. Later in the same year he sent an individual taken at the Biggarsberg in Natal.

A ♂ example sent by Mr. A. D. Millar in June 1888 was taken by him among grass along a brook near the Halfway House at Botha's Hill, Natal.

Localities of *Thymelicus Wallengrenii*.

A. South Africa.
   E. Natal.
      b. Upper Districts.—Estcourt (*J. M. Hutchinson*). Biggarsberg (*J. H. Bowker*). Botha's Hill (*A. D. Millar*).
   F. Zululand.—Napoleon Valley (*J. H. Bowker*).

342. (5.) Thymelicus Barberæ, (Trimen).

♀? *Cyclopides Barberæ*, Trim., Trans. Ent. Soc. Lond., 1873, p. 120, pl. i. f. 11 [♀].

Exp. al., (♂) 1 in. 1–2½ lin.; (♀) 1 in. 4 lin.

♂ *Glossy brownish-fuscous, with an ochreous tinge, spotted with white.* Fore-wing: a narrow, elongate spot at extremity of discoidal cell; beyond middle, a very irregular row of eight spots, of which the first three touch each other, and form a short costal and subapical stripe, the fourth and fifth (nearest hind-margin) are small and almost obsolete, and the seventh (immediately below disco-cellular spot) the largest. Hind-wing: a scarcely visible paler spot marks extremity of discoidal cell. Cilia varied with fuscous and white. UNDER SIDE.—*Hind-wing and costal and apical border of fore-wing hoary-grey, varied with brownish.* Fore-wing: a small whitish spot immediately above disco-cellular spot; five minute white marks on costal edge, of which the fourth adjoins first three spots of transverse row; fourth and fifth spots of transverse row merged in *an oblique apical marking of hoary-grey*. Hind-wing: an elongate white mark in discoidal cell before middle is scarcely separated from *a large irregular white marking occupying extremity of cell*, and extending to a hoary-grey hind-marginal suffusion; this marking is joined, at its upper portion, by *an oblique white stripe from costa about middle*; between median and submedian nervures *an ill-defined longitudinal white stripe*, extending from near base to join hind-marginal hoary-grey near anal angle.

♀ *All the spots rather larger; especially* (in *fore-wing*) the disco-cellular spot, which is broader and reniform or subreniform, and the fourth and fifth spots of transverse row; *and* (in *hind-wing*) the disco-cellular spot, which is quite distinct, though suffused and tinged with yellowish. Under side as in ♂.

This species is at once distinguished from all its congeners in South Africa by the remarkable disposition of the white markings of its under side, which indeed are unlike those presented by any member of the *Hesperidæ* that I can call to mind.

I have named this curious butterfly in honour of my friend Mrs. Barber, of Highlands, near Grahamstown, whose kind and valuable aid in working out the Rhopalocerous fauna I have had such frequent occasion to acknowledge. A single ♀ specimen was taken by Mrs. Barber in October 1871, "among long grass and rushes near water," in the Stormbergen, a range of mountains forming the boundary between the Queenstown and Albert Divisions. The only other Cape Colony example I have seen is a ♂ sent from Burghersdorp in 1883 by Dr. D. R. Kannemeyer; but ten years previously I received a fine pair captured in the Transvaal by Mr. H. Barber. The insect must be exceedingly local, as the four specimens mentioned are the only ones that have come under my notice.

Localities of *Thymelicus Barbera*.

I. South Africa.
   B. Cape Colony.
      b. Eastern Districts.—Stormbergen (*Mrs. Barber*). Burghersdorp, Albert District (*D. R. Kannemeyer*).
   K. Transvaal.—Special locality not noted (*H. Barber*).

Genus PAMPHILA.

Pamphila, Fabricius, " Illiger's Mag., vi. p. 287 (1807)."
   „   Westw. (part), Gen. Diurn. Lep., ii. p. 521 (1852).
   „   Trim. (part), Rhop. Afr. Aust., ii. p. 297 (1866).
   „   Speyer, Stett. Ent. Zeit., 1878, p. 183, and 1879, p. 489.
Hesperia, Plötz, Stett. Ent. Zeit., 1882, p. 314.

IMAGO.—*Head* broad, sometimes broader in ♂ than in ♀; *palpi* usually short,—middle joint broad,—wider anteriorly, densely clothed with stiff hair with the appearance of having been evenly clipped,—terminal joint usually rather short and thick, conical, often scarcely showing beyond hair of middle joint ; *antennæ* variable in length, stout,—club well marked, rather thick, cylindrical, more or less elongate, almost always ending in a short, slender, more or less pronounced hook ; tuft near base of each antenna rather short.

Thorax very robust, about as long as abdomen, clothed with short hair, which is usually less dense on the back. *Wings* rather small, thick. *Fore-wings* prolonged apically, especially in ♂ ; apex itself more or less acute ; costa nearly straight, or slightly hollowed beyond middle ; hind-margin in some species (more markedly in ♂) moderately concave inferiorly ; discoidal cell always more than half, and usually two-thirds the length of the wing, pointed superiorly at its extremity,— the closing nervules (middle and lower disco-cellular) being inwardly oblique, and usually well developed ; lower radial nervule originating considerably nearer to third median nervule than to upper radial, and usually so curved in conjunction with lower disco-cellular as to look like a fourth median nervule ; first median nervule variable as to distance of its origin from that of second nervule, sometimes originating about half-way between latter and base, but in many species much nearer to second nervule ; in ♂ sometimes a lower discal linear groove, situated as in some species of *Thymelicus*, or farther from base, shorter, moderately oblique inwardly, between submedian nervure and first median nervule. *Hind-wings* rather short, apically rounded, always more or less produced, and commonly lobate, at anal angle (especially in ♂) ; costa more or less strongly lobate and fringed with hair basally ; discoidal cell short, seldom extending to middle,—its closing nervule very slender, outwardly oblique ; radial nervule reduced to a mere line or wanting ; costal nervure ending at or just before apex ; subcostal nervure branching considerably before extremity of cell; median nervules long, the second and third given off very near each other. *Legs* rather long, stout; femora with rather thin longish hair beneath; appendage to fore-tibiæ rather large; middle and hind tibiæ strongly spurred, usually sparsely spinose beneath,—the latter (and usually also the former) superiorly with long hair; tarsi long and thick, strongly spinulose beneath.

Abdomen of variable length, but usually rather shorter than inner margin of hind-wings, and rarely slightly exceeding it.

LARVA.—Rather thick, smooth; head large, with the two lobes prominent. (Characters of *P. Comma*, Linn.—the type of the genus—as figured by Hübner and Duponchel.)

This genus is more numerous in species than any other in the family, and has by several authors been regarded as the typical one—*Hesperia*. Out of the very large number—between 500 and 600—of species referred to it, about two-thirds are peculiar to the New World—seventy having been described from North America alone. From its Ethiopian Region about sixty species have been recorded, and about forty-five from the Oriental Region. The Palæarctic Region is exceedingly poor, having yielded but twelve species, and the Australian Region (as far as known) is no richer.

It is with some hesitation that I have admitted the South-African *Callicles*, Hewits., and *Morantii*, Trim., into this genus, because in the former the tip of the antenna, though curved and acuminate, cannot be said to form a hook at all; and in the latter, the palpi (especially the terminal joint) are considerably longer and narrower than usual, and porrected forwards. Some latitude in respect of these organs is, however, recognised by entomologists in the case of *Pamphila*,—*P. Alcides*, Herr.-Schäff., from Asia Minor, having non-falcate antennæ, and *P. Buchholzi*, Plötz, from West Africa, possessing[1] long porrect, pointed, almost dependent palpi. The three larger South-African species—*Erinnys*, Trim., *Dysmephila*, Trim., and *Fiara*, Butl.—I was disposed, with Mr. Butler, to place in the genus *Proteides*; but a more thorough examination of their structure has convinced me that the greater proportional length of their antennæ is the only distinctive feature that they present.

The fourteen known South-African species may be arranged as follows, viz.:—

- A. Antennæ of moderate length; their club short, curved, and pointed at tip, but not hooked.—*P. Callicles*.
- B. Antennæ of moderate length; their club long, with a moderately long, gently-curved hook at tip; palpi long and porrect.—*P. Morantii*.
- C. Antennæ of moderate length; their club rather shorter and

[1] See Plötz, Stett. Ent. Zeit., 1882. p. 330.

thicker than in B., and with its terminal hook more bent; palpi normal.—*P. Zeno.*

D. Antennæ short; their club as in C.—*P. lugens.*
E. Antennæ very short; their club short and thick, with the terminal hook exceedingly small, short, and slender, and consisting of a single joint.—*P. hottentota.*
F. Antennæ rather long; their club rather long and stout, with the terminal hook rather long and sharply recurved.—*P. Moritili* and *P. Ayresii.*
G. Antennæ rather short; their club as in F.—*P. Monasi, P. Borbonica, P. Fatuellus,* and *P. Mohopaani.*
H. Antennæ long; their club as in F. and G.—*P. Erinnys, P. dysmephila,* and *P. Fiara.*

The colouring of most of them is dull; *Callicles* (with clearly-defined pale-yellow spots on a brown ground above and black-ringed white spots on a yellow ground beneath), *Morantii,* and *Zeno* (with warm ochre-yellow macular bands on a brown ground above), and *Erinnys* (with colouring like the last on the hind-wings above, and a whitish varied bronzy-glossed under-side), being the only species of a somewhat gayer aspect.

The most widely distributed species are *Hottentota* and *Mohopaani,* both of which occur near Cape Town occasionally, and are recorded from both Tropical belts. *Callicles* is known from such remote points as Damaraland and Somaliland, but in South Africa seems to have been met with only at Delagoa Bay and Natal. *Borbonica* inhabits Natal and Transvaal as well as Madagascar and the Mascarene Islands. All the remaining species belong to the south-eastern side, and six of them are known to occur within the Southern Tropic. The four which appear to be confined to South Africa are *Monasi* (limited to Natal), and *Erinnys, Dysmephila,* and *Fiara,* all of which inhabit Kaffraria and Natal, and penetrate as far as the eastern districts of the Cape Colony. The rarest of the South-African species are *Callicles, Morantii, Zeno, Lugens,* and *Ayresii.*

343. (1.) **Pamphila Callicles,** (Hewitson).

♀ *Cyclopides Callicles,* Hewits., Descr. New Sp. Hesp., ii. p. 42, n. 6 (1868); and Exot. Butt., v. pl. 59, ff. 10, 11 (1874).

Exp. al., (♂) 1 in. 1 lin.; (♀) 1 in. 2½ lin.

♂ *Glossy-brown, with straw-yellow terminal disco-cellular spot and irregular series of discal spots in both wings; cilia very pale yellowish, in fore-wing mixed with brownish except at posterior angle. Fore-wing:* basal fourth, and costal and inner-marginal border to beyond middle, rather closely irrorated with yellow; terminal disco-cellular spot elongate transversely, narrow and constricted superiorly; discal series of nine spots highly irregular,—the first three very small, forming a short incurved costal streak about midway between end of cell and apex,—the fourth and fifth also very small, situated nearer to hind-margin,—the sixth larger, subquadrate, in a straight line beneath first three,—the seventh and ninth large, longitudinally elongate, in a slightly oblique line beneath disco-cellular one (from which the seventh is separated only by median nervure),—and the eighth minute, between seventh and ninth, and in one specimen united to latter. *Hind-wing*: discoidal cell and nearly all inner-marginal area below and beyond it clothed with yellow hairs; terminal disco-cellular spot shaped similarly to that in fore-wing, but not so elongate; in discal series six spots, of which the first and sixth are very small, rounded, but the others narrow and elongated longitudinally, contiguous, the second and fourth projecting considerably beyond the rest. UNDER SIDE.—*Hind-wing, and all fore-wing except inner-marginal area up to a little distance from hind-margin, bright chrome-yellow; spots arranged as on upper side, but very sharply and distinctly black-edged; neuration beyond discal spots, and a hind-marginal edging line, black*. *Fore-wing*: inner-marginal fuscous area extending narrowly to posterior angle, and slightly encroaching on lower edge of discoidal cell. *Hind-wing*: close to base, between costal and subcostal nervures, a small black spot; another, rather farther from base, between median and submedian nervures; submedian and internal nervures black; all six spots of discal series equally distinct, contiguous,—and a seventh (separate) spot between costal nervure and first subcostal nervule.

Head above black, with a few creamy-white hairs intermixed; on vertex a median longitudinal line and a conspicuous spot next each eye, as well as a posterior half-ring round eyes, creamy-white; *palpi* above black with a few yellowish hairs intermixed, beneath creamy-white tinged with yellow superiorly,—the small terminal joint all black; *antennæ* black, ringed at joints with creamy-yellowish (conspicuously so beneath),—the club very slightly acuminated and recurved, dull-red at tip. *Thorax* above fuscous, densely clothed with yellowish hair, especially posteriorly; beneath creamy-whitish; legs yellow. *Abdomen* fuscous above, very conspicuously and sharply ringed with straw-yellow on segmental incisions; beneath creamy-whitish, with a row of elongate black spots on each side.

♀ *Like ♂ ; but both ground-colour and spots paler, and the yellow irroration in fore-wing and yellow pubescence in hind-wing very much duller, paler, and sparser*. *Fore-wing*: disco-cellular spot larger superiorly; seventh spot of discal row for the greater part not in line with, but beyond disco-cellular spot,—ninth spot in one example considerably enlarged and elongated longitudinally. UNDER SIDE.—As in ♂ ; but inner-marginal fuscous area paler in fore-wing, and in one example extending up to hind-marginal edge above posterior angle as far as first median nervule.

The markings on the under side of the hind-wings of this very distinct species are much like those of the European *Cyclopides Steropes* (W. V.), but narrower, less numerous, rather duller, and not so strongly black-edged.

Though apparently very scarce in collections, *Callicles* has so wide a known range that it is probably not rare in favourable localities. Damaraland seems to be its favourite country; the first examples that I saw were taken in that territory by the late Mr. C. J. Andersson and Mr. H. Hutchinson. Hewitson records the butterfly from Angola. Colonel Bowker in 1880 took a ♀ example in Upper Natal, and Mrs. Monteiro sent a ♂ from Delagoa Bay in 1883. I have also received a ♂ (without special locality) in a collection made by Mr. F. C. Selous in the country between Transvaal and the Zambesi. Butler in 1885 noted (*Proc. Zool. Soc. Lond.*, 1886, p. 775) two specimens from Somaliland.

Localities of *Pamphila Callicles*.

I. South Africa.
  E. Natal.
    b. Upper Districts.—Valley of Tugela and Mooi Rivers (*J. H. Bowker*).
  II. Delagoa Bay.—Lourenço Marques (*Mrs. Monteiro*).

II. Other African Regions.
  A. South Tropical.
    a. Western Coast.—Damaraland (*C. J. Andersson* and *H. Hutchinson*). "Angola (*Rogers*)."—Hewitson. "Congo: Kinsembo."—Butler.
    b1. Eastern Interior.—Special locality not noted, but South of Zambesi (*F. C. Selous*).
  B. North Tropical.—"Somaliland (*Thrupp*)."—Butler.

344. (2.) Pamphila Morantii, Trimen.

♀ *Pamphila Morantii*, Trim., Trans. Ent. Soc. Lond., 1873, p. 122.
Pamphila Ranoha, Westw., App. Oates' Matabele Land, p. 353, n. 74 (1881).

PLATE XII. fig. 3 (♂).

Exp. al., (♂) 1 in. 2 lin.; (♀) 1 in. 3 lin.

♂ Very dark brown, with fulvous or ochreous-yellow bars and spots; bases with ochre-yellow scaling and short hairs. *Fore-wing*: basal half of costa broadly clouded with ferruginous or yellow-ochreous; a moderate-sized subquadrate spot at extremity of discoidal cell; on disc an irregular, inwardly-oblique, inferiorly wider band from costa to submedian nervure, interrupted subcostally on lower radial nervule, separated from cellular spot only by median nervure, and emitting a narrow ray along submedian nervure towards base; in the ferruginous-marked examples there is a tinge of that colour over the upper separate part of this discal band. *Hind-wing*: a rather large obliquely-transverse discal patch, commencing abruptly widely on second subcostal nervule, and terminating narrowly on basal portion of median nervules. *Cilia* broad ochreous-yellow, in some examples tinged with ferruginous in fore-wing.

UNDER SIDE.—*Hind-wing and costal border and wide apical area of fore-wing ferruginous or yellow-ochreous; in the ferruginous-tinted examples the inner-marginal fold of hind-wing (except along its inferior margin) is yellow-ochreous.* Fore-wing: ground of lower half of wing dull-fuscous; upper part of discal band obliterated; lower part and disco-cellular spot paler than on upper side. Hind-wing: *three small black spots before middle,* viz., one in discoidal cell close to base, another at extremity of cell, and the third between median and submedian nervures; *a strongly-curved discal row of five similar spots* between costal and submedian nervures, two being above the interrupting disco-cellular fold and three below it. Cilia agreeing in colour with the hind-margin. Above, the head and palpi, and the prothorax clothed with ferruginous- or yellow-ochreous hairs; thorax and abdomen dark-brown, the former with some yellow hairs posteriorly, and the latter with the segmental incisions and a terminal tuft fulvous- or ochreous-yellow. Beneath, the palpi, breast, and legs, and the base of abdomen ochreous-yellow (in some examples tinged with ferruginous); median band and terminal part of abdomen pale-yellow, with a row of small black spots on each side.

♀ Like ♂, but all the ochre-yellow markings narrower, especially in fore-wing, where discal band is widely interrupted (from upper radial to third median nervule), and is much reduced inferiorly, leaving disco-cellular terminal spot isolated. UNDER SIDE.—Fuscous ground larger, and markings reduced in fore-wing.

Allied to *Lepenula*, Wallengr., but with much darker ground-colour on upper side, and readily known by possessing small black spots on the under side of the hind-wings, and by the acuminate and slightly hooked club of the antennæ. From the East-African *Herilus*, Hopff.,[1] which is also nearly related, it may be recognised by wanting the yellow disco-cellular spot and sub-inner-marginal stripe on the upper side of the hind-wings, and by having fewer and quite differently situated black spots on the under side.

I was disposed to keep separate *Ranoha*, West., in consequence of its somewhat larger and paler upper-side markings, and paler yellow under side, but finding that the Natalian *Morantii* varied in both these characters in both sexes (the under side, especially in a ♂ and two ♀ s, being as pale in tint as in *Ranoha* from the Upper Limpopo), I have decided that the two cannot be held distinct.

This is a rare species; I have seen only seven Natalian examples, three of which were of the decided ferruginous tinting on the under side, and one of an intermediate colour. An injured ♂ was captured at D'Urban by my Kafir collector on 5th February 1867, but I could not identify it as conspecific with the ♀ (taken in June 1869 at Pinetown by Mr. W. Morant) on which I subse-

---

[1] Peters' *Reise nach Mossamb.*, Ins., p. 419, t. xxvii. ff. 7, 8 (1862).

quently founded the species. Colonel Bowker has of late years sent at intervals five specimens in all (two ♂s, three ♀s) from Pinetown and Northdene, a few miles from D'Urban. Of these, one ♂ was taken at the beginning of June 1879, and the other was reared on 8th July 1883, from a pupa found in the rolled-up leaf of a tree at Northdene. Mr. A. D. Millar informs me that he had (1888) only seen two specimens near D'Urban; he forwarded one (♀) for identification.

Localities of *Pamphila Morantii*.

I. South Africa.
 F. Natal.—D'Urban. Pinetown and Northdene (*W. Morant* and *J. H. Bowker*).
 K. Transvaal.—Upper Limpopo (*F. C. Selous*).

II. Other African Regions.
 A. South Tropical.
  *a*1. Western Interior.— Ehanda, between Northern Ovampoland and Ombuella (*A. W. Eriksson*).
  *b*1. Eastern Interior.—" Zambesi: near Victoria Falls (*F. Oates*)."—Westwood.

345. (3.) Pamphila Zeno, Trimen.

♀ *Pamphila Zeno*, Trim., Trans. Ent. Soc. Lond., 3rd Ser., ii. p. 179 (1864); and Rhop. Afr. Aust., ii. p. 301, n. 194 (1866).

PLATE XII. fig. 2 (♀).

Exp. al., (♂) 1 in. 4 lin.; (♀) 1 in. 4½–5 lin.

 ♂ *Dark-brown, with orange-yellow spots. Fore-wing:* along costa, from near base to rather beyond middle, a yellow stripe; a large irregularly-shaped, disco-cellular spot, emitting a thin ray along median nervure to near base; beyond middle an interrupted, irregular, transverse row of nine spots, of which the first three form one marking close to costa, the fourth and fifth are small and nearest of the row to hind-margin, the sixth and seventh largest and (with the small eighth and large ninth) contiguous, the ninth being above submedian nervure and in a straight line below spot in cell; basal area densely scaled with yellow. *Hind-wing:* same pattern as fore-wing: disco-cellular spot smaller, divided by fold; transverse row composed of six contiguous spots, commencing some distance from costa, and reaching to submedian nervure,—the first spot much the smallest, and second spot the largest. *Cilia* brown; orange-yellow at posterior and anal angles. UNDER SIDE.—*Fore-wing:* spots as above, but much paler; *costal and apical area yellow with a greenish tinge. Hind-wing: spots uniting with costal and hind-marginal colouring to form a pale-yellow ground-colour,* on which the portions dark on upper side take the form of a broad median and very narrow submarginal macular fuscous band,—the latter incomplete band joining a wide blackish space at anal angle, which narrows gradually to base along inner-marginal fold; inner margin itself bordered with ground-colour; three small fuscous spots form a short transverse row near base.

♀ *Similar, but duller and paler throughout.* Fore-wing: costal stripe wanting, but represented by some yellow irroration.

This is a near relative of the widely spread Indo-Malayan *P. Augias*, (Linn.), but the ♂ wants the discal badge on the fore-wings of the latter, as well as the extension of the yellow costal band almost to apex, and the evenness and continuance of the discal row of spots (which forms a band) and its radiation on nervules to hind-margin. The ♀ *Zeno* also wants in the fore-wings the costal yellow beyond middle of *Augias* ♀, and her discal row of spots differs in the same way, though in a less degree, as noted with regard to the ♂ s. On the under side both sexes of the South-African species are of a paler yellow throughout, both as respects the ground-colour and the markings; the latter are better defined; and the fuscous spots near the base of hind-wings are a character wanting in *Augias*.

Colonel Bowker discovered this rare species in Kaffraria Proper in 1863, and sent me a ♂ and two ♀ specimens. He also captured a ♀ at Pinetown, Natal, in May 1879. Mr. A. D. Millar, forwarding a fine ♂ for identification, writes (June 1888) that he took the species numerously at the foot of the Howick Falls of the Umgeni, near Maritzburg, but that it is seldom seen on the coast. He observed one example only at Pinetown. At Oxford, in 1867, I noted one example in the Rev. H. Rowley's Zambesian collection.

Localities of *Pamphila Zeno*.

I. South Africa.
  D. Kaffraria Proper.—Bashee River (*J. H. Bowker*).
  E. Natal.
    a. Coast Districts.—Pinetown (*J. H. Bowker*).
    b. Upper Districts.—Howick, near Maritzburg (*A. D. Millar*).
II. Other African Regions.
  A. South Tropical.
    b. East Coast.—Zambesi River (*Rev. H. Rowley*).—Coll. Hope Oxon.

346. (4.) Pamphila Hottentota, (Latreille).

? ♀ *Papilio Niso*, Linn., Mus. Lud. Ulr. Reg., p. 339, n. 157 (1764); and Syst. Nat., i. 2, p. 796, n. 270 (1767).
♂ *Hesperia Hottentota*, Latr., Encyc. Meth., ix. p. 777, n. 133 (1823).
♂ ♀ *Hesperia Letterstedti*, Wallengr., K. Sv. Vet.-Akad. Handl., 1857; Lep. Rhop. Caffr., p. 49, n. 3.
♂ ♀ *Pamphila Letterstedti*, Trim., Rhop. Afr. Aust., ii. p. 300, n. 193 (1866).
♂ *Pamphila Hottentota*, Staud., Exot. Schmett., i. pl. 99 (1888).

PLATE XI. fig. 8 (♂); fig. 8a (♀).

Exp. al., (♂) 1 in. 2–2½ lin.; (♀) 1 in. 1–3½ lin.

♂ *Pale dull greyish-brown, with nearly entire hind-wing, and all but the wide apical hind-marginal area of fore-wing, suffused with ochre-*

ous-yellow. Fore-wing: on subcostal nervules, rather before midway between extremity of discoidal cell and apex, two very small and indistinct (sometimes quite evanescent) yellowish spots, forming a very short slender oblique mark. *Hind-wing:* costal border more or less brownish-grey towards apex. *Cilia* brownish-grey, becoming whitish about posterior and anal angles. UNDER SIDE.—*Hind-wing, and all fore-wing except a wide dull-whitish inner-marginal border, pale dull greyish-yellow with a faint greenish tinge, marked with some scarcely paler yellowish grey-edged spots. Fore-wing:* subcostal spots better defined than on upper side, and beginning a discal row of six, the third and fourth of which stand together beyond the rest, and the fifth and sixth obliquely nearer base. *Hind-wing:* discal row composed of elongate confluent spots, extending from costal to submedian nervure, sharply angulated on second subcostal nervule near hind-margin; whole of inner-marginal fold much paler, inclining to dull-whitish.

♀ *Rather darker, with a general yellowish tinge, but without any direct yellow suffusion; fore-wing with a discal row of small distinct pale-yellowish spots arranged as on under side of ♂. Fore-wing:* usually a third (uppermost) minute spot at subcostal commencement of discal row, also an eighth rather indistinct spot immediately above submedian nervure. *Hind-wing:* a short, faint, median discal row of three dull-yellowish elongate marks, representing in part that of under side. UNDER SIDE.—*Like that of ♂, but considerably yellower, and with all the markings much more distinct. Fore-wing:* inner-marginal and (especially) central area of a much darker grey. *Hind-wing:* inner-marginal fold greyer.

VARIETY A. ♂ and ♀ (*Letterstedti*, Wallengr.).

♂ Rather darker, without ochre-yellow suffusion, but with more or less of a pale dull greenish-yellow tinge, especially near bases. *Fore-wing:* often a decided greenish-yellow irroration basally and along costal border. UNDER SIDE.—Much yellower, nearer in tint to typical ♀, and with the markings almost as distinct. *Hind-wing:* inner-marginal fold grey, darker about anal angle,—only the inner edge yellowish.

♀ Like typical ♀, except that the discal spots are usually larger and yellower and better defined, especially in hind-wing, and that the yellow irroration from bases usually extends pretty widely over inner-marginal areas, as well as along costa of fore-wing. UNDER SIDE.—Like typical ♀, but central and inner-marginal area of fore-wing and inner-marginal fold of hind-wing darker.

(*Hab.*—All Eastern South Africa, and as far westward as Knysna, Cape Colony.)

In *Rhopalocera Africæ Australis* I treated the more prevalent form just described as typical (as indeed it is of *Letterstedti*, Wallengr.), noting that the rather smaller form, with ochreous-yellow suffusion, characteristic of the Cape Peninsula and the Western Districts, was probably *Hottentota*, Latr. As fresh reference to Latreille's brief description (of the upper side only) convinces me that he had before him a specimen of the ochre-yellow suffused ♂, that form now properly becomes typical of the species.

Mr. Aurivillius (*K. Sv. Vet.-Akad. Handl.*, 1882, xix. p. 125) has shown much ground for the opinion that *Niso*, Linn.—of which he reproduces two figures from Clerck's unpublished *Icones*[1]—is a Western ♀ of this species. But Linné's description is so absolutely inadequate for identification (treating the insect as probably only a variety of the European *Nisoniades Tages*), and Clerck's figures, except as regards the head and antennæ, are so unlike the butterfly concerned, that I cannot feel sufficient certainty to warrant my adopting Mr. Aurivillius' conclusion.

All the specimens I have taken about Cape Town, the few that have been met with in the neighbouring Western Districts, at Port Elizabeth, and near Grahamstown, are of the typical form. Möschler (*Verh. K.K. Zool.-Bot. Gesellsch. Wien*, 1883, p. 287) notes that at Baziya, in Kaffraria Proper, the ♂ sometimes exhibits the ochre-yellow ground-colour above; and I have received an intermediate example from the Bashee River.

This is an active and pugnacious little species, with habits quite like those of *Pamphila Sylvanus* and *P. Comma* in Europe. It haunts by preference little hollows and kloofs in hillsides; the ♂ often taking possession of some tall flower in an open position, and darting at every insect which approaches his perch. It is not common near Cape Town, but I took it numerously at Knysna and in Natal, and commonly near Grahamstown. I have met with it at various stations in South Africa from the end of September to the end of March, and took it at D'Urban, Natal, on 3d August 1865. I captured the paired sexes twice at Knysna, and on one occasion at Mapumulo in Natal; and Colonel Bowker sent me a pair taken *in copulâ* at King William's Town. Mr. W. S. M. D'Urban noted the butterfly as abundant in British Kaffraria, occurring from August to May.

I noted a ♂ in a collection made on the Gold Coast in 1873 by Lieutenant Bourke, R.N.

Localities of *Pamphila Hottentota*.

I. South Africa.
   B. Cape Colony.
      a. Western Districts.—Cape Town. Vogel Vley, Tulbagh District. Palmiet River, Caledon District (*T. D. Butler*). Knysna—[*Var.* A.].
      b. Eastern Districts.—Port Elizabeth. Uitenhage (*S. D. Bairstow*). Grahamstown. King William's Town (*J. H. Bowker*). Hope Town (*J. H. Bowker*).
      c. Orange Free State.—Special locality not noted (*C. Hart*).
   D. Kaffraria Proper.—Bashee River (*J. H. Bowker*)—[*Var.* A. and intermed.]. Heads of St. John's River (*J. H. Bowker*).
   E. Natal.
      a. Coast Districts.—D'Urban, Verulam, and Mapumulo—[*Var.* A.]. Pinetown (*J. H. Bowker*)—[*Var.* A.].
      b. Upper Districts.—Maritzburg and Greytown—[*Var.* A.]. Biggarsberg and Rorke's Drift (*J. H. Bowker*)—[*Var.* A.]. Estcourt (*J. M. Hutchinson*).

---

[1] Mr. Aurivillius wrote to me in 1881, that although the types of *Pap. Spio* and *Pap. Niso*, L., were unfortunately lost, he felt "quite certain that these" (Clerck's) "figures are delineated after the true types of Linné."

F. Zululand.—Napoleon Valley (*J. H. Bowker*)—[*Var.* A.]. St. Lucia Bay (the late *Colonel H. Tower*).
H. Delagoa Bay.—Lourenço Marques (*Mrs. Monteiro*).
K. Transvaal.—Potchefstroom District (*T. Ayres*)—[*Var.* A.].

II. Other African Regions.
B. North Tropical.
a. Western Coast.—Gold Coast (*J. Bourke*). "Gambia River (*Moloney*)."—Shelley.

347. (5.) Pamphila Monasi, *sp. nov.*

Exp. al., (♂) 1 in. 1–2 lin.; (♀) 1 in. 3–4 lin.

♂ *Pale dull-brown, with some sparse yellowish scaling over basal and inner-marginal areas; in both wings a discal series of transparent spots; cilia greyish, mixed with brown in fore-wing, and with yellowish-brown on hind-wing.* Fore-wing: longitudinally-elongated transparent spot in upper part of discoidal cell at a little distance before extremity; discal series of five spots arranged in the usual positions, viz., two minute ones transversely on subcostal nervules, about midway between extremity of cell and apex, and three in a very oblique series between lower radial and first median nervules, the first being minute, the second larger, and the third of a good size (the two latter elongate longitudinally and excised externally); yellowish scaling best developed on costa near base, more faintly and narrowly along inner margin. Hind-wing: spots of discal series small, situated transversely between second subcostal and first median nervules, the first and third projecting considerably beyond the second and fourth; sparse yellowish scaling and hairs prevalent over discoidal cell and inner-marginal area. UNDER SIDE.—*Hind-wing and costal border, and rather wide apical area of fore-wing, rather dull greenish-yellow.* Fore-wing: transparent spots as on upper side; ground-colour dark-grey, becoming much paler at and near posterior angle. Hind-wing: transparent spots considerably smaller than on upper side, and bounded externally by dark-brownish, sometimes almost obsolete; inner-marginal fold grey; an ill-defined grey spot marking extremity of cell.

♀ Quite like ♂, except that the yellowish scaling is more inconspicuous in both wings. UNDER SIDE.—Yellow much duller, and with a more decidedly greenish cast. Hind-wing: transparent spots on disc larger and better defined.

Although so much smaller, this species is nearly related to *P. Mohopaani*, (Wallengr.), and to *P. Borbonica*, (Boisd.). The ♀ bears a very considerable resemblance, in size and colour as well as in pattern, to the ♀ *P. Hottentota*, (Latr.), but may easily be known by her quite transparent discal spots, without any yellow tint, and duller under side, with small vitreous spots on the disc instead of large contiguous elongate yellow spots. The ♂, being closely similar to the ♀, is of course wholly different above from the spotless or almost spotless, paler and more or less yellow-suffused ♂ *Hottentota*.

I captured both sexes of this inconspicuous little Hesperide at D'Urban, Natal, in February 1867, and have since received a few examples from the late Mr. M. J. M'Ken and from Colonel Bowker, captured in the same locality. It is so very like the ♀ *Hottentota* in flight and appearance, that no doubt it is constantly mistaken for that butterfly, and so overlooked by collectors.

Localities of *Pamphila Monasi*.

I. South Africa.
   E. Natal.
      a. Coast Districts.—D'Urban.

348. (6.) Pamphila lugens, Hopffer.

♂ *Pamphila lugens*, Hopff., "Monatsb. K. Akad. Wissensch. Berl., 1855, p. 643;" and Peters' Reise n. Mossamb., Ins., p. 418, t. xxvii. ff. 5, 6 (1862).

Exp. al., (♂) 1 in. 3 lin.; (♀) 1 in. 4 lin.

♂ *Very dark brown, with a bronzy gloss in some lights; without markings of any kind;* cilia brown, but greyish externally in both hind-wing and lower half of fore-wing. UNDER SIDE.—Not so dark, especially in fore-wing, which has a very faint tinge of dull ochreous-yellow; in both wings a paler very ill-defined median shade, more apparent in fore-wing, which has also a very narrow similar paler shade along hind-margin.

Head and *body*, with *legs*, all dark-brown, above with a slight admixture of bronzy-yellow hairs. *Palpi* beneath with a considerable admixture of pale-yellowish hairs; *antennæ* black above, yellowish-white narrowly barred with black beneath.

♀ *Duller and paler*. *Fore-wing*: a discal series of five very indistinct minute transparent spots, viz., a subcostal curved row of three some way beyond extremity of discoidal cell, and two larger sublinear ones, situated one above, the other below second median nervule (the lower one nearer base). UNDER SIDE.—Paler than in ♂; transparent spots of fore-wing as on upper side.

Hopffer mentions that the ♂ when in fine condition exhibits on the under side of the fore-wings a scarcely perceptible discal series of five minute spots paler than the ground-colour. He records three ♂ s from Querimba, collected by the Peters' Expedition.

I have seen only two specimens—one of each sex—of this very sombre little butterfly; they were received at the South-African Museum in 1886 in a small series collected at Delagoa Bay by Mrs. Monteiro.

Localities of *Pamphila lugens*.

I. South Africa.
   II. Delagoa Bay.—Lourenço Marques (*Mrs. Monteiro*).
II. Other African Regions.
   A. South Tropical.
      b. East Coast.—" Querimba (*Peters*)."—Hopffer.

349. (7.) Pamphila Moritili, (Wallengren).

♀ *Hesperia Moritili*, Wallengr., K. Sv. Vet.-Akad. Handl., 1857; Lep. Rhop. Caffr., p. 49, n. 4.
Pamphila? Moritili, Trim., Rhop. Afr. Aust., ii. p. 305, n. 199 (1866).
Hesperia Neba, Hewits., Ann. and Mag. Nat. Hist., 4th Ser., xix. p. 84 (1877).

PLATE XII. fig. 4 (♂).[1]

Exp. al., (♂) 1 in. 1½–3 lin.; (♀) 1 in. 2–3 lin.

♂ *Glossy brownish-black, with small transparent spots in fore-wing and opaque yellowish-white ones in hind-wing.* Fore-wing: in discoidal cell, near extremity, two elongate spots, one above the other, the lower rather the larger; a strongly incurved discal row of five spots, of which the first and second are together and subcostal, very small, between end of cell and apex,—the third and fourth also very small between lower radial and second median nervules; and the fifth considerably larger, quadrate or subquadrate between second and first median nervules, and immediately beneath lower cellular spot; besides these vitreous spots, a small, rather elongate yellowish-white spot about middle, immediately above submedian nervure; from base to middle of inner margin a yellowish-white edging, widening towards its extremity. *Hind-wing*: rather darker than fore-wing, especially in basal half; a curved discal series of five small yellowish-white spots, between second subcostal nervule and submedian nervure,—the first being the smallest and the last the largest, and somewhat diffused. *Cilia* of fore-wing brown, mixed with whitish at posterior angle; of hind-wing, brown mixed with whitish, but entirely white round anal-angular lobe. UNDER SIDE.—*Much paler, dark-brown with a strong bronzy or æneous gloss (in hind-wing shifting to violaceous).* Fore-wing: transparent spots as on upper side; yellowish-white spot just above submedian nervure larger; inner-marginal yellowish-white streak wanting, but a similar streak along costa from base to about middle; basal and inner-marginal area darker and duller than rest of wing. *Hind-wing*: an ill-defined median whitish shade, more or less markedly exhibiting the violaceous gloss, on the outer edge of which the small discal spots corresponding to those of upper side are only just perceptible, except the

---

[1] In this figure the head is not large enough, the projection near anal angle of the hind-wings is too acute, and the pink stain on the under side of the hind-wings is too strong an expression of the faint violaceous gloss in nature.

fifth, which is rather larger than on upper side, and conspicuous; beyond the spots and also along inner-marginal fold the dark-brown surface is unmarked, except by a white line just before hind-marginal edge, between first median nervule and anal angle; about base some sparse whitish scaling.

Head and thorax very robust, the former quite as broad as the latter. Antennæ with a well-developed and strongly recurved hook; black above, half-ringed with yellowish-white beneath, and with basal incrassation of club also marked beneath with yellowish-white. Beneath, the palpi and front of the breast are conspicuously yellowish-white; rest of breast greyish. Abdomen blackish, with complete segmental incision rings of yellowish-white, very thin above, but wider laterally and beneath.

♀ Like ♂, but rather paler and duller. Fore-wing: about subcostal and median nervures, near base, a little dull-yellowish scaling. Hind-wing: spots of discal series yellower, slightly larger,—usually a minute additional spot between the subcostal nervules. UNDER SIDE.—Much paler. Fore-wing: costal yellowish-white extending to beyond middle; spots larger. Hind-wing: basal whitish scaling and median whitish shade, as well as discal spots (except the last), better developed; white hind-marginal line widened to a streak between first median nervule and submedian nervure.

This strongly-built little species, which in marking is not unlike the much larger P. Borbonica (Boisd.) in the fore-wings, may be recognised by its large head, very dark ground-colour, inner-marginal yellowish-white edging on upper side of the fore-wings, rather well-developed discal series of yellowish-white spots in hind-wings, and very glossy partly violaceous-shot under side. In the latter feature it much resembles P. Fatuellus, Hopff.,—especially the ♀. I was enabled to identify Neba, Hewits., with this species by examining the seven specimens in the Hewitson Collection at the British Museum.

I took a ♂ near D'Urban, Natal, in February 1867, and Colonel Bowker, Mr. W. D. Gooch, the late Mr. M. J. M'Ken, and Mr. A. D. Millar have all from time to time sent a specimen or two captured in that locality. As my last-named correspondent writes, however, the butterfly appears to be decidedly rare in that part of Natal. The paired sexes were taken by Colonel Bowker on the 13th August 1878. A single ♂, taken at Humbe, on the Cunene River, in October 1887, by Mr. Eriksson, has the vitreous spots decidedly larger and the under side paler and greyer than usual.

Localities of *Pamphila Moritili*.

I. South Africa.
  E. Natal.
    a. Coast Districts.—D'Urban. Pinetown (*W. Morant*).
  II. Delagoa Bay.—Lourenço Marques (*Mrs. Monteiro*).
  K. Transvaal.—Lydenburg District (*T. Ayres*).

II. Other African Regions.
   A. South Tropical.
      a 1. Western Interior.—Mossamedes: Humbe, Cunene River (*A. W. Eriksson*).

350. (8.) Pamphila Ayresii, sp. nov.

PLATE XII. fig. 1 (♀).[1]

Exp. al., (♂) 1 in.; (♀) 1 in. 1–2 lin.
Nearly allied to *Moritili*, Wallengr.

♂ *Pale greyish-brown; fore-wing with small vitreous spots, hind-wing with a dull yellowish-white discal streak.* Fore-wing: vitreous spots arranged as in *Moritili*, but only one (the upper) spot in discoidal cell near extremity, and the last spot in discal series (immediately above submedian nervure), wanting; some faint and sparse yellowish irroration in basal area, but no distinct inner-marginal streak. *Hind-wing*: discal yellowish-white streak very ill-defined, diffused externally, scarcely curved, lying between second subcostal and first median nervules. *Cilia* as in *Moritili.* UNDER SIDE.—*Hind-wing (except on inner-marginal fold) and narrow costal and moderately-wide apical hind-marginal border of fore-wing hoary-grey, closely and finely hatched or striolated with blackish.* Fore-wing: vitreous spots as on upper side; hind-marginal border narrowing to a point on first median nervule; ground-colour very much paler throughout inner-marginal area. *Hind-wing*: inner-marginal fold brownish-grey, not striolated.

♀ Like ♂; but in fore-wing two disco-cellular vitreous spots, and a small yellowish-white spot at end of discal series, just above submedian nervure, as in *Moritili.*

Besides the peculiar striolated hoary-grey under side of the wings, the much paler upper side and smaller head serve to distinguish this butterfly from *Moritili*, Wallengr.

I have dedicated the species to Mr. Thomas Ayres, the well-known ornithological observer, who has made very extensive collections of insects during his long residence in South Africa, and from whom a ♀ example (the first seen) was obtained by the South-African Museum in 1879. Until quite recently (1888) I had met with no other specimens, but, in a fine collection formed in the interior of South-Western Tropical Africa by Mr. A. W. Eriksson, I have found a pair, noted as captured at Omrora River, and a single ♂, noted as taken at Ehanda, between Northern Ovampoland and Ombuella, early in August 1887.

---

[1] In this figure the club of the antennæ is represented as too long; its real form is like that shown in fig. 4 (*Moritili*, Wallengr.). The vitreous spots of the fore-wings are too near the base, and the subcostal ones are not shown with sufficient distinctness; while the yellowish-white streak in the hind-wings is too curved.

Localities of *Pamphila Ayresii*.
I. South Africa.
  K. Transvaal.—Lydenburg District (*T. Ayres*).
II. Other African Regions.
  A. South Tropical.
    a 1. Western Interior.—Omrora River (*A. W. Eriksson*).

351. (9.) Pamphila Borbonica, (Boisduval).

Hesperia Borbonica, Boisd., Faune Ent. Madag., &c., p. 65, n. 3, pl. 9, ff. 5, 6 (1833).
Hesperia Fatuellus, Wallengr. (*nec* Hopff.), K. Sv. Vet.-Akad. Handl., 1857; Lep. Rhop. Caffr., p. 48, n. 1.
♂ ♀ *Pamphila Borbonica*, Trim. [part], Rhop. Afr. Aust., ii. p. 303, n. 197 (1866).

Exp. al., (♂) 1 in. 4–6 lin.; (♀) 1 in. 6–6½ lin.

♂ *Glossy brown, widely suffused from bases with olivaceous; with vitreous spots. Fore-wing:* in upper part of discoidal cell, near extremity, a small vitreous spot; beyond middle a transverse row of seven spots (of which the first three are minute, subcostal, and forming one narrow streak,—the other four larger, separate, the sixth being the largest, and the seventh, which is just above submedian nervure, tinged with yellowish), abruptly angulated inwardly on first median nervule. UNDER SIDE.—*Paler; hind-wing and costal and apical border of fore-wing olivaceous-yellow. Fore-wing:* spots as above. *Hind-wing:* beyond middle a row of three small white, thinly fuscous-ringed spots, between first subcostal and third median nervules; inner-marginal fold fuscous-grey superiorly, widening to anal angle.

♀ *Similar; vitreous spots larger.* UNDER SIDE.—As in ♂, but slightly duller in tint.

Cilia, in both sexes, throughout dull brownish-white.

The South-African examples differ from the typical form in their yellower under side, with larger white discal spots, and in possessing the small vitreous disco-cellular spot of the fore-wing.

I took two or three examples near D'Urban, Natal, in February 1867, and Colonel Bowker has also met with it there, and has further sent me two ♂ s, captured respectively at Pinetown in June and at Umbilo in December 1879. It seems to be decidedly much scarcer on the Natal coast than either of its near allies—*P. Fatuellus*, Hopff., and *P. Mohopaani*, Wallengr.

This butterfly was abundant in Mauritius during my visit in July 1865, and I captured many specimens at flowers in gardens.

Localities of *Pamphila Borbonica*.
I. South Africa.
  E. Natal.
    a. Coast Districts.—D'Urban. Pinetown and Umbilo (*J. H. Bowker*).
  K. Transvaal.—? Locality "(*N. Person*)."—Wallengren.

II. Other African Regions.
 A. South Tropical.
  bb. Eastern Islands.—Madagascar: Murundava (*Grevé*). Mauritius. "Bourbon."—Boisduval.

352. (10.) Pamphila Fatuellus, Hopffer.

♂ *Pamphila Fatuellus*, Hopff., "Monatsb. K. Akad. Wissensch. Berl., 1855, p. 643, n. 25."
♂ *Pamphila Fatuellus*, Hopff., Peters' Reise Mossamb., Ins., p. 417, pl. xxvii. ff. 3, 4 (1862).
Pamphila Borbonica, Trim. [part], Rhop. Afr. Aust., ii. p. 303, 304 (1866).

Exp. al., (♂) 1 in. 5–6 lin. ; (♀) 1 in. 6–7 in.

♂ *Very dark-brown, with exceedingly obscure dull-olivaceous basal suffusion ; fore-wing with a discal series of very small vitreous spots;* cilia dull-whitish, obscured with brownish in upper two-thirds of fore-wing. *Fore-wing*: vitreous spots usually only five, viz., a curved subcostal row of three minute ones and two larger ones (of which the lower is considerably the greater, and transversely elongate), respectively immediately above and below second median nervule near its origin ; very rarely two other minute spots, viz., one between lower radial and third median nervules, and the other just above submedian nervure slightly beyond middle. UNDER SIDE.—*Hind-wing and costal border and wide apical area of fore-wing dull-brown, with an ochreous-yellow tinge, and with a slight brassy gloss.* Fore-wing: costal border from base to beyond middle sparsely scaled with olivaceous-yellow ; spots as on upper side, but with the two additional ones almost always present (the lower one invariably). *Hind-wing*: a curved discal series of three minute whitish spots (sometimes very indistinct), lying between first subcostal and first median nervules ; just before this a very vague and ill-defined median dull-whitish fascia,—in some specimens all but obsolete.

♀ *Duller and paler than ♂, with more developed dull-olivaceous basal suffusion ; the vitreous spots in fore-wing considerably enlarged, always seven in number, and the lowest of them rather conspicuously white, and only semi-transparent.* UNDER SIDE.—Considerably paler than in ♂, with the submetallic gloss more pronounced and usually sub-violaceous, and with some hoary clouding on hind-margins ; *median whitish in hind-wing better developed, and in some specimens conspicuous.*

In my former work I gave this form as synonymous with *Borbonica*, Boisd., but fuller material has convinced me of its clear claim to be held a distinct species. As far as the ♂ sex is concerned, there is no chance of confounding the two butterflies, *Fatuellus* being much darker, with very much smaller (and usually fewer) vitreous spots in the fore-wings, and wanting on the under side the conspicuous olivaceous-yellow of *Borbonica*. The ♀ is, however, very like that of *Borbonica* on the upper side, but has less olivaceous suffusion from bases and a darker ground-colour, while on the under side it differs more widely than does the ♂ in respect of its want of the olivaceous-yellow, especially in those specimens where the sub-violaceous gloss and the whitish median fascia of the hind-wings are best expressed. Moreover, the shape of the wings is in both sexes different from that observable in *Borbonica*,—the fore-wings being less produced and blunter at the apex, and the hind-wings at the anal angle.

I first met with this species at D'Urban, Natal, in August 1865, and subsequently in the summer of 1867 captured a good many examples there. Colonel Bowker has sent me a number of specimens from the same neighbourhood, and Mr. A. D. Millar notes the species as common there. It flits about grass and low plants, and constantly visits flowers, *Vinca rosea* being its favourite in the Botanic Gardens at D'Urban.

Localities of *Pamphila Fatuellus*.

I. South Africa.
   E. Natal.
      a. Coast Districts.—D'Urban. Avoca and Pinetown (*J. H. Bowker*).
      b. Upper Districts.—Estcourt (*J. M. Hutchinson*).
   II. Delagoa Bay.—Lourenço Marques (*Mrs. Monteiro*).
II. Other African Regions.
   A. South Tropical.
      b. Eastern Coast.—"Querimba (*Peters*)."—Hopffer.

353. (11.) **Pamphila Mohopaani**, (Wallengren).

   ♀ *Hesperia Mohopaani*, Wallengr., K. Sv. Vet.-Akad. Handl., (1857); Lep. Rhop. Caffr., p. 48.
   ♂ ♀ *Pamphila Micipsa*, Trim., Trans. Ent. Soc. Lond., 3rd Ser., i. p. 290 (1862).
   ♂ ♀ *Pamphila Mohopaani*, Trim., Rhop. Afr. Aust., ii. p. 304, n. 198 (1866).

Exp. al., (♂) 1 in. 6–$7\frac{1}{2}$ lin.; (♀) 1 in. 7–$9\frac{1}{2}$ lin.

♂ *Glossy-brown, with a dense clothing of dull olivaceous-yellow hairs and scales over basi-inner-marginal area of fore-wing and all hind-wing except a broad costal and apical and narrow hind-marginal border; a well-marked discal series of vitreous spots in fore-wing, and a few small ones in hind-wing. Fore-wing:* apical half sparsely scaled with olivaceous-yellow; *in discoidal cell, close to extremity, two small vitreous spots, one above the other;* discal series of spots much as in *Borbonica*, but not so sharply angulated, the spots considerably smaller (but not so small as in *Fatuellus*, Hopff.), and, *in place of the lowest one, a narrow straight obliquely transverse streak* from first median nervule to submedian nervure. *Hind-wing:* discal spots minute, often indistinct, varying in number from one to five, lying between first subcostal and first median nervules. *Cilia* throughout dull-whitish, but browner in upper half of fore-wing. UNDER SIDE.—*Paler; hind-wing and narrow costal and broad apical border of fore-wing more or less coloured with pale olivaceous-yellow—in some specimens with a greyish tinge. Fore-wing:* male badge (streak between first median nervule and submedian nervure) not so distinct as on upper side, but bounded externally by two diffused indistinct whitish marks; vitreous spots as on upper side. *Hind-wing:* in upper part of discoidal cell, close to extremity, a small rounded white spot; discal spots more distinct and usually fully represented; occasionally a sixth spot between costal nervure and first subcostal nervule.

Head, thorax, and *abdomen* superiorly with a rather dense clothing of dull olivaceous-yellow hairs; inferiorly dull-white.

♀ Like ♂ ; but *the fore-wing with larger vitreous spots, and with two additional spots (one minute, immediately below sixth spot of discal row,—the other larger, outwardly acuminate, immediately above submedian nervure), in place of the straight oblique streak of the ♂* ; olivaceous-yellow scales and hairs of an obscurer tint. UNDER SIDE.—As in ♂, except that in fore-wing the two spots that replace the ♂ badge are distinct and well defined.

This is a very close ally of the widely-distributed Indian and Indo-Malayan *Mathias*, Fab., which has been recorded by Mr. Butler as inhabiting Aden, and I think it very likely, when a full series from Africa and Oriental localities can be compared, that *Mohopaani* will be recognised as merely a larger form of *Mathias*.

It is distinguishable without difficulty from its South-African allies, *Borbonica*, Boisd., and *Fatuellus*, Hopff., by possessing two disco-cellular vitreous spots in the fore-wings, and by the strong olivaceous-yellow tinting over both wings; while the ♂ has the further distinction of the linear discal sexual badge in the fore-wings.

Mohopaani seems to be pretty generally distributed throughout Southern Africa, and its occurrence has been noted in such widely remote parts of the continent, that it will probably be found to range throughout Tropical Africa. I have taken it in the neighbourhood of Cape Town and Knysna, and met with it numerously at D'Urban in Natal. Its habits are precisely those of its near congeners;—it is swift on the wing, but constantly arrests its flight to settle on flowers—its special resort in the neighbourhood of Cape Town being the flowers of the common cultivated vetch. On the Natal Coast it appears to be on the wing throughout the year. Colonel Bowker captured the paired sexes on 28th March 1880.

Localities of *Pamphila Mohopaani*.

I. South Africa.
   B. Cape Colony.
      a. Western Districts.—Cape Town. Knysna.
   D. Kaffraria Proper.—Tsomo and Bashee Rivers (*J. H. Bowker*).
   E. Natal.
      a. Coast Districts.—D'Urban. Verulam. Pinetown (*J. H. Bowker*).
      b. Upper Districts.—Estcourt (*J. M. Hutchinson*).

F. Zululand.—Napoleon Valley (*J. H. Bowker*).
H. Delagoa Bay.—Lourenço Marques (*Mrs. Monteiro*).
K. Transvaal.—Upper Limpopo (*F. C. Selous*).

II. Other African Regions.
  A. South Tropical.
    a. Western Coast.—Damaraland (*J. A. Bell*).
  B. North Tropical.
    a. Western Coast.—Gold Coast (*J. Bourke*). "Gambia River (*Moloney*)."—Shelley.

354. (12.) Pamphila Erinnys, Trimen.

Pamphila Erinnys, Trim., Trans. Ent. Soc. Lond., 3rd Ser., i. p. 290 (1861); and Rhop. Afr. Aust., ii. p. 303, n. 196, pl. 6, f. 8 [♂], (1866).

Exp. al., (♂) 1 in. 11 lin.—2 in. 1 lin.; (♀) 2 in.—2 in. 3 lin.

♂ *Dark-brown, with vitreous and yellow-ochreous spots. Fore-wing:* from base along inner-margin a clothing of short yellow hairs; a small, quadrate, vitreous spot in and near extremity of discoidal cell; three or four small similar spots form a short oblique streak from costa, between middle and apex; below them, from origin of third median nervule to submedian nervure, a row of three larger, narrow, vitreous spots (the lowest tinged with yellow) inclining towards middle of inner margin. *Hind-wing:* clothed with yellow hairs from base; *an angulated, transverse, macular, yellow-ochreous band* from near costa to middle of submedian nervure,—the first two spots smaller, separate, and one or the other of them often obsolete; a yellow-ochreous space at anal angle, extending narrowly along hind-margin. UNDER SIDE.—*Paler, and with a greenish-bronzy surface-gloss. Fore-wing:* a whitish mark on costa, above disco-cellular vitreous mark; a dentate submarginal, pale yellow-ochreous streak from costa to first median nervule; other markings as above; in apical region, between dentate streak and vitreous spots, an irroration of bluish scales. *Hind-wing: two white (or pale dull-ochreous) macular bands across central region*, close together, and partly connected by whitish nervures; some scattered whitish spots at base; anal-angular and marginal yellow-ochreous marking paler and narrower than above; parallel to hind-margin, except near anal angle, an ill-defined dentate whitish stripe; a dark purplish-brown cloud over anal-angular space.

♀ Similar; wings blunter. *Hind-wing:* yellow-ochreous median band broader, less curved. UNDER SIDE.—*Hind-wing:* some well-marked bluish irroration between central bands and submarginal stripe.

Cilia, in both sexes, in fore-wing dull-brown, tinged with yellow-ochreous about posterior angle, and in hind-wing yellow-ochreous tinged with brown near apex.

Head, thorax, and *abdomen* (in basal half) densely clothed above with pale olivaceous-brown and greyish hairs; abdomen beyond hairy coating dark-brown above, with conspicuous ochreous-yellow incision rings (becoming white laterally), and a terminal paler ochreous-yellow tuft.

A *variation* occurs in a ♀ taken at Highlands, near Grahamstown, by Mrs. Barber, the hind-wings presenting on the upper side an additional short broken yellow-ochreous bar a little before the ordinary discal one.

This fine and handsome species was discovered in Kaffraria Proper by Colonel Bowker in 1863; he noted it as occurring in deep shade on the banks of forest streams in January, settling soon after being roused on rocks among moss and ferns. It flies about sunset, as a rule, but on damp and cloudy days will visit flowers much earlier in the afternoon. In the Botanic Gardens at D'Urban, Natal, on 2d February 1867, I took easily with my fingers several specimens which were at rest, with wings all erect, on the trunks of trees. At Highlands, near Grahamstown, in January and February 1870, I noticed that the butterfly chiefly frequented *Petunia* and *Campanula* flowers, and Mrs. Barber observed one visiting the flowers of a *Crotalaria* during heavy rain. Mr. A. D. Millar informs me that he has often taken it at D'Urban on the blossom of the honeysuckle. Colonel Bowker has met with it on the Natal coast from December to April.

Localities of *Pamphila Erinnys*.

I. South Africa.
  B. Cape Colony.
    b. Eastern Districts.—Grahamstown.
  D. Kaffraria Proper.—Bashee River (*J. H. Bowker*).
  E. Natal.
    a. Coast Districts.—D'Urban. Pinetown (*J. H. Bowker*).
  F. Zululand.—Etshowe (*A. M. Goodrich* and *T. Vachell*).

355. (13.) Pamphila dysmephila, Trimen.

♀ *Pamphila dysmephila*, Trim., Trans. Ent. Soc. Lond., 1868, p. 96, n. 2, pl. vi. f. 10.

Exp. al., (♂) 1 in. 8–10 lin.; (♀) 1 in. 9–10 lin.

♂ *Pale-brown, the fore-wing beyond middle and the hind-wing along hind-margin with a purplish gloss; both wings densely clothed from base with dull yellowish-ochreous scales and hairs; fore-wing with a discocellular and two discal vitreous spots, and a white border on upper half of hind-margin.* Fore-wing: yellow-ochreous scales and (inferiorly) hairs covering basal third of wing, and thence extending narrowly along costa to apex and widely along inner margin to posterior angle; disco-cellular vitreous spot of moderate size, longitudinally elongate, sub-terminal, situated immediately above origin of second median nervule; two discal vitreous spots more quadrate, transversely elongate, obliquely placed between third and first median nervule, so that the lower (larger) one is not far below and beyond disco-cellular spot; hind-marginal white border moderately wide, rather diffused on its inner edge, extending from apex almost to second median nervule. *Hind-wing*: yellow-ochreous coating of scales and hairs generally distributed except on the margins. *Cilia* brownish-white throughout. *Hind-wing, and narrow costal border and wide apical area of fore-wing, pale brownish-grey, with a slight tinge of ochre-yellow, and a very faint violaceous gloss.* Fore-wing: vitreous spots as on upper side; in discoidal cell a very small imperfect fuscous annulet with a whitish centre just above vitreous spot; a subapical series of five similar annulets, acutely angulated on upper radial nervule—the two lower ones in an oblique line with the two discal vitreous spots; inner-marginal border very pale yellowish-grey. *Hind-wing*: an irregular, strongly-curved discal series of seven or eight very small dull-whitish fuscous-edged spots (varying in shape from circular to cuneiform), angulated just beneath second subcostal nervule; a sub-basal curved series of four similar less distinct spots; inner-marginal fold darker.

Head and *body* above very densely clothed with yellowish-brown hair; beneath of the same pale brownish-grey as the greater part of the wings.

Antennæ above whitish, but *with the club (except the hook) pure silvery-white*; beneath creamy, the club brownish, but with a white bar just before the hook, which is reddish.

♀ *Like ♂, but fore-wings without white hind-marginal border, and with smaller disco-cellular and discoidal vitreous spots; cilia browner; no purplish gloss.* UNDER SIDE.—*Hind-wing*: a conspicuous, straight, narrow, pure-white streak from base to hind-margin, running along subcostal nervure and its second nervule, and bordered inferiorly by dark-brown shading off into ground-colour.

Antennæ above blackish, half-ringed with white,—club with a white bar just before the hook; beneath as in ♂.

LARVA.—" Elongated; deep-green on the head and thoracic segments, light-green abdominally; head with two small jet-black spots, looking like eyes. Feeds on the dwarf wild date-palm (*Phœnix reclinata*), drawing the leaves together, and forming a silk-lined incomplete tube, sometimes six or seven inches in length."—J. H. Bowker, 23d March 1881.

Colonel Bowker adds that he was indebted to Mr. A. D. Millar, of D'Urban, for the discovery of this larva and its food-plant; and Mr. Millar has lately (1888) written to me that he had bred about a dozen *Dysmephila* from a small date-palm in his garden. He notes that the larvæ feed on the leaves above the tube that they have formed, retiring to the tube when not feeding; and that they are most difficult to discover among the folds of the leaves.

PUPA.—Elongate, cylindrical, abruptly acuminate at tip of abdomen. Shining reddish-brown, paler beneath. Attached to leaf by the tail, and almost concealed in the channel or imperfect tube formed by the larva. Length about one inch.

Colonel Bowker sent five pupa-skins of this species, enveloped in the folded palm-leaves.

Dysmephila is a near ally of *Cerymica*, Hewits., from Old Calabar. Only the ♂ of the latter is figured (*Exot. Butt.*, iv. pl. 57, ff. 20, 21) and described; the same sex of *Dysmephila* differs in being considerably smaller, and in the fore-wings having a conspicuous white hind-marginal border, the vitreous spots three instead of four in number, and white instead of pale-yellow, and wanting an oblique linear grey masculine badge just before lowest vitreous spot.[1]

Colonel Bowker discovered this fine species in Kaffraria Proper, capturing a single ♀ in March 1864 on the banks of the Bashee River; it was visiting flowers about sunset on a dark cloudy evening. Writing from D'Urban on the 28th May 1884, he observes, "This Skipper is a regular night-bird, and comes out with the owls and the bats. At this season it is almost dark at 5.45 p.m., and on these last two evenings I have taken five or six on the wing, chasing each other about a small tree. When in full chase they make a louder humming noise with their wings than most hawk-moths; I could often hear them a good ten yards off, when it was too dark to see them except when they rose above the horizon line." Mr. A. D. Millar writes that he has never captured this dusk-loving butterfly. The only Cape specimen that I have seen was taken near King William's Town by Miss F. Bowker in 1873. Colonel Bowker has taken it at D'Urban in December as well as in the winter months.

Localities of *Pamphila dysmephila*.

I. South Africa.
  B. Cape Colony.
    b. Eastern Districts.—Pembroke, near King William's Town (*Miss F. Bowker*).
  D. Kaffraria Proper.—Bashee River (*J. H. Bowker*).
  E. Natal.
    a. Coast Districts.—D'Urban and Pinetown (*J. H. Bowker* and *H. C. Harford*).

356. (14.) Pamphila Fiara, (Butler).

♀ *Proteides Fiara*, Butl., Trans. Ent. Soc. Lond., 1870, p. 503, n. 3.
♂ *Proteides Fiara*, Staud., Exot. Schmett., i. pl. 99 (1888).

Exp. al., (♂) 2 in.—2 in. 2 lin.; (♀) 2 in.—2 in. 3 lin.
Allied to *Dysmephila*, Trim.

♂ Rather dull ochreous-brown; hind-wing darker, with a median space of dark ochre-yellow. *Fore-wing*: on costa near apex two or three faint dull-creamy dashes at extremities of subcostal nervules. *Hind-wing*: median ochre-yellow filling discoidal cell (except close to base), and extending beyond and below it over about half the length of median nervules, but not reaching submedian nervure. *Cilia* very dull brownish-white. UNDER SIDE.—Hind-wing and narrow costal border and broad apical hind-marginal area of fore-wing pale dull brownish-grey, with a tinge of yellowish and a slight violaceous surface-gloss, rather sparsely speckled with blackish. *Fore-wing*: basal half of inner-marginal border

---

[1] I believe that a large West-African species which I noted in the British Museum collection will prove to be the ♀ *Cerymica*; on the under side of the hind-wings the white stripe is much broader than in ♀ *Dysmephila*, and the neuration is whitish.

pale dull-yellowish, glossy. *Hind-wing*: about centre a very faint yellowish cloud, not large, and without any defined outline.

Head and *body* as in *Dysmephila*, but abdomen whiter beneath. *Antennæ wholly white*, except a brownish outer line and chestnut-red terminal hook.

♀ *Similar, but paler. Fore-wing*: an ill-defined yellow-ochreous space or suffusion from base to about middle, not reaching above discocellular fold or beneath submedian nervure. *Hind-wing*: central yellow-ochreous space duller, more diffused. UNDER SIDE.—Paler; in both wings hind-margin clouded with hoary grey (on which the fuscous irroration is more apparent), and a purplish discal submacular ray (more distinct and larger in hind-wing) across basal portions of median nervules.

The fore-wings are in both sexes, and especially in the ♂, much more produced and pointed apically than in *Dysmephila*, and the total absence of vitreous markings is a notable distinctive character. In general colouring the ♂ has some resemblance on the upper side to *Hesperia Keithloa*, Wallengr.

This large and dull-tinted species, conspicuous only in its white antennæ, was first made known to me in 1874 by Mr. W. D. Gooch, who sent one of each sex, taken at Spring Vale, not far from D'Urban, Natal. Mr. Butler's type (from the description evidently a ♀) is noted with the locality "Kaffraria" only; but as the insect has occurred as far south as the Kowie River (Port Alfred) in the Cape Colony, there can be little doubt that it inhabits the intermediate Coast country between that place and Natal, including Kaffraria Proper. I have seen only seven examples; and Mr. A. D. Millar informs me that the butterfly is rather scarce at D'Urban, "usually appearing about dusk in search of food." A single ♂, taken in the same locality early in 1884, was sent to me by Colonel Bowker.

Localities of *Pamphila Fiara*.

I. South Africa.
  B. Cape Colony.
    b. Eastern Districts.—Port Alfred, Kowie River, Bathurst District (*H. Becker*).
  D. Kaffraria Proper.—Locality not noted specially. "(Druce)."—A. G. Butler.
  E. Natal.
    a. Coast Districts.—D'Urban (*W. D. Gooch, J. H. Bowker, A. D. Millar*). Pinetown (*B. Ayres*).
    b. Upper Districts.—Maritzburg (*A. Windham*).

GENUS ANCYLOXYPHA.

Ancyloxypha, Felder, Zool.-Bot. Gesellsch. Wien, 1862, p. 477.
Apaustus (Hübner, 1816), Plötz (part), Stett. Ent. Zeit., 1884, p. 151.

IMAGO.—Closely allied to *Pamphila*. *Antennæ* rather long, with a well-marked rather short club, ending in a well-curved slender hook; *palpi* closely appressed to face—the basal and middle joints broad and densely clothed with squamose hair,—*the terminal joint long, slender, naked, erect (slightly curved backward), rising considerably above head, and suddenly and sharply acuminate at tip.* Thorax moderately robust. Abdomen as long as, or rather longer than, inner margin of hind-wings.

Plötz (*loc. cit.*) refers seventy-one species to *Apaustus*, and among them seventeen African forms, of which I have been able to examine only two, viz., *Nothus*, Fab., from West Africa, and *Philander*, Hopff., from the East and South-East. The former is the type of Butler's genus *Ceratrichia*,[1] and appears to be widely different from *Apaustus* (= *Ancyloxypha*, Feld.), alike in its very long scarcely hooked antennæ, short terminal joint of palpi, comparatively slender body, and rounded hind-wings. The type of *Ancyloxypha*, is stated by Felder to be the North-American *Numitor*, Fab., and the few species that he describes are from Tropical South America.

The very remarkable structure of the terminal joint of the palpi at once distinguishes the members of this genus. I referred to it in describing[2] one of the three now known South-African species— *Mackenii*, Trim.—as a doubtful member of the genus *Pamphila*. An approach to this character is exhibited by the European species of *Thymelicus*. The South-African species are all blackish above, and are more or less white on the under side of the hind-wings; *Mackenii* is the smallest, and the ♂ has no white spots above, but a white edging and fringe to part of the hind-wings; *Philander* has large spots, some transparent and some opaque-white; and *Producta* has all the spots transparent, except one opaque-white spot in the fore-wings. The abdomen in *Mackenii* is tipped, and in *Philander* belted with white above.

These species are in South Africa only found on the south-east coast, but *Mackenii* (which is not uncommon at Natal) is recorded also from Angola; and *Philander* (which has occurred at Delagoa Bay) was originally discovered at Querimba, and has also been noted from Angola. *Producta* is only known to me by three specimens from Delagoa Bay and one from Kimberley.

357. (1.) Ancyloxypha Mackenii, Trimen.

♂ *Pamphila? Mackenii*, Trim., Trans. Ent. Soc. Lond., 1868, p. 95, pl. vi. f. 8.

Exp. al., (♂) 1 in. 1–2 lin.; (♀) 1 in. 1½–2½ lin.

♂ *Dull sooty-black; hind-wing with greater part of hind-margin edged with pure-white; cilia of fore-wing dark-brownish, becoming whitish about posterior angle, of hind-wing brownish about apex, but thence pure-white.* Fore-wing: usually two very indistinct dull-greyish discal spots, the upper between first and second median nervules, the lower immediately above submedian nervure. *Hind-wing*: hind-marginal white

[1] Cat. Fab. Diurn. Lep., p. 274 (1869).   [2] Trans. Ent. Soc. Lond., 1868, p. 95.

edging beginning below second subcostal nervule, abruptly widened between second and first median nervules, and emitting a more or less acute projection just below submedian nervure. UNDER SIDE.—*Hind-wing, and (less conspicuously) narrow costal and moderately broad apical margin of fore-wing, whitish, finely hatched and clouded with dark-brown, irrorated with yellow in parts, and shot faintly with violaceous.* Fore-wing : discal spots better marked than on upper side, but never well-defined. *Hind-wing :* basal area with a round white spot between subcostal nervules; quite across middle a straight rather narrow white fascia (with edges ill-defined), which crosses the extremity of discoidal cell ; dark-brown with yellow scaling most developed beyond this fascia, and itself succeeded by some ill-defined violaceous lunules ; about middle of hind-margin a narrow whitish cloud.

Head and *body* above blackish,—the head and thorax clothed with mixed yellow and blackish hairs,—*the abdomen with the last four segments shining pure-white*, and with an anal tuft of blackish hairs; beneath white. *Palpi* mixed blackish and yellow above, white beneath; *antennæ* black above, white-barred laterally, and whitish beneath,—the club with a conspicuous white bar beneath.

♀ *Duller, not so blackish as ♂ ; in fore-wing the two discal spots considerably larger, well-defined ; occasionally a third much smaller spot between second and third median nervules.* Fore-wing : between extremity of cell and apex a faint trace of minute spots placed transversely,—in one example fully developed in a curved subcostal line of three. *Hind-wing :* pure-white edging with less-developed expansions. UNDER SIDE.—*Fore-wing :* lower discal spots conspicuous, subcostal ones always more or less distinct; yellow scaling much more developed hind-marginally, especially at apex. *Hind-wing :* much whiter in basal area, so that median white fascia is barely separable on its inner edge.

Abdomen not shining-white terminally, but only with thin white incision rings, well-marked laterally, but almost obsolete above.

Though smaller and with broader wings, and wanting altogether the white central bar and transparent spot in the hind-wings, this Hesperid is nearly related to *Philander*, Hopff. I named it, in 1868, after the late Mr. M. J. M'Ken, who discovered it in Natal, and from whom I first received it. In June and August 1865, and subsequently in February and March 1867, I took several examples about D'Urban ; they frequented the edges of woods, flitting rapidly about bushes, and when settled holding all the wings erect. I took one specimen on flowers of *Lantana* in the Botanic Gardens. Colonel Bowker and Mr. A. D. Millar have both met with the species pretty frequently in the same vicinity.

Localities of *Ancyloxypha Mackenii*.

I. South Africa.
   E. Natal.
      a. Coast Districts.—D'Urban.

II. Other African Regions.
   A. South Tropical.—" Angola (J. J. Monteiro)."—Druce.

358. (2.) Ancyloxypha Philander, Hopffer.

♂ *Pamphila Philander*, Hopff., "Monatsb. Kön. Akad. Wissensch. Berlin, 1855, p. 643;" and Peters' Reise nach Mossamb., Ins., p. 416, t. xxvii. ff. 1, 2 (1862).

Exp. al., (♂) 1 in. 3 lin. [Hopff.]; (♀) 1 in. 4 lin.

Blackish-brown, with transparent and pure-white opaque spots. Fore-wing: at extremity of cell two transparent spots, of which the lower is larger and elongate; a little beyond and above these, a curved subcostal row of three very small transparent spots; below the two first-named spots, a large subquadrate transparent spot, between first and second median nervules, and a smaller quadrate transparent spot (nearer to hind-margin) between second and third median nervules; below the largest transparent spot, a conspicuous subquadrate pure-white opaque spot, between first median nervule and submedian nervure; basal area with some yellowish-grey scaling. *Hind-wing*: a pure-white, rather broad, central bar, extending from second subcostal nervule almost to inner margin about middle, and enclosing in its upper portion (in outer half of cell) a large sub-triangular transparent spot. Cilia brownish-white, becoming pure-white about posterior angle in both wings. UNDER SIDE.—*Fore-wing with costal margin and apical hind-marginal area brownish-ochreous; hind-wing pure-white bordered with brownish-ochreous, and with a large blackish-brown ochre-tinged irregular patch on inner-marginal area near anal angle. Fore-wing*: spots as on upper side. *Hind-wing*: costal border narrow, terminating some distance before apex; hind-marginal border moderately broad, terminating abruptly between first median nervule and submedian nervure; irregular dark patch broadest inferiorly (but leaving a narrow even white border on anal angular projection), growing narrower superiorly, rising to a point between origin of second and third median nervules; near base an elongate fuscous spot between median and submedian nervures; above cell a series of two or three fuscous dots, one beyond the other, the outermost between the subcostal nervules.

Judging from Hopffer's figure of the ♂, and a single ♀ before me, the sexes do not differ, except in the blunter, more rounded wings of the ♀, and her somewhat enlarged dark patch on the under side of the hind-wing; the white bar on the upper side of the hind-wing is also somewhat narrower in the ♀.

The under side of the palpi, thorax, and abdomen, and the sides and middle of the back of the latter, are white in both sexes.

Hopffer (*loc. cit.*) notes two ♂s from Querimba. The only specimen I have seen is a ♀ from Delagoa Bay, which was recently (August 1884) forwarded to me for determination by Mr. H. Grose Smith.

Localities of *Ancyloxypha Philander*.

I. South Africa.
    II. Delagoa Bay.—Lourenço Marques (*Mrs. Monteiro*).

II. Other African Regions.
  A. South Tropical.
    a. Western Coast.—"Angola (*Pogge*)."—Dewitz.
    b. Eastern Coast.—"Querimba."—Hopffer.

359. (3.) **Ancyloxypha producta**, *sp. nov.*

Exp. al., (♂) 1 in. $5\frac{1}{2}$-$6\frac{1}{2}$ lin.

♂ *Blackish-brown, with transparent spots; fore-wing with an inferior opaque pure-white spot, and a thin white longitudinal streak between it and base.* Fore-wing, about middle, a large irregularly-quadrate spot, crossed in its upper part by median nervure and bounded inferiorly by first median nervule; beyond it, a much smaller spot, between second and third median nervules; above and slightly beyond the large spot another rather smaller elongate spot; and beyond this last-named spot a subcostal curved row of three very small spots; opaque white spot conspicuous, roughly trigonate, below large central transparent spot and somewhat nearer to base; white longitudinal streak not quite touching this white spot or quite extending to base; basal and costal area irrorated with yellowish-grey. *Hind-wing:* a central, rather obliquely-transverse, rather narrow, transparent bar lying between second subcostal nervule and submedian nervure,—its lower part constricted, sub-opaque, and interrupted between median and submedian nervures; hairs of basal and inner-marginal area yellowish-grey, becoming whitish along a line between submedian and internal nervures. Cilia whitish, becoming white about posterior angle in both wings. UNDER SIDE.—*Hind-wing, and costal border and apical hind-marginal area of fore-wing, hoary-grey with a violaceous gloss.* Fore-wing: spots as on upper side, but white longitudinal streak wanting. *Hind-wing:* transparent bar as on upper side, but its lower constricted part joining a rather wide longitudinal white ray from base, which ends acutely beyond middle; anal-angular area dark-brown with a very narrow external edging of white; an elongate ill-defined whitish spot between first median nervule and submedian nervure; between costal and subcostal nervures two small faint fuscous spots, one beyond the other; a discal row of four or five similar spots, beginning near costa and thence skirting outer edge of transparent bar.

Head and *body* above blackish,—the *head* and *thorax* clothed with mixed greyish, dull yellow-ochreous, and blackish hairs,—the *abdomen* with half-rings of white; beneath white.

Notwithstanding its much larger size and very much longer wings—the fore-wing being produced apically and the hind-wing at anal angle—this species is very like *Philander*, Hopff., and the single example in the Hewitson Collection is referred with a ? to that butterfly. The resemblance is chiefly on the upper side, but *Producta* there presents in the fore-wing a longitudinal white streak in the basal region, which is wholly wanting in *Philander*. On the under side the hoary-grey violet-glossed hind-wing is very different from the pure-white brown-bordered surface in *Philander*, although the general pattern of the marking is similar. In outline of wings *Producta* resembles *Pamphila borbonica*, Boisd., but the elongation is more pronounced in the former. The palpi, thorax, and abdomen are white beneath, as in *Philander*, but the sides and middle of the back of the abdomen are not white, there being only fine white lines marking the segmental divisions.

The four examples I have seen appear to be ♂ s, viz., the one in the Hewitson Collection above mentioned; another kindly lent to me by Mr. H. Grose Smith; a third received in August 1884 by the South-African Museum; and a fourth captured at Kimberley on 5th March 1887 by Mr. H. L. L. Feltham. The three former examples were collected by Mrs. Monteiro at Delagoa Bay.

Localities of *Ancyloxypha producta*.

I. South Africa.
  B. Cape Colony.
    c. Griqualand West.—Kimberley (*H. L. L. Feltham*).
II. Delagoa Bay.—Lourenço Marques (*Mrs. Monteiro*).

GENUS ABANTIS.

Abantis, Hopffer, "Monatsb. K. Akad. Wissensch. Berlin, 1855, p. 643;" and Peters' Reise n. Mossamb., Zool., Ins., p. 414 (1862).
Leucochitonea, Wallengren, Sv. Vet.-Akad. Handl., 1857; Lep. Rhop. Caffr., p. 52.
Leucochitonea, Trimen, Rhop. Afr. Aust., ii. p. 306 (1866).
Abantis and *Sapœa*, Plötz, Stett. Ent. Zeit., 1884, p. 388, and 1885, p. 35.

IMAGO.—*Head* not quite as broad as thorax, shortly but densely pilose above, a compact tuft (very short in some species) near base of each antenna; *palpi* with basal and middle joints broad, densely and compactly scaly and hairy,—terminal joint usually short (very short in *Zambesina*, longest in *Levubu*), slender, blunt, clothed with very short appressed hairs, horizontally porrected; *antennæ* short, rather thick, with a rather suddenly-formed, elongate, cylindrical, bluntly-tipped club, sharply (in *Levubu* gently) recurved.

Thorax robust (less so in *Levubu*), shorter than abdomen; above scaly, with close down anteriorly, and longish hair laterally and posteriorly,—beneath with dense short hair. *Fore-wings* rather narrow, elongate apically; costa nearly straight (in ♀ more arched at base than in ♂); hind-margin very slightly concave just above posterior angle; inner margin nearly straight, very slightly hollowed beyond middle; discoidal cell narrow, about two-thirds length of wing; middle and lower disco-cellular nervules well developed, inwardly oblique; lower radial originating just midway between upper one and third median nervule; first median nervule given off considerably nearer to base than to second nervule. *Hind-wings* in ♂ (except in *Lerubu*) more or less prominent inferiorly about anal angle; costa short, almost straight beyond strong basal curve; subcostal nervure branched at some distance before extremity of discoidal cell; disco-cellular and lower radial nervules arranged as in fore-wing, but more attenuated, and the former less oblique. *Legs*[1] of moderate length and thickness; hind-femora inferiorly, and middle and hind tibiæ superiorly more or less hairy; appendage to fore-tibiæ narrow, elongate (in *Tettensis* ♂ very small, and in one example apparently wanting[2]); middle and hind tibiæ smooth, their terminal spurs rather long, but second pair on hind-tibiæ in some species (*Venosa* and *Zambesina*) very small; tarsi closely spinulose beneath.

Abdomen tapering in ♂, and a little shorter than inner margin of hind-wings; broad and blunt at extremity (and considerably shorter) in ♀.

Since I have been able to compare specimens of the exceedingly rare *Abantis Tettensis* with *Lerubu*, Wallengr., and the few other species which appear to have been rightly placed in the genus *Leucochitonea*, Wallengr., I consider that the two reputed genera cannot properly be separated, and should be therefore united under Hopffer's earlier name. *Lerubu*, Wallengren's type of *Leucochitonea*, certainly diverges more from *Tettensis* than the other species I include under *Abantis*, but the differences it presents (noted above) are insufficient for generic separation. I agree with Herrich-Schäffer (*Corr.-Blatt. Zool.-Min. Ver. Regensb.*, 1868, pp. 32 and 46) that neither the *Nireus*, Cram. (= *Arsalte*, Linn.), group—though strikingly like *Lerubu* in general aspect—nor the other American species described and figured by Felder (*Reise der Novara, Lep.*, iii. pp. 524–525, pl. lxxiv.), can rightly be placed in *Leucochitonea*; the latter have indeed been by most subsequent writers referred to the genus *Pythonides*. Mr. Hewitson also issued (*Exot. Butt.*, vols. iv. and v.) two plates of numerous so-called species of *Leucochitonea*, which cannot be retained in that genus, or consequently in *Abantis*.

Plötz (*Stett. Ent. Zeit., loc. cit.*) created the genus *Sapæa* for *A.* (*Leucochitonea*) *bicolor*, mihi, and afterwards added to it (1886) *A. Zambesina* (Westw.), placed by its describer in the genus *Oxynetra*, Feld., but a near ally of *A.* (*Leucochitonea*) *paradisea*, Butl. These species and *Venosa*, mihi, n. sp., are unquestionably inseparable generically from *Tettensis*, Hopff.

Four of the five known South-African species have been met with in the South-Tropical belt, the exception being *A. bicolor*, which appears

---

[1] The fore-coxæ in the ♂ *Lerubu* have a conspicuous tuft of very long black hairs springing close to the junction with the femur, and directed posteriorly.

[2] Hopffer (*op. cit.*) notes that this appendage was wanting in the ♀ *Tettensis* he described.

to be limited to Kaffraria Proper, Natal, and Zululand. *Tettensis* and *Lerubu* have been taken on the Vaal River, in Griqualand West, and *Venosa* in the Eastern Transvaal. *Paradisea* is recorded from Natal, Zululand, and Delagoa Bay, and appears to be less rare than its congeners.

The species are very different in colouring and pattern. *Tettensis* has the fore-wings blackish with large whitish markings, and the hind-wings white with a few black spots, and faintly margined with pale-red beneath. *Venosa* is pale brownish-grey tinged with ochre-yellow, and has all the neuration black above. *Bicolor* is pale ochre-yellow with a black border, and has four large black spots in the fore-wings. *Paradisea* is shining bluish-black with transparent and white (or pale-yellowish) spots and hind-wing patch; the body being spotted with orange and white. *Lerubu* is of the purest white, with a white spotted black border and the ends of nervules black; its upper and under surfaces scarcely differ.

360. (1.) Abantis Tettensis, Hopffer.

♀ *Abantis Tettensis*, Hopff., "Monatsb. K. Akad. Wissensch. Berl., 1855, p. 643;" and Peters' Reise Mossamb., Ins., p. 415, pl. xxvi. ff. 16, 17 (1862).

Exp. al., (♂) 1 in. 3–4 lin.; (♀) 1 in. 5 lin.

♂ *Blackish, with large yellowish-white semi-transparent basal and discal markings in fore-wing, and a hind-marginal series of small spots in both wings; hind-wing white, with black spots and hind-marginal border.* Fore-wing: neuration black; two white longitudinal stripes from base, the upper one narrow, along costal border to a little beyond middle—the lower one broad, occupying space between median nervure with part of its first nervule and submedian nervure to a little before middle; in discoidal cell, crossed by fold, a large white spot very near extremity; a superiorly strongly-curved, inferiorly sinuate discal row of eight spots, of which the five upper are small and elongate (the first and second almost linear), and form a continuous subapical curved bar; the sixth is small, subquadrate, separate, between second and third median nervules; and the seventh and eighth are much larger, and in a direct line between disco-cellular spot and submedian nervure; seven spots in hind-marginal series, the lowest bisected by black line of intra-neural fold. *Hind-wing:* field white up to costal edge, but more or less obscured with blackish along inner-marginal border; a very small black spot in discoidal cell at base, and another near base between costal and subcostal nervures; slightly before middle a curved transverse row of three widely-separated rather large spots; a submarginal row of six small spots, mostly touching the diffused blackish inner edge of hind-marginal black border (which becomes very broad in anal-angular area); six hind-marginal white spots, much smaller than those in forewing, especially the upper four; a more or less developed longitudinal black streak running from the lowest of the median row of spots to the base. *Cilia* dull-grey, except about anal angle of hind-wing, where it is pure white, and much longer than elsewhere. UNDER SIDE.—*Forewing with white markings almost wholly confluent, only the neuration and a hind-marginal line being black; hind-wing much as above, but hind-marginal border without any inward blackish diffusion; both wings with a pale-red costal edging.* Fore-wing: at posterior angle the lowest (bisected) spot of hind-marginal series distinct, black-edged inwardly; some faint greyish traces of the boundaries of the principal markings; costal red edge bright at base and extending faintly to apex. *Hindwing*: a conspicuous red edging all along inner margin; costal red edging paler; and usually some red of the same paler tint immediately preceding sharply-defined black hind-marginal border, and also extending over the space between median nervure with its first nervule and submedian nervure; all the black spots very sharply defined; white hind-marginal spots larger and better defined than on upper side.

Head: above black, with three white spots in front, one between antennæ, and two (very conspicuous) just behind antennæ; eyes externally edged with white; *palpi* black above, with a white spot at end of middle joint, ochre-yellow beneath,—short terminal joint black, with a yellow median streak beneath; *antennæ* black, the club ochre-yellow beneath. *Thorax* above black; the collar ochre-yellow, the pterygodes edged with white, and with an external white spot superiorly; beneath black, with white spots frontally and larger ones and tufts of white hair laterally and posteriorly; *legs* ochre-yellow, the femora half-black basally. *Abdomen* above black, with distinct white incision half-rings, and an ochre-yellow anal tuft; beneath white.

♀ (Figured by Hopffer, *op. cit.*) Like ♂, but with blunter, more rounded wings; some brownish-yellow suffusion about bases and margins of both wings; hind-wing without submarginal black spots and blackish suffusion. Under side as in ♂, but submarginal series of spots in hind-wing only represented by the two lowest, which are reduced in size.

Of this very distinct and exceedingly rare species I have seen only three examples of the ♂. Hopffer founded the species and genus on a single ♀ brought by the Peters Expedition from the Zambesi. Two examples from Angola are recorded by Mr. W. F. Kirby in his Catalogue of the late Mr. Hewitson's Collection. The spotting and the under-side tinting of the hindwings are very peculiar, strongly recalling the aspect of some of the smaller species of *Acræa*.

Colonel Bowker took the three ♂s above mentioned in November 1871 at Klipdrift (now Barkly), on the Vaal River, in Griqualand West; they were flying about rapidly on a hill-top, in company with their snow-white congener, *A. Lerabu* (Wallengr.).

Localities of *Abantis Tettensis*.

I. South Africa.
 B. Cape Colony.
  c. Griqualand West.—Barkly, Vaal River (*J. H. Bowker*).
II. Other African Regions.
 A. South Tropical.
  a. Western Coast.—"Angola."—W. F. Kirby.
  b1. Eastern Interior.—"Tette, Zambesi (*Peters*)."—Hopffer.

361. (2.) Abantis venosa, sp. nov.

Exp. al., (♂) 1 in. 6 lin.

♂ *Pale dull-grey, tinged with ochrey-yellow (especially hind-wing); neuration black throughout; a common linear black hind-marginal edging; fore-wing with a few transparent spots, of which two in middle are large.* Fore-wing: two small whitish spots at base, one costal, the other (black bordered) at the origin of median nervure; about middle two large, inwardly sub-acuminate, outwardly excavate-truncate transparent spots, divided only by median nervure,—the upper one in discoidal cell near its extremity rather shorter longitudinally than the lower one, which is bounded inferiorly by first median nervule; immediately beyond these large spots, between second and third median nervules, a very much smaller transparent spot; on costa, between third subcostal and upper radial nervules, a tripartite transparent marking of moderate size; all these transparent markings more or less completely blackish-edged. *Hind-wing:* ochrey-yellow tinge strongest on costa, but diffused over entire area; linear black hind-marginal edging becoming greatly wider and diffuse on anal-angular prominence. *Cilia* short and blackish, but at anal angle of hind-wing long and ochrey-yellow. UNDER SIDE.—*Very much paler; hind-wing white, with a black hind-marginal border; neuration not black, except to a limited extent in hind-wing.* Fore-wing: no blackish edging to transparent markings; basi-inner-marginal area tinged with pale creamy-yellowish; a basal ochrey-yellow mark between median and submedian nervures; costal, submedian, and internal nervures, and nervules near hind-marginal border, black; border narrow and fuscous apically, but becoming broader and darker inferiorly, particularly on anal-angular prominence.

Head and *body* much rubbed in the two specimens before me, but evidently black *above* with creamy-yellowish spots, and ochrey-red prothoracic collar, metathoracic tufts and anal tuft; and *beneath* with pale ochrey-yellow palpi, propectus and legs,—side of thorax blackish spotted with white, and abdomen white edged with blackish.

VAR. A. (♂) *Ochrey-yellow tinge more pronounced;* transparent markings of fore-wing not (or very faintly and partially) blackish-edged, and the smallest of the three middle ones much reduced or obsolete; *hind-wing without inferior diffused widening of hind-marginal border*. UNDER SIDE.—White much more restricted, confined to inner-marginal inferior half of wing; black hind-marginal border obsolete superiorly, and almost so inferiorly, except on anal-angular prominence, where it is very much narrower. (*Hab.*—Ehanda, interior of Western South-Tropical Africa.)[1]

The discovery of this species is due to Mr. C. F. Palmer, who forwarded the two rather worn specimens above described from Eureka, near Barberton, Eastern Transvaal, with the note that one of them was captured on 4th April 1888 about teak trees growing on a stony ridge. Although very different in the colouring and marking of the wings, and considerably larger, this species is in form, and apparently to a large extent in the marking of the head and body, nearly related to *A. Tettensis*. The complete black neuration of the upper side at once distinguishes it from the other South-African species of the genus.[2]

Localities of *Abantis venosa*.

I. South Africa.
    K. Transvaal.—Barberton (*C. F. Palmer*).

II. Other African Regions.
    A. South-Tropical.
        *a*1. Western Interior.—Ehanda (*A. W. Eriksson*)—[*Var.* A.].

362. (3.) Abantis bicolor, (Trimen).

♂ *Leucochitonea bicolor*, Trim., Trans. Ent. Soc. Lond., 3rd Ser., ii. p. 180 (1864); and Rhop. Afr. Aust., ii. p. 307, n. 201, pl. 6, f. 1 (1866).

Exp. al., (♂) 1 in. 5–6 lin.; (♀) 1 in. 6–7 lin.

♂ *Very pale ochreous-yellow, with black margins: fore-wing with large black spots.* Fore-wing: costa with a black border, narrowing to apex, where it joins the broad, interiorly irregularly-dentate, hind-marginal black band; inner margin also narrowly bordered with black; close to base, between median and submedian nervures, a large spot; a little beyond it two large spots form an oblique fascia, interrupted on first median nervule, between costal border and submedian nervure beyond middle; a large irregular costal spot or patch between middle

---

[1] Besides the two examples of the Variety here described, I have one ♂ from the same locality, and have seen two others in Mr. Jameson's collection from the Um-Vuli (a tributary of the Zambesi, in Mashunaland), in which all the transparent spots of the fore-wings are wanting, the white of the under side of the hind-wings and the blackish border at anal angle being also almost obsolete. In the existing dearth of material I am unable to decide whether these examples belong to a distinct species: the markings of head and body appear to be identical with those of the Variety.

[2] I noted a large (possibly ♀) specimen of this species in Mr. H. Grose Smith's collection in October 1886. I omitted to record the locality, but believe that it was Delagoa Bay.

and apex reaches second median, and its outermost angle joins hind-marginal border on first median nervule, leaving an oblique subapical band of yellow, crossed by four dark nervules. *Hind-wing*: spotless; inner margin broadly black-bordered throughout; a moderately wide hind-marginal border, broader towards anal angle, radiating slightly on nervules, and emitting a long black ray between first median nervule and submedian nervure, where it is also marked with two yellow dots close to hind-margin; sometimes two other similar dots more towards middle point of hind-margin. UNDER SIDE.—*All of a deeper yellow, with only very narrow blackish edgings, excepting on inner margin of hind-wing, where, however, the border is narrower, shorter (only reaching anal angle by a thin streak), and enclosing interiorly a yellow stripe;* neuration in both wings finely black beyond middle. *Fore-wing*: spots forming fascia only represented; spot near base and beyond middle only indicated by some fuscous scales. *Cilia* blackish above and below, with a few whitish hairs at anal angle of hind-wing.

Head: above black, with a frontal ochre-yellow spot between palpi and a white spot behind it; also a white spot before and an ochre-yellow spot behind base of each antenna, and a white spot on vertex; *palpi* above black with a whitish-yellow spot at end of middle joint, beneath ochre-yellow,—terminal joint entirely black; *antennæ* black. *Thorax* above black; collar with three ochre-yellow spots, pterygodes with one, and a yellowish-white tip; two ochre-yellow spots on each side (? also some on back); and two tufts of pale yellowish hair in the black posterior pubescence; beneath black, the middle broadly ochre-yellow in front and narrowly yellowish with two lateral whitish tufts posteriorly; an ochre-yellow spot on each side in front; *legs* black,— the middle and hind femora ochre-yellow in front, and the hind ones bearing a long ochre-yellow tuft.

Abdomen above black, with a terminal ochre-yellow tuft, and with four broad conspicuous ochre-yellow transverse lateral stripes which narrow to points dorsally without meeting; beneath paler ochre-yellow with a black stripe on each side.

♀ *Much duller; the ground-colour tinged throughout with brownish, and the dark markings (especially the borders) diminished and not so black.* Fore-wing: costal border narrowly fuscous to beyond middle, not touching first subcostal spot; this spot more distinctly separate than in ♂ from the spot obliquely below and beyond it; hind-marginal border much reduced and narrowed, especially inferiorly, but radiating on nervules throughout; inner-marginal border almost linear beyond middle. *Hind-wing*: hind-marginal and inner-marginal borders greatly reduced, especially in anal angular area; the ochre-yellow dots in the former increased in number and size to a series of from five to seven rather ill-defined spots. UNDER SIDE.—As in ♂, but duller and rather darker.

Thorax with three pairs—anterior, mesial, and posterior—of ochre-yellow spots above, besides those on the collar. *Abdomen* with the lateral stripes enlarged, quite confluent on each side, and their narrowed dorsal points meeting.

Colonel Bowker discovered this well-marked species in Kaffraria Proper in 1863, and the only specimen (a ♂) which the collection he presented in the following year to the South-African Museum contained was described by me in 1864 and figured (*op. cit.*) in 1866. Two ♂s subsequently received from him were noted as taken in January; they were very swift on the wing, frequenting the edge of a forest; one was captured when perched on a twig of *Hibiscus*. Two other ♂s were secured by Colonel Bowker before leaving the Bashee River; and more recently (in 1879 and 1883) he has sent four ♀s from Pinetown in Natal, one of which is marked as captured on the 5th April. The insect is evidently very rare: I have seen only four besides the nine examples just mentioned, viz., a pair in the collection from Natal sent by Colonel Bowker to the Colonial and Indian Exhibition of 1886, a ♀ taken in Zululand by Captain Goodrich, and a ♂ lately (1888) sent to me for examination by Mr. A. D. Millar of D'Urban. The latter kind correspondent informs me that he has taken but two specimens during his long and successful researches in the vicinity of D'Urban.

Localities of Abantis bicolor.

I. South Africa.
  D. Kaffraria Proper.—Bashee River (*J. H. Bowker*).
  E. Natal.
    a. Coast Districts.—D'Urban (*A. D. Millar*). Pinetown (*J. H. Bowker*).
  F. Zululand.—Etshowe (*A. M. Goodrich*).

363. (4.) Abantis paradisea, (Butler).

♂ *Leucochitonea paradisea*, Butl., Trans. Ent. Soc. Lond., 1870, p. 499; and Lep. Exot., p. 167, pl. lix. f. 8 (1874).
♂ *Hesperia (Oxynetra) Namaquana*, Westw., Thes. Ent. Oxon., p. 183, pl. xxxiv. f. 10 (1874).
♂ *Leucochitonea paradisea*, Staud., Exot. Schmett., i. pl. 100 (1888).[1]

Exp. al., (♂) 1 in. 7–8½ lin.; (♀) 2 in. 0½ lin.

♂ *Glossy black, with a slight submetallic dark-green lustre: forewing with rather large semi-transparent white spots near base and inner margin, and vitreous ones superiorly and subapically; hind-wing with a very large semi-transparent white rounded patch; in both wings the semi-transparent markings are sometimes tinged with pale ochre-yellow; cilia short and black, except about anal angle of hind-wing, where they are longer and mixed with whitish. Fore-wing:* at base, below subcostal nervure, a silky-white spot; three spots in basal area arranged in a triangle,—the upper one small (in discoidal cell), and the two lower ones (between median nervure and its first nervule and sub-

---

[1] Dr. Staudinger (*op. cit.*, p. 299) suggests that this species and *L. bicolor*, Trim., probably belong to the genus *Abantis*.

median nervure) nearly equidistant from it—the outer of the two very much larger than the inner; in discoidal cell, very near extremity, a large vitreous spot, variable in shape, broader inferiorly; a much elbowed discal row of five spots, of which the first three are small, forming a nearly straight continuous subcostal transverse streak midway between end of cell and apex; the fourth is rather large, subquadrate, just below the upper three, between third and second median nervules; and the fifth (the largest in wing) is elongated longitudinally, and situated immediately below the large disco-cellular spot, between second and first median nervules; below this last spot, above submedian nervure, a small semi-transparent white spot. *Hind-wing:* white patch roughly rotundate in outline, leaving a black border tolerably wide at base, narrow on costa, wide on hind-margin and inner margin, and very broad in anal-angular area; it is crossed by strongly black-clouded subcostal and median nervures and their nervules, dividing it very unequally into three large and three small sections—the portion between third and second median nervules being minute, and sometimes obsolete. UNDER SIDE.—*As on upper side, but the semi-transparent markings all of a purer white.* Hind-wing: a white spot at base on costal prominence; a conspicuous white longitudinal ray close to inner-marginal edge; median nervure and the origins of its nervules not black-clouded.

Head above black, with three white spots down middle line, one in front and one behind base of each antenna, and a linear one behind each eye; palpi black, with a white spot on middle joint; antennæ black; beneath white at base, but the palpi (except small terminal joint) ochre-yellow. *Thorax* above black, with six orange-yellow spots, viz., one on each side of collar, two (larger and longer) on pterygodes; and two (smaller and paler) on mesonotum; with two white dots marking middle of collar, and two white spots the bases of pterygodes; and posteriorly clothed with black hairs, but (superiorly) with two conspicuous tufts of white; beneath pale ochre-yellow in front on middle, and laterally black with three white spots; *legs* black, the middle and posterior tibiæ white externally, and the latter tufted with white hair. *Abdomen* above blue-black, with an ochre-yellow anal tuft, and marked laterally with four orange-red transverse bars, confluent inferiorly, but acutely pointed superiorly,—the points of the two sets of bars not meeting dorsally; beneath pure white, with a black longitudinal streak on each side.

♀ *Like ♂, but the four spots in basal and inner-marginal area of fore-wing ochre-yellow, and the white space in hind-wing strongly tinged with the same colour.* UNDER SIDE.—The same spots in fore-wing ochre-yellow, but much paler; in hind-wing the subcostal nervure and its nervules, as well as the median, without black clouding.

The type specimen (a Natal ♂) described and figured by Butler has the three sub-basal spots (but not the small outer inner-marginal one) ochre-yellow as in the ♀. I have not seen any ♂ s having this character so strongly developed, but several exhibit a yellowish tinge over the spots in question, as well as over the large white patch in the hind-wings,—notably a specimen from Zululand and another (smaller than usual) from Ehanda, to the northward of Ovampoland.

The only specimen of the ♀ that has come under my notice was captured at Etshowe, in Zululand, by Captain Goodrich, of the 27th Inniskilling Fusiliers; it is much larger than the ♂, with the wings (especially the hind-wings) blunter and subtruncate.

A. paradisea is a very handsome species, but is surpassed in beauty by its near congener *A. Zambezina* (Westw.),[1] of which a single ♂ was brought from the Zambesi many years ago by the Rev. H. Rowley, but which in 1887 was taken rather numerously by Mr. Eriksson in the country between Northern Ovampoland and Ombuella. Mr. Walter Morant in October 1869 sent me a specimen of *Paradisea*, with the information that it was captured with outspread wings on a young Syringa tree at Pinetown, Natal, and that the butterfly occurred there in the months of March, April, and May. Colonel Bowker has forwarded examples taken in the same locality on 5th April, and nearer D'Urban in the middle of February and in June; and he notes that the wings are always held either open or not more than half closed, and that the species much affects loquat trees. He adds that a number of the pupæ were found at Umzinto, on the coast of Natal, in the rotting stumps of a wooden fence. Mr. A. D. Millar, of D'Urban, writes that Pinetown is the only locality known to him for this species, and that he had caught several fine examples there towards the end of April, including ♂ and ♀ *in copulâ*. He notes that the ♀ had "orange markings in place of the white ones of the ♂," and describes the ♂ s as "settling on twigs and leaves, darting about and returning to the same spot."

In 1878 I received from Mr. H. Barber a single specimen captured by him not far from Shoshong (Bamangwato).

Localities of *Abantis paradisea*.

I. South Africa.
  E. Natal.
    a. Coast Districts.—D'Urban (*J. H. Bowker*). Pinetown (*W. Morant, B. Ayres, J. H. Bowker,* and *A. D. Millar*).
  F. Zululand.—Etshowe (*A. M. Goodrich* and *T. Vachell*).
  H. Delagoa Bay.—Lourenço Marques (*Mrs. Monteiro*).

II. Other African Regions.
  A. South Tropical.
    a1. Western Interior.—Omrora and Ehanda, northward of Ovampoland (*A. W. Eriksson*).
    b1. Eastern Interior.—Bamangwato Country (*H. Barber*).

---

[1] *Thes. Ent. Oxon.*, p. 183, pl. xxxiv. f. 9 (1874). This lovely insect is smaller than *Paradisea*, and the ground-colour of the wings is on both surfaces brilliantly glossed with metallic blue and green; the spots of the fore-wing are mostly larger; and the collar and large posterior tufts of the thorax, as well as the anal tuft of the abdomen, are rich deep red with a tinge of carmine, while the sides of the abdomen are pure white bordered with black. In the type specimen figured by Westwood—of which I made a description in 1867—the sides of the abdomen were dull-yellowish, having probably become discoloured. I have not yet met with the ♀ of this species.

364. (5.) Abantis Levubu, (Wallengren).

♂ *Leucochitonea Levubu*, Wallengr., K. Sv. Vet.-Akad. Handl., 1857; Lep. Rhop. Caffr., p. 52.
„ „ Trim., Rhop. Afr. Aust., ii. p. 306, n. 200 (1866).

Plate XII. fig. 5 (♂).[1]

Exp. al., (♂) 1 in. 6–7 lin.; (♀) 1 in. 7½ lin.

♂ *Pure white, with narrow black margins.* Fore-wing: costa and hind-margin rather narrowly edged throughout; a submarginal black streak, slightly inclining towards hind-marginal black, and becoming obsolete about or rather above submedian nervure; before this, a curved black streak from costa beyond middle extends to third median nervule, where it joins submarginal streak; *all nervules, between curved streak and costa and hind-margin, clouded with black*, as well as (more slightly) those in lower half of wing, near hind-margin; by these black markings a hind-marginal series of seven, and a curved subapical series of six white spots (the latter much more elongate than the former spots) are completely separated from the white field. *Hind-wing:* hind-margin with a narrow edge, becoming broad at anal angle, whence it radiates on nervules (chiefly along submedian nervure, which is clouded throughout). Under side.—*Scarcely differing from upper side.* Hind-wing: inner margin edged with black; a black ray between it and submedian nervure; median nervules black towards hind-margin, especially the first, between which and submedian nervure is a short black ray; *a broad black clouding conspicuously marks basal half of costal nervure.* Cilia white, black-spotted at ends of nervules; at anal angle of hind-wing longer than elsewhere.

Head above black, with the following white spots, viz., one frontal, one central and vertical, one posterior to base of each antenna; palpi above black, with a conspicuous white spot at base of middle joint, and a smaller one at extremity of that joint; beneath, pure white, except terminal joint, which is black. *Thorax* above black, clothed with long white hair laterally and posteriorly; collar with a median white spot and a larger ochre-yellow one on each side; pterygodes conspicuously deep ochre-yellow in basal half, but beyond that white and hairy; beneath white frontally, mesially, and posteriorly, but ochre-yellow laterally; legs black, with the front of the tibiae and the terminal half of the front of the femora of the middle and hind pairs pure white. *Abdomen* above black, with a black anal tuft, beneath and on the sides pure white.

♀ Like ♂, but the black streaks and outer neuration more finely marked, so that the separated hind-marginal and subapical spots are all rather larger. Ochre-yellow shoulder-patches on thorax paler.

---

[1] This figure is much too yellow, the white ground being in nature quite pure; the inner-marginal shading in the hind-wings is too dark; and the rich ochre-yellow shoulder-patches are very inadequately shown.

This beautiful insect bears a strong superficial resemblance to the South-American *Arsalte*, Linn. (= *Niveus*, Cram.), which, with *Petrus*, Hübn., and other New-World allies, has been by authors misplaced in the genus *Leucochitonea*, differing as they all do very markedly in their much more slender antennæ with a thin very gently-curved club, more hirsute palpi, smaller thorax, shorter abdomen, and much blunter fore-wings. *Levubu* is unlike every other South-African member of the *Hesperidæ*, and must be highly conspicuous in its pure-white black-edged livery when on the wing. It seems to present scarcely any variation, but the ♂ s lately (1888) received from Northern Ovampoland—where they were taken by Mr. Eriksson—have the black markings thicker in the fore-wings. Only two ♀ s have come under my notice,—one taken on the Upper Limpopo by Mr. F. H. Barber in 1875, and the other by Mr. H. Barber in Matabeleland in 1878 or 1879.

Though having a pretty wide range through Tropical South Africa, *Levubu* does not appear to have been met with abundantly in any locality. Its most southern station known to me is Griqualand West, where (at Klipdrift, on the Vaal River) Mrs. Barber and Colonel Bowker both took a few specimens. The latter, in November 1871, found the butterfly only about a particular hill-top, keeping to a space of limited extent; it was exceedingly swift on the wing, and looked in flight like a bit of burnished silver. Mr. Eriksson's Ovampoland examples are noted as taken in November 1887 and January 1888. Mr. T. Ayres has forwarded six specimens from the district of Potchefstroom, Transvaal.

Localities of *Abantis Levubu*.

I. South Africa.
  B. Cape Colony.
    c. Griqualand West.—Vaal River; Barkly (*Mrs. Barber* and *J. H. Bowker*).
  K. Transvaal.—Upper Limpopo (*H. Barber, F. H. Barber*, and *F. C. Selous*). Potchefstroom District (*T. Ayres*).

II. Other African Regions.
  A. South Tropical.
    a. Western Coast.—Damaraland (*J. A. Bell*).
    a1. Western Interior.—North-East Damaraland: Omaramba—Oamatako (*A. W. Eriksson*). North Ovampoland: Omrora (*A. W. Eriksson*).
    b1. Eastern Interior.—Matabeleland (*H. Barber*). Tati River (*J. L. Fry*). " Sakasusi or Dry River (*Oates*)."—Westwood.

Genus CAPRONA.

Caprona, Wallengren, K. Sv. Vet.-Akad. Handl., 1857; Lep. Rhop. Caffr., p. 51.
Caprona, Trim., Rhop. Afr. Aust., ii. p. 308 (1866).

IMAGO.—*Head* densely clothed above with long scales and short hairs; tuft of hairs near base of each antenna slender; *palpi* as in *Abantis*, but very short terminal joint rather broader and more distinctly hairy and scaly; *antennæ* as in *Abantis*, but comparatively shorter, and with the club rather more abruptly formed and more rounded at tip.

Thorax moderately robust, about as long as abdomen, with rather sparse long hair above laterally and posteriorly. *Fore-wings* with costa as in *Abantis*; apex acute but not produced; hind-margin moderately dentate, slightly hollowed just below apex, and again between first median nervule and submedian nervure, the intervening space being convexly prominent; posterior angle prominent; these hind-marginal characters more pronounced in the ♀; inner margin concave mesially (more so in *Pillaana* than in *Canopus*); neuration mainly as in *Abantis*, but disco-cellular nervules less oblique (the lower one considerably longer and curved inwardly), and discoidal cell longer. *Hind-wings* rather short, except inner-marginally; costa short and straight after prominent basal lobe; hind-margin angulated between first and second subcostal, and more prominently between second and third median nervules; anal angle moderately lobed; inner-marginal border clothed with long hair; cilia very long on anal-angular lobe; neuration as in *Abantis*. *Legs*[1] with femora all thinly hairy inferiorly; middle and hind tibiæ with long sparse hair superiorly; spur-like appendage on fore-tibiæ well developed, as well as second pair of spurs on hind-tibiæ.

Abdomen shaped as in *Abantis*, but bearing dorsally, on posterior part of segments one to three, long, sparsely-set, radiating, erect hairs.

The shape of the wings, which is unlike that of any other South-African genus of *Hesperidæ*, best distinguishes *Caprona* from *Abantis*, and at the same time exhibits considerable resemblance to that found in the South-American genus *Helias*; indeed, one species of the latter, *H. Lacæna*, Hewits. (as remarked by Mr. Butler in *Ent. M. May.*, 1870, p. 98), strikingly resembles *C. Canopus* both in outline and colouring of the wings, the pattern also nearly agreeing, except in the smaller size of the transparent spots in *H. Lacæna*. The exceedingly blunt, evenly-thick club of the antennæ, however, well distinguishes *Caprona* not only from *Helias*, but—judging from the figure and description—also from the Indian genus *Darpa*, Moore,[2] in which the dentation of the hind-margins is more pronounced.

The only known species are the type *Pillaana*, Wallengr., in which the ground-colour is greyish-brown with paler markings, and *Canopus*, *mihi*, which is white with ferruginous and blackish markings,—both species bearing also transparent spots. *Pillaana* is much the rarer, being known only by a very few specimens, respectively obtained in Natal, Bechuanaland, Damaraland, and Matabeleland; while *Canopus*, with a less extended range (not being known to me to occur in

---

[1] The fore-coxæ in the ♂ *Pillaana* (but not in *Canopus*) have a tuft of long black hairs situated and directed exactly as in *Abantis Lerubu*, Wallengr.

[2] *Proc. Zool. Soc. Lond.*, 1865, p. 781, pl. xlii. fig. 2. The type, *D. Hanria*, Moore, is described as having "the usual second pair of spurs on the hind-tibiæ invisible."

any tropical locality) is not uncommon on the eastern side of South Africa, especially in Natal. I have not seen living *Pillaana*, but observed in Natal that *Canopus* frequented the borders of woods and often visited the flowers of Labiate plants, and that it held the wings fully expanded when settled.

365. (1.) Caprona Pillaana, Wallengren.

Caprona Pillaana, Wallengr., K. Sv. Vet.-Akad. Handl., 1857; Lep. Rhop. Caffr., p. 51.
♀ „ „ Trim., Rhop. Afr. Aust., ii. p. 308, n. 202 (1866).

PLATE XII. fig. 6 (♂), 6a (♀).[1]

♂ *Greyish-brown, varied with darker and with paler transverse markings; fore-wing with transverse vitreous markings, hind-wing with semi-transparent dingy-whitish median fascia.* Fore-wing: just about middle a narrow, transverse, sinuated, submacular vitreous stripe, finely black-edged, from costa to submedian nervure, interrupted on first median nervule, between which and submedian nervure are a minute and a small spot in succession; a minute separate vitreous black-edged spot just beyond stripe, between origins of second and third median nervules; a little beyond vitreous stripe a submacular blackish irregular stria from first median nervule to submedian nervure; on costa, about midway between vitreous stripe and apex, a short narrow rather irregular oblique vitreous streak, crossed by three nervules and immediately succeeded by a fuscous cloud; immediately beyond this a whitish costal transverse mark, which is the beginning of an almost imperceptible waved pale discal streak, bounding exteriorly the blackish irregular stria; a rather narrow paler hind-marginal edging. Hind-wing: basal area dark-brown, with a tinge of ferruginous superiorly; median semi-transparent fascia broad, extending from costa to below median nervure and its first nervule (where it ends abruptly, but emits two small acute projections), and traversed by two parallel unequal greyish-brown streaks, traces of a discal pale streak continuous of that of fore-wing; ground-colour beyond this tinged with ferruginous; hind-marginal pale streak as in fore-wing, but inclining to whitish at anal angle. Cilia of the ground-colour black-spotted at extremities of nervules. UNDER SIDE.—*Much paler, especially in hind-wing, where dull-whitish (slightly grey-clouded at base) occupies area to beyond middle.* Fore-wing: vitreous markings as on upper side; a hind-marginal pale cloud succeeds sinuated discal streak. Hind-wing: streaks traversing median fascia almost obsolescent; an obsolescent similar streak crossing disk beyond them.

♀ *Like ♂, but paler; the discal darker and paler markings more*

---

[1] In fig. 6a the under-side colouring is not accurately given, the fore-wing being much too dark and the hind-wing too yellow.

apparent. Fore-wing: vitreous markings broader; a sub-basal denticulate blackish transverse streak; at apex a dark-brown spot. *Hind-wing:* semi-transparent fascia broader, prolonged on its inner side almost to inner margin. UNDER SIDE.—Considerably paler, nearly all the surface being dull-whitish, except an interrupted submarginal brownish fascia. *Fore-wing:* costal commencement and inner-marginal termination of submarginal fascia ferruginous-brown. *Hind-wing:* no grey at base.

This butterfly, the type of Wallengren's genus *Caprona*, appears to be exceedingly rare. I have seen only four examples—a ♀ (described in *Rhopalocera Africæ Australis*), taken in Damaraland by Mr. John A. Bell as long ago as 1862; another ♀, captured on the Upper Limpopo, not very far south of Bamangwato, by Mr. F. Barber in 1875; a ♂ found at Pinetown, Natal, by Colonel Bowker in 1879; and a ♂ taken by Mr. F. C. Selous, on the Makloutse River, not far south of Tati, in Southern Matabeleland, in 1882. Its obscure colouring may, however, lead to its being overlooked by most collectors.

Localities of Caprona Pillaana.

I. South Africa.
    E. Natal.
        a. Coast Districts.—Pinetown (*J. H. Bowker*).
    L. Bechuanaland.—Upper Limpopo (*F. W. Barber*).

II. Other African Regions.
    A. South Tropical.
        a. Western Coast.—Damaraland (*J. A. Bell*).
        b1. Eastern Interior.—Makloutse River, South Matabeleland (*F. C. Selous*).

366. (2.) Caprona Canopus, Trimen.

Caprona Canopus, Trim., Trans. Ent. Soc. Lond., 3rd Ser., ii. p. 180 (1864); Rhop. Afr. Aust., ii. p. 309, n. 203, pl. 6, f. 2 [♂], (1866).
♂ *Caprona Canopus*, Staud., Exot. Schmett., i. pl. 100 (1888).

Exp. al., (♂) 1 in. 3–4½ lin.; (♀) 1 in. 5½–6½ lin.

♂ *Semi-transparent creamy-white, with vitreous bands and spots; base in both wings and apical area of fore-wing ferruginous-ochreous, varied with blackish marks. Fore-wing:* basal patch rather broad, well-defined, irrorated with fuscous atoms; touching it, in discoidal cell a small subvitreous spot, separated by a fuscous-ochreous line from a subvitreous band of three rather large contiguous spots between subcostal and submedian nervures; a broad median vitreous band of six spots (of which the two costal are sublinear, the two central large and subquadrate, and the two lower small and irregular in shape) extends from costa to submedian nervure, and is edged internally by a thin fuscous-ochreous line, and externally by the apical patch, which encloses a small, round, vitreous spot between third and second median nervules;

colouring nearer apex itself pale-ferruginous, enclosing a short vitreous streak of four spots on costa, and two pale indistinct lines, hind-marginal and submarginal; a wavy streak of fuscous-ochreous reaches inner margin, leaving anal angle white; on costa, between apical and central vitreous some ill-defined, elongate, blackish marks; continuous of transverse line of apical vitreous, three or four blackish spots. *Hind-wing*: basal fuscous-ochreous as in fore-wing; first subvitreous band of fore-wing continued to submedian nervure or almost to inner margin; outer band irregular, narrow, merged with inner on costa, and only reaching first median nervule; cell closed by a fuscous-ochreous streak, succeeded by a transverse macular row of the same hue quite across wing to inner margin, where it much widens, and is sometimes joined to an outer, parallel, incomplete, ochreous shade; inner-marginal region densely fringed with long hairs; a hind-marginal fuscous-ochreous bounding line from second median nervule to anal angle. *Cilia* in fore-wing brownish as far as first median nervule, thence white to posterior angle, with fuscous spots at ends of nervules; in hind-wing broader (especially long about anal angle) white, with small fuscous spots at ends of median nervules only. UNDER SIDE.—*White purer; basal colouring wholly absent. Fore-wing*: apical colouring very much paler, obsolescent on inner side, not varied with dark spots. *Hind-wing*: vitreous markings indistinct; ochreous stripes almost obsolete, *but a conspicuous round black spot in fold of inner margin near anal angle*, marking termination of outer one.

Antennæ white, with black club. *Body* above blackish, clothed with mixed ochreous-yellow and greyish hairs, beneath white; *abdomen* above with thin white incision rings, a white anal tuft, and thin erect tufts of hairs.

♀ Quite like ♂, except for its somewhat paler basal patches and larger apical patch in fore-wing.

The peculiar and strongly-contrasted colouring and marking of this species at once distinguish it from its dull-tinted congener, *C. Pillaana*, and a further character of distinction is its much less excised inner margin of the fore-wings.

Colonel Bowker discovered this beautiful *Caprona* in Kaffraria Proper in the year 1863, and forwarded several fine specimens to the South-African Museum. On the coast of Natal it is by no means rare: I took my first specimen at D'Urban on 23rd June 1865, and subsequently (February and March 1867) became well acquainted with the insect in that neighbourhood. It is not a rapid flyer, but has a rather fluttering motion on the wing; it frequently lights on flowers, and sometimes on the under side of leaves, holding all the wings expanded. It seems to appear throughout the year, as Colonel Bowker took several in August 1878, and Captain Goodrich others in Zululand during October and November 1886. As noted in my former book (ii. p. 310), Colonel Bowker observed that, when in flight, this butterfly made a sharp creaking or buzzing noise. I failed to detect any sound of the kind in the living examples to which I listened in Natal; but Colonel Bowker's observation is confirmed by his niece, Mrs. Bailie, who informed me that in 1869 she was attracted to a specimen of *Canopus* by the clicking sound it made (in a locality about sixteen miles east of King William's Town), and followed the insect until she succeeded in capturing it. It is not improbable that this sound may be peculiar to the ♂ when courting the ♀.

Localities of *Caprona Canopus*.

I. South Africa.
  B. Cape Colony.
    b. Eastern Districts.—"King William's Town District."—Mrs. Bailie.
  D. Kaffraria Proper.—Bashee River (*J. H. Bowker*).
  E. Natal.
    a. Coast Districts.—D'Urban. "Lower Umkomazi."—J. H. Bowker.
    b. Upper Districts.—Tunjumbili, Tugela River.
  F. Zululand.—Etshowe (*A. M. Goodrich* and *T. Vachell*).
  K. Transvaal.—Potchefstroom District (*T. Ayres*).

Genus PTERYGOSPIDEA.

Pterygospidea, Wallengren, K. Sv. Vet.-Akad. Handl., 1857; Lep. Rhop. Caffr., p. 53.
Nisoniades (Hübner, 1816), Trimen, Rhop. Afr. Aust., ii. p. 310 (1866).

IMAGO.—*Head* not quite as wide as thorax; tuft near base of each antenna rather full; *palpi* short, broad, densely scaly and hairy; terminal joint conical, short or very short [except in *Djælælæ* and *Kobela*, not projecting much beyond short hairs of middle joint]; *antennæ* of moderate length, with a more or less gradually-formed, elongate, fusiform club, tapering to a point, more or less bent or elbowed, but (except in *Flesus* and some intimately allied species) not hooked.

Thorax moderately stout, considerably shorter than abdomen. *Fore-wings* rather broad; costa scarcely arched beyond basal curve; hind-margin entire or moderately denticulate (in *Djælælæ*, *Motozi*, *Nottoana*, *Phyllophila*, and *Flesus*, more or less excised or hollowed inferiorly between second or first median nervule and submedian nervure); inner margin straight; costal nervure situated at some distance from costa, except at its extremity—at or a little beyond middle—where it abruptly curves upward; discoidal cell about two-thirds length of wing; lower radial originating a little nearer to upper one than to third median nervule; first median nervule given off much nearer to base than to second median nervule; in ♂ *Nottoana* a long costal groove or fold from near base to about middle. *Hind-wings* rounded, not (or very slightly) produced about anal angle; hind-margin usually slightly or moderately dentate (in *Djælælæ* decidedly so, and with a marked excavation between second subcostal and third median nervules). *Legs* of moderate length, rather stout, more or less hairy; terminal spurs on middle and hind tibiæ long and strong,—the second pair on the latter also well developed; tarsi more or less spinulose beneath; spur on fore-tibiæ rather large; *in male*, fore-tibiæ sometimes (*Djælælæ*) with a long external fringe of broad scale-like hairs, extending also to first joint of tarsi, or (*Mokeezi*) with a dense coating of short hair, somewhat lengthened at extremity of tarsi; hind-femora at base beneath and hind-tibiae throughout above in some species (*Djalala* and *Mokeezi*) bearing very long bristly hairs, forming a tuft directed posteriorly; or hind-tibiae and first joint of hind-tarsi (the latter enlarged to thickness of tarsus) tufted superiorly throughout with long dense softer hair (*Motozi* and *Nottoana*); or, lastly, hind-tibiae with two fringes, superior and inferior, of fine long silky hair (*Flesus*).

Abdomen moderately stout, variable in length, but always shorter— usually much shorter—than inner margin of hind-wings; in ♂ hollowed and hairy in basal part beneath; in ♀ not blunt and broad at tip as in *Caprona*.

It is difficult to separate this genus from *Nisoniades*—represented by *Tages*, Linn., and *Marloyi*, Boisd., in Europe, and by a considerable number of species in North America—except by the length and flexure of the antennal club, and the shorter and less hairy palpi. The South-African species have been dispersed in several Hübnerian genera,— Plötz, for instance,[1] makes *Motozi* an *Ephyriades*, *Nottoana* an *Antigonus*, and *Djalala*, *Kobela*, *Mokeezi*, and *Flesus* species of *Tagiades*,—but I have not succeeded in discovering among them any distinctions of generic value. The remarkable variation as regards the secondary sexual characters of the ♂, as in the case of *Pyrgus* (to which genus *Pterygospidra* presents considerable affinity), does not afford any constant criterion for grouping the species, these badges differing in such closely-allied forms as *Flesus* and *Nottoana*, while the ♀s throughout offer no tangible points of distinction.

Djalala approaches nearest to *Nisoniades* in respect to palpi and antennae, but is singular in the toothed and excised hind-margin of the hind-wings (which resembles to a slight extent that found in *Caprona*);[2] its small size, purplish-glossed black-brown upper side and reddish under side render it easily recognised. All the South-African species possess some transparent markings in the fore-wings, but these are very small in *Djalala*, *Kobela*, and the ♂s of *Nottoana*, *Motozi*, and *Flesus*, while in both sexes of *Mokeezi* they are enlarged into two oblique bars, and of a light-yellow tint. The ground-colour is in all of some shade of brown (in *Nottoana* much darker in the ♂ than in the ♀). In general aspect *Kobela* is more like a *Nisoniades* than any of the others.

The species known to inhabit South Africa are sylvan; they rest with fully-expanded wings, chiefly on leaves—*Flesus* and *Mokeezi* almost always, and *Nottoana* by preference, on the under surface of the latter. I have frequently taken *Djalala* and *Motozi* settled on the ground.

[1] *Jahrb. Nass. Ver. für Nat.*, 1884, pp. 59, 85, 94, &c.
[2] The larger *P. crosula* and *P. angulata*, Feld., from the Oriental Region, are figured as possessing a very similar hind-marginal outline.

Only *Djælæla* and *Nottoana* are known to extend as far to the south-west as Knysna in Cape Colony. The former and *Flesus* are most widely distributed over the African continent, but the only other known to me to occur within the tropical boundary is *Motozi*. From Natal I have obtained all but *Kobela*; this last, with *Motozi* and *Mokeezi*, extend into the eastern border of Cape Colony.

I am not acquainted with many species of this genus from other countries; but several near allies of *Flesus* occur in the Indo-Malayan area, and some related to *Mokeezi* in the same region and in Tropical Africa.

367. (1.) Pterygospidea Kobela, Trimen.

♂ ♀ *Nisoniades Kobela*, Trim., Trans. Ent. Soc. Lond., 3rd Ser., ii. p. 180 (1864); and Rhop. Afr. Aust., p. 312, n. 205, pl. 6, f. 4 [♀], (1866).

Exp. al., (♂) 1 in. $5\frac{1}{2}$–9 lin.; (♀) 1 in. 7–$10\frac{1}{2}$ lin.

♂ Dusky blackish-brown, here and there clouded with paler scales; with small vitreous, and indistinct larger black spots; in both wings a subterminal disco-cellular spot, and a transverse discal row of spots. *Fore-wing:* cellular spot encloses a vitreous dot,—often wanting; discal row of spots elbowed on discoidal nervules, composed of nine spots, the three first of which (next costa) are minute and always vitreous, the fourth and fifth black only and nearer margin than the rest, the others more or less indistinctly vitreous-centred; between first median nervule and submedian nervure, before middle, a large indistinct black spot; beyond discal spots and also along hind-margin some faint-yellowish scaling. *Hind-wing:* spots without vitreous centres; discal row more regular than in fore-wing, of six or seven spots. UNDER SIDE.—*Rather paler, more glossy; spots more distinct, but smaller. Fore-wing:* inner margin bordered with dull-greyish; spot before middle obsolete. *Hind-wing:* a double streak closing cell; spots of transverse row bounded outwardly and often centred with dull-yellowish scales. *Cilia* of fore-wing fuscous, indistinctly mixed with dull greyish-yellow, more apparent near anal angle; of hind-wing conspicuously *dull yellowish-white interrupted with fuscous on nervules*,—paler beneath.

♀ Paler, browner; spots of fore-wing (especially lower ones of discal row) always with vitreous or subvitreous centres,—that in discoidal cell sometimes containing a second (inferior) vitreous spot; all spots more apparent than in ♂.

Antennæ fuscous above; half-ringed with creamy-yellow beneath, and with the club (except its dull-red tip) of the same colour.

Palpi beneath with some pale-yellowish hairs mixed with the fuscous ones.

However indistinct the spots in general may be in the ♂, the three minute vitreous ones commencing the discal row in the fore-wings are always well marked.

VAR. A. (♂ and ♀.)—*Darker; the black spots almost indistinguishable; the vitreous spots of fore-wing, on the contrary, very distinct* (except, in ♂, the two lowest of discal row); always two disco-cellular vitreous spots in ♂, confluent into one elongate mark in ♀. *Cilia* with more distinct pale marks in fore-wing, and of a clearer creamy-white in hind-wing; beneath in both wings the paler portions are wider and whiter.

Hab.—Transvaal.

In marking this rather large and very dark species much resembles the considerably smaller *P. Nottoana*, Wallengr.; but, irrespective of size, it differs in its robuster structure, blunter and more rounded wings, tendency to vitreous or subvitreous centres to the dark spots on the fore-wings of the ♂, and much smaller development of lower discal vitreous spots in the ♀.

Colonel Bowker discovered this Hesperid in 1862, frequenting thick forest on the Bashee River in Kaffraria Proper; he reported it as being more numerous in the following year, but restricted to certain localities. In 1872 he sent five specimens from the Perie Bush in the King William's Town district. The occurrence of the variety above described in the Eastern Transvaal renders it most probable that the butterfly inhabits the intermediate tracts of country, but I have seen no specimens from either Natal or Zululand.

Localities of *Pterygospidea Kobela*.

I. South Africa.
  B. Cape Colony.
    b. Eastern Districts.—Perie Bush, King William's Town District (*J. H. Bowker*).
  D. Kaffraria Proper.—Bashee River (*J. H. Bowker*).
  K. Transvaal.—Lydenburg District (*T. Ayres*—Var.).

368. (2.) Pterygospidea Djælælæ, Wallengren.

♂ *Pterygospidea Djælælæ*, Wallengr., K. Sv. Vet.-Akad. Handl., 1857; Lep. Rhop. Caffr., p. 54, n. 5.
Nisoniades Umbra, Trim., Trans. Ent. Soc. Lond., 3rd Ser., i. p. 289 (1862).
Nisoniades Djælælæ, Trim., Rhop. Afr. Aust., ii. p. 311, n. 204 (1866).

PLATE XII. fig. 7 (♀).[1]

Exp. al., (♂) 1 in. 1–3 lin.; (♀) 1 in. 1½–3½ lin.

♂ *Very dark glossy purple-brown, varied with some almost black patches, and very sparsely sprinkled with yellowish-white scales; fore-wing with very small discoidal vitreous spots. Fore-wing*: a very large indistinctly defined roughly-triangular darker patch rather before middle,—its wide base on inner margin, its apex on costal nervure; on outer edge of this patch two minute subcostal vitreous spots—the lower one just within discoidal cell—and three other small vitreous

---

[1] This figure does not sufficiently exhibit the excision of the hind-margins—in fore-wing just above posterior angle, and in hind-wing between second subcostal and third median nervules. On the upper side of the fore-wing, the subapical vitreous streak of three small spots is obscured; it should be quite distinct.

spots (of which the middle one is twice as large as the others) in an irregular line between base of third median nervule and just below first median nervule; a much smaller roughly sub-triangular subcostal darker patch beyond middle—its base on third median nervule—outwardly edged by a thin oblique slightly curved streak composed of three minute vitreous spots; at apex and posterior angle a good-sized darker mark, with traces of a connecting submarginal dark fascia between the two; close to hind-margin a black streak, immediately succeeded by an indistinct yellowish-white one. *Hind-wing:* an indistinct broad darker fascia about middle; beyond it traces of a more indistinct dark streak; at apex a darker mark extending along upper third of hind-margin. *Cilia* blackish, with pale-greyish internervular interruptions, indistinct in fore-wing but well defined in hind-wing. UNDER SIDE.—*Ferruginous-orange, with blackish striæ indicating roughly the outline of the darker patches of upper side. Fore-wing:* vitreous spots as on upper side, but with fuscous edges, rarely an additional small one in discoidal cell immediately below ordinary one; a fuscous line marking extremity of cell; continuous of vitreous streak of three subcostal spots, two fuscous spots between upper radial and third median nervules; inner-marginal border ashy-grey. *Hind-wing:* a thin terminal disco-cellular streak; base ashy-grey; an interrupted sub-basal stria, and a less widely interrupted but submacular and irregular discal stria from near costa to submedian nervure.

Head (with palpi and antennæ) and *body* above like upper side of wings; *palpi* beneath dull ferruginous-ochreous; *antennæ* beneath white, except black apical half of club. *Thorax* and *abdomen* beneath dull-brown, *legs* also dull-brown, *except tibia and tarsus of front pair, which are snow-white externally; the tibia and first joint of tarsus bearing a dense snow-white fringe.*

♀ Like ♂, the discal spots of fore-wings rather larger. UNDER SIDE.—Orange paler.

Antennæ not white beneath, but with whitish half-rings. *Fore-legs* not white on tibia and tarsus, but wholly brown and without fringe.

A large ♂, taken by Mr. C. F. Palmer near Barberton, Transvaal, on 12th April 1888, varies in having the upper side more uniformly dark and with a more apparent purplish gloss, while the under side is very much darker, the ferruginous-orange only appearing as an upper discal patch (small and ill defined in hind-wings), the rest of the surface being not much paler and of the same colour as the upper side.[1]

In the ordinary ♂ it occasionally happens that the two minute subcostal spots about middle are obsolescent, or (very rarely) absent entirely.

[1] Mr. A. G. Butler (*Proc. Zool. Soc. Lond.*, 1888, p. 81) observes that ♀ specimens received from Wadelai (about 3° N. lat. in Central Africa) "show no trace of the ochraceous colouring on the under surface which characterises the ♂ specimens." The ordinary South-African ♀ s that I have seen present, on the contrary, a somewhat paler (but quite as decided) ferruginous-orange under side than the ♂ s. It is not impossible that the Tropical specimens may prove to constitute a distinct race or variety.

The ♀ seems to be very seldom met with: the specimen figured was taken *in copula* on 3d September 1878, by Colonel Bowker, at D'Urban, Natal.

I took this curious little butterfly rarely at Plettenberg Bay, Cape Colony, in February 1858, and also near Grahamstown in the same month of 1870. It is numerous in Kaffraria and Natal, and in the latter country (chiefly on the coast) I captured many specimens in the summer of 1867. It is tolerably active, frequenting hilly and bushy spots, and often settling with expanded wings on the ground or on low plants. I have noticed that it soon loses its glossy dark tints, becoming worn and dull; the ♂. too, not long retaining the remarkable snow-white fringe of hair and long scales on the front legs.

Djælælæ has an immense African range, and is recorded also to occur at Aden.

Localities of *Pterygospidea Djælælæ.*

I. South Africa.
  B. Cape Colony.
    a. Western Districts.—Plettenberg Bay.
    b. Eastern Districts.—Port Elizabeth (*J. L. Fry*). Between Zwartkop and Coega Rivers (*J. H. Bowker*). Grahamstown (*Mrs. Barber*). Mitford Park, Albany District. King William's Town (*W. S. M. D'Urban*). Lower Kei River (*J. H. Bowker*). Mooi Plaat, Stormbergen, Albert District (*D. R. Kannemeyer*).
  C. Orange Free State.—Special locality not stated (*C. Hart*).
  D. Kaffraria Proper.—Butterworth, Tsomo River, and Bashee River (*J. H. Bowker*). "Baziya (*Baur*)."—Möschler.
  E. Natal.
    a. Coast Districts.—D'Urban (*J. H. Bowker*). Verulam. Itongati River. Mapumulo. "Lower Umkomazi."—J. H. Bowker.
    b. Upper Districts.—Maritzburg. Greytown. Estcourt (*J. M. Hutchinson*). Colenso (*W. Morant*).
  F. Zululand.—Napoleon Valley (*J. H. Bowker*). Etshowe (*A. M. Goodrich* and *T. Vachell*). St. Lucia Bay (the late *Colonel H. Tower*).
  K. Transvaal.—Potchefstroom District (*T. Ayres*). Eureka, near Barberton (*C. F. Palmer*—VAR.). "Schoman's Farm (*Person*)."—Wallengren.

II. Other African Regions.
  A. South Tropical.
    a. Western Coast.—"Angola."—Butler.
    b. Eastern Coast.—"Somaliland (*Thrupp*)."—Butler.
    b1. Eastern Interior.—Matabeleland: Shashani River (*F. C. Selous*).
  B. North Tropical.
    a. Western Coast.—"Gambia River (*Moloney*)."—G. E. Shelley.
    b1. Eastern Interior.—"Wadelai (*Emin Bey*)."—Butler. Abyssinia: "Shoa (*Antinori*)."—Oberthür.

IV. Asia.
  A. Southern Region.—Arabia: "Aden (*Yerbury*)."—Butler.

369. (3.) **Pterygospidea Motozi,** Wallengren.

♀ *Pterygospidea Motozi*, Wallengr., K. Sv. Vet.-Akad. Handl., 1857; Lep. Rhop. Caffr., p. 53.

♀ *Nisoniades Pulo*, Trim., Trans. Ent. Soc. Lond., 3rd Ser., i. p. 404 (1862).
♀ *Nisoniades Motozi*, Trim., Rhop. Afr. Aust., ii. p. 313, n. 206, pl. 6, f. 3 (1866).

Exp. al., (♂) 1 in. 2–3 lin.; (♀) 1 in. 4–5½ lin.

♂ *Pale-brown; fore-wing with a broad dark-brown median fascia and some small discal vitreous spots; hind-wing with indistinct darker discal spots; in both wings a submarginal indistinct submacular darker streak; cilia of the ground-colour, fading into greyish-white externally. Fore-wing:* median fascia very broad on costa, where it encloses a triangular mark (apex downward) of ground-colour, terminating in a small narrow subterminal disco-cellular vitreous spot; inner edge of fascia not well defined, except between first median nervule and submedian nervure, where it is bounded by an indistinct pale linear mark; outer edge well defined, strongly elbowed between upper radial and third median nervules, bounded subcostally by the usual slightly oblique streak of three small vitreous spots, and between third and first median nervules by two small vitreous spots (of which the lower is elongate vertically and considerably larger). *Hind-wing:* a good-sized subterminal disco-cellular darker spot, sometimes with an indistinct pale centre; two indistinct smaller sub-basal spots, and another below larger disco-cellular one; seven spots of discal series forming an irregular streak, and containing sometimes traces of paler centres. UNDER SIDE.—*Pale ochre-yellow; vitreous spots as above, but with diffused dark-grey borders; submarginal macular streak of both wings, and all spots of hind-wing (the latter with more distinct pale centres), dull-greyish; hind-margin bounded by a slender blackish line. Fore-wing:* immediately beyond vitreous spots a rather narrow dark-grey discal fascia with ill-defined edges.

Antennæ fuscous with thin white rings, and a broad white ring at base of club; *palpi* brown above, pale-yellow beneath.

♀ *More varied than in ♂, some hoary-grey clouding more or less prevalent basally and discally, and the less strongly-defined dark fascia of fore-wing mixed with paler ferruginous-brown; vitreous spots of fore-wing greatly enlarged (especially the lowest discal one, which is sub-quadrate); subterminal disco-cellular spot in hind-wing enlarged, rounded, vitreous. Fore-wing:* line of inner edge of median fascia between first median nervule and submedian nervure by two small black spots; enlarged subterminal disco-cellular vitreous spot very variable in shape and size, and surmounted by an also variable very small vitreous spot; submarginal streak usually more developed and externally clouded slightly with ferruginous-brown. *Hind-wing:* submarginal streak widened apically into a good-sized border. UNDER SIDE.—*Variable in tint, but almost always much duller and browner than in ♂; the yellow being usually much suffused with brownish, and its brighter portions in most specimens confined to median discal and submarginal spots between the enlarged darker markings; vitreous spots as on upper side.*

The ♂ varies somewhat in depth of colouring on the upper side, and in one example from Southern Matabeleland the dark fascia of the fore-wings scarcely appears. In the ♀, besides the variations noted above, the vitreous spots are sometimes very much reduced in size, and in three examples (two from Natal and one from Delagoa Bay) the disco-cellular one in the hind-wings is obsolete as in the ♂. These examples, as well as several of the more ordinary pattern, exhibit in the fore-wings two minute subvitreous spots on the disc immediately below the largest vitreous spot.

Although so much larger and paler, and without marked excision of the hind-margins of the wings, the ♂ of this species is evidently not distantly allied to *P. Djalælæ*, Wallengr.; while the ♀ is more like the ♀ *P. Nottoana*, Wallengr., owing to its well-developed vitreous spots.

In August 1865 I took three ♀s in the Botanic Garden at D'Urban, Natal; they settled on the ground with expanded wings; and Colonel Bowker has since forwarded specimens captured in that neighbourhood during July and August. In the summer of 1867 (February and March), I met with several of both sexes, chiefly in the coast districts; they frequented bushes at the edge of woods, frequently settling on the leaves. Mr. A. D. Millar informs me that the butterfly is numerous at Sydenham, near D'Urban. In Kaffraria Proper Colonel Bowker found it but rarely; and single specimens have occurred in the eastern districts of the Cape Colony.

Localities of *Pterygospidea Motozi*.

I. South Africa.
  B. Cape Colony.
    b. Eastern Districts.—Pluto's Vale, Albany District (*W. S. M. D'Urban*). Uitenhage. Lower Kei River (*J. H. Bowker*).
  D. Kaffraria Proper.—Bashee and Tsomo Rivers (*J. H. Bowker*).
  E. Natal.
    a. Coast Districts.—D'Urban. Verulam. Pinetown, Avoca, and Mouth of Tugela River (*J. H. Bowker*).
    b. Upper Districts.—Greytown. Estcourt (*J. M. Hutchinson*). Between Tugela and Mooi Rivers (*J. H. Bowker*).
  H. Delagoa Bay.—Lourenço Marques (*Mrs. Monteiro*).
  K. Transvaal.—Upper Limpopo (*F. C. Selous*).

II. Other African Regions.
  A. South Tropical.
    b1. Eastern Interior.—South Matabeleland. Makloutse River (*F. C. Selous*).

370. (4.) **Pterygospidea Mokeezi**, Wallengren.

Pterygospidea Mokeezi, Wallengr., K. Sv. Vet.-Akad. Handl., 1857; Lep. Rhop. Caffr., p. 54.

♂ ♀ *Hesperia Amapondа*, Trim., Trans. Ent. Soc. Lond., 3rd Ser., i. p. 405 (1862).

♂ ♀ *Nisoniades Mokeezi*, Trim., Rhop. Afr. Aust., ii. p. 316, n. 210, pl. 6, f. 5 [♂], (1866).

Exp. al., (♂) 1 in. 8–9 lin. ; (♀) 1 in. 9–10½ lin.

♂ *Dark-brown; fore-wing with two subvitreous, pale yellow-ochreous, transverse stripes.* Fore-wing: in basal portion and along inner margin irrorated with dull-yellowish; a rather wide oblique stripe, irregularly indented externally and internally, from costa slightly before middle to submedian nervure not far before posterior angle, where it ends almost in a point; a subapical much shorter and narrower stripe, touching neither costa nor hind-margin, and ending abruptly on third median nervule. *Hind-wing:* irrorated with yellowish scales, excepting along costa and hind-margin; at extremity of discoidal cell a sub-reniform yellow-ochreous spot; in some specimens a discal submarginal series of from five to seven elongate yellow-ochreous spots rarely distinct; inner-marginal area with brownish-yellow hairs. *Cilia* of fore-wing brown, of hind-wing pale-yellow. UNDER SIDE.—*Paler; yellow markings as above, but with fewer scales than on upper side; a very limited yellowish irroration at bases.* Fore-wing: a pale yellowish-grey inner-marginal border, joining extremity of longer transverse stripe. *Hind-wing:* spot at extremity of cell larger, more conspicuous; in cell, a more or less distinct, small, yellow spot; in some specimens the discal row of yellow spots is more or less distinct, and there is also a pair of small spots of the same colour below terminal disco-cellular spot, between first median nervule and submedian nervure.

Head above dark-brown, with the following pale-yellow markings, viz., a transverse streak in front, another between bases of antennæ, a spot behind each antenna, and an outer half-ring round each eye; *palpi* black above, pale-yellow beneath; *antennæ* black, the club pale-yellow beneath. *Thorax* and *abdomen* above dark-brown, with bronzy-yellow hair; beneath pale-brown,—the former with a pale-yellow median longitudinal stripe in front. *Legs* pale yellow-ochreous, shaded with brown on femora and tibia.

♀ *Like ♂, but slightly paler; semi-transparent stripes of fore-wing rather broader.*

Both sexes are subject to the same amount of variation as regards the yellow spotting of the hind-wings beyond the middle; but a single ♂ taken near D'Urban, Natal, by Mr. W. D. Gooch, presents on the under side, in addition to the spots above described, two small subcostal ones, one just above costal nervure before middle, and the other just below it about middle. This ♂, as well as two others from the same locality and one from the Kei River in the Cape Colony—two of the four with the yellow discal spottings of the hind-wings undeveloped —is remarkable for having a minute subvitreous ochre-yellow spot in the fore-wings between and near the origins of the second and third median nervules, and about midway between the two subvitreous stripes.[1]

---

[1] In a fifth (♀) specimen, taken by myself at Tunjumbili, Natal, this additional spot in the fore-wings is considerably larger. This example has the stripes of the fore-wings decidedly broader than usual; in the hind-wings, however, the disco-cellular terminal spot is a mere line, and on the upper side almost obsolete.

This species is allied to *P. Galenus*, (Fab.),[1] from Tropical Western Africa, but differs widely in the simple two-striped instead of many-spotted pattern of the fore-wings, and wholly distinct markings and colouring of the hind-wings. It is a strictly sylvan insect, and apparently more prevalent in high-lying woods than in those situated at low levels. It was scarce at D'Urban during the summer of 1867; but Mr. W. D. Gooch and Colonel Bowker have taken a good many specimens in that neighbourhood, and Mr. A. D. Millar writes that he has found it numerous in bush-lands along the Natal coast. At Tunjambili, I found it very abundant on the 7th and 8th March, flitting actively about the undergrowth in open glades of the hill-forest, and often lighting suddenly on the under side of leaves. In Kaffraria Proper, Colonel Bowker noted the butterfly as of rare occurrence; but in the King William's Town district, Mr. W. S. M. D'Urban took it in abundance in the forest on the hill above Frankfort. *Mokeezi* is a very handsome warmly-tinted "Skipper," and conspicuous on the wing. Some of Colonel Bowker's specimens are recorded as taken in November and December; Mr. D'Urban's were captured on 28th March.

Localities of *Pterygospidea Mokeezi*.

I. South Africa.
  B. Cape Colony.
    b. Eastern Districts.—Frankfort, near King William's Town (*W. S. M. D'Urban*). Lower Kei River (*J. H. Bowker*).
  D. Kaffraria Proper.—Bashee River (*J. H. Bowker*).
  E. Natal.
    a. Coast Districts.—D'Urban. Isipingo (*J. H. Bowker*).
    b. Upper Districts.—Tunjambili, Umvoti County.
  F. Zululand.—Etshowe (*A. M. Goodrich* and *T. Vachell*).
  K. Transvaal.—Lydenburg District (*T. Ayres*).

371. (5.) Pterygospidea Nottoana, Wallengren.

♂ *Pterygospidea Nottoana*, Wallengr., K. Sv. Vet.-Akad. Handl., 1857; Lep. Rhop. Caffr., p. 54, n. 4.
♂ ♀ *Nisoniades Sabadius*, Trim.; Rhop. Afr. Austr., ii. p. 315, n. 208 (1866).

Exp. al., (♂) 1 in. $3\frac{1}{2}$–5 lin.; (♀) 1 in. $3\frac{1}{2}$–6 lin.

♂ *Very dark glossy brown, tinged more or less obscurely with ferruginous; in both wings a subterminal disco-cellular spot, a discal series of spots, and a hind-marginal series, all black, more or less indistinct; in fore-wing, the first and second (and rarely also the third) spots of discal series with a minute vitreous centre; cilia not quite so dark as ground-colour.* Fore-wing: disco-cellular mark very oblique, composed of two (sometimes quite separated) spots; nine spots in discal row, of which the seventh (between first and second median nervules) is the largest, and the eighth and ninth are almost as small as spots first to fifth; two small black spots before middle, one above the other, between first median nervule and submedian nervure. Hind-wing: disco-cellular spot and seven spots of discal series very indistinct; two spots before middle, one above and the other below disco-cellular spot. UNDER

---

[1] For figure of this species, see Staudinger, *Exot. Schmett.*, i. pl. 100 (1888).

SIDE.—*Considerably paler; darker spots of fore-wing scarcely perceptible. Hind-wing:* darker spots less indistinct, occasionally pretty well defined, especially the large first spot in discal row and the spot preceding it between costal and subcostal nervures.

Head above almost black, with a frontal transverse ochre-yellow streak and also a vertical one; on inner edge of each eye a small ochre-yellow spot; *palpi* above with mixed black and ochre-yellow hair, beneath all ochre-yellow with a tinge of rufous; *antennæ* black, with the club creamy-yellow beneath. *Body* above black, beneath ashy-grey; front of thorax beneath dull ochre-yellow in the middle; *legs* ochre-yellow with tarsi rather paler.

♀ *Very much paler, dull pale reddish-brown with a slight violaceous gloss; fore-wing with large disco-cellular and discal vitreous spots. Fore-wing:* two subterminal disco-cellular vitreous spots obliquely placed, contiguous, the lower and outer one larger; above first of these spots a smaller subcostal one; small vitreous spots at beginning of discal row larger than in ♂, three or four in number; vitreous spots of lower part of discal row four—the first rather small, between third and second median nervules, the second very large between second and first median nervules and just below lower disco-cellular spot, and the third and fourth very small, one above the other, between first median nervule and submedian nervure; all these vitreous spots more or less edged narrowly with blackish. *Hind-wing:* spots as in ♂, but more distinct on the paler ground-colour. UNDER SIDE.—*Paler; vitreous spots of fore-wing without dark edges; darker spots of hind-wing usually less distinct; in both wings a dull-whitish linear hind-marginal edging. Hind-wing:* inner-marginal area paler, inclining to dull-whitish.

Head and *body* very much paler than in ♂, greyish-white beneath; *palpi* beneath and *legs* merely tinged with ochre-yellow.

This species is closely allied to *P. Sabadius*, (Boisd.), of Mauritius, but differs, as far as the ♂ sex is concerned, in its much darker and less rufous colour, very much larger black spots (two good-sized ones instead of a single small one in discoidal cell of fore-wing), and fuscous instead of yellow-ochreous cilia; as well as in the possession of two small vitreous spots at the beginning of the discal row of the fore-wing. I have not seen the ♀ *Sabadius*, but judging from Boisduval's figure and description (*Faune Ent. de Madag.*, &c., p. 62, pl. 9, f. 2), the cellular and lower discal vitreous spots in the fore-wing of *Nottoana* ♀ are considerably larger, and the two lowest of the latter are only represented by a fuscous mark in *Sabadius*.

I found this butterfly rarely in the Knysna district of the Cape Colony, firstly in October and November 1858, and again from the middle of February to April 1859. I did not meet with it at all numerously in Natal, but captured occasional ♂ s about D'Urban during February and at the beginning of April 1867. Colonel Bowker has taken several examples in the same neighbourhood in August; and Mr. A. D. Millar informs me that the butterfly is found numerously at Sydenham near D'Urban. It is quite a woodland species, keeping much about a particular spot, but usually singly; it haunts bushes, often settling with expanded wings on leaves, sometimes on their under surfaces. I have taken it on the flowers of *Lentana* and *Acacia*. The ♂ s, like *Djælælæ*, soon lose their dark glossy freshness and grow worn and reddish; they occur much more frequently than the ♀ s.

Localities of *Pterygospidea Nottoana*.

I. South Africa.
  B. Cape Colony.
    a. Western Districts.—Knysna and Plettenberg Bay.
    b. Eastern Districts.—Grahamstown (*Mrs. Barber*). Bathurst (*Mrs. Barber*) and Tharfield (*Miss M. L. Bowker*).
  D. Kaffraria Proper.—Tsomo River (*J. M. Bowker*).
  E. Natal.
    a. Coast Districts.—D'Urban.
  K. Delagoa Bay.—Lourenço Marques (*Mrs. Monteiro*).

372. (6.) Pterygospidea phyllophila, (Trimen).

♀ *Nisoniades phyllophila*, Trim., Trans. Ent. Soc. Lond., 1883, p. 362.

PLATE XII. fig. 8 (♀).

Very closely allied to *P. Nottoana*, Wallgrn.
Exp. al., (♀) 1 in. 5–7 lin.

♀ *Pale dull reddish-brown with fuscous spots; fore-wing also with disco-cellular and discal vitreous spots; all arranged as in Nottoana, and quite similar in relative size and shape.* Fore-wing: dark edges of vitreous spots better marked; two dark spots between upper and lower vitreous spots of discal row sometimes much enlarged and elongated longitudinally. *Hind-wing:* dark spots more distinct than in *Nottoana*; along hind-margin, except near apex, a narrow irroration of whitish scales, more developed towards anal angle. *Cilia* in fore-wing reddish-brown, except in slight concavity just above posterior angle, where it is white; in hind-wing brown apically, but elsewhere white. UNDER SIDE.—*Hind-wing white, broadly bordered with pale reddish-brown costally and apically.* Fore-wing: at base a little whitish suffusion; fuscous inferior spot before middle faint, but marked with two sub-vitreous dots. *Hind-wing:* costal brown border very narrow at base, but broadening before middle, and very broad in apical area; lowest dark spot before middle (and often also fifth spot of discal row) obsolete; a short brown hind-marginal edging between first median nervule and submedian nervure.

Body and *palpi* beneath white,—the latter tinged with ochre-yellow at tip of middle joint; *legs* white.

Although the conspicuous white field of the under side of the hind-wings gives this form such a distinct appearance, and makes it resemble a small *Flesus*, Fab., I am inclined to the opinion that it may prove to be inseparable from *Nottoana*, especially as one or two of the few known Cape ♀ s of the latter exhibit a slight tendency to a whitish suffusion about the inner margin of the hind-wings. The ♂ *Nottoana* from Natal is a little larger, darker, and more heavily fuscous-spotted than most of the Cape ♂ s; and it is not impossible that *Phyllophila* may actually be merely the modified ♀ of this slight ♂ variation proper to Natal and Delagoa Bay, particularly as ♀ s of the Cape form have not reached me from those countries.

I have before me nine Natalian specimens, and one from Delagoa Bay. Six examples from the latter locality are in the Hewitson Collection in the British Museum. The specimen figured was captured by myself at D'Urban in February 1867; it was settling on leaves, and resting with outspread wings. Two of the seven specimens collected in the same locality by Colonel Bowker were taken on the 15th August 1878; they frequented leaves in the same manner.

Localities of *Pterygospidea phyllophila*.

I. South Africa.
   E. Natal.
      a. Coast Districts.—D'Urban.
II. Delagoa Bay.—Lourenço Marques (*Mrs. Monteiro*).

373. (7.) Pterygospidea Flesus, (Fabricius).

Papilio Flesus, Fab.,[1] "Sp. Ins., ii. p. 135, n. 621" (1781); Ent. Syst., iii. 1, p. 338, n. 286 (1793).
♂ *Papilio Ophion*, Dru., Ill. Nat. Hist., iii. pl. xvii. ff. 1, 2 (1782).
♂   „   „   Stoll, Suppl. Cram. Pap. Exot., p. 127, pl. xxvi. ff. 4, 4 c (1791).
? *Thymele Ophion*, Boisd., Faune Ent. Madag., &c., p. 63, pl. 9, f. 4 (1833).
♂ ♀ *Nisoniades Ophion*, Trim., Rhop. Afr. Aust., ii. p. 313, n. 207 (1866).
? *Tagiades insularis*, Mab., Ann. Soc. Ent. France, 5, vi. p. 272, n. 21 (1876).

Exp. al., (♂) 1 in. 8–10 lin.; (♀) 1 in. 9 lin.—2 in.

♂ Dull pale greyish-brown (*darker in some specimens*), *with more or less indistinct discal blackish spots and submarginal bluish-white irroration; fore-wing with small vitreous spots, incompletely and diffusedly blackish-edged.* Fore-wing: about middle, close to costa, a vitreous spot; below it, in discoidal cell near extremity, two very obliquely placed rather widely separated vitreous spots; discal series of vitreous spots composed of a superior subcostal very oblique slightly curved row of three very small (or even minute) ones about midway between

---

[1] In 1881 I examined the Fabrician type of *Flesus* in the Banksian Collection at the British Museum, and can confirm Mr. Butler's identification of it with Drury's *Ophion*, of which it is a small ♂.

extremity of cell and apex, and an inferior inwardly-oblique row of two, immediately adjacent to lower and outer disco-cellular spot, one just above and the other just below second median nervule; between three superior and two inferior discal vitreous spots, in some specimens two additional minute vitreous spots, one above and the other below lower radial nervule; immediately before disco-cellular and inferior discal vitreous spots, a rather broad, ill-defined blackish fascia, extending to submedian nervure; before middle, between median and submedian nervures, a short, narrow, obliquely transverse blackish streak, of which the portion below origin of first median nervule is marked with two minute bluish-white spots on its inner edge; bluish-white submarginal irroration always sparse, very variable in development, sometimes scarcely perceptible, but occasionally forming an ill-defined fascia. *Hind-wing:* a strongly-curved, somewhat irregular discal series of seven blackish spots, very variable in distinctness, and in some examples obsolete; bluish-white irroration better expressed than in fore-wing, most developed at anal angle and along lower half of hind-margin, but usually also shown to some extent before discal series of spots. UNDER SIDE.—*Fore-wing with blackish markings quite obsolete; hind-wing pure white, with a broad costal-apical brown border.* *Fore-wing:* vitreous spots as on upper side, but without blackish edges; submarginal irroration, without bluish tinge, confined to inferior disc below vitreous spots, sometimes forming two ill-defined whitish spots between first median nervule and submedian nervure. *Hind-wing:* base bluish-white; costal border narrow at base but broad beyond middle, and terminating rather abruptly on hind-margin a little below second subcostal nervule; along lower edge of costal border a series of three conspicuous good-sized black spots, the first and second respectively before and about middle, and both below costal nervure, the third (first of discal series) between subcostal nervules; remaining spots of discal series very much smaller than the first (two or more of them sometimes obsolete), but very conspicuous on the pure-white ground; close to hind-margin a submacular, more or less interrupted, narrow brown border, best developed on each side of first median nervule. *Cilia* pale-brown, mixed with white in fore-wing below first median nervule, in hind-wing (to a greater degree) below second subcostal nervule.

Head and *body* brown above, white beneath. *Head* with seven small pure-white spots above, viz., two in middle in front, one behind base of each antenna, and three at back (one in middle and one on each side); *palpi* brown above, with a small white spot at extremity of middle joint,—beneath pure white, with terminal joint and tip of middle joint black; *antennæ* black, with club beneath creamy-yellowish. *Legs* white,—the tibiæ and tarsi faintly tinged with brownish. *Abdomen* above and laterally with segmental incisions very finely marked with pure white.

♀ *Like* ♂, *but rather paler, and with all the vitreous spots of fore-wing larger*. *Fore-wing:* disco-cellular enlarged vitreous spots touching or almost touching, and not rarely confluent. In both sexes the size of the disco-cellular and discal vitreous spots of the fore-wing varies considerably, and in some ♂ s they are even minute.

Boisduval (*op. cit.*) described as *Ophion* Madagascar specimens which, except for their smaller size, he could not distinguish from Drury's West-African type-form; but M. Mabille (*loc. cit.*) has separated the Malagasy insect as *T. insularis*. From the latter author's full description of the two examples he possessed, I am disposed to conclude that Boisduval was right in regarding the insular and continental specimens as belonging to the same species, especially as the butterfly offers considerable variation in Africa itself;[1] but not having seen any examples from Madagascar, I cannot determine the point with certainty.

Flesus has several near allies in the Indo-Malayan Sub-Region, of which the nearest are the Indian and Malaccan *P. Gana* (Moore), and the Javan *Japetus* (Cram.); in both these species, however, there is a conspicuous broad white black-spotted lower hind-marginal border on the *upper side* of the hind-wings.

I first made the acquaintance of this widely-distributed Ethiopian species in life at D'Urban, Natal, in June 1865, and again took it there in the following August; and in February and March 1867 captured a good many specimens there and at two other localities on the Natal coast. It is active and rapid in flight; the white under side of the wings renders it then conspicuous, but it has the habit of settling constantly and very abruptly on the under side of leaves, so as suddenly to disappear from view altogether. When thus settled, the wings are held fully expanded. I only on one occasion took an example on a flower.

Localities of *Pterygospidea Flesus*.

I. South Africa.
  E. Natal.
   a. Coast Districts.—D'Urban. Verulam. Itongati River. Morewood's Bay (*J. H. Bowker*). "Lower Umkomazi."—(*J. H. Bowker*).
   F. Zululand.—Etshowe (*A. M. Goodrich* and *T. Vachell*). St. Lucia Bay (the late *Colonel H. Tower*).
   H. Delagoa Bay.—Lourenço Marques (*Mrs. Monteiro*).

II. Other African Regions.
  A. South Tropical.
   a. Western Coast.—"Angola (*Pogge*)."—Dewitz. "Chinchoxo (*Falkenstein*)."—Dewitz.

---

[1] Stoll's figure (*op. cit.*) of a "Sierra Leone" specimen represents the discal series of black spots and the hind-marginal brown border on the under side of the hind-wings as much more developed than in any South-African examples which I have seen.

B. North Tropical.
  a. Western Coast. — "Camaroons (*Buchholz*)." — Plötz. Accra: "Aburi (*Weigle*)." — Möschler. Ashanti. — Coll. Brit. Mus. Gold Coast (*E. Bourke*). Sierra Leone.—Coll. Mus. Oxon. "Gambia River (*Moloney*)."—G. E. Shelley.

Genus HESPERIA.

Hesperia, Fabricius (part), Ent. Syst., iii. 1, pp. 258 and 325 (1793).
Ismene, Swainson, Zool. Illust., i. pl. 16 (1820-21).
Ismene, Westwood, Gen. Diurn. Lep., ii. p. 514 (1852).
Rhopalocampta, Wallengren, K. Sv. Vet.-Akad. Handl., 1857; Lep. Rhop. Caffr., p. 47.
Ismene, Trimen, Rhop. Afr. Aust., ii. p. 317 (1866).
Hesperia, Butler, Cat. Fab. Diurn. Lep., p. 269 (1869); and Ent. M. Mag., 1870, p. 58.

IMAGO.—*Head* very broad, not narrower than thorax, densely hairy, but without extra-antennal tufts; *palpi* with basal and middle joints much swollen (especially the middle joint, which is egg-shaped and twice as long as the basal one), and very densely clothed with scales and hairs (of which the marginal hairs on each side beneath are rather longer than the rest),—terminal joint long, not shorter than middle one, very slender, straight, not acuminate, clothed with exceedingly fine, short, closely-appressed hairs, porrected horizontally; *antennæ* of moderate length or rather short, very gradually and moderately thickened from about middle, and towards extremity tapering to a fine point, and very strongly recurved into a long hook.

Thorax exceedingly stout, clothed with dense hair, longer than usual dorsally. *Fore-wings* more or less elongated apically; costa nearly straight, but slightly curved at base and apex; inner margin slightly concave; discoidal cell narrow, long or very long (from three-fifths to two-thirds the length of wing), closed by slender disco-cellular nervules in an inwardly-oblique line; lower radial nervule well developed, originating midway between upper one and third median nervule; first median nervule very long, much curved, given off not far from base at a great distance before origin of second. *Hind-wings* more or less prominently and broadly lobate about anal angle; basal protuberance of costa very prominent, fringed with hair; disco-cellular nervules obsolete or reduced to an extremely attenuated line; radial nervule also greatly reduced or obsolete. *Legs* rather long, stout; all femora inferiorly, and middle and hind tibiæ superiorly, more or less densely fringed with long silky hair; appendage to fore-tibiæ well developed; terminal spurs of middle and hind tibiæ, and second pair of spurs of the latter, rather long and stout; all tarsi strongly and closely spinulose beneath.

Abdomen thick, but tapering posteriorly, usually considerably shorter than inner margin of hind-wing, but in ♂ of some species almost as long; a terminal tuft in ♂.

LARVA.—Elongate, contracted on second segment. Head large, subtriangular, but rounded superiorly, with a depression in the middle of its upper edge.

PUPA.—Stout, rounded; abdomen tapering abruptly to a point. Head with frontal projection short, slender, and acute; eye-covers elevated, conical. Prothorax dorsally with a laterally projecting acute prominence, like that on head, but shorter. In loose irregular web in curled leaf; attached by the anal prominence (which is rather long, stout, and curved), and with a *free* silken girth, giving off from its middle point over the back a silken stay, attached to the leaf far away from the attachments of the ends of the girth.

(The above characters are taken from Mr. J. P. Mansel Weale's notes and drawings of the larva of *H. Forestan*, and from pupæ of that species and *H. Keithloa*.)

I have followed Mr. Butler (*op. cit.*) in adopting as the type of *Hesperia* the species which Fabricius in 1793 placed at the head of his *Hesperiæ Urbicolæ* (= the modern Family *Hesperidæ*), viz., the Indian *H. Exclamationis*, Fab. From this species the type of Swainson's genus *Ismene*, the Javan *I. ædipodea*, Sws., cannot be generically separated, the only differences presented by the latter being greater width of fore-wings, and (in the ♂ only) a sub-basal velvety patch on those wings. I had for some time supposed Wallengren's genus *Rhopalocampta* to be a natural one, but upon closer comparison of the three South-African species on which he founded it with others from different regions recognised as species of *Ismene*, Sws., I can find nothing warranting its being kept apart.

The genus *Hesperia* is readily recognised by the long, slender, terminal joint of the palpi, and the very gradually thickened and then attenuated antennæ terminating in a very long thin hook. The species composing it are all above the medium size in this Family, most of them being large (above 2 inches in expanse), and one, the West-African *Iphis*, Drury, quite a giant among the *Hesperidæ*, attaining an expanse of over 3 inches. The colouring of *Iphis*, and of its nearest congener, the considerably smaller *Juno*, Plötz, is very peculiar, viz., a shining indigo-black or very dark bronzy-green upper side, with a more decided submetallic and partly blue-shot under side, while the head and palpi and under side of end of abdomen are deep-red. West Africa also affords the most brilliant members of the genus, *H. Bixæ*, (L.), and *H. Chalybe*, (Westw.), in which the body and a space from the bases of the wings above are of an intense glittering blue shot with verditer-green. Most of the *Hesperiæ* are, however, of dull-brown and yellowish tints, often relieved on the under side of the hind-wings by a white band.

All the species are proper to the Old World; out of about seventy recorded, nearly half belong to India and other parts of the Oriental

Region. Only one species is known from the Palæarctic Region,[1] inhabiting Vladivostok on its extreme east. The Australian Region appears to have yielded sixteen, and the Ethiopian Region nineteen species. Of the last named, four appear to be peculiar to Madagascar, and one (*Arbogastes*, Guén.) to Madagascar and Réunion. The five found in South Africa are *Forestan*, Cram.; *Pisistratus*, Fab.; *Anchises*, Gerst.; *Keithloa*, Wallengr.; and *Unicolor*, Mabille. The first and second of these scarcely differ from each other on the upper side, but on the under side of the hind-wings the white band is in *Pisistratus* marked with three black spots. *Anchises* also presents this black-spotted white band, but has also two orange-red spots; while *Keithloa* has no white bands, but vivid orange-red spots. In *Unicolor*, which is considerably smaller and darker, there are no markings on either side of the wings. *Forestan* and *Pisistratus* range very widely through Tropical Africa, the former also reaching Madagascar and Mauritius; *Anchises* appears to be chiefly East-African, and has been recorded from Aden; *Unicolor* was first described from Congo specimens; and *Keithloa* is not known to me to have occurred beyond South-African limits. All five species inhabit the eastern side of South Africa; and *Forestan* and *Keithloa* penetrate as far to the south and west as Port Elizabeth in the Cape Colony. *Anchises* has been received from Delagoa Bay only; *Unicolor* from that place and Natal.

The three species I have observed in life (*Forestan*, *Pisistratus*, and *Keithloa*) have a rapid but *bustling* flight, reminding one, though much quicker, of that of such Noctuæ as *Plusia* and allied genera. Owing to their constant visits to flowers, they are not difficult to capture; when settled, they hold the wings erect, the hind pair being kept, however, a little more open than the fore-wings.

374. (1.) Hesperia Forestan, (Cramer).

Papilio Forestan, Cram., Pap. Exot., iv. pl. cccxci. ff. E, F (1782).
♂ ♀ *Ismene Florestan*, Trim. [part], Rhop. Afr. Aust., ii. p. 318, n. 213 (1866).

Exp. al., (♂) 2 in.—2 in. 1 lin.; (♀) 2 in. 2–3 lin.

♂ *Dull pale greyish-brown, darker in hind-wing; fore-wing with a slight tinge of yellowish-grey basally, hind-wing with a basi-median pale ochreous-yellow patch. Fore-wing:* paler about middle; some greyish-yellow hairs at base and on inner margin. *Hind-wing:* much darker (almost black) at and near anal angle; central and inner-marginal area thickly clothed with pale ochreous-yellow hairs, mixed with greyish near base. UNDER SIDE.—*Smoother, sometimes with a faint violaceous gloss.*

---

[1] Of this species, *H.* (*Ismene*) *Aquilina*, Speyer (*Stett. Ent. Zeit.*, 1879, p. 500) notes that it is one of the few *Hesperidæ* that exhibit a sexual distinction in the neuration; the second median nervule of the fore-wings originating in the ♂ nearer to the first than to the third, but in the ♀ just the reverse.

Fore-wing: a dull whitish space on middle of inner margin, ill defined on its edges, rising slightly above first median nervule. *Hind-wing: a broad, even, transverse pure white band from costa to submedian nervure,* where it narrows, and is abruptly and widely interrupted, ending in a separate elongate white spot close to inner margin beyond middle. *Cilia* on both sides grey, but *orange-yellow* on lobe of anal angle.

Head above ashy-grey, with an ochreous-yellow spot on each side between eye and base of antenna ; *palpi* with terminal joint wholly black, but with middle joint above black-tipped with ochreous-yellow, —beneath ochreous-yellow, with a conspicuous outer black longitudinal streak ; *antennæ* black. *Thorax* above ashy-grey mixed with greenish ; collar edged anteriorly with an orange-yellow line ; beneath ashy-grey, except mesially in front, where it is orange-yellow ; *legs* fuscous, the coxæ, femora, and tibiæ with long dense tufts of orange-yellow hair, the latter (and also the fore-tarsi) more or less striped longitudinally with creamy-yellow in front. *Abdomen* above dark-grey with whitish, laterally and inferiorly black with white, segmental rings ; a median longitudinal ochre-yellow stripe beneath.

♀ like ♂, but the ground-colour rather darker.

LARVA.—" Pale-yellow, with purplish or crimson transverse bands. Head brick-red or yellowish, with two frontal rows of black spots—the upper row of six, the lower of five ; mandibles black. Each segment with a deep crimson or purplish transverse median band ; the yellow ground between these bands crossed by two narrow streaks of the same colour. Anal segment with three somewhat angulated black marks. Above spiracles a series of rather elevated yellowish spots. Legs black ; pro-legs deep yellow or orange.

" About an inch in length, but variable in size. Feeds on *Robinia pseud-acacia* (and probably on *Erythrina caffra*), fastening the pinnæ of a leaf together with silk and feeding in the shelter thus formed."— J. P. Mansel Weale, *in epist.*

PLATE II. fig. 6.

PUPA.—Pale greyish-green, covered with a dense chalky-white efflorescence ; first three dorsal segmental incisions of abdomen yellow. Under surface with nine small black spots, viz., one on each eye-case, one (rather smaller) on middle line of haustellum, two (not so widely apart as those on eyes) on middle pair of legs, two (very minute) on third pair of legs, not so widely apart as the two preceding them, and two (larger than the rest and sub-rhomboidal) on wing-covers, a little more widely apart than those on eyes. Spiracles linear, black. Head with an acute, median, superior, small, short, slender black projection ; a similar projection, rather dorsal than lateral, on each side of collar of thorax, and a little below it, on shoulder, three black dots arranged triangularly. Anal projection blackish.

Length ¾ inch; greatest width (across base of abdomen and wing-covers) ¼ inch.

(Described from a King William's Town example sent to me by Mr. Mansel Weale, from which the imago emerged on 24th March 1873.)

I have also reared a ♂ imago from a similar pupa received from Colonel Bowker (D'Urban, Natal) in March 1881.

PLATE II. fig. 6a.

Mr. Weale wrote in 1873 that his *Robinia* trees at Brooklyn, near King William's Town, were covered with the larvæ of *Forestan*, which completely riddled the clusters of leaves within which they fed.

This well-known species appears to range throughout the Ethiopian Region. Its closest ally, which accompanies it in Madagascar and Réunion, is *Arbogastes*, Guen. (= *Margarita*, Butl.),[1] but it is also very nearly related to *Pisistratus*, Fab., which has a similarly wide continental distribution, but does not seem to occur in Madagascar or the Mascarene Islands. In South Africa it inhabits only the eastern side, not having, to my knowledge, been taken westward of Port Elizabeth, and is commoner on the Natal coast than elsewhere. In January and February 1867 I took a good many examples about D'Urban; and also met with the species in Mauritius in July 1865, and near Grahamstown in February 1870. Its flight is very swift, but *bustling*, reminding one of that of the *Plusiæ* and other diurnal members of the Noctuæ tribe; it keeps steadily to flowers (in Natal preferring the cultivated *Lantana*), appearing both in the heat of the day and a little after sunset. The wings are elevated when the insect is settled, the hind-wings being held slightly apart from the fore-wings.

Localities of *Hesperia Forestan*.

I. South Africa.
   B. Cape Colony.
      b. Eastern Districts.—"Port Elizabeth."—S. D. Bairstow. Grahamstown. King William's Town (*J. P. Mansel Weale*).
   D. Kaffraria Proper.—Bashee River (*J. H. Bowker*).
   E. Natal.
      a. Coast Districts.—D'Urban.
      b. Upper Districts.—Estcourt (*J. M. Hutchinson*).
   F. Zululand.—Etshowe (*A. M. Goodrich* and *T. Vachell*).
   K. Transvaal.—Barberton, De Kaap (*C. F. Palmer*).

II. Other African Regions.
   A. South Tropical.
      a. Western Coast.—Damaraland (*J. A. Bell*). "Angola (*Pogge*)."—Dewitz.
      *a*1. Western Interior.—Otiembora (*A. W. Eriksson*).
      b. Eastern Coast.—"Querimba (*Peters*)."—Hopffer.
      bb. Eastern Islands.—"Madagascar."—Saalmüller. "Réunion."—Guenée. Mauritius.

---

[1] Figured in Grandidier, *Hist. Madag.*, xix. Atlas, pl. 51, ff. 5, 5a (1885), and distinguished by its sub-metallic greenish-grey basal suffusion on the upper side (shading off into hoary-grey in hind-wings), and inferiorly much narrower white band on the under side of the hind-wings.

B. North Tropical.
 a. Western Coast.—Accra: "Aburi (*Weigle*)."—Möschler. "Gambia River (*Moloney*)."—G. F. Shelley. "Senegal."—Hopffer.
 b1. Eastern Interior.—" Monbuttu (*Emin Pasha*)."—Butler. Abyssinia: "Shoa (*Antinori*)."—Oberthür. "Nubia."—Hopffer.

375. (2.) Hesperia Pisistratus, Fabricius.

Hesperia Pisistratus, Fab., Ent. Syst., iii. 1, p. 345, n. 311 (1793).
Rhopalocampta Valmaran, Wallengr., K. Sv. Vet.-Akad. Handl., 1857; Lep. Rhop. Caffr., p. 48.
Ismene Florestan, ? Var., Trim., Rhop. Afr. Aust., ii. p. 319 (1866).
Ismene Pisistratus, Westw. [part], Oates' Matabele Land, App., p. 352 (1881).
Ismene Pisistratus, Staud., Exot. Schmett., i. pl. 98 (1888).

PLATE XII. fig. 10 (♂).

Exp. al., (♂) 2 in. 1–1½ lin.; (♀) 2 in. 3–3½ lin.
Closely allied to *H. Forestan*, (Cram.).

♂ Not differing from *H. Forestan*, except that cilia of hind-wing on anal-angular lobe is of a duller and paler yellow. UNDER SIDE.— *White band of hind-wing differently shaped from that in Forestan, being much narrower on costa and much wider on submedian nervure, and marked inferiorly with three conspicuous black spots*, viz., two small rounded ones on outer edge between third and first median nervules, and the third larger, elongated, piercing it upwards from submedian nervure.

Head above with a much larger and paler yellow lateral spot on each side (the antennæ springing from the centre of it), and with also a median longitudinal yellow mark; *palpi* above with paler and wider yellow tufts at extremity of middle joint, beneath much paler, creamy-white tinged with yellow on middle joint. *Legs* with ochreous-yellow tufts and whitish stripes. *Abdomen* beneath with a white or creamy middle longitudinal stripe.

♀ Like ♂, but darker in ground-colour.

The distinctive characters above noted appear to be very constant[1] in both sexes, and fully warrant the separation of *Pisistratus* from *Forestan*. From some pencil notes and outline drawings made by Mr. W. D. Gooch near D'Urban, Natal, I gather that the larva is very like that of *Forestan*, but the front view of the head shows a third (inferior) transverse row of three spots; and the transverse bands are described as black, and two on each segment, while the parallel lines between each two pairs of bands are also black and enclose an orange spiracular spot. It is also noted that the food-plant is a wild species of *Indigofera*. Outlines given of the pupa exhibit no appreciable difference

---

[1] Professor Westwood, however, observes (App. to Oates' *Matabele Land*, 1881, p. 352) that " the black mark extending from near the anal margin on the under side of the hind-wings was more divided than usual in Drury's specimen, figured by Jones, so as to have led Fabricius to describe the hind-wings as four-spotted."

from that of *Forestan*, and the under-side black spots appear to be identical in size and position. The pupal state lasted in February for ten days.

In flight and habits there is nothing to distinguish this butterfly from *Forestan*. In Natal the two forms are often found on the wing together, and in February 1867 I on several occasions captured both on the same bush of *Lantana* near D'Urban. *Pisistratus*, however, was the rarer, and it seems to be generally scarcer in South Africa. Colonel Bowker did not send any specimens from Basutoland, but informed me that the species was not uncommon at Maseru in the autumn, appearing about a fortnight later than *Forestan*.

Localities of *Hesperia Pisistratus*.

I. South Africa.
  B. Cape Colony.
    d. Basutoland.—" Maseru."—J. H. Bowker.
  E. Natal.
    a. Coast Districts.—D'Urban. Pinetown (*J. H. Bowker*).
    b. Upper Districts.—Colenso (*W. Morant*).
  F. Zululand.—Etshowe (*A. M. Goodrich*). St. Lucia Bay (the late Colonel H. Towers).
  H. Delagoa Bay.—Lourenço Marques (*Mrs. Monteiro*).
  K. Transvaal.—Potchefstroom District (*T. Ayres*). Marico and Upper Limpopo Rivers (*F. C. Selous*).
II. Other African Regions.
  A. South Tropical.
    a. Western Coast.—Damaraland (*J. A. Bell*).
    *a*1. Western Interior.—Omrora River, to North of Ovampoland (*A. W. Eriksson*).
    *b*1. Eastern Interior.—Near Bamangwato (*H. Barber*). Makloutse River (*J. L. Fry*).
  B. North Tropical.
    a. Western Coast.—" Accra (*Buchholz*)."—Plötz. Gold Coast (*E. Bourke*).

376. (3.) Hesperia Keithloa, (Wallengren).

♀ *Rhopalocampta Keithloa*, Wallengr., K. Sv. Vet.-Akad. Handl., 1857; Lep. Rhop. Caffr., p. 48.
♂ ♀ *Ismene Stella*, Trim., Trans. Ent. Soc. Lond., 3rd Ser., i. p. 287 (1862).
♂ ♀ *Ismene Keithloa*, Trim., Rhop. Afr. Aust., ii. p. 317, n. 212 (1866).

PLATE XII. fig. 9 (♂).[1]

Exp. al., (♂) 1 in. 11 lin.—2 in; (♀) 2 in.—2 in. 2 lin.

♂ *Dull ochreous-brown. Fore-wing:* an orange edging on costa at base; a very slight ochre-yellow suffusion over basi-inner-marginal area. *Hind-wing:* much darker than fore-wing, especially at anal angle, *clothed with orange hairs except near costa and hind-margin; cilia orange from anal angle for more than half the extent of hind-margin,* fading thence into greyish. UNDER SIDE.—*Tinged with bronzed-green,*

---

[1] In this figure the orange cilia and discal space and the under-side orange-red marks of the hind-wings are not nearly bright enough in colour.

and with more or less of a violaceous gloss. Fore-wing: on inner margin, from base to about middle, a dull-whitish space, not rising above first median nervule. *Hind-wing: before anal angle, between first median nervule and submedian nervure, an orange-red marking, divided by a transverse black streak;* below this mark an orange spot on inner margin; orange cilia as above.

Head and *body* above much darker and browner than in *Forestan* or *Pisistratus;* the spots of head as in *Forestan*, but with an additional median frontal spot, and all orange-red; *abdomen* with thin whitish median rings and orange-yellow incision rings. Beneath, *palpi*, median front of *thorax*, spots and tufts of *legs*, and median stripe of *abdomen* all orange-red.

♀ Like ♂, but slightly paler.

This very handsome and distinct species is readily known by its want of any white band on the under side of the hind-wings and by the presence there of the vivid orange-red spots near the anal-angular lobe. In the former respect it is allied to the Malagasy *Ratek*, (Boisd.),[1] but this species has the under side of the hind-wings without marking of any kind. Perhaps the nearest known ally of *Keithloa* is the Socotran species named *Jucunda* by Butler (*Proc. Zool. Soc. Lond.*, 1881, p. 179, pl. xviii. f. 8). This is very like *Keithloa* on the upper side, but wants the orange discal pubescence in the hind-wings; and on the under side of the same wings it has, in addition to two orange-red marks near anal angle, some median marks of the same colour, the lowest and largest of which bears a black spot.

I have the skins of two pupæ sent to the Museum by Colonel Bowker from D'Urban in August 1881, from one of which the imago emerged in the following October. They are very like the pupa of *Forestan*, being covered with a dense white efflorescence, but their colouring beneath this is dark-red instead of pale-greenish.

Unlike its two South-African congeners, *Forestan* and *Pisistratus*, this butterfly has rather a limited range, and I have not found any record of its occurrence in Tropical Africa; from its presence, however, at Delagoa Bay, it is very likely to extend into Mozambique. It is numerous on the Natal Coast, where its habits quite agree with those of its congeners just mentioned. I took many specimens about D'Urban in February 1867; and met with the species singly at Port Elizabeth in January, and at King William's Town in February 1878. Colonel Bowker, writing from D'Urban on 24th May 1887, observed: "*Keithloa* is very common here now, and I have been noticing the curious behaviour of the sexes. The ♀ darts away from a flowering-tree the species frequents and settles on the ground, closely followed by the ♂; after a little she rises slowly, keeping her wings constantly fluttering, while the ♂ circles round her; and when they reach about a yard above the ground, off they

---

[1] This species was, I believe, erroneously catalogued as South-African in my earlier work (ii. p. 317). The specimen then described was ticketed "Natal" on the authority of Mr. E. L. Layard; but I have little doubt that it was actually taken in Madagascar, and afterwards by accident was misplaced among the South-African allied species.

go to the flowers of the tree for a drink, but soon return to go through the same evolutions."

Localities of *Hesperia Keithloa*.

I. South Africa.
  B. Cape Colony.—Port Elizabeth. Grahamstown (*Mrs. Barber*). King William's Town.
  D. Kaffraria.—" Baziya (*Baur* and *Hartmann*)."—Möschler.
  E. Natal.
    a. Coast Districts.—D'Urban. Pinetown and Mouth of Tugela River (*J. H. Bowker*). " Lower Umkomazi."—J. H. Bowker.
    b. Upper Districts.—Maritzburg (*S. Windham*).
  II. Delagoa Bay.—Lourenço Marques (*Mrs. Monteiro*).

377. (4.) Hesperia Anchises, (Gerst.).

♂ ♀ *Ismene Anchises*, Gerst., Gliederth.-Fauna d. Sansibar-Gebiet., p. 374, n. 29, pl. xv. ff. 6, 6a (1873).
Ismene Taranis, Hewits., Ann. and Mag. Nat. Hist., 4th Ser., vol. xviii. p. 347 (1876).

Exp. al., 2 in.—2 in. $3\frac{1}{2}$ lin.

" Uniformly brown; fore-wing lighter and with a greyish tinge, hind-wing, except at base, darker pitch-brown; bases with bluish slate-grey hair; in hind-wing hair along inner margin light grey-brown, and near anal angle even hoary; *cilia* of fore-wing brown basally but narrowly chalk-white outwardly,—of hind-wing broad and pure chalk-white, edged with light golden-yellow at the anal-angular projection. UNDER SIDE.—*Hind-wing:* chalk-white band ending rather farther from inner margin than from hind-margin, elongately and bluntly triangular, its broader base (towards inner margin) bearing a larger round deep black spot and a smaller acute one; beyond this two orange-red scaled spots, of which the inferior one is smaller, immediately at anal angle, and the superior one larger, more curved anteriorly and longitudinally parallel to the outline of the superior side of anal-angular projection.

" *Head* above slate-grey; border of eyes above and in front reddish-yellow, behind almost vermilion; *palpi* orange-yellow with black terminal joint; *antennæ* black. *Thorax* above slate-grey, beneath bluish-grey on the sides. *Fore-legs* with femur above blackish, beneath with a tuft of reddish-yellow hair; *middle and hind legs* with tibiæ white-scaled at tip and bearing vivid fox-red hair beneath. *Abdomen* above brownish-grey, beneath chalk-white on each side, in the middle spotted with reddish-yellow."

Not having seen specimens of this Hesperid, I have above adapted Gerstaecker's description (*op. cit.*). It should be added that his figure shows the white under part of the abdomen to be crossed by four black streaks. *Anchises* combines to some extent the markings on the under side of the hind-wings of *Pisistratus* and *Keithloa*, having both a white black-spotted band and two orange-red spots.

The species was discovered by Dr. Kersten (on Van Der Decken's East-African Expedition) at Lake Jipé, and in 1876 was renamed *Taranis* by Hewitson from Zanzibar specimens. I have included it in the South-African list on the strength of a coloured drawing of a Delagoa Bay butterfly forwarded to me in 1884 by Mr. H. Grose Smith, which appeared to represent the ♀ *Anchises*. The drawing differed from Gerstaecker's figures in the following particulars, viz., the hind-wings on the upper side duller and paler, and on the under side having the upper ochre-yellow spot smaller and on the hind-marginal edge (instead of some little way before it), and the lower spot more elongate and along hind-marginal edge (instead of only touching the edge just at anal angle); the cilia on anal-angular lobe white instead of yellow; the under side of the abdomen less regularly and distinctly banded with black and white.

Localities of *Hesperia Anchises*.

I. South Africa.
  II. Delagoa Bay.—" Lourenço Marques (*Mrs. Monteiro*)."—H. G. Smith.
II. Other African Regions.
  A. South Tropical.
    b. Eastern Coast.—" Zanzibar."—Hewitson.
    b1. Eastern Interior.—" Lake Jipé (*Kersten*)."—Gerstaecker. " Victoria Nyanza."—Butler.
IV. Asia.—Arabia: " Aden (*Yerbury*)."—Butler.

378. (5.) Hesperia unicolor, (Mabille).

Ismene unicolor, Mab., Ann. Soc. Ent. de France, Sér. 5, vii. p. xxxix. n. 47 (1877); and Bull. Soc. Zool. de France, 1877, p. 230.

Exp. al., (♂) 1 in. 9 lin.; (♀) 1 in. 10 lin.

♂ *Dark-brown, without markings, but with a slight violaceous gloss; bases (especially towards inner margin in fore-wing) with a bronzy tinge;* a coating of bronzy-yellowish hairs (thickest towards inner margin) over basi-median and lower discal area of hind-wing, but leaving free a broad costal, apical, and hind-marginal border; cilia dull-whitish in fore-wing, white in hind-wing, UNDER SIDE.—*Much paler, the violaceous gloss much stronger. Fore-wing:* inner margin with a still paler ill-defined border. *Cilia* as on upper side.

Head and *body* above brown, clothed with bronzy hair (intermixed, on head and front of thorax with grey and shining green hairs); an ochre-yellow streak inwardly edging each eye; *palpi* clothed with mixed pale ochre-yellow and black hairs, and with a black exterior streak beneath; *antennæ* black with the tip dark-red. *Thorax* beneath grey, but with a pale ochre-yellow median space in front. *Legs* brown, with some yellowish scales and hair. *Abdomen* beneath with a median dull pale-yellowish stripe and indistinct segmental half-wings.

♀ Like ♂.

This sombre-tinted Hesperid seems to differ from *H. Libeon*, (Druce),[1] a native of Angola, solely in wanting the " indistinct small

---
[1] *Proc. Zool. Soc. Lond.*, 1875, p. 416.

white spot near the anal angle" on the under side of the hind-wings. Should the two prove to be the same species, Druce's name will have priority. Mabille's insect is recorded as a native of Congo; his descriptions above cited apply well to the South-African specimens I have seen.

Mr. H. Grose Smith made this butterfly known to me by a coloured drawing of a Delagoa Bay specimen which he forwarded in 1884, and I subsequently examined a pair from that locality in his fine collection. In February 1885 I received a fine ♂ captured on the 14th of that month by Colonel Bowker at Northdene near D'Urban, Natal, and in May 1887 two ♀s arrived from the same neighbourhood, also just taken by my colleague. The first of these was secured at garden flowers in the evening in company with *Keithloa*. Colonel Bowker has met with no other examples, and Mr. A. D. Millar writes that only one has come under his notice.

Localities of *Hesperia unicolor*.

I. South Africa.
   E. Natal.
      a. Coast Districts.—D'Urban (*J. H. Bowker*).
   II. Delagoa Bay.—Lourenço Marques (*Mrs. Monteiro*).
II. Other African Regions.
   A. South Tropical.
      a. Western Coast.—"Congo."—Mabille.

APPENDIX I.

*ADDITIONAL SPECIES, RECENTLY ASCERTAINED
TO BE SOUTH-AFRICAN.*

VOL. I.

FAMILY **NYMPHALIDÆ**.
SUB-FAMILY ACRÆINÆ.
GENUS ACRÆA.

379. (3A.) Acræa Machequena, H. G. Smith.

Acræa Machequena, H. G. Smith, Ann. and Mag. Nat. Hist., 5th Ser., xix. p. 62 (1887).

Exp. al., (♂) 1 in. 9 lin.—2 in.; (♀) 2 in. 2–3 lin.

♂ *Pale yellowish-red, semi-transparent ; fore-wing with broad apical area transparent grey; hind-wing with basal, discal, and submarginal black spots.* Fore-wing: spotless; yellowish red bounded by costa as far as extremity of discoidal cell, inner margin to posterior angle, and disco-cellular nervules,—below discoidal cell it extends more towards hind-margin, presenting a somewhat convex outer edge as far as posterior angle. Hind-wing: base narrowly blackish; in discoidal cell two spots, of which one (close to base) is smaller than the other (about middle); a minute terminal disco-cellular spot; three spots below cell, near base, of which the two lower ones (close to inner margin) are smaller than the upper one; about middle, a sinuated discal series of eight rather small spots (of which the third and fourth are sometimes obsolescent, and the sixth is nearer base than the rest); hind-margin itself narrowly edged with fuscous-grey, which is immediately preceded by a series of six good-sized internervular hastate black spots, each of which is bounded outwardly by a smaller spot of the ground-colour interrupting the fuscous-grey edging. UNDER SIDE.—Reddish areas very much paler, and with a glossier surface; spots of hind-wing as on upper side.

♀ *Reddish colouring wanting; fore-wing with a slight tinge of brownish-yellow (chiefly along nervures) over inner half; hind-wing dull-whitish.* Hind-wing: basal and discal spots reduced in size,—the basal disco-cellular spot and the first (and sometimes also the second) spot of the discal series obsolete; submarginal hastate spots considerably enlarged and lengthened, the spots bounding these externally ochre-yellow; hind-marginal fuscous-grey edging broader, diffused. UNDER SIDE.—Almost colourless; only submarginal black spots of hind-wing well marked.

This species is a very near ally of *A. Ranavalona*, Boisd.,[1] a native of Madagascar. It differs from the latter, as far as the ♂ is concerned, in having the red area much paler and yellower, and in the fore-wings of very much greater extent; in both sexes there is no sub-basal black spot just below costal nervure, the submarginal black spots of the hind-wings are much larger, and the adjoining hind-marginal rufous spots of a much duller tint, while the spots of the discal series are considerably smaller, especially in the ♀, which entirely lacks the large and conspicuous first (costal) spot of the ♀ *Ranavalona*.[2]

The curious arrangement of the hastate submarginal black spots of the hind-wings, with their externally adherent reddish spots interrupting the actual grey edging of the hind-margin, readily distinguishes *Machequena* from all other known South-African *Acræa*. In tint and general aspect the ♂ is not unlike *A. Neobule*, Doubl., but the absence of markings in the fore-wings, and black, laterally ochreous-spotted (instead of wholly ochreous) terminal half of the abdomen above, at once mark it as a different species. The spotless fore-wings recur in some examples of the variable *A. Cerasa*, Hewits., but the latter has no submarginal spots at all in the hind-wings.

Mrs. Monteiro discovered this *Acræa* at Delagoa Bay early in the year 1886, and sent me a rough sketch of it in March, and specimens later on. She met with a good many examples of both sexes, but did not note anything peculiar in the habits of the species.

---

[1] *Faune Ent. de Madag.*, &c., p. 30, pl. 6, ff. 3, 4, 5 (1833).

[2] C. Ward in Part II. of his *African Lepidoptera*, p. 9, states that he had recently (1874) received pairs taken *in copulâ* of ♂ and ♀ *Ranavalona*, " in no way differing" (that is, the ♀ agreeing with Boisduval's *Ranavalona* ♂), and of the ♀ *Ranavalona*, Boisd., "with a ♂ only differing in being rather smaller;" and he on this account separates the latter under the name of *A. Manandaza*. [His description here is identical with that previously published in the *Entom. Monthly Mag.*, 1872, ix. p. 147.]

The occurrence of a red-tinted ♀ resembling the ♂ is accordant with several similar cases in the same genus; but I incline to the belief (in the absence, however, of the specimens with which Mr. Ward dealt) that some error was made in respect to the alleged pair in which the ♂ only differed in size from the pale ♀. In Ward's figures (*op. cit.*, pl. vii. ff. 1, 2) the larger "♀" has a good deal of reddish suffusion in the hind-wings, while in the smaller "♂" this is quite absent.

Locality of *Acræa Machequena*.
I. South Africa.
II. Delagoa Bay.—Lourenço Marques (*Mrs. Monteiro*).

380. (3b.) **Acræa Igola,** *sp. nov.*

Exp. al., (♂) 1 in. 11 lin.; (♀) 2 in.

♂ *Brick-red, in hind-wing spotted and bordered with black; apical area of fore-wing transparent, colourless, bordered with blackish and crossed by black-clouded nervules.* Fore-wing: red area semi-transparent, filling discoidal cell, and occupying all space below cell to inner marginal edge as far as posterior angle; a rather narrow blackish costal border, becoming very much broader at apex; hind-marginal border below upper radial nervule not much wider than costal border, inwardly somewhat excavated between nervules, and diminishing to a point at posterior angle; at base a short black longitudinal mark between median and submedian nervures, and a longer black streak along inner margin to before middle; an ill-defined narrow terminal disco-cellular black mark. *Hind-wing:* hind-marginal black border of moderate width, emitting inwardly short acute sublinear nervular dentations and very short internervular ones; black spots of moderate size, distinct, rounded, viz., one in cell near base; an irregular sub-basal curved series of four, of which the second is in cell; two small ones obliquely placed at upper part of extremity of cell; and an irregular median discal series, of which the superior ones (except that next costa) are very small and partly obsolete, but the three lower large and conspicuous, and forming a straight row from below second median to internal nervule; base with a rather wide blackish suffusion; inner-marginal border pale-yellowish beyond middle. UNDER SIDE.— *Exceedingly glossy; red in fore-wing extremely pale; hind-wing pale dull-yellowish with a very faint reddish tinge.* Fore-wing: costal, apical, and hind-marginal border thinly scaled with pale-yellow crossed by blackish-clouded nervules. *Hind-wing:* hind-marginal border grey, rather closely scaled with pale-yellow, crossed by blackish nervules, and also by a series of short internervular black streaks not reaching hind-margin; spots more conspicuous than above,—three additional ones present, viz., two near base on costa and inner margin respectively, and a very small one on inner margin about middle.

Abdomen above black, with a superior lateral series of six small white spots, and an inferior lateral white streak,—beneath yellowish-white.

♀ *Red replaced by very pale yellowish; borders not so dark.* Fore-wing: base widely suffused with fuscous, especially below cell, where suffusion reaches' to beyond middle; submedian nervure and median nervure with its first nervule blackish-clouded. *Hind-wing:* spots larger, especially upper ones of discal series; basal blackish suffusion considerably wider; hind-marginal border broader, emitting nervular rays inwardly. UNDER SIDE.—*Fore-wing:* border as in ♂; basal fuscous suffusion and yellowish area beyond both paler than on upper side. *Hind-wing:* whiter than in ♂, without reddish tinge; black spots larger; hind-marginal border broader, but its internervular submarginal dark marks less defined.

Abdomen with larger white spots.

This *Acræa* has at the first glance much the appearance of a dwarf *Horta*, but the blackish border of the fore-wings and broad unspotted black border of the hind-wings readily distinguish it. The ♂ seems most nearly related to *A. Peneleos*, Ward,[1] from Cameroons and Old Calabar, but has much shorter and blunter fore-wings, with a very large red area in place of only two longitudinal red streaks on the inner margin near the posterior angle; and in the hind-wings has the inferior discal spots differently shaped and arranged. The ♀ comes very close to a Madagascar *Acræa* described and figured by Saalmüller[2] as *A. Boscæ*, but differs in its heavy basal blackish suffusion of both fore and hind wings, and the form and disposition of the black spots of the hind-wings. In the last-mentioned character the ♀ *Igola* agrees much better with the ♀ *Masamba*, Ward,[3] which is, however, a much larger insect, without basal black suffusion, and with very hyaline fore-wings, which bear a dusky spot in the discoidal cell. The ♂ *Masamba* has a small field of red in the fore-wings, only filling basal half of cell and not rising above first median nervule, and the discal black spots of the hind-wing are very much larger, especially that next to costa; moreover the fore-wings are as much elongated apically as in the ♂ *Peneleos*.

Major H. D'Aguilar discovered this butterfly in Zululand during May 1886, and communicated to me through Colonel Bowker the two specimens (♂ and ♀) from which the above description is made. He wrote that the ♂s appeared about the middle of May, flying over a yellow-flowered climbing plant in thick forest near Etshowe; they kept usually about twenty-five feet from the ground, and were not scarce. The only ♀ observed sat low down and was easily captured; this example occurred near the end of May, when Major D'Aguilar left the neighbourhood; but the wings of a second ♀ were found by him in the same spot.

Locality of *Acræa Igola*.

I. South Africa.
   F. Zululand.—Etshowe (*Major H. D'Aguilar*).

---

[1] *Ent. M. Mag.*, viii. p. 60 (1871); and *African Lepidoptera*, i. p. 7, pl. vi. ff. 3, 4 (1873), [♂].
[2] *Lep. Madag.*, i. p. 76, pl. i. f. 3, [♀].
[3] *Ent. M. Mag.*, ix. p. 3. (1872); and *African Lepidoptera*, ii. p. 10, pl. 7, ff. 3, 4 (1874). See also Saalmüller, *op. cit.*, p. 75, pl. 3, f. 32, for a dull-reddish tinted ♀ of this Madagascar species.

381. (11A.) Acræa Acrita, Hewitson.

Acræa Acrita, Hewitson, Exot. Butt., iii. pl. 8 (Acræa iii.), f. 18 [♂].

Exp. al., (♂) 2 in. 4–6 lin.

♂ *Deep ochreous-red, with a ferruginous tinge, spotted with black; wide apical area of fore-wing ochreous-yellow, paler on its inner side; cilia white.* Fore-wing: a good-sized ovate black spot in outer half of discoidal cell, and another (more elongate) at extremity of cell; in an oblique line below latter spot towards posterior angle, two other spots, the upper one large, rounded, above first median nervule,—the lower one smaller, narrower, below that nervule; sometimes a small round spot below median nervure, some way before origin of the first nervule; costa very finely black-edged from about middle; apex narrowly or broadly black-bordered; hind-margin narrowly black-edged from end of apical border to posterior angle; boundary between ochreous-red and apical ochreous-yellow areas not sharply defined, the latter beginning about extremity of cell. *Hind-wing:* paler on its margins; base in the middle very narrowly irrorated with blackish; a small spot in discoidal cell close to base; an irregular sub-basal row of four spots, of which the first and fourth (respectively next costa and inner margin) are very small and indistinct, and the second is in cell; immediately beyond upper disco-cellular nervule a rather small rounded spot; an exceedingly irregular discal row of six moderate-sized spots, of which the first and fifth are nearer base than the rest; a rather broad black hind-marginal border enclosing seven large spots of the ground-colour, but rather paler. UNDER SIDE.—*Hind-wing and moderately broad apical area of fore-wing with a ground of clear yellowish-creamy, largely obscured in the former by broad dull lake-red, in the latter by narrow reddish-ochreous, internervular rays.* Fore-wing: black-edging everywhere linear; black spots as on upper side; nervules crossing apical creamy space very finely black. *Hind-wing:* except in discoidal cell for about two-thirds of its length from base, and in a more restricted space between median and submedian nervures, the internervular lake-red rays leave only a creamy clouding on neuration and round spots; latter as above, but three small additional ones close to base; spots of hind-marginal border clear yellowish-creamy, and so enlarged that black is reduced to a very distinct and well-defined inner and outer edging to, and nervular lines of separation between, them.

♀ *Dusky-brown, with the subapical space beyond extremity of discoidal cell conspicuously paler;* all spotting similar to that of ♂; colouring of under side much duller.

There is much probability that the ♂ of this striking and distinct *Acræa* is in life of considerably brighter colouring than appears in my description, which was made from Hewitson's types and from a ♂ captured in 1883 by Mr. Selous. I include the species in this Appendix, on the strength of several examples from Delagoa Bay which I saw in Mr. H. Grose Smith's collection in 1886. Its nearest described allies appear to be *A. Caldarena*, Hewits., and *A. Stenobea*, Wallengr., but it differs greatly from both these congeners by its deep-red upper side (in the ♂ sex), and want of basal dusky suffusion and subapical oblique series of spots in the fore-wings, as well as by the disposition of the discal spots, and the existence of large spots of the ground-colour in the hind-marginal border on the upper side, in the hind-wings. The wide unbroken ochre-yellow apical area of the fore-wings in the ♂ recalls the still larger space in *A. Anemosa*, Hewits.,—a congener in other respects so totally different in marking.[1]

This species is very rare in collections, and would appear to be so, or possibly extremely local, in nature. Mr. Hewitson's types (a ♂ and a ♀) were from the Zambesi,[2] and the only specimen (a very fine ♂) sent me by Mr. Selous was taken in Mashunaland, an elevated region not far to the south of that river. The six Delagoa Bay specimens (three ♂s, three ♀s) above mentioned were, I believe, captured by Mrs. Monteiro. A solitary example is recorded by M. C. Oberthür (*Études d'Entomologie*, iii. p. 24, 1878) as having been taken by M. Raffray in Zanguebar.

Localities of *Acræa Aerita*.

I. South Africa.
   II. Delagoa Bay.—Lourenço Marques (*Mrs. Monteiro* ?).
II. Other African Regions.
   A. South Tropical.
      b. Eastern Coast.—"Zanguebar: Tchouacka (*Raffray*)."—Oberthür "Zanzibar."—Kirby. Cat. Coll. Hewitson.
      b1. Eastern Interior.—Zambesi.—Coll. Hewitson. Mashunaland (*F. C. Selous*).

Sub-family NYMPHALINÆ.

Genus HARMA.

382. (2.) Harma Coranus, H. G. Smith.

♂ ♀ *Cymothoe Coranus*, H. G. Smith, Ann. and Mag. Nat. Hist., Feb. 1889, p. 133.

Exp. al., (♂) 2 in. 5–6 lin.; (♀) 2 in. 6 lin.

♂ *Cream-colour; a submarginal sharply-dentated black streak of unequal thickness, and a hind-marginal narrow fuscous border; common*

---

[1] A much closer ally of *Aerita*, but exhibiting one or two more features of alliance with *Caldarena*, has lately (1888) been sent to me by Mr. A. W. Eriksson from Tropical South-Western Africa. This species presents, however, in the fore-wings, besides the very broad apical black of *Caldarena*, a subapical *white* patch.

[2] In Kirby's Catalogue of the Hewitson Collection, p. 51, Zanzibar is given as the locality of *Aerita* (of which five examples are recorded); but Zambesi was the locality given me by the late Mr. Hewitson for his two type specimens, and is that noted in his published description above cited.

median streak of under side dimly perceptible; nervules defined with black towards hind-margin. *Fore-wing:* submarginal streak thickened on points of internervular dentation, and at each point having a small acute projection towards hind-margin; fuscous border much widened at apex, sending a rather wide projection for a little distance along costa, and also a short streak or ray along subcostal nervure,—also abruptly widened between lower radial and second median nervules, so as to join black submarginal streak, which here curves inward considerably; costa narrowly edged with black; base with a narrow grey suffusion, somewhat wider on inner margin. *Hind-wing:* submarginal streak more acutely dentated than in fore-wing, and more distinct from hind-marginal fuscous border; a very narrow basal grey suffusion, but prolonged outwardly over a small (inferior) part of discoidal cell and below median nervure and base of its first nervule; inner-marginal groove and a small space beyond its extremity (before anal angle) brownish-grey, paler in the middle. UNDER SIDE.—*Almost as pale as upper side to about middle, where a narrow (almost linear) dark-brown streak crosses from costa of fore-wing to just before anal angle of hind-wing;* immediately beyond streak a very pale brownish-grey fascia, rather narrow, and sharply dentate externally on nervules; this is closely succeeded by narrow stripe of the same colour, exhibiting corresponding dentations; submarginal line indistinctly represented, except that the apices of the dentations form a series of very small rather ill-defined blackish spots. *Fore-wing:* the following fine fuscous markings before middle, viz., a highly irregular, unsymmetrical, transverse figure in discoidal cell just before origin of first median nervule; two similar longer less irregular figures (open inferiorly), one immediately before, the other immediately beyond extremity of cell; external edge of the outer of these two markings interruptedly prolonged by a very fine broken and irregular line, a little before dark-brown streak, almost to submedian nervure; an annulet, succeeded by a very fine transverse line, just below inner irregular disco-cellular figure; a small diffused fuscous mark in cell, near base; a very slight ochrey-red tinge along edge of hind-margin. *Hind-wing:* disco-cellular fine linear figures arranged as in fore-wing, but the inner one shaped like a somewhat oblique 8, and the two outer ones considerably straighter; two transverse irregular lines (completely and widely interrupted by disco-cellular space) ending on submedian nervure.

♀ *Fuscous, with a common pure-white transverse discal band, closely succeeded by an internervular series of rather small, mostly sagittate-lunulate white marks.* *Fore-wing:* beyond extremity of discoidal cell, between subcostal nervure and third median nervule, a well-defined moderately broad white bar (crossed by two black nervules), the lower end of which is externally only divided from discal white band by a slender curved black streak; discal band inwardly only slightly excavate between nervules, but outwardly very deeply, so as to present very sharp nervular dentations,—superiorly the band consists of two very narrow attenuated rather indistinct separate sagittate marks, but from lower radial nervule is continuous, widening on inner margin; upper three of succeeding series of spots subapical, slightly diffused, not lunulate. *Hind-wing:* discal band much narrowed on costa, but broader than in fore-wing from costal to submedian nervure,—its inner edge even but somewhat convex inwardly, its outer edge not nearly so sharply dentated as in fore-wing; white spots of succeeding series much smaller than in fore-wing, scarcely lunulate, except the upper two, which are almost obsolete. On the fuscous ground there are darker traces of the submarginal markings and (in the fore-wing) of discocellular ones. UNDER SIDE.—Area before median streak very palegreyish (becoming whitish along inner-marginal border of hind-wing); beyond median streak follows whitish, representing band of upper side, but its external edge and the succeeding white spots very vague; hindmarginal border rather widely pale-greyish tinged (more decidedly in fore-wing) with ochrey-yellow; fine lineolar markings before middle much as in ♂, but broader in figure, and much less conspicuous. *Forewing:* costal bar whitish, immediately before transverse median streak, which is not nearly so dark as in ♂. *Hind-wing:* median streak with a distinct curve inward superiorly.

This *Harma* combines many characters of *H. Alcimeda*, (Godt.), Variety A., and *H. Cænis*, (Drury), the ♂ being on the whole nearer to that sex of the former, while the ♀ is distinctly intermediate between those of the two species named. The ♂ *Coranus* is on the upper side very near Var. A. of the ♂ *Alcimeda*, both in tint of ground-colour and in markings, but it has much less grey suffusion at the bases; presents no trace (except in one example the faintest indication near inner margin of hind-wings) of the inner submarginal lunulate or dentate black streak; and wants altogether the hind-marginal orangeochreous spots; on the under side the resemblance to Var. *Alcimeda* is also close, but difference is observable in the paler ground before middle with more defined linear black figures; the darker (and in hindwings straighter) median transverse streak; the very much broader and better defined brownish-grey fascia immediately succeeding the median streak; and the very much fainter ochrey-red tinge on the hind-margins. The much larger size and considerably less prominence of the apical production of the fore-wings and the anal-angular one of the hind-wings, readily distinguish the ♂ *Coranus* from its smaller ally, and approximate it to the ♂ *Cænis*, to which it is further related by the absence of the hind-marginal orange-ochreous spots. It is separated from the latter by its want of the inner submarginal dentate or lunulate streak, and the more regular and less strong dentation of the outer one; while on the under side it has no trace of the peculiar very pale greenish tint of *Cænis;* the common median streak is deflected inwardly near the costa of the fore-wings, instead of running directly onward; and the outer submarginal streak is much fainter and far more regular in its dentation.

The single specimen of ♀ *Coranus* is no larger than a full-sized ♀ of the Var. A. of *Alcimeda*: the features in which it differs from the latter and approximates the ♀ *Cænis* are, *on the upper side*, (1) the absence of the disco-cellular white transverse mark in the fore-wings, and (2) the greater development of the subcostal white bar beyond the cell; (3) the greater width of the discal white band on inner margin of the fore-wings, and (4) the absence in both wings of the orange hind-marginal spots; while, *on the under side*, (5) it has none of the ferruginous tint, but is of a grey even paler than in *Cænis*, and (6) has the white band much wider and outwardly very vaguely defined. The characters in which it approaches the ♀ *Alcimeda* Var. are its small size and the shortness and comparative bluntness of the indistinct black submarginal spots of the upper side, which, although hastate, are only about half as long, and not nearly so acute, as in *Cænis*.

This is the insect mentioned in the text (p. 312) as taken by Colonel Bowker in Natal. I hesitated to describe it from a solitary and imperfect ♂ example; but having since received a perfect ♂ and a passable ♀ from Zululand (where they were taken by Captain A. M. Goodrich, of the Royal Inniskilling Fusiliers), I am now able to give its distinctive characters. Colonel Bowker's specimen was captured at Pinetown in June 1883; in sending it to me, he remarked on its large size and other differences from *Alcimeda*, and noted that it had the same flight and habits as the latter. Captain Goodrich took his specimens at Etshowe,—the ♀ in November 1886, and the ♂ in April 1887; he informed me that there was nothing noticeable in their behaviour to distinguish them from *H. Alcimeda*, Var. A., which occurred in the same locality.

Localities of *Harma Cænides*.

I. South Africa.
    E. Natal.—Pinetown (*J. H. Bowker*).
    F. Zululand.—Etshowe (*A. M. Goodrich*).

II. Other African Regions.
    A. South Tropical.
       b. Eastern Coast.—"Mombasa (*Last*)."—H. G. Smith.

GENUS CHARAXES.[1]

383. (14A.) **Charaxes Violetta**, H. G. Smith.

    ♂ ♀ Charaxes Violetta, H. G. Smith, Ent. M. Mag., xxi. p. 247 (1885); and Rhop. Exot., pt. i, pl. i. ff. 1–3 (1887).

"*Upper side.*—♂ Anterior wings dark brown, suffused slightly with violet, with a curved row of violet-blue spots across the middle

---

[1] In October 1885 Mrs. Monteiro (then at Delagoa Bay) sent me a coloured drawing of a *Charaxes* which she had taken, with the note that it had previously been forwarded to

of the wings, and a submarginal row of similar spots from near the costa to the inner margin, the lower half of the two rows becoming confluent, the two submarginal spots near the apex nearly white. Posterior wings with a broad central band of violet-blue, suffused with white from the second subcostal nervule to the abdominal fold near the anal angle; above this band are two pairs of violet-blue spots, a submarginal row of seven small spots, and a row of elongated spots on the margin on each side of the nervules, all violet-blue suffused with white. ♀ Anterior wings with a broad curved band across the centre of the wings from the costa to the inner margin, and two white transverse spots near the apex. Posterior wings: the inner half, from near the base, white, suffused with violet, a submarginal row of small white spots, and a marginal white line intercepted by the nervures, both suffused with violet.

"*Under side.*—Both wings as in *Ch. Cithæron*, except that the central black line across both wings, which is broadly bordered on the outside with white, is straight and continuous, not irregular and interrupted as in *Cithæron*. This species on the upper side has a general resemblance to *Cithæron*; it is more violet-blue, and is smaller in size, particularly the female, which is not so large as the male *Cithæron*, while the under side of both sexes is very distinct from *Cithæron*.

"Exp. 3⅛ inches.

"*Hab.*—Delagoa Bay."

In the absence of any examples of this close ally of *Cithæron*, Feld., I give Mr. H. Grose Smith's original description from the Journal quoted. Judging from the figures of the upper side of the ♂ and the under side of both sexes more recently (*op. cit.*) published by Mr. Smith, it appears that in the ♂ the violet-blue spots are in their deeper tint more like those of the ♂ *Xiphares*, Cram., although the violet-blue band of the hind-wings is, as regards width and the white suffusion in its inferior part, more like that of *Cithæron*. It is also noteworthy that on the under side, although the thin blue-black transverse striæ

Mr. H. Grose Smith, adding, "Mr. Smith says, 'The blue band is much wider than in *Etesipe*, and the tails are larger;'—when he sees the original I shall know if the difference is due to my hurried sketch." This drawing represented a ♂ *Charaxes* evidently very close to, if not identical with, *C. Etesipe* of Godart, a beautiful West-African species, which has a blue-shot black upper side, bearing a common submarginal series of pale-blue spots, and also a few white spots on the fore-wings, and a strikingly handsome under side of creamy-white and ferruginous markings on a grey ground further varied with black spots. Before returning the drawing to Mrs. Monteiro, I noted that the greater width of the blue submarginal macular band was represented as in the hind-wings only. This feature is characteristic in a more marked degree of the Madagascar representative of *Etesipe* named by Hewitson *Cacuthis* (*Exot. Butt.*, iii. pl. 32, ff. 12, 13, 1863); and, as far as I could judge from the drawing referred to, the Delagoa Bay form must be intermediate between *Etesipe* and *Cacuthis*. I learn from Mr. Grose Smith that he has not found any example of this butterfly in the collections which reached him from Delagoa Bay; but Mrs. Monteiro's drawing leaves no doubt of the occurrence of the species in that locality. It is especially desirable to procure the ♀ of the Delagoa Bay insect, as that sex differs greatly on the upper side in the West-African and Madagascar allies.

(and especially the long median one common to both fore and hind wings) are in both sexes much straighter and more continuous than in either *Cithæron* or *Xiphares*—their white edgings (in the ♂ throughout, and in the ♀ as regards the hind-wings and sub-basal part of the fore-wings) approximate *Violetta* more to *Xiphares*. This remark does not apply to the under side of the fore-wings in the ♀ *Violetta*, where the broad white band externally bounding the median blue-black stria is very like that found in the ♀ *Cithæron*. The considerably smaller size of both sexes in comparison with the two large allies in question is an important distinction of *Violetta*.

Locality of *Charaxes Violetta*.

I. South Africa.
II. Delagoa Bay (? Mrs. Monteiro).

384. (17.) Charaxes Azota, Hewitson.

♀ *Philognoma Azota*, Hewits., Ent. M. Mag., xiv. p. 82 (1877); and
♂ *Charaxes Azota*, op. cit., p. 181 (1878).

"*Upper side*.—Male black. Both wings with the outer margin broadly bright brick-red (colour of *C. Protoclea*). Anterior wing with the costal margin rufous, the band of the outer margin marked by black spots, and divided as it reaches the apex into two bands of separate spots, one of which is on the margin. Posterior wing dentated, the red margin covering half the wing, and marked near the anal angle by two minute black spots centred with white.

"*Under side*.—Red-brown. Both wings with a submarginal series of minute white spots. Anterior wing with four bands in the cell and a spot beyond it; crossed at the middle by a band of glossy colour bordered with brown, the apex marked by two white spots. Posterior wing crossed near the base by a band of red-brown, and at the middle and beyond it by broad glossy bands.

"Exp. $3\frac{3}{10}$ inches.

"This fine species was taken by Mr. Thelwall at Nyassa. It is closely allied to *C. Protoclea*, and is the ♂ of a butterfly from the collection of the Monteiros, which (having lost the posterior part of the hind-wings) I described as *Philognoma Azota*."—(Hewitson, *loc. cit.*)

"*Upper side*.—Female dark brown. Both wings crossed by a very broad white band tinted with lilac, commencing on the anterior wing at the third median nervule, where it is tinted with orange, and becoming broader in the posterior wing, and reaching to the abdominal fold. Both wings with the outer margin broadly rufous. Anterior wing with a bifid orange spot near the middle of the costal margin, and a broad submarginal rufous band. Posterior wing with one tail.

"*Under side*.—Rufous-brown. Both wings crossed as above by a common band, but narrower, tinted with yellow, and marked near its internal border by several brown spots. Anterior wing with several dark-brown spots in and below the cell, the bifid spot and rufous band as above; two pale spots near the apex, and a black spot at the anal angle, and a submarginal series of white spots.

"Exp. $3\frac{8}{10}$ inches.

"*Hab.*—Delagoa Bay: Monteiro."—(Hewitson, *loc. cit.*)

Not having seen any example of this species, I can only give the above descriptions of the sexes by Hewitson, which were kindly furnished to me by Mr. W. F. Kirby, and which I have also since had an opportunity of consulting in the Journal in which they appeared. Mr. Hewitson did not figure the insect, nor could I find it mentioned in the printed Catalogue (1879) of his Collection. On consulting Mr. Kirby, he wrote in reply (19th February 1888), "I am sorry to say that *Charaxes Azota* does not seem to be in his [Hewitson's] Collection. As it was only described shortly before his death, I think it probable that the specimens were never incorporated at all."

C. Protoclea, Feisth.,[1] to which Hewitson states *Azota* to be closely allied, is a native of the Western Coast of North Tropical Africa; it was originally recorded from the Casamanza, but specimens in the Hewitson Collection are ticketed Calabar and Camaroons. Another, but much smaller, West-African ally is *C. Anticlea*, (Drury),[2] recorded from Sierra Leone and (Hewitson Collection) as far south as Angola. All three species appear to agree in the possession of a broad rufous or brick-red hind-marginal border,—a feature which should readily distinguish *C. Azota* from any other hitherto known South-African *Charaxes*.

Localities of *Charaxes Azota*.[3]

I. South Africa.
   II. " Delagoa Bay (*Monteiro*)."—Hewitson.
II. Other African Regions.
   A. South Tropical.
      b. Eastern Interior.—" Nyassa (*Thelwall*)."—Hewitson.

---

[1] *Ann. Soc. Ent. de France*, 1850, p. 260.
[2] *Illustr. Nat. Hist.*, iii. pl. xxvii. ff. 5, 6 (1782).
[3] I think that *Charaxes Calliclea*, H. G. Smith (Ann. and Mag. Nat. Hist., Feb. 1889, p. 130), recorded from Mombasa, East Africa, is identical with *C. Azota*,—judging from the published description only.

VOL. II.

FAMILY LYCÆNIDÆ.
GENUS LYCÆNA.

385. (15A.) Lycæna Pephredo, *sp. nov.*

Exp. al., (♂) 1 in. 3½–5 lin.; (♀) 1 in. 5 lin.

♂ *Dark greyish-brown, rather glossy; a very slender fuscous hind-marginal edging line; cilia very narrowly dark-brown next to wing, but beyond this conspicuously white; hind-wing tailless.* UNDER SIDE.—*Rather pale brownish-grey, with a slight hoary cast (faintly tinged with bluish in hind-wing) over basal areas;* ordinary markings arranged as in *Niobe,* Trim., but mostly rather blurred and faint—*especially discal series of spots, which in fore-wing is sometimes almost, or even entirely, obsolete;* a white transverse stripe as in *Niobe* (but much better developed, and *in hind-wing broad, sinuate, and conspicuous*), externally bounding discal series of spots; hind-margins narrowly whitish, with a very fine bounding brownish line interrupted on internervular folds; cilia white throughout, except that on the half next to wing there are a nervular series of very thin brownish marks. *Fore-wing:* terminal disco-cellular lunule constant, tolerably well marked; discal series of spots indistinct, always more or less obsolete inferiorly,—in one example reduced to three, in another to two, of the superior spots,—and in one case wanting wholly; the adjoining white stripe narrow, submacular, obsolescent about first median nervule. *Hind-wing:* ordinary three subbasal ocelli wanting in two examples, and obsolescent in a third; ordinary ocellus at costal commencement of discal series well developed in four examples, but obsolescent in the fifth,—terminal ocellus of series pretty well defined in three examples,—remaining spots of series dark-brownish or brownish-grey, without distinct whitish edging inwardly, and in two examples faintly marked; *white stripe outwardly bounding discal series very conspicuous, sinuated, continuous, widening inferiorly,* immediately succeeded by a series of brown lunules, most pronounced in the concavity of its lower curve; between first and second median nervules, near hind-margin, a sublinear black spot densely scaled with greenish-silvery; terminal disco-cellular lunule thin and dull, but rather conspicuous from its broad white edging.

♀ *Like* ♂, *except that in hind-wing cilia are linear pale-brownish nervular interruptions.* UNDER SIDE.—As in ♂, but nearly all the markings more distinct, especially the brown submarginal macular series.

This *Lycæna* on the upper side most resembles *L. ignota*, Trim., but is much darker, and presents no trace of the ordinary hind-marginal spot near the anal angle of the hind-wings; moreover, the pure white of its cilia makes a conspicuous distinction. This last feature gives it something of the aspect of *L. Methymna*, Trim., but the broad nervular interruptions of brown, so constant in the latter variable species, are wanting. On the under side *Pephredo* differs altogether from *Ignota*, as well in its ashy-grey ground-tint as in the remarkable development of its discal white band and the irregular obsolescence or failure of many of its ordinary markings. It looks, indeed, like a "sport" of *Niobe*, Trim., and as such I should probably have regarded any solitary example, until further material had demonstrated that the upper side was the same in both sexes, and dark greyish-brown instead of violaceous in the ♂, and varying from violaceous to reddish-brown in the ♀.

I owe the knowledge of this addition to the South-African butterfly fauna to Mr. C. W. Morrison, who, in October and December 1888, most kindly sent me from Estcourt, Natal, the five ♂ s and one ♀ of which I have given the above description. Mr. Morrison informs me that he took about a dozen of this species, and that it had a more sustained flight and was more wary than *L. Methymna*, which occurred in the same locality.

Locality of *Lycæna Pephredo*.

I. South Africa.
 E. Natal.
  b. Upper Districts.—Estcourt (*C. W. Morrison*).

GENUS LYCÆNESTHES.

386. (7.) **Lycænesthes Mahota**, H. G. Smith.

Lycænesthes Mahota, H. G. Smith, Ann. and Mag. Nat. Hist., 5th Ser., xix. p. 65 (1887).

"*Upper side.*—Both wings orange-brown. Anterior wings with the base, costal margin, the upper part of the cell, and the exterior margin broadly dark-brown. Posterior wings with the base, the costal, exterior, and inner margins, two spots near the anal angle, and an interrupted submarginal line dark-brown.

"*Under side.*—Both wings greyish-brown, lighter in the middle, with the orange colour showing through, crossed with several bands of white and two white submarginal lines. Posterior wings with two spots near the anal angle, both irrorated with silver, the spot farthest from the angle bordered on three sides with orange.

"Expanse $1\frac{1}{8}$ inch."

Mr. Smith kindly showed me his types of this species, and I ascertained that they agreed with a ♂ from Old Calabar belonging to the South-African Museum. The pale-orange field is much more restricted in the hind-wings than in the fore-wings, lying indeed outside the discoidal cell, and not rising above second subcostal, or extending much below first median nervule. The under side bears a singular likeness to that of *Lycæna Telicanus*, but is still closer to that of a West-African congener, *Lycænesthes Lychnides*, Hewits. In the latter species the ♂ is wholly dark-brown on the upper side, but the ♀ has a small orange mark near the inner margin of the fore-wings. Two allied Gaboon *Lycænesthes*, named by Hewitson respectively *L. Lusones* and *L. Leptines*, have similar but darker under sides, and very restricted orange markings on the upper side; they are also considerably smaller than *L. Mahota*.

Localities of *Lycænesthes Mahota*.

I. South Africa.
II. Delagoa Bay.—Lourenço Marques (*Mrs. Monteiro*).

II. Other African Regions.
B. North Tropical.
a. Western Coast.—Old Calabar (the late *D. G. Rutherford*).

GENUS DEUDORIX.

387. (4A.) Deudorix Dinomenes, H. G. Smith.

♂ ♀ *Deudorix Dinomenes*, H. G. Smith, Ann. and Mag. Nat. Hist., 5th Ser., xix. p. 65 (1887).

"♂ *Upper side.*—Copper-red, paler than in *Dinochares*" (= *Licinia*, vol. ii. p. 111, No. 176 *supra*) "and more glossy. Anterior wings with a brown apex.

"*Under side.*—Anterior wings darker and redder than in *Dinochares*, the lower portion down to the inner margin orange. Posterior wings with three basal spots larger than in *Dinochares*, and the lines forming the bands on both wings broader and redder; the outer portion of the posterior wings from beyond the middle irrorated with white.

"♀ *Upper side.*—Dull blue, more grey than in *Dinochares*, in other respects resembling it; but on the posterior wings is a marginal black spot between the caudal spot and the lobe.

"*Under side.*—The spots and lines redder and broader than in *Dinochares*.

"Expanse 1¼ inch."

I saw Mr. Smith's types of this *Dendorix* in 1886, and having been disappointed in obtaining any specimens except one much broken ♂ from the Northern Transvaal, I give his description of both sexes. The ♂ is readily distinguished from the other orange-red species of the genus in South Africa by the much greater extent of red in the fore-wings, leaving, in fact, only a tolerably broad blackish apex, and by the more shining and golden tint of that colour.

The Transvaal ♂ received by me presents a variation in the direction of *Licinia*, Mabille, having a narrow costal and linear hind-marginal blackish edging in the fore-wings.

This butterfly was one of Mrs. Monteiro's numerous discoveries at Delagoa Bay; she wrote to me of its capture in 1886.

Localities of *Deudorix Dinomenes*.

I. South Africa.
    H. Delagoa Bay.—Lourenço Marques (*Mrs. Monteiro*).
    K. Transvaal.—Marico River (*F. C. Selous*).

APPENDIX II.

ADDITIONS AND CORRECTIONS.

VOL. I.

PREFACE, p. vii.—Under "II. Other African Regions. A. South Tropical," was accidentally omitted, after "*a*1. Western Interior,"—

b. Eastern Coast. *bb*. Islands. *b*1. Eastern Interior.

RHOPALOCERA, p. 27, line 4.[1]

Besides *Parnassius*, three other genera of *Papilioninæ*, viz., the East-Siberian *Luehdorfia*, the Australian *Euryeus*, and the South-American *Euryades*, present the horny abdominal pouch in the ♀. As mentioned above (vol. iii. p. 192, *note*), there is good evidence to show that, in the case of *Parnassius*, this appendage is a temporary one, resulting from the hardening of a fluid substance yielded by the ♂ during coition.

RHOPALOCERA, p. 30, note 2.

Dr. A. Speyer (*Stett. Ent. Zeit.*, 1878, p. 168) mentions the peculiar attitude of the Hesperide *Thanaos Tages* when quite at rest, and states that Zeller had long before (*Isis*, 1847, p. 288) noticed it both in *Tages* and in newly-emerged *H.* (*Pyrgus*) *Malvarum*, Ochs.

RHOPALOCERA, p. 39, line 1.

The Abyssinian representative of *Papilio Merope* is *P. Antinorii*, Oberthür (*Ann. Mus. Stor. Nat. Genova*, xv. p. 131, pl. ix. f. 4 ♀).

I have recently received from my kind friend and correspondent, M. Oberthür, a pair of another most interesting and distinct species of the *Merope* group, viz., *P. Humbloti*, Oberth., discovered in the island of Grand Comoro. In this species the sexes resemble each other even more closely than in *Meriones*, Feld., of Madagascar, the ♂ having along the costa of the fore-wings a moderately wide black bar in the same position as the broader one of the ♀; while the hind-wings in both sexes (in complete distinction from all other forms of the group) present a broad, unbroken, inwardly deeply dentated, black *hind-marginal border* (*not* a discal band). The tail is short and black, and in the ♀ has a narrow yellowish edge quite at the tip.

---

[1] This correction applies also to my remark on the subject under the Sub-Family *Acræinæ*, p. 129, line 30.

RHOPALOCERA, p. 40, note 1.

Terias Hecabe should be withdrawn from the list of species extending beyond the African Region, as I think the South-African representative distinct, and have named it *T. Butleri.* (See vol. iii. p. 23.) On the other hand, *Lycæna Telicanus* is to be added to this list, so that the number, thirteen, remains unaltered.

Family NYMPHALIDÆ.

Sub-Family DANAINÆ.

Genus DANAIS, p. 51, line 14.

I overlooked a third African species, viz., *D. formosa*, Godman (*Proc. Zool. Soc. Lond.*, 1880, p. 183, pl. xix. f. 1), brought from the mainland opposite Zanzibar by Mr. J. T. Last. It is related to the variety of *D. Limniace*, mentioned in the text, but has the singular feature of a large rufous-ochreous basal patch occupying nearly half the area of the fore-wings, while in the hind-wings the basal third is whitish crossed by strongly black-clouded nervures.

It is most interesting to find that this remarkable *Danais*, which combines to some extent the coloration of the *Limniace* and *Chrysippus* groups, is unquestionably mimicked (as M. Oberthür has pointed out in *Ann. Soc. Ent. de France*, 1886, p. cxiv.) by the larger *Papilio Rex*, Oberth. (*loc. cit.*, and *Études d'Ent.*, liv. 12, pl. i. f. 2, 1888), which inhabits the same part of East Africa, and is a very isolated form, recalling both *P. Antimachus* and *P. Antenor*, but not nearly related to either.

Danais Chrysippus, p. 54, line 10 from bottom of page.

Mr. Meldola (*Ann. and Mag. Nat. Hist.*, 5th Ser., i. p. 158, 1878) notes the significant fact that in Australia, where *D. Chrysippus* does not occur, the ♀ of *Argynnis Niphe* (var. *inconstans*) is like the ♂.

Amauris Echeria, p. 59, line 21.

Plötz (*Stett. Ent. Zeit.*, xli. p. 189, 1880) notes five examples taken at Victoria, on the mainland, just opposite Fernando Po, by Dr. Buchholz, but does not state whether they differed from typical *Echeria*.

Additional locality :—Transvaal : Barberton (*J. P. Cloete*).

The single specimen (♂) received from Mr. Cloete is of the typical yellow-spotted form.

Sub-Family SATYRINÆ.

Ypthima Asterope, p. 67.

Additional localities in Western Districts of Cape Colony :—Prieska, Orange River, Victoria West (*F. Purcell*), and Spectakel, Namaqualand (*L. Péringuey*).

Additional localities on Western Coast of North Tropical Africa :—" Cameroon Mountains : Bonjongo," Niger : " Abo," and Accra : " Aburi (*Buchholz*)." —Plötz.

ADDITIONS AND CORRECTIONS.

Cænyra Hebe, p. 69.

Fig. of ♂ (typical), Staud., Exot. Schmett., pl. 82 (1887).

Genus PHYSCÆNEURA, p. 71, line 25.

A singular intermediate form between *P. Panda* and *P. Leda* is described and figured by Mr. Godman (*Proc. Zool. Soc. Lond.*, 1880, p. 183, pl. xix. ff. 2, 3) as *P. Pione*, from a specimen obtained by Mr. Last on the Gnuru Hills, opposite Zanzibar. In this species the white of the upper side is less developed than in *Leda*, leaving a brown inner-marginal border from base to beyond middle in the fore-wings, and a costal narrow one in the hind-wings; the ocelli, too, are mostly visible. On the under side it is much nearer *Panda*, being everywhere striated except for a small discal space in each wing immediately before the ocelli. Only a single example occurred in Mr. Last's collection.

Physcæneura Panda, p. 71.

Fig. *Maniola Panda*, Staud., Exot. Schmett., pl. 81 (1887).

Pseudonympha sp., near *P. Natalii*, Boisd., p. 82, *note*.

Judging from the descriptions given, I think Mr. Selous' Tropical South-African species is identical with the *Yphthima Bera* of Hewitson (*Ent. M. Mag.*, 1877, p. 107), from Lake Nyassa, from which *Neocænyra duplex*, Butler (*Proc. Zool. Soc. Lond.*, 1885, p. 758), a native of Somaliland, does not appear to be separable. Mr. A. W. Eriksson has lately (1888) sent me fourteen specimens taken on the Okavango River.

Pseudonympha Hippia, p. 82.

Four more examples of this rare butterfly have come under my notice; one was captured by Mr. H. L. L. Feltham on the summit of Table Mountain on the 15th January 1888, and three were taken by Mr. R. M. Lightfoot on the same mountain, but at the lower elevation of about 2300 feet, on the 1st February 1889.

Mycalesis Safitza, p. 105, and *Mycalesis perspicua*, p. 107.

There is considerable ground for believing that the variety occurring in both these species in which the under-side ocelli are very greatly reduced or almost obsolete is the winter or dry-season form. My attention was directed to this view by Mr. L. de Nicéville, who sent me his interesting paper on the analogous cases observed by him among Indian *Satyridæ* of the genera *Mycalesis*, *Ypthima*, and *Melanitis* (*Journ. Asiat. Soc. Bengal*, lv. p. 229, 1886). In *Safitza* and *Perspicua*, all the specimens which bear dates of capture by myself or others in Eastern South Africa have been carefully referred to, and I find that all but one Zululand *Safitza*, ticketed March 1887 (which has small ocelli), and a Natal one ticketed May 1884 (which has them moderately developed), bear out M. de Nicéville's Indian experience that the summer or wet-season specimens have the under-side ocelli fully developed, while the winter or dry-season specimens have them either greatly reduced or obsolete. It is worth noting that at Knysna, on the south coast of Cape Colony, where the year is not sharply divisible into a wet summer and a dry winter season, but where the rainfall is more irregularly distributed, the long series of *Safitza* which I captured consisted almost entirely of specimens with either obsolete or very small ocelli.

Additional locality of *M. Safitza* in Eastern Districts of Cape Colony:— Tharfield, Kleinemond River (*Miss M. L. Bowker*).

Additional locality of *M. perspicua* in Transvaal:—Eureka, near Barberton (*C. F. Palmer*).

Melanitis diversa, p. 116.

Fig. of ♀, *Gnophodes Parmeno*, Staud., Exot. Schmett. pl. 78 (1886).

Lethe Indosa, p. 121.

Fig. of ♂, *Lethe dendrophilus*, var. *albo-maculatus*, Staud., Exot. Schmett., pl. 78 (1886).

Additional localities :—Coast Districts of Natal: Umzinto (*J. H. Bowker*). Zululand: Etshowe (*A. M. Goodrich* and *T. Vachell*).

Meneris Tulbaghia, p. 125.

Fig. of ♀, Staud., Exot. Schmett., pl. 57 (1885).

Larva and Pupa, p. 127.

I found a full-grown larva of this species at Rondebosch, near Cape Town, on the 19th November 1885, and append the following description of it, viz.:—

Ochre-yellow, with a broad conspicuous median dorsal blackish stripe, narrowing toward tail; on each side a supra-spiracular waved rather indistinct waved dusky-grey stripe; spiracles ringed with blackish; all the legs and the under surface of a very much paler and duller ochre-yellow. Head dark-red, set sparsely but generally with short stiff black bristles; body generally (including legs and two short acute hindward-pointing projections at tail) set sparsely with short whitish bristles,—those on the body planted in regular successive transverse lines, which are closer together on the hinder part of each segment.

Rather broad and flattened dorsally, tapering gradually toward the tail from the tenth segment; head globose,—the next adjoining segment somewhat constricted. Length, 2¼ inches.[1]

This exceedingly sluggish larva was resting near the top of a wooden fence (on which I had previously discovered two pupæ); it seemed about to pupate, but did not do so until the 25th November. The butterfly (a ♀) emerged on 26th December.

The fence in question divided a public road from a piece of ground bare of vegetation in the immediate neighbourhood of the fence, except for a bank recently planted with the "Kweek" grass (*Stenotaphrum glabrum*),—the plant mentioned in the text as conjectured to be a food-plant of *Tulbaghia*. Long and careful search on this bank, however, failed to produce another larva.

The following description of the pupa is made from three specimens, viz., the two found suspended on the fence on 14th and 19th November respectively, and that of which the larva pupated on 25th November.

Pale sandy-yellowish, with a generally-distributed pinkish-white bloom; semi-transparent; wing-cases very finely and indistinctly striolated with short grey lines. Numerous small spots and dots of black, of which the following are the principal, viz., a median longitudinal dorsal abdominal series, of which

---

[1] It remains to be seen whether the larva varies irregularly or sexually, or perhaps locally, as regards green or ochre-yellow colouring. Mrs. Barber mentions that her pale-green larva was much yellower in a younger stage, and certainly all the young ones hatched from the eggs laid by the Cape Town ♀ mentioned in the text were sandy-yellowish.

the two spots at base of abdomen are much larger than the others; also another abdominal series on each side of back, with a larger suffused spot at base; two spots on each blunted side of thorax; three across back of thorax in front; two at extremity of head, and one on lower edge of each eye-case; three, at unequal distances, along leg-cases; on each wing-case, near frontal margin, one at base, and two about middle, and also a straight transverse series of six minute spots beyond middle; on each side of abdomen four series of small or minute spots; at base of abdomen beneath, immediately beyond end of wing-cases, a median elongate roughened double blackish streak, apparently marking the case of the termination of the maxillæ. Anal foot-stalk black, flattened beneath and hollowed, sending off two ridges on under surface of end of abdomen. Length, 1 in. 4 lin.; greatest depth, from hollow of back to meeting-line of wing-cases in front of breast, 7½ lin.; greatest width, 5 lin.

Attached to a small quantity of strong reddish-brown silk, in one case near the top of the fence, and in the other near the bottom close to the ground. The upper one was in the depression or channel formed by the over-lapping of one board over the other; it was quite unconcealed, and indeed conspicuous on the black of the fence. The lower one, on the contrary (a thicker and rounder individual, with all the black markings stronger), was not only partly sheltered by a grass (*Briza major*), but also in close approximation to some spider's web and the bud-sheath of a *Pinus;* in this case (as in the one at Seapoint recorded in the text) the pale colour and black specklings of the pupa rendered it decidedly inconspicuous.

The smaller pupa from the top of the fence produced a ♂ *Tulbaghia* on the 8th December; the larger one died, but I conclude, from its size and agreement in stronger black markings with the pupa resulting from the larva above described, that it was that of a ♀.

Examination of the living larva confirmed my judgment that *Meneris* is properly placed with the Sub-Family *Satyrinæ*.

Sub-Family ACRÆINÆ.

Acræa Rabbaiæ, p. 133.

Additional locality :—Zululand : Etshowe (*A. M. Goodrich*).

A single ♂ in the collection of Captain Goodrich was taken in this locality in February 1887.

Acræa Neobule, p. 137.

Additional locality on West Coast of North Tropical Africa :—" Niger : Abo (*Buchholz*)."—Plötz.

Acræa Violarum, p. 141.

Additional locality in Transvaal :—Barberton (*C. F. Palmer*).

Acræa Nohara, p. 142.

Additional locality in Transvaal :—Barberton (*C. F. Palmer*).

Acræa Petræa, p. 144.

Fig. of ♂, Staud., Exot. Schmett., pl. 33 (1885).

Acræa Aglaonice, p. 151.

Additional locality :—Zululand : Etshowe (*A. M. Goodrich*).
The single ♀ in Captain Goodrich's collection was taken in March 1887. It differs from those described in the text by the possession of a good-sized space of white in the lower central area of the hind-wings, extending from the basal to the hind-marginal blackish.

Acræa Anemosa, p. 157.

Fig. of ♂, Staud., Exot. Schmett., pl. 33 (1885).

Acræa Barberi, p. 162.

Additional locality in Cape Colony :—Prieska, Orange River, Victoria West District (*F. Purcell*).
The single ♂, collected in November 1887 by Mr. Purcell, agrees with the variation (recorded in the text) from Griqualand West.

Acræa Encedon, p. 163.

Fig. of Var. A. (*Lycia*, Fab.), *Acræa Braunei*, Staud., Exot. Schmett., pl. 33.
A remarkable *Aberration* of this Variety, taken at Pinetown, Natal, by Major D'Aguilar, in March 1886, has been sent to me by Colonel Bowker. In this example (a ♂) the fore-wings are obscured with sooty black, altogether obliterating subapical bar, and leaving only a paler dusky space along inner margin and in middle of wing; black spots enlarged, lengthened, diffused. The hind-wings have a broad dusky-black border, pierced deeply by the much-elongated, outwardly-acuminate, discal spots, which unite with the internervular dusky rays; the basal and terminal disco-cellular spots are also much enlarged and elongated. On the under side similar peculiarities are observable, but in both fore and hind wings the hind-marginal border is yellowish, and in the hind-wings the elongated discal spots are more conspicuous on the clear ground.
A less remarkable "sport" of the typical or rufous form of this species is a ♂ taken at Etshowe, Zululand, by Mr. T. Vachell, in which the discal spots of the fore-wings are greatly elongated towards base.
Additional locality in Transvaal :—Barberton (*C. F. Palmer*).
Additional localities on the Western Coast of North Tropical Africa :— "Cameroons and Abo (*Buchholz*)."—Plötz.

Acræa Rahira, p. 166.

Additional localities in Cape Colony :—Western Districts: Stellenbosch (*L. Péringuey*). Eastern Districts : Tharfield, Kleinemond River (*Miss M. L. Bowker*).

Acræa Anacreon, p. 168.

In March 1888 I received from Mr. J. M. Hutchinson a number of specimens of the pupa of this species (from larvæ collected at Estcourt, Natal), four of which were still alive and yielded the perfect insect. This pupa, though smaller and more slender, is not at all unlike that of *A. Horta*; the ground being white or creamy, with the outlines of wings and neuration strongly marked with black, and the abdomen having on each side a sub-dorsal and an inferior-lateral continuous chain-like series of orange-yellow spots in black rings. On back, two longitudinal black streaks from head, meeting and thickening on middle of thorax, and thence narrowing to a common terminal point on first abdominal segment. Eye-cases black superiorly; black on wing-cases most strongly developed near apex, along hind-margin, and along line of median nervure and its first nervule. Length, 7–8 lines.

Acræa Buxtoni, p. 170.

The Zanguebar *Acræa* subsequently described by Hewitson under this name (*Ent. M. Mag.*, xiv. p. 155, 1877) has nothing to do with *A. Buxtoni*, Butler, but is, as Mr. Distant has pointed out (*Proc. Zool. Soc. Lond.*, 1880, pp. 184–185), a close ally of *A. Horta*, and apparently identical with *A. insignis*, Dist., *loc. cit.*, pl. xix. f. 6.

Acræa Cabira, p. 173.

Fig. of ♂, Staud., Exot. Schmett., pl. 33 (1885).

Planema Esebria, p. 177.

Fig. of Var. A. (♂), *Acræa Esebria*, Staud., Exot. Schmett., pl. 33 (1885).
Additional locality on Western Coast of North Tropical Africa:—"Camaroons: Victoria (*Buchholz*)."—Plötz.
A ♂ of Var. A., taken by Captain Goodrich at Etshowe, Zululand, has the yellow markings much reduced, especially in the fore-wings, where the sub-apical bar is almost obsolete.

Planema Aganice, p. 180.

Additional locality:—Transvaal: Barberton (*C. F. Palmer*—VAR.).
The single ♂ received from Mr. Palmer is noted by him as captured on "1st April 1888, in gully on stony ridge, Eureka, near Barberton." It expands to the largest size of the ♂, viz., 2 in. 11 lin., and is a very marked variation from the typical form, presenting the following differences, viz., transverse pale-yellowish band much broader in both fore and hind wings, especially in the latter, where it is even broader proportionally than in the ordinary ♀; in the fore-wing the lowest of the component spots of the band is on the upper side as large as in the ♀, and on the under side only separated by first median nervule from next (largest) spot of the band.
Dewitz (*K. Leop.-Car.-Deutsch. Akad. Nat.*, xli. p. 190, 1879) states that the single specimen (sex not given) from Angola referred by him to *Aganice* had the band of the fore-wings much broader than in Hewitson's figure, and also than in a Camaroons specimen in the Berlin Museum.

Pardopsis punctatissima, p. 183.

Fig. *Acræa punctatissima*, Staud., Exot. Schmett., pl. 33 (1885).
Additional locality:—Zululand: Etshowe (*A. M. Goodrich* and *T. Vachell*).
Of the thirteen Zululand specimens collected by Captain Goodrich and Mr. Vachell, all but one had the spots smaller than usual, and in three ♂s and two ♀s these markings were very small.
Additional locality:—North Tropical Africa, Eastern Interior: Abyssinia; "Shoa (*Antinori*)."—Oberthür.

Sub-Family NYMPHALINÆ.

Atella Columbina, p. 193.

Fig. *Atella Phalanta,* Staud., Exot. Schmett., pl. 36 (1885).

Lachnoptera Ayresii, p. 196.

Additional locality :—Zululand : Etshowe (*A. M. Goodrich* and *T. Vachell*).

Eurema Hippomene, p. 204.

An Abyssinian ♂ kindly sent to me by M. Ch. Oberthür presents the peculiarity of a ferruginous-red suffusion in basal half of discoidal cell, and also a tinge of the same colour over the middle part of the pale-yellow band on the under side.

Eurema Schœneia, p. 207.

Additional locality :—North Tropical Africa : Eastern Interior ; Abyssinia ; "Shoa (*Antinori*)."—Oberthür.
Fig. of ♂ and ♀, *Vanessa Schœneia,* Oberth., Ann. Mus. Civ. Genova, xv. pl. ix. ff. 1, 2 (1883).
A ♂ example from Abyssinia, kindly sent to me by M. Oberthür, confirms that author's identification of *Schœneia* in the late Marquis Antinori's collection.

Junonia Cebrene, p. 210.

Fig. of ♂, *Junonia Œnone,* Staud., Exot. Schmett., pl. 37 (1885).
Additional locality in Cape Colony :—Western Districts : Van Wyk's Vley, Carnarvon District (*E. G. Alston*).

Precis Sophia, p. 221.

Fig. of ♂, Staud., Exot. Schmett., pl. 37 (1885).

Precis Cloantha, p. 222.

Fig. of ♂, Staud., *op. cit.,* pl. 38 (1885).
Additional locality in Transvaal :—Barberton (*C. F. Palmer*).
Additional locality on Western Coast of North Tropical Africa :—"Accra (*Buchholz*)."—Plötz.

Precis Octavia, p. 229.

Fig. of ♂, Staud., Exot. Schmett., pl. 38 (1885).
The intimate relation of this butterfly to *P. Sesamus,* Trim., referred to in the text, is further borne out by a ♂ example (very like the larger of the two Transvaal specimens mentioned on p. 231) received from Colonel Bowker. He notes it as captured close to Malvern, ten miles from D'Urban, Natal, on 11th September 1887 ; it was in company with two *P. Sesamus.* In this specimen the inner half of the red discal field is in both fore and hind wings obscured by violaceous-grey irrorations.

A second smaller ♂, almost exactly similar in colouring, was obtained in the same locality by Mr. C. Barker, and was received by me in January 1888.
Additional localities on Western Coast of North Tropical Africa:—Cameroons: "Mungo" and Niger; "Abo (*Buchholz*)."—Plötz.

Precis Sesamus, p. 231.

Precis Sesamus, Trim., Trans. Ent. Soc. Lond., 1883, p. 347.
Fig. of ♂, *Precis Amestris*, var. *Caffraria*, Staud., Exot. Schmett., pl. 38 (1885).

PUPA.—Prominences on head rather short, wide apart, minutely and unequally bifid at tip. Thorax dorsally elevated into a prominence with a somewhat blunted point, but laterally acutely so at base and also before middle of inner margin of wing-cases; a pair of acute tubercles on back of each thoracic segment, besides some minute intermediate ones. Abdomen with three dorsal series of similar tubercles,—the tubercles of the series on each side much larger and longer than those of the middle series; two lateral series of very much smaller tubercles.

Abdomen beyond fourth segment and lower median area of thorax reddish-brown; all dorsal area as far as end of fourth segment brassy-gilded, with numerous small brown spots, and with more burnished brassy-gilded spots on tubercles and their bases; a duller gilded dorsal stripe, narrowing to tip of abdomen; superior half of wing-cases rather dully gilded, and with a short brown streak (composed of three subquadrate spots) about middle of inner margin; head dully gilded beneath. Length 1 inch.

This description was made from two living pupæ sent to me from Estcourt, Natal, by Mr. J. M. Hutchinson in March 1888. They produced pure *Sesamus*, ♀ and ♂ respectively, on 14th and 15th March. I can detect no difference between the empty skins of these specimens and that of an *Octavia* pupa described on pp. 219 and 230.

An interesting (apparently ♀) example, rather nearer to *Sesamus* than any of the three specimens recorded on p. 233, was sent by Mr. C. Barker from Malvern, Natal, with the individual near *Octavia* mentioned above. This example retains in the fore-wings much of the basal blue irroration, and has barely a trace of red in the striation of the discoidal cell, while the upper part of the median fascia is pure-blue, but its lower part (inwardly bordering the macular red band) is dull reddish-violaceous; in the hind-wings, however, the red invades the violaceous or lower half of median band, and there is also a red disco-cellular stria. In both wings, however, the submarginal blue lunules are reduced as in *Octavia*. The under side is quite like that of *Sesamus*, except that the discal red is much more developed in the fore-wings, and also faintly indicated in the hind-wings.

Oberthür (*Ann. Mus. Civ. St. Nat. Genova*, xviii. pp. 721–722, 1883) records several examples intermediate between *Amestris*, (Drury), and *Octavia*, taken by the late Marquis Antinori in Shoa, Abyssinia, and is of opinion that these two forms can no longer be regarded as separate species.

Dewitz (*Berl. Ent. Zeitschr.*, xxix. p. 142, pl. 2, 1885) gives (fig. 5) true *Sesamus*; (figs. 2, 3, and 4) variations indicating more or less approach to *Octavia*; and (fig. 1) a specimen very much resembling the variation near *Octavia* depicted in my pl. 4, f. 4. These intermediate examples appear to be in the Berlin Museum; the only locality noted is that of fig. 3, viz., Liberia.

The evidence in this most interesting case certainly seems to indicate that *Octavia*, *Amestris*, and *Sesamus* are as yet incompletely segregated forms, and that fertile inter-crossing is not unfrequent among them.[1]

---

[1] *Precis Actia*, Distant (*Proc. Zool. Soc. Lond.*, 1880, p. 185, pl. xix. f. 7), seems to a large extent to be intermediate between *P. Sesamus* and *P. Archesia*. It has the outline

Precis Pelasgis, p. 236.

I received in May 1888 from Mr. A. D. Millar the cast skins of the larva and pupa of this species. Mr. Millar found about a dozen of the larvæ feeding on cultivated plants of *Coleus* at D'Urban, Natal, and reared two *Pelasgis*, which made their appearance on the 4th May.

The larva is described by him as entirely dark-brown. As far as can be judged from the skin cast before pupation, it does not differ much from that of *Octavia*, except that the head is black, instead of yellowish-brown with black marks, and the body-spines are shorter, more slender, and mostly yellowish brown instead of black.

The skin of the pupa strongly resembles those of *P. Sesamus* and *P. Octavia*, but is of a darker and browner tint, with the surface not so much granulated, and the tubercular points smaller. Mr. Millar does not state whether this pupa has any gilded ornamentation.

Precis Elgiva, p. 240.

Fig. of ♂, Staud., Exot. Schmett., pl. 37 (1885).

Precis Tugela, p. 241.

Additional locality:—Zululand: Etshowe (*A. M. Goodrich* and *T. Vachell*).

Of three Zululand examples, two were captured by Captain Goodrich respectively in November 1886 and January 1887. All three were ♂ s and had the discal band tinged with rufous, chiefly in its outer portion. The example captured in January was of a more uniform pale ochre-yellow on the under side than in the ♂ described in the text, most of the brown markings being much fainter, except the common dark-brown streak.

Salamis Anacardii, p. 244.

Fig. of ♂, Staud., Exot. Schmett., pl. 38 (1885).
Additional locality:—Zululand: Etshowe (*A. M. Goodrich* and *T. Vachell*).

Genus CRENIS, p. 248.

From notes and pencil outlines by Mr. W. D. Gooch, it is evident that the larva of this genus (it is uncertain whether *C. Natalensis* or *C. Boisduvali* was the species reared from Mr. Gooch's specimens) is very similar to that of *Precis*. The branched spines or horns borne on the head are, however, shown as much shorter, and the dorsal branched spines of the third, fourth, fifth, eleventh, and twelfth segments are much larger than the others. The second (first thoracic) segment is small and has very short spines; the thirteenth (anal) segment has two subdorsal and two lateral, all four simple, unbranched.

Crenis Boisduvali, p. 252.

Additional locality:—Zululand: Etshowe (*T. Vachell*).

(except that the hind-margins are almost without dentation) and apparently under-side colouring (though more uniform) of *Archesia*, and the upper-side striping is represented as more even and continuous than in that species, but the upper-side colouring is quite of the character of that exhibited by *P. Sesamus* and *P. Amestris*. Hab.—"Masassi, East Africa."

Crenis Morantii, p. 253.

Additional locality :—Zululand : Etshowe (*A. M. Goodrich*).

Crenis Rosa, Hewitson, p. 255.

♂ *Crenis Pechuelii*, Dewitz, Nov. Act. Leop.-Carol.-Deutsch. Akad. Nat., xli. p. 195, tab. xxvi. f. 1 (1879).
♂ *Crenis Pechuelii*, Staud., Exot. Schmett., i. p. 107, pl. 40 (1885). [Female noted in text.]

I overlooked the fact that *C. Pechuelii* is unquestionably the ♂ of Hewitson's earlier *C. Rosa*, owing partly to the dull colouring and small size of Dewitz's figure, which gives but little idea of the great beauty of the insect. Staudinger's figure is not much better, but renders more faithfully the warm, rich ochre-yellow of the under side of the fore-wings. Dewitz's types (six ♂ s) were taken by Pogge in Angola ("Lat. 10° S., long. 17-22° E."), but Staudinger mentions having received a pair from Vivi, on the Congo, and others from the Quango, a southern tributary of the Congo.

I have lately (1888) had the pleasure of receiving eleven fine ♂ s from Otjimbora, a locality within the Portuguese boundary of the interior of the province of Mossamedes, just to the north-east of Ovampoland, and situated apparently in about S. lat. 17° 8′, and E. long. 17° 27′. These specimens were taken by the well-known hunter and naturalist Mr. A. W. Eriksson, at dates between the 20th November and 3d December 1887.[1] I give from these examples the characters distinguishing the ♂ from the ♀.

♂ *Exp. al.*, 2 in. 5½–7½ lin. *Violaceous-blue brighter, submetallic, more or less shot with pink.* *Fore-wing :* Oblique fuscous bars between extremity of discoidal cell and apex wanting ; near apex, from fourth subcostal to lower radial nervule, a transverse row of three small, or very small, fuscous spots, of which the lowest is the largest ; hind-marginal fuscous border very much narrower, almost linear in some examples, its radiation on nervules and on subcostal nervure much thinner ; a submarginal series of very small fuscous spots, extending from costa to between second and first median nervules. *Hind-wing :* black spots of discal series smaller, not encircled by paler blue. UNDER SIDE.—As in ♀, but brighter ; the greenish-white markings more glossy, and sometimes with a tinge of violaceous-blue. *Fore-wing :* no trace of inner oblique fuscous bar or of spot below third median nervule ; costal-apical widening of greenish-white edging wanting. *Hind-wing :* median and premedian greenish-white transverse bands usually coalesce, more or less completely, about median nervure and the base of its first nervule, but in four examples they are completely separated, as in ♀.

This lovely Nymphaline has much the aspect of a magnified Lycænide, its shot-blue upper side strongly resembling that of various species of *Lycæna*, while the under side in tints and hind-wing ocellate spots strongly recalls that of the European *Chrysophanus Hippothoë*.

Additional localities in South Tropical Africa :—Western Interior.—Mossamedes Province : Otjimbora (*A. W. Eriksson*). "Angola (Pogge)."—Dewitz. Congo : "Vivi (*Pechuel-Loesche*) and Quango River (*Von Mechow*)."—Staudinger.

---

[1] Four of the nearly allied and almost equally beautiful *C. Benguelæ* were taken by Mr. Eriksson at the same time and place, and two other examples at Ehanda, considerably farther northward, in September 1887. In 1873 I made a description of two ♂ s of this *Crenis* among other butterflies collected at Kabenda, about forty miles north of the mouth of the Congo, by Lieutenant Larcom, R.N.

Eurytela Hiarbas, p. 258.

Fig. of ♂, Staud., Exot. Schmett., pl. 39 (1885).
Additional locality on Western Coast of Tropical North Africa:—Cameroons: "Victoria (*Buchholz*)."—Plötz.

Eurytela Dryope, p. 261.

Mrs. Monteiro sent me the wings of this butterfly from Delagoa Bay, and informed me that she had reared it from a green larva, closely set with green branched spines and bearing two brown horns on the head, which fed on the castor-oil plant.

Additional localities on Western Coast of Tropical North Africa:—Cameroons: "Victoria," and Accra: "Aburi (*Buchholz*)."—Plötz.

Hypanis Ilithyia, p. 264.

Colonel Bowker forwarded in 1887 a pair of small specimens of the typical form taken *in copulâ* at Malvern, near D'Urban, Natal, on the 6th September, by Mr. C. Barker. The ♂ of this pair has the under side of the hind-wings moderately dark ochrey-chocolate, with reduced (but distinct) ordinary black and white markings, while the ♀ has it ochre-yellow, with creamy bars and fully developed ordinary markings.

Additional localities on Western Coast of Tropical North Africa:—Cameroons: "Mungo" and Accra: "Aburi (*Buchholz*)."—Plötz.

Neptis Agatha, p. 270.

Fig., Staud., Exot. Schmett., pl. 50 (1885).
Additional locality:—Zululand: Etshowe (*A. M. Goodrich* and *T. Vachell*).

Diadema Misippus, p. 277.

Hypolimnas Alcippoïdes, Butler (*Ann. and Mag. Nat. Hist.*, 5th Ser., xii. p. 102, 1883), from Lake Nyanza, seems clearly to be merely a ♀ *Misippus* with the whitish suffusion on the disk of the hind-wings mentioned on p. 279. It is noted that a second form of ♀ like a small *Inaria* occurs in the same locality, while the ♂ is stated not to differ from the ordinary ♂ *Misippus*.

Additional localities on Western Coast of Tropical North Africa:—Cameroons: "Victoria," and also "Accra and Aburi (*Buchholz*)."—Plötz.[1]

Pseudacræa Tarquinia, p. 289.

Fig. of ♂, Staud., Exot. Schmett., pl. 49 (1885).
Additional locality:—Zululand: Etshowe (*A. M. Goodrich* and *T. Vachell*).[2]

---

[1] Mr. L. de Nicéville informs me that Mr. W. H. Edwards has recorded *Misippus* as having occurred in Florida.

[2] A ♀ taken by Mr. Vachell is smaller than usual, and has the spots of the fore-wings (except that between first and second median nervules) almost obsolete; while in the hind-wings the ochre-yellow patch is very narrow—the basal black being broader—and marked at extremity of discoidal cell with an obliquely transverse black streak.

Pseudacræa Delagoæ, p. 291.

An apparently very near ally of this species is *Panopea protracta*, Butler (*Ent. M. Mag.*, 1874, p. 164); but it is described as very much larger, viz., *exp. al.*, 3 in. 6 lin.

(*Hab.*—Cabinda [Congo] (*Monteiro*).

Pseudacræa imitator, p. 293.

Five examples (four ♂ s, one ♀) were received from Colonel Bowker during 1887, with the note that they were captured at Northdene, near D'Urban, Natal, in the first quarter of 1884.

Pseudacræa Trimenii, p. 296.

Since my account of this fine species was published, I have examined twenty-four additional specimens, nineteen being ♂ s and five ♀ s: all were taken in the D'Urban district of Natal (except two of the ♀ s, which were captured in Zululand),—sixteen ♂ s and two ♀ s by Colonel Bowker, three ♂ s and one ♀ by Mr. A. F. Evans.

Of this series, twelve ♂ s and four ♀ s are of the normal coloration; but three of the ♂ s exhibit a limited white lower discal suffusion on the upper side of the hind-wings. The remaining seven ♂ s and one (Zululand) ♀ exhibit more or less failure of the yellow-ochreous subapical bar of the fore-wings, thus approximating to the Variety A. (*P. Colvillei*, Butler) described in the text. Five of these ♂ s, all taken by Colonel Bowker at Northdene from January to May 1884, exhibit a complete and regular gradation in this character, from a much narrower but complete yellow-ochreous bar, through an attenuated bar obsolescent and whitish superiorly, to an example where only a small yellow-ochreous trace of the lower extremity of the bar remains.

Additional locality:—Zululand: Etshowe (*A. M. Goodrich* and *T. Vachell*).

Euphædra Neophron, p. 305.

A ♂ example taken near Delagoa Bay was received by the South-African Museum from Mrs. Monteiro in 1886.

Hamanumida Dædalus, p. 309.

Additional locality in Transvaal:—Barberton (*J. P. Cloete* and *C. F. Palmer*).

Harma Alcimeda, p. 312.

Additional locality for Variety A.:—Zululand: Etshowe (*A. M. Goodrich* and *T. Vachell*).

Charaxes Zoolina, p. 318.

VARIETY A.—♂ *Fuscous borders much broader; their enclosed spots of ground-colour more or less obliterated. Fore-wing: basal fuscous greatly developed, filling discoidal cell up to terminal transverse stripe; lower basal area (in one example) tinged with dull fulvous-ochreous. Hind-wing: base narrowly fuscous; median dusky streak deepening into fuscous superiorly; narrow longitudinal brown streak of under side, between median and sub-*

median nervures, also diffusedly represented on upper side. (*Hab.*—Delagoa Bay: Lourenço Marques (*Mrs. Monteiro*).—Three examples.)

A ♂ of this Variety was taken at Etshowe, Zululand, by Mr. T. Vachell in 1886–87.

A near ally of *Zoolina* is *C. Betsimisaraka*, Lucas, a native of Madagascar.[1] Another allied form (varying in the direction of a great *reduction* of the dark markings) is *C. Kahldeni*, Homeyer and Dewitz (*Berl. Ent. Zeitschr.*, 1882, p. 381), from "Pungo Andonga, Angola (*A. von Homeyer*)." It differs in the very great attenuation of the dark markings, especially those of the hind-margin of the hind-wings, which on the upper side are almost obsolete in both sexes, according to the description and accompanying figures.

Additional locality (of *C. Zoolina*):—North Tropical Africa. Eastern Interior. Abyssinia: "Shoa (*Antinori*)."—Oberthür.

Charaxes Neanthes, p. 320.

Fig. of ♂, Staud., Exot. Schmett., pl. 58 (1885).

Besides the Malagasy *Betanimena*, Lucas[2] (erroneously given as *Betsimisaraka* of the same author on p. 321, line 21), two other close allies of *Neanthes* appear, from the description and figures, to be *C. Homeyeri*, Dewitz, and *C. Ehmckii*, Dewitz and Homeyer (*Berl. Ent. Zeitschr.*, 1882, p. 382), from "Pungo Andonga, Angola (*A. V. Homeyer*)." Both these forms have a broad whitish suffusion from the base of the wings (exceedingly wide in the hind-wings of *C. Homeyeri*).

Additional locality (of *C. Neanthes*):—North Tropical Africa. Eastern Interior. Abyssinia: "Shoa (*Antinori*)."—Oberthür.

Charaxes Varanes, p. 321.

Fig. of ♂, *Palla Varanes*, Staud., Exot. Schmett., pl. 60 (1885).

Pupa, p. 324.—In June 1887 I received from Colonel Bowker two living pupæ of this butterfly, from which I obtained a ♂ perfect insect on June 19 (pupation 18th May), and a female on June 26th. These pupæ were semi-transparent, of a uniform bright-green, with a wax-like surface; six abdominal spiracles on each side, and six caudal tubercles, black. They were very thick and swollen abdominally, but became gradually slenderer and narrower anteriorly. Head very shallowly bifid. Inferior outline, along middle line of head and thorax, almost straight; dorsal median line of thorax and inner-marginal edges of wing-cases somewhat prominently but smoothly ridged or keeled. Length, about 1 inch; greatest width and also greatest depth—across third abdominal segment—½ inch.

The band of light reddish-brown silk to which these pupæ were attached completely encircled the slender stem of *Cardiospermum halicacabum*, on which the insects were suspended. Colonel Bowker found the larvæ feeding on this widely-spread tropical climber near D'Urban, Natal, and called my attention to the general resemblance borne by the pupæ to the younger somewhat heart-shaped inflated seed-vessels of the plant, which hang similarly, though by a much longer pedicel.

Additional locality on Western Coast of Tropical North Africa:—Cameroons: "Victoria (*Buchholz*)."—Plötz.

---

[1] Figured in Grandidier's *Hist. Phys. Nat. etc. de Madag.*, xix. Lep., Atlas i., pl. 21, ff. 2, 2a (1885).

[2] Figured in Grandidier's *Hist. Phys. Nat. etc. de Madag.*, xix. Lep., Atlas i., pl. 23, ff. 5, 6, 7 (1885).

Charaxes Jahlusa, p. 325.

A ♂ example, captured at Eureka, near Barberton, Eastern Transvaal, by Mr. C. F. Palmer, on the 12th May 1888, is very near the Zambesi specimen mentioned in the text, but on the upper side the ground-colour is generally brighter, and not obscured near apex of fore-wings, and the black hind-marginal border of the hind-wings is rather broader; while on the under side the colouring is more vivid, and the silvery submarginal marks in upper part of fore-wings are almost replaced by an enlargement of the adjacent salmon-red marks.

Mrs. Barber wrote to me in 1887 that Mr. Bourchier Bowker had noticed this butterfly in remarkable numbers in a piece of ground covered with *Pappea Capensis* (one of the Family *Sapindaceæ*), which she believes to be one of the insect's food-plants.

Additional locality in Eastern Districts of Cape Colony:—Junction Drift, Commadagga, Somerset East District (*Mrs. Barber*).

Charaxes Candiope, p. 327.

The Malagasy ally mentioned in a note on p. 328 is *C. Antamboulu*, Lucas. It is figured in Grandidier's *Hist. Phys. Nat. etc. de Madag.*, vol. xix. Lep. Atlas i., pl. 23, ff. 3, 4 (1885).

Charaxes Pelias, p. 331.

Additional locality in Western Districts of Cape Colony:—Waterfall, Tulbagh District (*W. Thomas*).

The single example kindly presented to me by Mr. Wyan Thomas was taken on or about 12th February 1888, settling on the stem of a pear-tree.

Charaxes Saturnus, p. 334.

Fig. of ♂, *Charaxes Pelias*, Staud., Exot. Schmett., i. pl. 58 (1885). Additional locality:—Delagoa Bay: Lourenço Marques (*Mrs. Monteiro*).

Charaxes Brutus, p. 335.

Additional locality:—Delagoa Bay: Lourenço Marques (*Mrs. Monteiro*).

Charaxes Castor, p. 338.

♂ *Exp. al.*, 3 in. 7–8 lin.

Like ♀, but common macular band much narrower, and of a deeper tint of ochreous-yellow. *Fore-wing*: hind-marginal internervular yellowish marks and adjacent white marks in cilia much smaller, especially the latter. *Hind-wing*: only three submarginal blue lunules, lying between submedian nervure and third median nervule; both tails (but more especially that at end of third median nervule) considerably shorter. (*Hab.*—Delagoa Bay—two examples; and Barberton, Transvaal—one example.)

Compared with Cramer's figure (C.) of the upper side, these South-African specimens differ in having the common macular band considerably wider throughout, and without perceptible ferruginous edging, and also in presenting a mixture of white and greenish in the hind-marginal lunules (a touch of ochre-yellow in the two upper ones) quite as in the ♀.

Feisthamel's fine figure quoted in the text represents an apparent ♀ from the River Casamanza, a little to the south of the Gambia. The Delagoa Bay ♀ s differ from it in the same characters as those which have just been noted as distinguishing the ♂ s from Cramer's figure of a ♂ from the Guinea Coast; but the Casamanza specimen presents no trace of any ferruginous edging to the common macular band.

The ♂ specimen from Barberton has the common macular band rather narrower and of a deeper tint than in the two ♂s from Delagoa Bay, and possesses in the fore-wing a small diffuse ferruginous-ochreous spot in the upper part of the discoidal cell near its extremity. Its captor, Mr. C. F. Palmer, notes that it was taken on a teak tree, on a "kopje" on the stony ridge at Eureka, on the 24th April 1888.

Additional locality on Western Coast of North Tropical Africa:—Cameroons: " Victoria (*Buchholz*)."—Plötz.

Charaxes Achæmenes, p. 340.

Mr. C. F Palmer has forwarded two ♂s from Eureka, near Barberton; one of them is noted as captured at the same station as *C. Castor*, on the 12th May 1888.

Additional locality in Western Interior of South Tropical Africa:—Omrora (*A. W. Eriksson*).

Charaxes Ethalion, p. 342.

Fig. of ♂, *Charaxes Ephyra*, Staud., Exot. Schmett., pl. 58 (1885).
Additional locality:—Transvaal: De Kaap, near Barberton (*C. F. Palmer*).

Charaxes Phæus, p. 344.

In 1886 the South-African Museum acquired two ♀ examples taken at Delagoa Bay by Mrs. Monteiro, and I have recently (October 1888) received a worn ♂ captured near Barberton, North-East Transvaal, by Mr. J. P. Cloete.

Charaxes Xiphares, p. 346.

Additional locality:—Transvaal: Eureka, near Barberton (*C. F. Palmer*).

VOL. II.

Family ERYCINIDÆ.
Sub-Family LIBYTHÆINÆ.

Libythea Laius, p. 5.

Additional locality :—Zululand : Etshowe (*A. M. Goodrich*).
The single specimen (♂) in Captain Goodrich's collection was ticketed "February 1887."

Family LYCÆNIDÆ.

Lycæna Osiris, p. 15.

Additional localities in Zululand and Transvaal :—Etshowe (*T. Vachell*); Eureka, near Barberton (*C. F. Palmer*).
The two specimens forwarded by Mr. Palmer were noted as captured on the 10th and 15th April 1888 respectively.

Lycæna Asopus, p. 16.

Additional locality on Western Coast of North Tropical Africa :—Accra : "Aburi (*Buchholz*)."—Plötz.

Lycæna Methymna, p. 27.

Additional locality in Natal :—Estcourt (*C. W. Morrison*).

Lycæna Mahallokoæna, p. 44.

Additional locality in Transvaal :—Eureka, near Barberton (*C. F. Palmer*).

Lycæna Lysimon, p. 45.

A very small ♀ (*exp.* 9 lin.) was taken at Estcourt, Natal, by Mr. C. F. Palmer, and sent for my inspection; all its under-side markings were very indistinct.
Additional localities :—"Lower Niger (*H. A. Forbes*)."—Godman and Salvin [*Knysna*, Trim.]. Southern Arabia : "Shaik Othman, Huswah, and Lahej (*Yerbury*)."—A. G. Butler [*Knysna*, Trim.].

Lycæna lucida, p. 47.

Additional localities :—Zululand : Etshowe (*A. M. Goodrich* and *T. Vachell*). Transvaal : Eureka, near Barberton (*C. F. Palmer*).

Lycæna Gaika, p. 50.

The paired sexes were taken at Northdene, near D'Urban, Natal, by Colonel Bowker on 11th June 1883.

Additional locality :—Southern Arabia : "Aden and Haithalkim (*Yerbury*)."—A. G. Butler.

Lycæna Trochilus, p. 52.

Additional localities :—Transvaal : Eureka, near Barberton (*C. F. Palmer*). Southern Arabia : "Aden, Huswah, and Lahej (*Yerbury*)."—A. G. Butler.

Lycæna Bœtica, p. 58.

Additional locality on Western Coast of North Tropical Africa :—" Accra (*Buchholz*)."—Plötz.

Lycæna Noquasa, p. 64.

Additional locality in Upper Districts of Natal :—Ulundi, sources of Bushman's River (*J. M. Hutchinson*).

Among the examples received from Mr. Hutchinson there are two ♀s, which in the hind-wings on the upper side present below (as well as above) the ordinary hind-marginal spot traces of bluish-white annulets, and in a third ♀ these inferior markings are developed into minute, black, bluish-ringed spots. Mr. Hutchinson wrote, in February 1888, that he found this species in hundreds about the vleys at Ulundi.

Lycæna Lingeus, p. 66.

Additional localities :—Cape Colony, Eastern Districts : Tharfield, Kleinemond River (*Miss M. L. Bowker*). North Tropical Africa : Western Coast : "Cameroon Mountains ; Bonjongo (*Buchholz*)," and "Accra ; Aburi (*Buchholz*)."—Plötz : "Lower Niger (*H. A. Forbes*)."—Godman and Salvin. Eastern North Tropical Interior : Abyssinia ; "Shoa (*Antinori*)."—Oberthür.

Lycæna Telicanus, p. 69.

Additional localities :—Cape Colony, Eastern Districts : Tharfield, Kleinemond River (*Miss M. L. Bowker*). Transvaal : Eureka, near Barberton (*C. F. Palmer*). Western Coast of North Tropical Africa : "Lower Niger (*W. A. Forbes*)."—Godman and Salvin [*Pulchra*, Murray].

Lycæna Jesous, p. 72.

Additional locality :—Zululand : Etshowe (*T. Vachell*).

Lycæna Macalenga, p. 74.

Additional locality :—Transvaal : Heidelberg (*C. Barker*).

Lycæna Moriqua, p. 75.

Additional locality :—Zululand, Etshowe (*A. M. Goodrich*).

Lycæna Natalensis, p. 77.

Additional locality :—Zululand, Etshowe (*A. M. Goodrich* and *T. Vachell*).

Lycæna Calice, p. 80.

Additional locality :—Zululand : Etshowe (*T. Vachell*).

Lycæna Thespis, p. 87.

Additional locality in Eastern Districts of Cape Colony :—Tharfield, Kleinemond River (*Miss M. L. Bowker*).

Lycæna Bowkeri, p. 88.

Additional locality in Upper Districts of Natal :—Estcourt (*C. W. Morrison*).

Lycænesthes Amarah, p. 94.

Additional locality :—Zululand : Etshowe (*A. M. Goodrich* and *T. Vachell*).
A ♀ taken by Mr. Vachell has on the under side all the dark edgings of the markings stronger than usual, that of the lower spots of the discal row in the fore-wings being diffusedly extended baseward.

Lycænesthes Sylvanus, p. 98.

♀ *Pseudodipsas Sylvanus*, Staud., Exot. Schmett., pl. 94 (1887).
Additional locality on Western Coast of North Tropical Africa :—" Cameroon Mountains : Bonjongo (*Buchholz*)."—Plötz.

Deudorix Antalus, p. 107.

Additional localities :—Transvaal : Eureka, near Barberton (*C. F. Palmer*). North Tropical Africa, Western Coast : Cameroons ; " Victoria (*Buchholz*)."—Plötz.

Deudorix Diocles, p. 108.

Fig. of ♂, *Deudorix Diocles*, Staud., Exot. Schmett., i. pl. 96 (1888).

Deudorix Licinia, p. 111.

Fig. of ♂ and ♀, *Deudorix Dinochares*, H. G. Smith, Ann. and Mag. Nat. Hist., 5th Ser., xix. p. 64, n. 7.
" ♀ *Upper side*.—Dull blue, shading broadly into brown on the costal and exterior margins, paler in the middle of the anterior wings. Posterior wings with a black caudal spot, lobe grey. *Under side*.—Grey with red markings as in the ♂." (H. G. Smith, *loc. cit.*)

Capys Alphæus, p. 115.

Fig. of ♂, *Capys Alphæus*, Staud., Exot. Schmett., i. pl. 95 (1887).
Additional locality in Western Districts of Cape Colony :—Worcester District : Hex River (*L. Péringuey*).

Hypolycæna Cæculus, p. 116.

Additional locality in Eastern Interior of South Tropical Africa:—Rampungu River, "about 15° S. lat." (*F. C. Selous*).

Hypolycæna Philippus, p. 118.

Fig. of ♀, *Hypolycæna Philippus*, Staud., Exot. Schmett., i. pl. 96 (1888).
Additional locality on Western Coast of North Tropical Africa:—Accra: "Aburi (*Buchholz*)."—Plötz.

Hypolycæna Buxtoni, p. 119.

Additional locality:—Zululand: Etshowe (*A. M. Goodrich* and *T. Vachell*).

Hypolycæna Lara, p. 123.

Fig. of ♂, *Hypolycæna Lara*, Staud., Exot. Schmett., i. pl. 96 (1888).
Additional localities:—Cape Colony, Eastern Districts: Tharfield, Kleinemond River (*Miss M. L. Bowker*). Transvaal: Eureka, near Barberton (*C. F. Palmer*).

Iolaus Silas, p. 127.

Fig. of ♂, *Iolaus Silas*, Staud., Exot. Schmett., i. pl. 95 (1887).
Additional locality:—Zululand: Etshowe (*T. Vachell*).

Iolaus Trimeni, p. 129.

Additional locality in Transvaal:—Eureka, near Barberton (*C. F. Palmer*).
Two worn ♂ s of this rare species reached me from Mr. Palmer, who notes that they were taken on "teak" trees on a stony ridge on the 16th April 1888.

Iolaus Bowkeri, p. 132.

Fig. of ♂, *Iolaus Bowkeri*, Staud., Exot. Schmett., i. pl. 95 (1887).
Additional localities:—Transvaal: Sheba Range (*C. F. Palmer*). Western Interior of South Tropical Africa: Ehanda and Okavango River (*A. W. Eriksson*).

Iolaus Mimosæ, p. 135.

Additional locality:—Transvaal: Eureka, near Barberton (*C. F. Palmer*).

Iolaus Aphnæoides, p. 137.

A very fine example of this rare species has been presented to me by Mr. F. C. Selous; it was taken on 1st November 1888 at Panda-ma-Tenka, about thirty-five miles due south of the Victoria Falls of the Zambesi. I have also seen a specimen in Mr. H. Grose Smith's collection, which was captured in the Trans-Kei Kaffrarian territory by Mr. F. N. Streatfeild.

Iolaus Pallene, p. 138.

Fig. *Sithon Pallene*, Staud., Exot. Schmett., i. pl. 95 (1887).
This curious species occurs sparingly in Natal. Colonel Bowker wrote to me in January 1888 that two were taken at Stanger (near the coast, a few miles to the south of the Tugela) by the Rev. Mr. Pettman; and in October Mr. C. W. Morrison acquainted me with his having captured it near Estcourt.
Additional locality in South Tropical Africa :—Okavango River (*A. W. Eriksson*).

Myrina ficedula, p. 141.

Fig. of ♂, *Myrina Silenus*, Staud., Exot. Schmett., i. pl. 95 (1887).
Additional locality :—Zululand : Etshowe (*A. M. Goodrich* and *T. Vachell*).

Myrina dermaptera, p. 144.

LARVA.—Dull pale olivaceous-green, except seventh segment, which is ferruginous; a conspicuous white median mark on sixth segment, and a similar one on eighth segment, both apparently extending partly on the ferruginous seventh segment.
This note of the colouring of the larva is made from a drawing (giving a dorsal view only) by Mrs. Barber of a full-grown specimen just before pupation, found near D'Urban, Natal. Mrs. Barber's figure of the pupa (dorsal view) resembles that by Captain Harford, noted in the text.

Aphnæus Hutchinsonii, p. 148.

♂ *Aphnæus Zanzibarensis*, H. G. Smith, Ann. and Mag. Nat. Hist., Feb. 1889, p. 136.
Since the description in the text was made I have had the satisfaction of examining two very fine males taken in the same locality as the type—Estcourt, Natal—by Mr. C. W. Morrison, and also a ♂ captured at Eureka, near Barberton, Transvaal, by Mr. C. F. Palmer, and so am enabled to amend that description in the following particulars, viz. :—

Exp. al., 1 in. $4\frac{1}{2}$–$6\frac{1}{2}$ lin.
Upper side blue, not metallic, bright and pale, with scarcely any violaceous tinge; cilia white with black nervular interruptions (very slender on hindwing). *Fore-wing:* costa at base not yellow-ochreous, but with a dull purplishred stain. *Hind-wing:* tails black, white-tipped,—that on first median nervule short, but that on submedian nervure nearly thrice as long, viz., $2\frac{1}{2}$ lines. UNDER SIDE.—Ground-colour dull-olivaceous (in the Transvaal example much yellower). *Fore-wing:* a submarginal dark ferruginous streak much as in hindwing, but thinner. *Hind-wing:* four small silvery spots in outer area very minute and indistinct; submarginal ferruginous streak very dark and angulated towards base on submedian nervure; hind-marginal edging black streak becomes orange on anal-angular lobe on both sides of origin of long tail; immediately before edging a series of five ill-defined diffused small white spots dotted externally with silvery.
Head black (not reddish-brown); thorax beneath ferruginous-brown, with an anterior long white mesial mark, another posterior one, and three white spots on each side; legs with ferruginous-brown femora (hind pair inferiorly edged and tipped with white) and white tibiæ and tarsi—the tibiæ with a black ring at base, and the tarsi with black tips.
The upper side much resembles that of *A.* (*Spindasis*) *Somalina*, Butl. (*Proc. Zool. Soc. Lond.*, 1885 (publ. 1886), p. 764, pl. xlvii. f. 5), but the blue

is in both wings more limited, and the subapical white spots of the fore-wing are differently arranged. The under side is altogether different in the two species.

Mr. Morrison most liberally allowed me to retain one of the two specimens above mentioned; it is noted as captured on 4th November 1888. He informs me that he took what appeared to be a ♀ on the same date, not differing from the ♂ except in its larger size and squarer wings. During the four seasons of Mr. Morrison's acquaintance with this brilliant species, he has not met with it before 15th October or after 7th November.

Aphnæus Natalensis, p. 150.

Fig. of ♂, *Aphnæus Natalensis*, Staud., Exot. Schmett., i. pl. 95 (1887).

Aphnæus Masilikazi,[1] p. 154.

Additional locality in Zululand :—Etshowe (*A. M. Goodrich* and *T. Vachell*).

Aphnæus Ella, p. 154.

Additional locality in Transvaal :—Eureka, near Barberton (*C. F. Palmer*).
The single specimen (♂) received from Mr. Palmer is noted by him as having been captured about "teak" trees on a stony ridge on 12th April 1888.

Aphnæus Phanes, p. 156.

Additional locality :—Transvaal : Eureka, near Barberton (*C. F. Palmer*).

Aphnæus pseudo-zeritis, p. 160.

♂ *Chloroselas Esmeralda*, Butl., Proc. Zool. Soc. Lond., 1885 [publ. 1886], p. 765, pl. xlvii. f. 4.

Additional localities :—Natal: D'Urban (*A. D. Millar*). Eastern Interior of South Tropical Africa (*F. C. Selous*). Eastern Interior of North Tropical Africa : "Somaliland : Bunder Maria (*Yerbury*)."—Butler.

♀ *Wholly dark-brown, without any trace of blue* ; orange spot at anal angle of hind-wing as in ♂. Under side as in ♂. (Specimen received from Mr. A. D. Millar, who took it near D'Urban in December 1887.)

On careful comparison of two ♂s taken by Mr. Selous—which quite agree with Mr. Butler's description of *C. Esmeralda*—and of three very fine ♂s

---

[1] Mr. A. G. Butler (*Ent. M. Mag.*, xx. pp. 250–251, 1884) has described two near allies of this species—one from Lake Nyassa (*A. Nyassæ*) and the other from the Victoria Nyanza (*A. Victoriæ*). I have not seen the former, but a ♀ example of the latter has occurred in a collection recently formed in the Western Interior of South Tropical Africa by Mr. A. W. Eriksson, having been captured on the Omrora, "a river between Ovaquenyama and Ombuela," to the north of Ovampoland, in November 1887. It is very like *Masilikazi* ♀ on the upper side, but on the under side the fore-wing exhibits a very strongly flexuose (instead of straight) subapical fascia, the lower portion of which is connected with the median fascia by an oblique similarly-coloured mark lying between second and third median nervules —and in the hind-wing the long fascia angulated before anal angle is much more flexuose and distinctly composite of six unequal portions, while the short outer subapical fascia is distinctly composed of two very unequal portions, a small part on costa projecting beyond and almost apart from the rest of the marking. These and some minor distinctive markings noted by Mr. Butler fully warrant the separation of *A. Victoriæ* from *A. Masilikazi*.

taken near D'Urban by Mr. Millar, with the type of *A. pseudo-zeritis*, I have come to the conclusion that *Esmeralda* is identical with *Pseudo-zeritis*. The type of the latter is considerably worn, and the metallic spots on its under side are more brassy than silvery, while the fulvous-yellow tinge over the disco-cellular and costal part of the fore-wings is more pronounced, and there is pronounced brownish fuscous clouding over both basal and lower discal areas of the hind-wings. Mr. Selous' two examples have on the under side a paler ground-colour, wholly wanting any dusky clouding. The D'Urban specimens, on the contrary, are beneath of a darker greyer tint than the others, presenting also a small but very pronounced fuscous cloud on the middle disc of the hind-wings, and having all the silvery spots very brilliant. The greenish hue which replaces in certain lights the blue of the upper side is well exhibited in all the specimens except the type of *Pseudo-zeritis*. There are two linear tails on the hind-wing, respectively on the first median nervule and the submedian nervure—the latter tail being the longer of the two.

It has been rightly pointed out by Mr. Butler that the subcostal nervure of the fore-wings has only three nervules, instead of four (rarely five) as in *Aphnæus*; but this is the only important feature of his new genus *Chloroselas*, approximating it to *Chrysorychia*. The butterfly under notice is, however, much less robust in structure than *Chrysorychia*, and has much longer and more hirsute palpi; and it is on the whole so thorough an *Aphnæus* that I hesitate to separate it generically on account of its wanting one branch of the subcostal nervure of the fore-wings.

In December 1887 Mr. Millar had the good fortune to fall in with this exquisite little species near D'Urban. He wrote to me that he took about a dozen examples "flying round acacia trees at Clare Estate, Sydenham. They perched both on twigs and leaves, from which they darted at one another, and were very swift on the wing, usually returning to the same spot. They had not, so far as I am aware, been seen before near D'Urban."

Chrysorychia Harpax, p. 162.

Fig. of ♂, *Axiocerces Perion*, Staud., Exot. Schmett., i. pl. 94 (1887).

Additional locality in Eastern Districts of Cape Colony:—Tharfield, Kleinemond River (*Miss M. L. Bowker*). Additional locality in Transvaal:—Eureka, near Barberton (*C. F. Palmer*).

Chrysorychia Amanga, p. 165.

Additional localities:—Natal: "D'Urban."—A. D. Millar. Transvaal: Eureka, near Barberton (*C. F. Palmer*).

The specimens received from Mr. Palmer are noted as having been taken on 14th March and 17th April 1888; they were settling on grass and low plants about "kopjes," on a stony ridge.

Zeritis Lycegenes, p. 175.

Additional locality in Upper Districts of Natal:—Ulundi, Weenen County (*J. M. Hutchinson*).

It gave me great pleasure to receive three ♂s and a ♀ of this very rare species from Mr. Hutchinson in February 1888; and I am thus enabled to supplement my description of the solitary specimen previously known to me in the following particulars, viz.:—

Exp. al., (♂) 11 lin.—1 in. ¼ lin.; (♀) 11½ lin.—1 in. 1 lin.

It is clear, on comparison with the ♀ received from Mr. Hutchinson, that the smaller example described in the text is also of that sex. The larger ♀

is somewhat duller, and the black spots of the fore-wing are proportionally smaller, but the hind-marginal black edging has in both wings (but especially in the fore-wing) more prominent internervular projections. ♂ *Orange-red deeper, more metallic ; black hind-marginal edging in both wings broader in apical half ; bases slightly dusky*—in one example more strongly so. UNDER SIDE.—As in ♀, but the colouring generally deeper.

The specimens here noted were captured in January 1888; and Mr. Hutchinson wrote that the butterfly was very plentiful for some months at one spot on a stony hillside at Ulundi.

Zeritis Thysbe, p. 181.

Fig. of ♂, *Axiocerces Thysbe*, Staud., Exot. Schmett., i. pl. 94 (1887).

Zeritis Thero, p. 186.

Additional locality in Eastern Districts of Cape Colony :—Grahamstown (*Miss M. L. Bowker*).

Zeritis Sardonyx, p. 188.

Additional localities in Cape Colony :—Western Districts : Ookiep, Namaqualand (*G. Warden*) ; Prieska, Victoria West (*F. Purcell*).

Mr. Warden sent seven examples from Ookiep. Mr. Purcell's single specimen (♂) was captured in November 1887.

Zeritis Argyraspis, p. 189.

Additional locality in Western Districts of Cape Colony :—Prieska, Orange River, Victoria West (*F. Purcell*).

Zeritis Wallengrenii, Var. A., p. 193.

Additional locality in Western Districts of Cape Colony :—Prieska, Orange River, Victoria West (*F. Purcell*).

The specimens (two ♂s, two ♀s) collected by Mr. Purcell nearly resemble the somewhat divergent examples of this Variety noted in the text as sent from the neighbourhood of Grahamstown by Mrs. Barber and Colonel Bowker, but in both sexes the orange is more largely developed in the fore-wing, especially in the basal area. It is interesting to find this species occurring at the same season (November 1887) and in the same locality as its close but larger and more brilliant ally, *Z. Argyraspis*.

Zeritis Aranda, p. 198.

Additional localities :—Orange Free State : Parijs, Upper Vaal River (*E. G. Alston*). Natal : Upper Districts ; Estcourt (*C. W. Morrison*). Zululand : Etshowe (*T. Vachell*).

The single ♀ sent by Mr. Alston from the Free State resembles the two large and dark examples described on p. 199, footnote 1.

Zeritis Pierus, p. 202.

Additional locality in Eastern Districts of Cape Colony :—Grahamstown (*Miss M. L. Bowker*).

Zeritis Taïkosama, p. 203.

Additional locality in Upper Districts of Natal :—Estcourt (*C. W. Morrison*).

Zeritis Molomo, p. 205.

Additional locality in Cape Colony:—Western Districts: Van Wyk's Vley, Carnarvon (*E. G. Alston*).

Pentila tropicalis, p. 211.

Fig. of ♂, *Pentila tropicalis*, Staud., Exot. Schmett., i. pl. 94 (1887).
Additional locality :—Zululand : Etshowe (*A. M. Goodrich* and *T. Vachell*).

D'Urbania Amakosa, p. 215.

Fig. of (?) ♀, *D'Urbania Amakosa*, Staud., Exot. Schmett., i. pl. 94 (1887).
Additional locality in Natal :—Upper Districts : Ulundi, Weenen County (*J. M. Hutchinson*).

The six ♂ s sent by Mr. Hutchinson are like those noted from Kraai River (p. 216, *note*), but two of them have the orange stripe of the hind-wings broader; the four ♀ s accompanying them are large, and with the orange bands extremely broad, and in the fore-wing (as in the Grahamstown examples) prolonged almost to the base by the broad inner marginal orange border.

Mr. Hutchinson wrote (January 1888) that he took this butterfly numerously at Ulundi—one pair being *in copulâ*—in the early part of January; the ♂ s were much more plentiful than the ♀ s, and kept closely to the rocks which the species frequents, but the ♀ s were more wandering, and flew longer without settling. He also found the pupæ, which he mentions as occurring exactly as described in the text.

D'Urbania Saga, p. 219.

Additional locality in the Western Districts of Cape Colony :—Ceres, Tulbagh District.

In October 1887 (12th to 15th) I had the pleasure of observing this interesting species in life, at the upper end of Michell's Pass, close to the village of Ceres. This station lies very much lower than that of the original discovery of the insect, being only about 1700 feet above the sea; and the appearance of the butterfly so much earlier in the season at the new locality is probably traceable to the warmer climate there. The eight ♂ s that I captured expanded from 1 in. 4½ lin. to 1 in. 6½ lin.; they were fresh from the chrysalis, and the disco-cellular ochreous-yellow markings of the fore-wing in them are more apparent than described in the text, the inner marking more or less faintly extending to the base, and an additional similar elongate diffused mark appearing below the median nervure. The spots of the discal series vary as to size and distinctness, and in five examples the three subapical spots in the fore-wing are entirely obsolete.

I found the habits of this butterfly to agree with Mr. Péringuey's account; it sat very close on the sides or top of rocks and stones, and settled again very speedily after being disturbed. It was rather more wary than I expected, several times evading the sweep of the net, and in this way reminded me of some of the species of *Zeritis*. When at rest it is scarcely discernible, owing to the very great resemblance of the under-side colouring to that of the rock surfaces. I saw no example of the ♀, which probably makes her first appearance a little later than the ♂.

D'Urbania Aslauga,[1] p. 220.

Fig. of ♂ (as ♀), *Liptena Aslauga*, Staud., Exot. Schmett., i. pl. 94 (1887).

Alæna Amazoula, p. 223.

Fig. of ♂ *Alæna Amazoula*, Staud., Exot. Schmett., i. pl. 33 (1885).

Genus DELONEURA, p. 224.

I find that a second species of this genus has been described by Plötz (*Stett. Ent. Zeit.*, xli. p. 204, n. 189, 1880), viz., *D. marginata*, from two ♀ specimens collected by Buchholz respectively at Victoria (Cameroons) and Agovè. It is stated to be ochre-yellow, with the base, costa, and hind-margin broadly brown, and the cilia clay-yellow above but light-brown beneath. The outline of the wings is noted to be pretty much as in *D. immaculata*, except that in the hind-wings the anal angle is acute, and excised on the inner-marginal side.

Arrugia Protumnus, p. 228.

Fig. of ♂ *Arrugia Protumnus*, Staud., Exot. Schmett., i. pl. 94 (1887).

Arrugia brachycera, Variety, p. 230.

On the 15th January 1888, Mr. H. L. L. Feltham captured ten examples of this Variety on the summit of Table Mountain. A ♀ which he kindly presented to me expands 1 in. 4½ lin.; its upper-side colouring is less dark than the ♂ s, but equally dull.

Arrugia Basuta, p. 231.

Additional locality:—Orange Free State, Parijs, Upper Vaal River (*E. G. Alston*).

Lachnocnema Bibulus, p. 235.

Fig. of ♂ *Lucia Bibulus*, Staud., Exot. Schmett., i. pl. 94 (1887).

A grass sent to me by Colonel Bowker in January 1888, as in all probability the food-plant of this butterfly, has been kindly determined by Mr. Mac-Owan as a species of *Sorghum*.

---

[1] A near ally of *D. Aslauga* has been described and figured by Mr. W. F. Kirby, from Ashanti, viz., *Teriomima* (?) *Hildegarda* (*Ann. and Mag. Nat. Hist.*, xix. p. 367, 1887; and *Rhop. Exot.*, April 1888, p. 16, pl. Lycæn. iv. ff. 7, 8). It is distinguished on the upper side by its much duller, paler tint and broad hind-marginal blackish border in both wings. On the under side the orange spots are larger, and the outer ones more sharply sagittiform.

ERRATA.

VOLUME I.

Page 28, foot-note 1, line 1, *for* "*Teracolus*," *read* "*Anthocharis*."
„ 32, foot-note 1, line 1, *for* "Mr.," *read* "Mrs."; line 3, *for* "*arachnoïdes*," *read* "*arachoïdes*."
„ 52, line 14 from foot, *for* "♂," *read* "♀."
„ 54, line 14 from foot, *for* "♂," *read* "♀."
„ 91, line 10 from foot, *for* "Bul.," *read* "Butl."
„ 127, line 15 from foot, *for* "*cornuta*," *read* "*ferruginea*."
„ 141, line 18 from foot, *for* "pale, spotted," *read* "pale-spotted."
„ 149, line 17 from foot, *omit* first "Zanzibar."
„ 172, line 6 from foot, *for* "coasts," *read* "coast."
„ 207, line 15, *for* "germinate," *read* "geminate."
„ 264, line 21, for "♂," *read* "♀."
„ 266, line 5 from foot, *for* "me," *read* "one."
„ 296, line 8, *for* "♀," *read* "♂."
„ 300, line 15 from foot, *dele* "pl. vi. f. 3."
„ 321, line 21, *for* "*Betsimisaraka*," *read* "*Betanimena*."
„ 343, line 13 from foot, *for* "in," *read* "on."
„ 352, line 12 (excluding foot-note) from foot, *before* "tibia" at end of line, *insert* "front."
„ 354, line 17, *for* "*Aglaonice*," *read* "*Aglaonice*."

VOLUME II.

Page 26, line 19, *for* "Hoch," *read* "Hock."
„ 106, line 24, *for* "spines," *read* "species."
„ 166, line 4 from foot (exclusive of foot-note), *for* "*Zeritis*," *read* "*Chrysorychia*."
„ 188, line 11, *for* "District," *read* "Districts."
„ 203, line 16, *for* "Swellingdam," *read* "Swellendam."
„ 209, line 1, *for* "♂ s," *read* "♀ s."

INDICES.

SYSTEMATIC INDEX TO THE RHOPALOCERA

DESCRIBED IN VOLUME III.

	PAGE
Family PAPILIONIDÆ	1
Sub-Family *PIERINÆ*	3
Genus PONTIA, Boisduval	7
P. Alcesta, (*Cram.*)	8
Genus TERIAS, Swainson	10
T. Brigitta, (*Cram.*)	14
T. Zoë, *Hopff.*	16
T. floricola, *Boisd.*	19
T. Æthiopica, *Trim., sp. n.*	21
T. Butleri, *Trim., sp. n.*	23
T. Desjardinsii, *Boisd.*	24
T. regularis, *Butl.*	26
Genus MYLOTHRIS, Butler	28
M. Agathina, (*Cram.*)	30
M. Trimenia, (*Butl.*)	33
M. Rueppellii, (*Koch*)	34
Genus PIERIS, Schrank	37
P. Saba, (*Fab.*)	40
P. Thysa, *Hopff.*	44
P. Pigea, *Boisd.*	46
P. alba, *Wallengr.*	48
P. Simana, *Hopff.*	50
P. Charina, *Boisd.*	52
P. Spilleri, *Staud.*	54
P. Ogygia, *Trim.*	56
P. Zochalia, *Boisd.*	57
P. Mesentina, (*Cram.*)	59
P. Gidica, *Godt.*	64
P. Abyssinica, *Lucas*	66
P. Severina, (*Cram.*)	68
P. Hellica, (*Linn.*)	73
Genus HERPÆNIA, Butler	76
H. Eriphia, (*Godt.*)	77
Genus TERACOLUS, Swainson	80
T. subfasciatus, *Swains.*	92
T. Eris, (*Klug*)	93

	PAGE
Genus TERACOLUS—*continued.*	
T. Agoye, (*Wallengr.*)	98
T. Bowkeri, *Trim.*	100
T. Ione, (*Godt.*)	101
T. speciosus, (*Wallengr.*)	105
T. Jobina, *Butl.*	107
T. Phlegyas, *Butl.*	109
T. Regina, (*Trim.*)	111
T. Eunoma, (*Hopff.*)	114
T. Annæ, (*Wallengr.*)	114
T. Wallengrenii, *Butl.*	118
T. Auxo, (*Lucas*)	120
T. Topha, (*Wallengr.*)	123
T. Evenina, (*Wallengr.*)	126
T. simplex, *Butl.*	130
T. Achine, (*Cram.*)	131
T. Gavisa, (*Wallengr.*)	134
T. Antevippe, (*Boisd.*)	136
T. Halyattes, *Butl.*	139
T. Evippe, (*Linn.*)	140
T. Omphale, (*Godt.*)	142
T. Theogone, (*Boisd.*)	145
T. Antigone, (*Boisd.*)	148
T. Phlegetonia, (*Boisd.*)	151
T. Microcale, *Butl.*	154
T. Laïs, *Butl.*	155
T. Celimene, (*Lucas*)	157
T. Vesta, (*Reiche*)	160
Genus COLIAS, Fabricius	163
C. Electra, (*Linn.*)	165
Genus ERONIA, Boisduval	169
E. Cleodora, *Hübn.*	171
E. Leda, (*Boisd.*)	174
E. Buquetii, (*Boisd.*)	177
E. Argia, (*Fab.*)	179
Genus CALLIDRYAS, Boisduval	182
C. Florella, (*Fab.*)	185

	PAGE		PAGE
Sub-Family PAPILIONINÆ	191	Genus THYMELICUS—continued.	
Genus PAPILIO, Linnæus,	194	T. Lepenula, (Wallengr.)	300
P. Policenes, Cram.	201	T. Macomo, (Trim.)	302
P. Anthens, Cram.	205	T. niveostriga, (Trim.)	303
P. Porthaon, Hewits.	207	T. Wallengrenii, Trim.	304
P. Colonna, Ward	209	T. Barberæ, (Trim.)	306
P. Leonidas, Fab.	211	Genus PAMPHILA, Fabricius	307
P. Brasidas, Feld.	214	P. Callicles, (Hewits.)	309
P. Corinneus, Bertol.	217	P. Morantii, Trim.	311
P. Morania, Angas	220	P. Zeno, Trim.	313
P. Demoleus, Linn.	223	P. Hottentota, (Latr.)	314
P. ophidicephalus, Oberth.	229	P. Monasi, Trim., sp. n.	317
P. Constantinus, Ward	232	P. lugens, Hopff.	318
P. Euphranor, Trim.	235	P. Moritili, (Wallengr.)	319
P. Lyæus, Doubl.	237	P. Ayresii, Trim., sp. n.	321
P. Cenea, Stoll.	243	P. Borbonica, (Boisd.)	322
P. Echerioïdes, Trim.	255	P. Fatuellus, Hopff.	323
		P. Mohopaani, (Wallengr.)	324
Family HESPERIDÆ	259	P. Erinnys, Trim.	326
Genus CYCLOPIDES, Westwood	264	P. dysmephila, Trim.	327
		P. Fiara, (Butl.)	329
C. Metis, (Cram.)	266	Genus ANCYLOXYPHA, Felder	330
C. Malgacha, (Boisd.)	268	A. Mackenii, (Trim.)	331
C. Syrinx, Trim.	269	A. Philander, (Hopff.)	333
C. Ægipan, Trim.	271	A. producta, Trim., sp. n.	334
C. Meninx, Trim.	272	Genus ABANTIS, Hopffer	335
C. Willemi, (Wallengr.)	273	A. Tettensis, Hopff.	337
C. Lepeletierii, (Latr.)	274	A. venosa, Trim., sp. n.	339
C. Tsita, Trim.	276	A. bicolor, (Trim.)	340
C. inornatus, Trim.	277	A. paradisea, (Butl.)	342
Genus PYRGUS, Westwood	278	A. Levubu, (Wallengr.)	345
P. Vindex, (Cram.)	280	Genus CAPRONA, Wallengren	346
P. Dromus, Plötz	283	C. Pillaana, Wallengr.	348
P. Mafa, Trim.	284	C. Canopus, Trim.	349
P. Asterodia, Trim.	284	Genus PTERYGOSPIDEA, Wallengren	351
P. Agylla, Trim., sp. n.	286	P. Kobela, (Trim.)	353
P. Transvaaliæ, Trim., sp. n.	286	P. Djælælæ, Wallengr.	354
P. Diomus, Hopff.	287	P. Motozi, Wallengr.	356
P. Sataspes, Trim.	289	P. Mokeezi, Wallengr.	358
P. nanus, Trim., sp. n.	290	P. Nottoana, Wallengr.	360
P. Sandaster, Trim.	291	P. phyllophila, (Trim.)	362
P. Elma, Trim.	293	P. Flesus, (Fab.)	363
P. Mohozutza, (Wallengr.)	294	Genus HESPERIA, Fabricius.	366
P. Chaca, Trim.	296	H. Forestan, (Cram.)	368
P. Tucusa, Trim.	297	H. Pisistratus, Fab.	371
Genus THYMELICUS, Herrich-Schäffer	299	H. Keithloa, (Wallengr.)	372
		H. Anchises, (Gerst.)	374
		H. unicolor, (Mab.).	375

ADDITIONAL SPECIES IN APPENDIX I.

Family NYMPHALIDÆ.
  Sub-Family *ACRÆINÆ.*
    Genus ACRÆA.
      A. Machequena, *H. G. Smith* . . . 377
      A. Igola, *Trim., sp. n.* . 379
      A. Acrita, *Hewits.* . . 381
  Sub-Family *NYMPHALINÆ.*
    Genus HARMA.
      H. Coranus, *H. G. Smith* 382

Genus CHARAXES.
  C. Violetta, *H. G. Smith* 385
  C. Azota, *Hewits.* . . 387
Family LYCÆNIDÆ.
  Genus LYCÆNA.
    L. Pephredo, *Trim., sp. n.* 389
  Genus LYCÆNESTHES.
    L. Mahota, *H. G. Smith* . 390
  Genus DEUDORIX.
    D. Dinomenes, *H. G. Smith* . . . 391

INDEX.

SYNONYMS (including the names of recognised *Varieties*) are in italic type. After each Species-Name follows (between brackets) that of the Genus to which the Species is referred in this Work.

A.

	PAGE
ABANTIS	III. 335.
Abyssinica (Pieris)	III. 40, 66.
Acara (Acræa)	I. 132, 159.
Acca (Lycæna)	II. 42.
Achæmenes (Charaxes)	I. 340.
Acheloïa (Hypanis)	I. 264.
Achine (Teracolus)	III. 87, 131.
Achine (Teracolus)	III. 134, 142.
ACRÆA	I. 131.
ACRÆA	I. 175, 182; II. 222.
ACRÆIDÆ	I. 128.
ACRÆINÆ	I. 128.
Acrita (Acræa)	III. 381.
Acronycta (Acræa)	I. 153.
Acte (Teracolus)	III. 143.
Actiaca (Acræa)	I. 142.
Ægipan (Cyclopides)	III. 271.
Æthiopica (Terias)	III. 21.
Æthon (Zeritis)	II. 176.
Aganice (Planema)	I. 180; III. 399.
Agapenor (Papilio)	III. 201, 205.
Agatha (Neptis)	I. 270.
Agathina (Mylothris)	III. 30.
Agathina (Mylothris)	III. 33, 34.
Agave (Lycæna)	II. 72.
Aglaonice (Acræa)	I. 132, 151; III. 398.
Aglaspis (Zeritis)	II. 194.
Aglatonice (Salamis)	I. 244.
Agoye (Teracolus)	III. 83, 98.
Agrippina (Pieris)	III. 68.
Agylla (Pyrgus)	III. 286.

	PAGE
ALÆNA	II. 222.
Alba (Pieris)	III. 48.
Alcesta (Pontia)	III. 8.
Alcides (Myrina)	II. 141.
Alcimeda (Harma)	I. 312.
Alcippus (Danais)	I. 51, 53.
ALCYONEIS	I. 209.
Aleurona (Callidryas)	III. 185.
Aliena (Terias)	III. 24.
Almeida (Zeritis)	II. 200.
ALOEIDES	II. 167.
Alphæus (Capys)	II. 113.
Amakosa (D'Urbania)	II. 215; III. 417.
Amanga (Chrysorychia)	II. 164.
Amaponda (Pterygospidea)	III. 358.
Amarah (Lycænesthes)	II. 94; III. 411.
Amata (Teracolus)	III. 85.
AMAURIS	I. 56.
AMBLYPODIA	II. 114, 146.
Amelia (Teracolus)	III. 89.
Amestris (Precis)	I. 231.
Amina (Teracolus)	III. 157.
Amphimalla (Acræa)	I. 149.
Amytis (Teracolus)	III. 140.
Anacardii (Salamis)	I. 244.
Anacreon (Acræa)	I. 132, 168; III. 398.
Anchises (Hesperia)	III. 374.
ANCYLOXYPHA	III. 330.
Anemosa (Acræa)	I. 132, 157.
Angolensis (Teracolus)	III. 140.
Annæ (Teracolus	III. 86, 114.

	PAGE		PAGE
Anta (Deudorix).	II. 107.	Barberæ (Lycæna)	II. 14, 56.
Antalus (Deudorix).	II. 107.	Barberæ (Thymelicus)	III. 306.
Anteupompe (Teracolus)	III. 86.	Barberi (Acræa)	I. 132, 162; III. 398.
Antevippe (Teracolus)	III. 88, 136.	Barklyi (Zeritis)	II. 208.
Antharis (Papilio)	III. 205.	Basuta (Arrugia)	II. 231.
Anthedon (Euralia)	I. 282.	*Batikeli* (Deudorix)	II. 107.
Anthemenes (Papilio)	III. 217.	BELENOIS	III. 37.
Antheus (Papilio)	III. 197, 205.	*Bellua* (Acræa)	I. 155.
ANTHOCHARIS	III. 80.	*Benigna* (Lycæna)	II. 75.
ANTHOPSYCHE	III. 80.	BIBLIDES	I. 185.
Antigone (Teracolus)	III. 87, 148.	*BIBLIS*	I. 258, 261, 264.
APAUSTUS	III. 330.	Bibulus (Lachnocnema)	II. 235; III. 418.
Aphnæoïdes (Iolaus)	II. 137; III. 412.	Bicolor (Abantis)	III. 340.
APHNÆUS	II. 146.	*Boguensis* (Pieris)	III. 68.
APHRISSA	III. 182.	*Bohemani* (Teracolus)	III. 92.
Aranda (Zeritis)	II. 198; III. 416.	Boisduvali (Crenis)	I. 252.
Arcas (Chrysophanus)	II. 91.	*Bolina* (Diadema)	I. 277.
Arceusia (Melanitis)	I. 113.	Boöpis (Junonia)	I. 217.
Archesia (Precis)	I. 134.	Borbonica (Pamphila)	III. 309, 322.
Arethusa (Teracolus)	III. 140.	*Borbonica* (Pamphila)	III. 323.
Argenteostriatus (Cyclopides)	III. 272.	Bowkeri (Iolaus)	II. 132.
Argia (Eronia)	III. 179.	Bowkeri (Leptoneura)	I. 98.
Argillaceus (Teracolus)	III. 160, 161	Bowkeri (Lycæna)	II. 87, 88.
Argynnides (Charaxes)	I. 325.	Bowkeri (Teracolus)	III. 83, 100.
Argyraspis (Zeritis)	II. 189.	Brachycera (Arrugia)	II. 230; III. 418.
ARRUGIA	II. 226.	Brasidas (Papilio)	III. 197, 214.
Aslauga (D'Urbania)	II. 220; III. 418.	Brigitta (Terias)	III. 13, 14.
Asopus (Lycæna)	II. 16.	*Brutius* (Charaxes)	I. 335.
Asteris (Lycæna)	II. 24.	Brutus (Charaxes)	I. 335.
Asteris (Lycæna)	II. 18, 20, 21, 23, 26.	*Brutus* (Papilio)	III. 243.
Asterodia (Pyrgus)	III. 284.	Buquetii (Eronia)	III. 177.
Asterope (Ypthima)	I. 66.	Butleri (Terias)	III. 23.
ATELLA	I. 188.	Buxtoni (Acræa)	I. 132, 170; III. 399.
ATERICA	I. 307.	Buxtoni (Hypolycæna)	II. 119.
Aurigineus (Teracolus)	III. 89.	*Buxtoni* (Teracolus)	III. 109.
Aurota (Pieris)	III. 59.		
Auxo (Teracolus)	III. 86, 120.	C.	
Axina (Acræa)	I. 147, 148.	Cabira (Acræa)	I. 132, 173.
AXIOCERSES	II. 161.	Cæculus (Hypolycæna)	II. 116.
Ayresii (Lachnoptera)	I. 196.	Cærulea (Euryphene)	I. 306.
Ayresii (Pamphila)	III. 309, 321.	*Caffer* (Aphnæus)	II. 150.
Azota (Charaxes)	III. 387.	*Caffra* (Acræa)	I. 159.
		Caffra (Mycalesis)	I. 105.
B.		Caffrariæ (Lycæna)	II. 14, 23.
Bætica (Lycæna)	II. 58.	*Cajus* (Charaxes)	I. 335.
Balfouri (Charaxes)	I. 323.	Caldarena (Acræa)	I. 132, 149.
Bankia (Melanitis)	I. 113.	Calice (Lycæna)	II. 80.
		Calice (Lycæna)	II. 82.

INDEX.   429

Callicles (Pamphila) . III. 308, 309.
CALLIDRYAS . . . . III. 182.
Candiope (Charaxes) I. 327; III. 407.
Canissus (Iolaus) . . II. 137.
Canopus (Caprona) . . III. 349.
CAPRONA . . . . . III. 346.
CAPYS . . . . . . II. 112.
CARCHARODUS . . . III. 278.
Carduelis (Pyrameis) . I. 200.
Cardui (Pyrameis) . . I. 200.
CARTEROCEPHALUS . III. 264.
Cassina (Leptoneura) . I. 102.
Cassiopea (Eronia) . . III. 179.
Cassius (Pseudonympha) I. 89.
Cassius (Pseudonympha) I. 82.
Cassus (Leptoneura) . I. 100.
Cassus (Leptoneura) . I. 102.
Casta (Teracolus) . . III. 126.
Castor (Charaxes) . I. 338; III. 407.
Catharina (Lycæna) . II. 31.
CATOPSILIA . . . . III. 182.
Cebrene (Junonia) . . I. 210.
Cebrene (Teracolus) . III. 140.
Celæus (Lycæna) . II. 18, 20, 24, 27.
Celimene (Teracolus) . III. 157.
Cenea (Papilio) . . . III. 198, 243.
Cerasa (Acræa) . . . I. 132, 139.
Ceres (Iolaus) . . . II. 134.
Ceres (Terias) . . . III. 19.
Ceryne (Precis) . . . I. 224.
Chaca (Pyrgus) . . . III. 296.
Chaka (Aphnæus) . . II. 154.
CHARAXES . . . . I. 315.
Charina (Pieris) . . . III. 52.
Cheles (Cyclopides) . . III. 273.
Chersias (Thymelicus) . III. 300.
CHLOROSELAS . . . III. 414.
Chrysantas (Zeritis) . II. 177.
Chrysaor (Zeritis) . . II. 172.
Chrysippe (Danais) . . I. 51.
Chrysippus (Danais) I. 51; III. 394.
CHRYSOPHANUS . . . II. 90.
CHRYSORYCHIA . . . II. 161.
Cinadon (Charaxes) . I. 329.
Cinerescens (Teracolus) III. 115.
Cissus (Lycæna) . . II. 14, 31.
Cithæron (Charaxes) . I. 344.
Clelia (Junonia) . . . I. 214.
Cleodora (Eronia) . . III. 171.

Cloantha (Precis) . . I. 222.
Clytus (Leptoneura) . I. 92.
Clytus (Leptoneura) . I. 96, 98.
CŒNYRA . . . . . I. 68.
COLIAS . . . . . . III. 163.
COLIAS . . . . . . III. 182.
Colonna (Papilio) . . III. 195, 209.
Columbina (Atella) . . I. 193.
Columbina (Atella) . . I. 189.
Colvillei (Pseudacræa) I. 296, 298 note.
Constantinus (Papilio) . III. 198, 232.
Cora (Hypanis) . . . I. 264.
Coranus (Harma) . . III. 382.
Corinneus (Papilio) . III. 198, 217.
CRENIS . . . . . I. 248; III. 402.
CRUDARIA . . . . . II. 167.
CYCLOPIDES . . . . III. 264.
Cydonia (Acræa) . . I. 177.
CYNTHIA . . . . . I. 198.
Cynthia (Acræa) . . . I. 173.

D.

Dædalus (Hamanumida) I. 309.
Damarensis (Teracolus) III. 130.
Danaë (Teracolus) III. 114, 115, 118.
DANAIDÆ . . . . I. 45, 47.
DANAINÆ . . . . I. 47.
DANAIS . . . . . I. 50; III. 394.
DANAIS . . . . . I. 56.
Dariaves (Deudorix) . II. 110.
DEBIS . . . . . . I. 118.
Deceptor (Euralia) . . I. 286.
Deidamia (Teracolus) . III. 126.
Deidamioïdes (Teracolus) III. 126.
Delagoæ (Pseudacræa) I. 291; III. 405.
Delegorguei (Lachnocnema) . . . . II. 235, 236.
DELONEURA . . . II. 224; III. 418.
Delphine (Teracolus) . III. 148, 151.
Demoleus (Papilio) . . III. 198, 223.
Dendrophilus (Lethe) . I. 120.
Dendrophilus (Lethe) . I. 121.
Dermaptera (Myrina) II. 144; III. 413.
Desjardinsii (Terias) . III. 13, 24.
Desjardinsii (Terias) . III. 26.
DEUDORIX . . . . . II. 105.
DIADEMA . . . . . I. 275.
DIADEMA . . . . . I. 281.

	PAGE
Dingana (Leptoneura)	I. 97.
Dinomenes (Deudorix)	III. 391.
Diocippus (Diadema)	I. 277.
Diocles (Deudorix)	II. 108.
Diomus (Pyrgus)	III. 287.
DIPSAS	II. 105.
Diversa (Melanitis)	I. 116.
Djælælæ (Pterygospidea)	III. 354.
Dolorosa (Lycæna)	II. 14, 41.
Dominicanus (Amauris)	I. 61.
Dorippus (Danais)	I. 51, 53.
Doubledayi (Acræa)	I. 132, 147.
Dromus (Pyrgus)	III. 283.
Drona (Terias)	III. 17.
Druceanus (Charaxes)	I. 329.
DRYAS	III. 169.
Dryope (Eurytela)	I. 261; III. 404.
D'Urbani (Lachnocnema)	II. 236.
D'Urbani (Pseudonympha)	I. 80.
D'URBANIA	II. 213.
Dysmephila (Pamphila)	III. 309, 327.

E.

Echeria (Amauris)	I. 57; III. 394.
Echerioides (Papilio)	III. 199, 255.
Edusina (Colias)	III. 165.
Eione (Teracolus)	III. 151.
Electra (Colias)	III. 165.
Elgiva (Precis)	I. 239.
Ella (Aphnæus)	II. 154.
Elma (Pyrgus)	III. 293.
Emolus (Lycæna)	II. 61.
Emolus (Lycænesthes)	II. 98, 100.
Encedon (Acræa)	I. 163; III. 398.
Encedonia (Acræa)	I. 163.
Eosphorus (Teracolus)	III. 98.
Epaphia (Pieris)	III. 40.
Ephyra (Charaxes)	I. 342.
EREBIA	I. 73.
Ericus (Lycæna)	II. 66.
Erinnys (Pamphila)	III. 309, 326.
Eriphia (Herpænia)	III. 77.
Eris (Teracolus)	III. 83, 93.
Erithalion (Charaxes)	I. 342.
ERONIA	III. 169.
Erone (Teracolus)	III. 105.
Erosine (Zeritis)	II. 186.

	PAGE
Erxia (Eronia)	III. 171.
ERYCINIDÆ	II. 1.
ERYCINIDÆ	II. 7.
Erylus (Hypolycæna)	II. 118.
Esebria (Planema)	I. 177; III. 399.
Esmeralda (Aphnæus)	III. 414.
Ethalion (Charaxes)	I. 342.
Eucharis (Teracolus)	III. 123, 148.
EUMENIDES	II. 7.
EUNICA	I. 248.
Eunoma (Teracolus)	III. 84, 114.
EUPHÆDRA	I. 302.
Euphranor (Papilio)	III. 198, 235.
Eupithes (Harma)	I. 312.
EURYMUS	III. 163.
EURYPHENE	I. 305.
EURYTELA	I. 256.
EURYTELIDÆ	I. 45, 185.
Eurytis (Atella)	I. 193.
Eusirus (Mycalesis)	II. 105.
Evadrus (Zeritis)	II. 195.
Evarne (Teracolus)	III. 120.
Evenina (Teracolus)	III. 87, 126.
Evenus (Mycalesis)	I. 105.
Evippe (Teracolus)	III. 88, 140.
Exole (Teracolus)	III. 134, 142.

F.

Fatuellus (Pamphila)	III. 309, 323.
Fatuellus (Pamphila)	III. 322.
Fenestrata (Acræa)	I. 151.
Ferax (Pyrgus)	III. 287.
Fiara (Pamphila)	III. 309, 323.
Ficedula (Myrina)	II. 141.
Flaminia (Teracolus)	III. 148.
Flavida (Pieris)	III. 42.
Flesus (Pterygospidea)	III. 363.
Florella (Callidryas)	III. 185.
Florestan (Hesperia)	III. 368, 371.
Floricola (Terias)	III. 19.
Floricola (Terias)	III. 24.
Forestan (Hesperia)	III. 368.
Friga (Teracolus)	III. 148.

G.

Gaika (Lycæna)	II. 14, 50.
Galathinus (Teracolus)	III. 148.

INDEX. 431

Gallenga (Pieris)	III. 54.
Gavisa (Teracolus)	III. 134.
Gidica (Pieris)	III. 64.
Glauca (Lycæna)	II. 14, 21.
GNESIA	I. 130, 131.
GNOPHODES	I. 111.
GODARTIA	I. 299.
Goochii (Neptis)	I. 273.
Gorgias (Hypolycæna)	II. 123.
Griqua (Lycæna)	II. 14, 79, 84.

II.

Hæmus (Mylothris)	III. 34.
Halyattes (Teracolus)	III. 88, 139.
HAMANUMIDA	I. 307.
HARMA	I. 310.
Harmonides (Teracolus)	III. 136.
Harpax (Chrysorychia)	II. 162.
Hebe (Cœnyra)	I. 69.
Hecate (Precis)	I. 238.
Helice (Pieris)	III. 73.
HELICONIDÆ	I. 45, 47.
Hellica (Pieris)	II. 40, 73.
Hero (Teracolus)	III. 131.
HERPÆNIA	III. 76.
HESPERIA	III. 366.
HESPERIA	III. 278, 307.
HESPERIDÆ	III. 259.
HESPERIINA	III. 259.
Hiarba (Eurytela)	I. 258.
Hiarbas (Eurytela)	I. 258.
Higinia (Pieris)	III. 40.
Hintza (Lycæna)	II. 14, 79.
HIPIO	I. 111.
Hippia (Pseudonympha)	I. 82 ; III. 395.
Hippia (Pseudonympha)	I. 84, 85.
Hippocoon (Papilio)	III. 244.
Hippocrates (Lycæna)	II. 14, 35.
Hippocrene (Teracolus)	III. 136.
Hippomene (Eurema)	I. 204 ; III. 400.
Hirundo (Hypolycæna)	II. 121.
Hoffmannseggii (Lycæna)	II. 69.
Horta (Acræa)	I. 132, 134.
Hottentota (Pamphila)	III. 309, 314.
Hutchinsonii (Aphnæus)	II. 148; III. 413.
Hyale (Colias)	III. 165, 167 note.
HYALITES	I. 130, 131.
Hyblæa (Callidryas)	III. 185.

Hybridus (Teracolus)	III. 143.
HYPANARTIA	I. 203.
HYPANIS	I. 263.
Hypatia (Acræa)	I. 155.
Hypatia (Pieris)	III. 40.
Hyperbioides (Pseudonympha)	I. 89.
Hyperbius (Pseudonympha)	I. 75.
Hyperides (Teracolus)	III. 131.
HYPOLYCÆNA	II. 114.
Hypopolia (Lycæna)	II. 14, 30.

I.

IDMAIS	III. 80, 85.
Ignifer (Teracolus)	III. 136.
Ignota (Lycæna)	II. 14, 39.
Igola (Acræa)	III. 379.
Ilithya (Hypanis)	I. 264.
Ilithyia (Hypanis)	I. 264 ; III. 404.
Ilythia (Hypanis)	I. 264.
Imitator (Pseudacræa)	I. 293 ; III. 405.
Immaculata (Deloneura)	II. 226.
Imperator (Teracolus)	III. 102
Inana (Pieris)	III. 46.
Inaria (Diadema)	I. 277.
Indosa (Lethe)	I. 121.
Injusta (Mycalesis)	I. 105.
Inornatus (Cyclopides)	III. 277.
Insularis (Pterygospidea)	III. 363.
IOLAUS	II. 125.
Iolaus (Hypolycæna)	II. 123.
Ione (Teracolus)	III. 84, 101.
Ione (Teracolus)	III. 105, 109, 111.
Irrorata (Pseudonympha)	I. 76.
ISMENE	III. 366.
Ismene (Melanitis)	I. 113.
Ithonus (Teracolus)	III. 136.

J.

Jahlusa (Charaxes)	I. 325 ; III. 407.
Jalone (Teracolus)	III. 102.
Jesous (Lycæna)	II. 72.
Jobates (Lycæna)	II. 14, 33.
Jobina (Teracolus)	III. 84, 107.
JUNONIA	I. 209.
JUNONIA	I. 243.

K.

Kama (Lycæna)	II. 16.
Keiskamma (Teracolus)	II. 123.
Keithloa (Hesperia)	III. 372.
Kersteni (Lycænesthes)	II. 96.
Knysna (Lycæna)	II. 45.
Kobela (Pterygospidea)	III. 353.

L.

Laches (Lachnocnema)	II. 235.
LACHNOCNEMA	II. 233.
LACHNOPTERA	I. 195.
Laïs (Teracolus)	III. 88, 155.
Laius (Libythea)	II. 5.
Lara (Hypolycæna)	II. 123.
Larydas (Lycænesthes)	II. 96.
Larydas (Lycænesthes)	II. 98.
Leda (Eronia)	III. 174.
Leda (Melanitis)	I. 112.
Lemnos (Lycænesthes)	II. 98.
LEMONIIDÆ	II. 1.
Leonidas (Papilio)	III. 197, 211.
Leonidas (Papilio)	III. 214.
Lepeletierii (Cyclopides)	III. 274.
Lepenula (Thymelicus)	III. 300.
LEPTONEURA	I. 91.
LEPTOSIA	III. 7.
Leroma (Zeritis)	II. 169.
LETHE	I. 119.
Letsea (Lycæna)	II. 14, 40.
Letterstedti (Pamphila)	III. 314.
LEUCOCHITONEA	III. 335.
Levubu (Abantis)	III. 345.
LIBYTHÆINÆ	II. 2.
LIBYTHEA	II. 4.
LIBYTHEIDÆ	II. 1, 2.
LIBYTHIDES	II. 1, 2.
Licinia (Deudorix)	II. 111; III. 411.
Limbata (D'Urbania)	II. 213.
Lingeus (Lycæna)	II. 66.
Liodes (Lycænesthes)	II. 100.
LIPTENA	II. 213.
Livida (Lycænesthes)	II. 103.
Lorduca (Pieris)	III. 60.
LOXURA	II. 140.
LUCIA	II. 233.
Lucida (Lycæna)	II. 14, 47.
Lugens (Pamphila)	III. 309, 318.
Lyæus (Papilio)	III. 198, 237.
Lyæus (Teracolus)	III. 148.
LYCÆNA	II. 11.
LYCÆNA	II. 90, 93.
LYCÆNESTHES	II. 93.
LYCÆNIDÆ	II. 7.
Lycegenes (Zeritis)	II. 175; III. 415.
Lycia (Acræa)	I. 163.
Lycoris (Teracolus)	III. 148.
Lygus (Acræa)	I. 153.
Lyncurium (Zeritis)	II. 174.
Lysimon (Lycæna)	II. 14, 45; III. 409.
Lysimon (Lycæna)	II. 50.

M.

Macalenga (Lycæna)	II. 72, 74.
Machequena (Acræa)	III. 377.
Mackenii (Ancyloxypha)	III. 331.
Mafa (Pyrgus)	III. 284.
Mahallokoæna (Lycæna)	II. 14, 44.
Mahota (Lycænesthes)	III. 390.
Maimuna (Teracolus)	III. 94, 95, note.
Malagrida (Zeritis)	II. 194.
Malagrida (Zeritis)	II. 189, 192.
Malatha (Pieris)	III. 40.
Malgacha (Cyclopides)	III. 268.
Manjaca (Acræa)	I. 170.
Marginalis (Euralia)	I. 282.
Marpessa (Neptis)	I. 272.
Mars (Zeritis)	II. 198.
Masilikazi (Aphnæus)	II. 152.
Melæna (Lycæna)	II. 79, 82.
MELANITIS	I. 111.
Melantha (Hamanumida)	I. 309.
Meleagris (Hamanumida)	I. 309.
Melicerta (Neptis)	I. 270.
MENERIS	I. 123.
Menestheus (Papilio)	III. 229.
Meninx (Cyclopides)	III. 272.
Merope (Papilio)	III. 243.
Mesentina (Pieris)	III. 59.
Messalina (Papilio)	III. 255.
Messapus (Lycæna)	II. 14, 42.
Methymna (Lycæna)	II. 14, 27.
Metis (Cyclopides)	III. 266.
Metophis (Lycæna)	II. 14, 54.
Micipsa (Pamphila)	III. 324.
Microcale (Teracolus)	III. 88, 154.

INDEX. 433

	PAGE
Mima (Euralia)	I. 284.
Mimosæ (Iolaus)	II. 135.
Mintha (Leptoneura)	I. 96.
Misippe (Diadema)	I. 277.
Misippus (Diadema)	I. 277; III. 404.
Mohopaani (Pamphila)	III. 309, 324.
Mohozutza (Pyrgus)	III. 294.
Mokeezi (Pterygospidea)	III. 358.
Molomo (Zeritis)	II. 205.
Monasi (Pamphila)	III. 309, 317.
Montana (Pseudonympha)	I. 82.
Morania (Papilio)	III. 198, 220.
Morantii (Crenis)	I. 253.
Morantii (Pamphila)	III. 308, 311.
Moriqua (Lycæna)	II. 72, 75.
Moritili (Pamphila)	III. 309, 319.
MORPHIDÆ	I. 45, 185.
Motozi (Pterygospidea)	III. 356.
MYCALESIS	I. 103.
Mycena (Melanitis)	I. 113.
MYLOTHRIS	III. 28.
MYRINA	II. 140.
MYRINA	II. 114.
MYSCELIA	I. 248.

N.

NAIS	II. 167.
Nais (Zeritis)	II. 181.
Namaquana (Abantis)	III. 342.
Namaquus (Aphnæus)	II. 158.
Nanus (Pyrgus)	III. 290.
Narica (Pontia)	III. 8.
Narycia (Pseudonympha)	I. 77.
Narycia (Pseudonympha)	I. 79.
Natalensis (Aphnæus)	II. 150.
Natalensis (Aphnæus)	II. 152.
Natalensis (Crenis)	I. 250.
Natalensis (Crenis)	I. 252.
Natalensis (Lycæna)	II. 72, 77.
Natalica (Acræa)	I. 132, 155.
Natalica (Acræa)	I. 153.
Natalica (Precis)	I. 238.
Nataliensis (Acræa)	I. 141.
Natalii (Pseudonympha)	I. 81.
Neanthes (Charaxes)	i. 320; III. 406.
Neba (Pamphila)	III. 319.
Nega (Iolaus)	II. 127.
Neita (Pseudonympha)	I. 79.
Neluska (Acræa)	I. 147.

	PAGE
Neobule (Acræa)	I. 132, 137.
Neophron (Euphædra)	I. 304.
NEPTIS	I. 268.
Niarius (Amauris)	I. 61.
Niobe (Lycæna)	II. 14, 36.
Nireus (Papilio)	III. 237.
Niso (Pamphila)	III. 314.
NISONIADES	III. 351.
Nohara (Acræa)	I. 132, 142.
Noquasa (Lycæna)	II. 58, 64; III. 410.
Norma (Ypthima)	I. 66.
Notobia (Lycæna)	II. 58, 62.
Nottoana (Pterygospidea)	III. 360.
Nycelus (Zeritis)	II. 195.
NYCHITONA	III. 7.
NYMPHALIDÆ	I. 45.
NYMPHALIDÆ	I. 185.
NYMPHALINÆ	I. 185.
NYMPHALIS	I. 315.

O.

Ochlea (Amauris)	I. 60.
Octavia (Precis)	I. 227; III. 400.
Octavia (Precis)	I. 229.
Œnone (Junonia)	I. 210, 214.
Ogygia (Pieris)	III. 56.
Olympusa (Lycæna)	II. 95.
Omphale (Teracolus)	III. 142.
Omphale (Teracolus)	III. 134.
Omphaloides (Teracolus)	III. 143.
Oneaa (Acræa)	I. 147.
Ophidicephalus (Papilio)	III. 198, 229.
Ophion (Pterygospidea)	III. 363.
Orbona (Pieris)	III. 40.
Orejus (Hypolycæna)	II. 118.
Orithyia (Junonia)	I. 217.
Orthrus (Zeritis)	II. 207.
Ortygia (Lycæna)	II. 14, 26.
Orus (Chrysophanus)	II. 91.
Osbecki (Zeritis)	II. 183.
Osiris (Lycæna)	II. 14, 15.
Otacilia (Lycænesthes)	II. 102.
Oxylus (Leptoneura)	I. 94.

P.

Palemon (Lycæna)	II. 66, 67.
Pallene (Iolaus)	II. 138; III. 413.
Palmus (Zeritis)	II. 185.

VOL. III. 2 E

PAMPHILA	III. 307.
PAMPHILA	III. 299.
Panda (Physcæneura) .	I. 71.
PANOPEA	I. 288.
PAPILIO	III. 194.
PAPILIONIDÆ . .	III. 1.
PAPILIONIDÆ . .	III. 191.
PAPILIONINÆ . .	III. 191.
Paradisea (Abantis) .	III. 342.
PARDOPSIS	I. 182.
PAREBA	I. 131.
Parmeno (Melanitis) .	I. 116.
PARNASSIIDÆ . .	III. 192.
Parrhasus (Salamis) .	I. 244.
Parsimon (Lycæna) . .	II. 14, 18.
Parsimon (Lycæna)	II. 20.
Parva (Lycæna) . . .	II. 52.
Pato (Pterygospidea) .	III. 357.
Patricia (Lycæna) . .	II. 14, 20.
Pechuelii (Crenis) . .	III. 403.
Pelarga (Precis) . . .	I. 226.
Pelasgis (Precis) . .	I. 236 ; III. 402.
Pelias (Charaxes) . .	I. 331 ; III. 407.
Pelias (Charaxes) . .	I. 334.
Pelopidas (Papilio) . .	III. 211.
PENTILA	II. 210.
PENTILA	II. 213.
Pephredo (Lycæna) . .	III. 389.
Perion (Chrysorychia) .	II. 163.
PERIPLYSIA	I. 71.
Perspicua (Mycalesis)	I. 107 ; III. 395
Petræa (Acræa) . . .	I. 132, 144.
Petalus (Arrugia) . .	II. 228.
Phæus (Charaxes) . .	I. 344 ; III. 408.
Phalanta (Atella) . .	I. 189, 193.
Phalantha (Atella) . .	I. 189.
Phanes (Iolaus) . . .	II. 156.
PHASIS	II. 167.
Phedima (Melanitis) .	2. 113.
Philander (Ancyloxypha)	III. 333.
Philippus (Hypolycæna)	II. 118.
Phlegetonia (Teracolus)	III. 151.
Phlegyas (Teracolus) .	III. 109.
PHŒBIS	III. 182.
Phosphor (Zeritis) . .	II. 179.
Phyllophila (Pterygospidea)	III. 362.
PHYSCÆNEURA . .	I. 71 ; III. 395.
PIERIDÆ	III. 1, 3.

PIERINÆ	III. 3.
PIERIS	III. 37.
PIERIS	III. 28, 76.
Pierus (Zeritis) . . .	II. 202.
Pierus (Zeritis) . .	II. 198, 200, 203.
Pigea (Pieris) . . .	III. 40, 46.
Pigra (Pieris) . . .	III. 48, 50.
Pillaana (Caprona) . .	III. 348.
PINACOPTERYX . . .	III. 37.
Pisistratus (Hesperia) .	III. 371.
Pitho (Lycæna) . . .	II. 87.
PLANEMA	I. 175.
Policenes (Papilio) . .	III. 197, 201.
Polinice (Hypanis) . .	I. 264.
Poliarnus (Papilio) . .	III. 201.
Pollux (Charaxes) . .	I. 338.
POLYOMMATUS . .	II. 11, 90.
Pompilius (Papilio) . .	III. 201.
PONTIA	III. 7.
PONTIA	III. 37.
Poppea (Mylothris) . .	III. 34.
Porthaon (Papilio) . .	III. 195, 207.
PRECIS	I. 219.
Procne (Teracolus) . .	III. 145.
Producta (Ancyloxypha)	III. 334.
Protea (Acræa) . . .	I. 177.
PROTOGONIOMORPHA .	I. 243.
Protumnus (Arrugia) .	II. 228.
Protumnus (Arrugia) .	II. 230, 231.
PSEUDACRÆA . . .	I. 288.
Pseudocale (Teracolus) .	III. 140.
PSEUDONYMPHA . . .	I. 73 ; III. 395.
Pseudo-zeritis (Aphnæus)	II. 160 ; III. 414.
PTERYGOSPIDEA . . .	III. 351.
PTYCHOPTERYX . . .	III. 80.
Pulchella (Terias) . .	III. 16.
Pulchra (Lycæna) . .	II. 69.
Pulsius (Zeritis) . . .	II. 186.
Punctatissima (Pardopsis)	I. 183 ; III. 399.
Puncticilia (Lycæna) .	II. 14, 29.
Pygmæa (Lycæna) . .	II. 50.
Pylades (Papilio) . .	III. 217, 220.
PYRAMEIS	I. 198.
PYRAMEIS	I. 203.
Pyrene (Callidryas) . .	III. 185.
PYRGUS	III. 278.
Pyrocis (Zeritis) . . .	II. 180.

INDEX.

R.

Rabbaiæ (Acræa) . . I. 132, 133.
Rahel (Terias) . . III. 14. 17.
Rahira (Acræa) . . . I. 132, 166.
Ranoha (Pamphila) . . III. 311.
Raphani (Pieris) . . III. 73.
Rechila (Papilio) . . III. 244.
Regina (Teracolus) . . III. 111.
Regularis (Terias) . . III. 26.
Rhadia (Callidryas) . . III. 185.
RHOPALOCAMPTA . . III. 366.
ROMALEOSOMA . . . I. 302.
Rosa (Crenis) . . . I. 255; III. 403.
Rosimon (Lycæna) . . II. 79.
Roxane (Teracolus) . . III. 131.
Rueppellii (Mylothris) . III. 34.
Rufosparsa (Callidryas) III. 185.
Rumina (Zeritis) . . II. 186.

S.

Saba (Pieris) . . . . III. 39, 40.
Sabacus (Pseudonympha) I. 85.
Sabacus (Pseudonympha) I. 88.
Sabadius (Pterygospidea) III. 360.
Sabrata (Pieris) . . . III. 44.
Saclava (Neptis) . . . I. 272.
Safitza (Mycalesis) . I. 105; III. 395.
Saga (D'Urbania) . II. 219; III. 417.
SALAMIS . . . . I. 243.
Salmoneus (Zeritis) . . II. 186.
Sandaster (Pyrgus) . . III. 291.
Sandaster (Pyrgus) . . III. 288.
SAPÆA . . . . III. 335.
Sardonyx (Zeritis) . . II. 188.
Sataspes (Pyrgus) . . III. 289.
Sataspes (Pyrgus) . . III. 288.
Saturnus (Charaxes) . I. 334.
SATYRIDÆ . . . I. 62.
SATYRINÆ . . . I. 62.
SCELOTHRIX . . . . III. 278.
Schœncia (Eurema) I. 207; III. 400.
Scamani (Hypolycæna) II. 119.
Sebagadis (Lycæna) . . II. 42.
Seis (Acræa) . . . . I. 137.
Senegalensis (Terias) . III. 21.
Serena (Acræa) . . . I. 170.
Sesamus (Precis) . I. 231; III. 401.
Severina (Pieris) . . III. 40, 68.

Syanzini (Acræa) . . I. 163, 164.
Sichela (Lycæna) . . II. 58, 62.
Sichela (Lycænesthes) . II. 100.
Sidus (Iolaus) . . . II. 130.
Silas (Iolaus) . . . . II. 127.
Silvius (Arrugia) . . II. 228.
Simana (Pieris) . . . III. 40, 50.
Simana (Pieris) . . . III. 46.
Simia (Precis) . . . I. 227.
Similis (Papilio) . . . III. 211.
Simonsii (Mycalesis) . I. 109.
Simplex (Teracolus) . III. 88, 130.
Simplicia (Ypthima) . III. 66.
Sipylus (Teracolus) . . III. 126, 134.
SITHON . . . . II. 105.
Siwani (Lycæna) . . II. 33.
Solandra (Melanitis) . I. 113.
Sophia (Precis) . . . I. 221.
Speciosus (Teracolus) . II. 84, 105.
Spilleri (Pieris) . . . III. 40, 54.
SPILOTHYRUS . . . III. 278.
SPINDASIS . . . . II. 146.
Spio (Pyrgus) . . . III. 280.
Stellata (Lycæna) . . II. 14, 49.
Stenobea (Acræa) . . I. 132, 153.
STEROPES . . . . III. 264.
Stictica (Pardopsis) . . I. 183.
Subfasciatus (Teracolus) III. 92.
Subfumosus (Teracolus) III. 148.
Subvenosus (Teracolus) . III. 134.
Suetonius (Zeritis) . . II. 202.
Swainsonii (Callidryas) III. 185.
Sybaris (Lycæna) . . II. 79, 85.
Sylvanus (Lycænesthes) II. 98.
SYNCHLOË . . . . III. 37.
SYRICHTHUS . . . . III. 278.
Syrinx (Cyclopides) . . III. 269.
Syrinx (Pieris) . . . III. 60.
Syrtinus (Teracolus) . III. 120.

T.

Taïkosama (Zeritis) . . II. 203.
Tantalus (Lycæna) . . II. 14, 38.
Taranis (Hesperia) . . III. 374.
Tarquinia (Pseudacræa) I. 289.
TELCHINIA . . . . I. 131.
Telicanus (Lycæna) . . II. 69.
TERACOLUS . . . . III. 80.

	PAGE
TERIAS	III. 10.
Tettensis (Abantis)	III. 337.
Theogone (Teracolus)	III. 88, 145.
Thero (Zeritis)	II. 186.
THESPIA	III. 80.
Thespis (Lycæna)	II. 87.
Thurius (Charaxes)	I. 346.
Thyestes (Charaxes)	I. 346.
THYMELICUS	III. 299.
Thyra (Zeritis)	II. 195.
Thysa (Pieris)	III. 40, 44.
Thysbe (Zeritis)	II. 181.
Thysbe (Zeritis)	II. 185.
Tibullus (Papilio)	III. 244.
TINGRA	II. 211.
Tisiphone (Leptoneura)	I. 92.
Tjoane (Chrysorychia)	II. 162.
Topha (Teracolus)	III. 86, 123.
Tragicus (Papilio)	III. 209.
Transvaaliæ (Pyrgus)	III. 286.
Trimeni (Iolaus)	II. 129; III. 412.
Trimeni (Teracolus)	III. 131.
Trimenia (Mylothris)	III. 33.
Trimenii (Eronia)	III. 171.
Trimenii (Pseudacræa)	I. 296; III. 405.
Trimenii (Pseudonympha)	I. 88.
Tritogenia (Herpænia)	III. 77.
Trochilus (Lycæna)	II. 14, 52.
Trophonius (Papilio)	III. 244.
Tropicalis (Pentila)	II. 211.
Tsita (Cyclopides)	III. 276.
Tsomo (Lycæna)	II. 58, 68.
Tucusa (Pyrgus)	III. 297.
Tugela (Precis)	I. 241; III. 402.
Tukuoa (Precis)	I. 226.
Tulbaghia (Meneris)	I. 125; III. 396.

U.

Umbra (Pterygospidea)	III. 354.
Unicolor (Hesperia)	III. 375.
URBICOLÆ	III. 259.

V.

Vaillantiana (Amauris)	I. 57.
Valmaran (Hesperia)	III. 371.
Varanes (Charaxes)	I. 321; III. 406.
Varia (Eronia)	III. 179.
Venosa (Abantis)	III. 339.
Vesta (Teracolus)	III. 89, 160.
Vindex (Pyrgus)	III. 280.
Vindex (Pyrgus)	III. 287.
Violarum (Acræa)	I. 141.
Violetta (Charaxes)	III. 385.

W.

Wahlbergi (Euralia)	I. 282.
Wakefieldii (Godartia)	I. 300.
Wallengrenii (Teracolus)	III. 118.
Wallengrenii (Thymelicus)	III. 304.
Wallengrenii (Zeritis)	II. 192; III. 416.
Willemi (Cyclopides)	III. 273.

X.

XANTHIDIA	III. 10.
Xiphares (Charaxes)	I. 346.

Y.

YPTHIMA	I. 65.

Z.

Zanzibarensis (Aphnæus)	III. 413.
Zaraces (Arrugia)	II. 231.
Zeno (Pamphila)	III. 309, 313.
ZERITIS	II. 167.
ZERITIS	II. 112, 161, 226.
ZERYTHIS	II. 167.
Zetes (Acræa)	I. 159.
Zeuxo (Zeritis)	II. 171.
Zochalia (Pieris)	III. 57.
Zoë (Terias)	III. 16.
Zoolina (Charaxes)	I. 318; III. 405.
Zorites (Zeritis)	II. 169.

LIST OF SPECIES FIGURED IN THE PLATES.[1]

PLATE X.

	VOL. III. PAGE
Fig. 1.—*Pontia Alcesta* (Cram.), ♀. *Hab.*—D'Urban, Natal (*J. H. Bowker*)	8
Fig. 2.—*Terias Desjardinsii*, Boisd., Var. ♂. *Hab.*—Lydenburg District, Transvaal (*T. Ayres*)	24
Fig. 3, 3a.—*Mylothris Rüppellii*, Koch, ♂ and ♀. *Hab.*—Bashee River, Kaffraria, and Kei River, Cape Colony (*J. H. Bowker*)	34
Fig. 4.—*Pieris Charina*, Boisd., ♂. *Hab.*—Knysna, Cape Colony (*R. Trimen*)	52
Fig. 5, 5a.—*Pieris Pigea*, Boisd., ♂ and ♀. *Hab.*—D'Urban, Natal (*R. Trimen*)	46
Fig. 6.—*Pieris Zochalia*, Boisd., ♀. *Hab.*—Grahamstown, Cape Colony (*R. Trimen*)	57

PLATE XI.

Fig. 1.—*Pieris Gidica*, Godt., ♂. *Hab.*—D'Urban, Natal (*R. Trimen*)	64
Fig. 2.—*Pieris Ogygia*, Trimen, ♀. *Hab.*—D'Urban, Natal (*M. J. M'Ken*)	56
Fig. 3.—*Teracolus Regina*, Trimen, Var. ♀. *Hab.*—Lydenburg District, Transvaal (*T. Ayres*)	111
Fig. 4.—*Teracolus Bowkeri*, Trimen, ♂. *Hab.*—Vaal River, Griqualand West (*J. H. Bowker*)	100
Fig. 5.—*Papilio Colonna*, Ward, ♂. *Hab.*—Delagoa Bay (*Mrs. Monteiro*)	209
Fig. 6.—*Thymelicus Lepenula*, Wallengr., ♂. *Hab.*—Potchefstroom District, Transvaal (*T. Ayres*)	300
Fig. 7.—*Thymelicus Wallengrenii*, Trimen, ♀. *Hab.*—Napoleon Valley, Zululand (*J. H. Bowker*)	304
Fig. 8, 8a.—*Pamphila Hottentota* (Latr.), ♂ and ♀. *Hab.*—Cape Town and Mapumulo, Natal (*R. Trimen*)	314

---

[1] The *habitat* of each of the specimens figured is given, with the name of the collector who is the authority for it.

PLATE XII.

	VOL. III. PAGE
Fig. 1.—*Pamphila Ayresii*, Trimen, ♀. *Hab.*—Lydenburg District, Transvaal (*T. Ayres*).	321
Fig. 2.—*Pamphila Zeno*, Trimen, ♀. *Hab.*—Bashee River, Kaffraria (*J. H. Bowker*).	313
Fig. 3.—*Pamphila Morantii*, Trimen, ♂. *Hab.*—Pinetown, Natal (*J. H. Bowker*).	311
Fig. 4.—*Pamphila Moritili*, Wallengr., ♂. *Hab.*—D'Urban, Natal (*R. Trimen*)	319
Fig. 5.—*Abantis Lerubu* (Wallengr.), ♂. *Hab.*—Potchefstroom District, Transvaal (*T. Ayres*).	345
Fig. 6, 6a.—*Caprona Pillaana*, Wallengr., ♂ and ♀. *Hab.*—Pinetown, Natal (*J. H. Bowker*), and Crocodile River, S. of Shoshong (*F. Barber*)	348
Fig. 7.—*Pterygospidea Djælælæ*, Wallengr., ♀. *Hab.*—D'Urban, Natal (*J. H. Bowker*).	354
Fig. 8.—*Pterygospidea phyllophila* (Trimen), ♀. *Hab.*—D'Urban, Natal (*R. Trimen*)	362
Fig. 9.—*Hesperia Keithloa*, Wallengr., ♂. *Hab.*—Tugela River Mouth, Natal (*J. H. Bowker*)	372
Fig. 10.—*Hesperia Pisistratus*, Fab., ♂. *Hab.*—D'Urban, Natal (*R. Trimen*)	371

END OF VOL. III.

PRINTED BY BALLANTYNE, HANSON AND CO.
EDINBURGH AND LONDON.

Plate 12

www.ingramcontent.com/pod-product-compliance
Lightning Source LLC
Chambersburg PA
CBHW022146300426
44115CB00006B/373